AF248912

Springer Series in Reliability Engineering

Series Editor

Professor Hoang Pham
Department of Industrial and Systems Engineering
Rutgers, The State University of New Jersey
96 Frelinghuysen Road
Piscataway, NJ 08854-8018
USA

Other titles in this series

The Universal Generating Function in Reliability Analysis and Optimization
Gregory Levitin

Warranty Management and Product Manufacture
D.N.P Murthy and Wallace R. Blischke

Maintenance Theory of Reliability
Toshio Nakagawa

System Software Reliability
Hoang Pham

Reliability and Optimal Maintenance
Hongzhou Wang and Hoang Pham

Applied Reliability and Quality
B.S. Dhillon

Shock and Damage Models in Reliability Theory
Toshio Nakagawa

Risk Management
Terje Aven and Jan Erik Vinnem

Satisfying Safety Goals by Probabilistic Risk Assessment
Hiromitsu Kumamoto

Offshore Risk Assessment (2nd Edition)
Jan Erik Vinnem

The Maintenance Management Framework
Adolfo Crespo Márquez

Human Reliability and Error in Transportation Systems
B.S. Dhillon

Complex System Maintenance Handbook
Khairy A.H. Kobbacy and D.N. Prabhakar Murthy

Hoang Pham

Editor

Recent Advances
in Reliability
and Quality in Design

 Springer

Hoang Pham, PhD
Department of Industrial and Systems Engineering
Rutgers, The State University of New Jersey
96 Frelinghuysen Road
Piscataway, NJ 08854-8018
USA

ISBN 978-1-84800-112-1   e-ISBN 978-1-84800-113-8

DOI 10.1007/978-1-84800-113-8

Springer Series in Reliability Engineering ISSN 1614-7839

British Library Cataloguing in Publication Data
Recent advances in reliability and quality in design. -
   (Springer series in reliability engineering)
   1. Reliability (Engineering)
   I. Pham, Hoang
   620'.00452
ISBN-13: 9781848001121

Library of Congress Control Number: 2008923784

Cover design: deblik, Berlin, Germany

Printed on acid-free paper

9 8 7 6 5 4 3 2 1

springer.com

*This book is dedicated to
Dr. Thad Regulinski on his 80th birthday
for his many years of contributions
in Reliability Engineering and profession!*

Preface

Growing international competition has increased the need for all engineers and designers to ensure the level of quality and reliability of their products before release, and for all manufacturers to produce products at their best reliability level at the lowest cost. This implies that the interest in reliability and quality will continue to grow for many years to come.

This book comprises 25 chapters, organized in five parts: System Reliability Computing, Reliability Engineering in Design, Software Reliability and Testing, Quality Engineering in Design, and Applications in Engineering Design. It aims to present the latest theories and methods of reliability and quality, with emphasis on systems design, models and applications. The subjects covered include reliability engineering, maintenance, quality in design, failure analysis, robust design, software reliability and engineering, engineering reliability in design, software development process and improvement, reliability computing, software measurements, software cost effectiveness, applications in reliability design, stress-strength probabilistic, statistical process control, stochastic process modeling, repairable systems, safety analysis, accelerated life modeling, probabilistic modeling and risk analysis. Each chapter will be written by active researchers and/or experienced practitioners with international reputations in the field and with the hope of bridging the gap between theory and practice in reliability and quality in design. Authors of many outstanding papers from the 12th *ISSAT Conference Proceedings of the International Conference on Reliability and Quality in Design* (2006) have been invited to expand their conference papers for contribution as chapters to this book.

The book consists of five parts. Part I of the book contains five papers, deals with different aspects of *System Reliability Computing*. Chapter 1 by Zeephongsekul describes in detail the characteristic of central limit theorem for a reliability measure, called gauge measure, which is based on a marked point process. It also discusses directions of applications related to this new reliability measure. Chapter 2 by Tian, Li, and Zuo discusses a recent advance in the modeling and reliability evaluation of multi-state k out of n systems and its applications in engineering. Chapter 3 by Zhang, Xie, and Tang discusses a method for the parameter estimation of Weibull distribution when there is no censoring using weighted least squares estimation. They also present a simple approximation formula to calculate the weights for a small sample of size. Chapter 4 by Nakagawa and Mizutani discusses several characteristics of periodic cumulative damage models where the total damage is additive. The derivations of obtaining the optimum replacement policies along with numerical examples are also discussed. Chapter 5 by Filus and Filus presents an overview and the development concepts of stochastic reliability modeling approaches. Some analytical description and application of stochastic dependences such as conditioning method and transformations method are also discussed. Chapter 6 by Liu and Mazzuchi discusses a comprehensive literature review on the various burn-in aspects respect to cost functions. The authors also discuss various cost optimization models and their warranty policies considering the concept of "per-item-output".

Part II of the book contains five papers, and focuses on *Reliability Engineering in Design*.

Chapter 7 by Elsayed and Zhang presents a predictive maintenance model addressing multiple imperfect maintenance actions and optimization procedures to determine the optimum system maintenance threshold level that achieves the maximum system availability. Chapter 8 by Lu and Wang presents a method to estimate the reliability and its confidence limits for the Weibull distribution, when there are only few or no failure data available. The Monte Carlo simulation technique with only three failure samples was also discussed to obtain the estimates of a two-parameter Weibull distribution. Chapter 9 by Xie and Wang presents an extended stress-strength interference analysis method to calculate the fatigue reliability under constant cyclic load with uncertainty in stress amplitude. For a specified cyclic load amplitude distribution, fatigue reliability can be calculated using the statistical average of the probabilities. Applications of the methods are also discussed to show the effect of load uncertainty on reliability analysis. Chapter 10 by Fukuda, Tokuno, and Yamada presents a method to evaluate the performance of the software systems considering real-time properties including time-dependent debugging activities using Markov process. Chapter 11 by Xie, Wang, Hao and Zhang provides a general review of the load-component strength interference relationship and then presents a time-dependent strength function to estimate the failure probability of series pipeline systems under randomly multiple load actions.

Part III of the book contains five papers, focuses on *Software Reliability and Testing*.

Chapter 12 by Folleco, Khoshgoftaar and Van Hulse discusses the impact of noise based on the incomplete measurement data on the evaluation of software quality imputation techniques including Bayesian multiple imputation, nearest neighbor imputation, decision tree imputation, and regression imputation. Chapter 13 by Kimura presents a linearized growth curve model and its parameter estimation using the method of two-parameter numerical differentiation. Chapter 14 by Hwang and Pham discusses a generalized time-delay software reliability model addressing the time required to identify and prioritize the detected faults before removing them optimal policies using the method of steps. Numerical examples based on software failure data are presented to illustrate the use of the proposed model when it is applied in practice. Chapter 15 by Lipton and Gokhale presents architecture-based software reliability analysis and optimization methods for software systems addressing interface failures on application reliability using simulated annealing approach. Chapter 16 by Fujiwara, Inoue and Yamada discusses various software reliability growth models considering the time-dependent behavior of the fault-detection rate functions and the characteristics of module composition of the software system. Several applications also discussed to illustrate the methods.

Part IV of the book contains four papers, focuses on *Quality Engineering in Design.*

Chapter 17 by Yamada and Takahashi presents a description of rubber product and defect phenomena and discusses several design of experiments based on quality engineering approaches to identify the causes as well as enhance the product's quality and the process productivity. Chapter 18 by Son and Savage discusses an integrated mean and tolerance economic design model consisting of the production cost and the expected loss of quality cost over a planned horizon at the customer's discount rate based on present worth of loss of quality. They also demonstrate the methods using an application in automotive industry.

Chapter 19 by Castagliola, Celano and Fichera discusses a logarithmic transformed EWMA chart that monitors a statistic that depends on the sample variance and presents sensitivity analysis of the economic-statistical design to the implementation of a S^2 Shewhart chart. Chapter 20 by Fukushima and Yamada aims to prevent project failures by developing the risk management methods based on real-world experience and software development practices. It also analyzes the effects of project management factors using the multiple linear regression technique.

Part V of the book contains five papers, on *Applications in Engineering Design.*

Chapter 21 by Wanpracha, Pham, Hwang, Liang and Pham discusses the state-of-the-art approaches such as support vector machine, natural language processing, classification regression tree etc. in data mining that may be applicable to analyzing complex categorizing text records. The chapter also discusses several research challenges and directions in analyzing text records and mining. Chapter 22 by Siu briefly discusses the needs of visually impaired people in using public toilets and then identifies several key areas that worth to consider in designing the facilities, using the concept *friendly, informative, safe,* and *hygienic.* Chapter 23 by Miller and Gupta discusses assurance cases for critical infrastructures with a concentration on reliability and safety for Supervisory, Control, and Data Acquisition systems and presents a risk management structure based on a goal-based assurance approach to improve the return on investment. Chapter 24 by Fukuda discusses various detecting driver's emotion perspectives and context-dependent approaches in terms of human errors due to the rapid and frequent changes in real-world environments. Chapter 25 by Pham discusses some recent research and modeling in the area of aging and mortality modeling in demography. The chapter also presents several common distribution functions and the force of mortality functions that used in the field.

All the chapters are written by 50 leading experts in the field in academia and industry. I am deeply indebted and wish to thank all of them for their contributions and cooperation. Thanks are also due to the Springer staff for their editorial work. I hope that the readers including engineers, teachers, scientists, postgraduates, researchers, and practitioners in the areas of both engineering and applied science, will find this book a state-of-the-references survey and a valuable resource for understanding the latest developments in reliability and quality and its applications in engineering design.

Hoang Pham
Piscataway, New Jersey
June 2007

Contents

Part III Software Reliability and Testing

**12 Software Fault Imputation in Noisy
and Incomplete Measurement Data**
Andres Folleco, Taghi M. Khoshgoftaar, Jason Van Hulse 255

**13 A Linearized Growth Curve Model
for Software Reliability Data Analysis**
Mitsuhiro Kimura 275

Contributors

P. Castagliola	Institut Universitaire de Technologie de Nantes, France
Giovanni Celano	University of Catania, Italy
W. Chaovalitwongse	Rutgers University, USA
Elsayed A. Elsayed	Rutgers University, USA
S. Fichera	University of Catania, Italy
Jerzy K. Filus	Oakton Community College, USA
Lidia Z. Filus	Northeastern Illinois University, USA
Andres Folleco	Florida Atlantic University, USA
Takaji Fujiwara	Fujitsu Peripherals Limited, Japan
Masamitsu Fukuda	Tottori University, Japan
Shuichi Fukuda	Tokyo Metropolitan Institute of Technology, Japan
Toshihiko Fukushima	Nissin Systems Co., Ltd., Japan
Swapna S. Gokhale	University of Connecticut, USA
Rashi Gupta	University of Missouri, USA
Guangbo Hao	Northeastern University, China
Jason Van Hulse	Florida Atlantic University, USA
Seheon Hwang	Rutgers University, USA
Shinji Inoue	Tottori University, Japan
Taghi M. Khoshgoftaar	Florida Atlantic University, USA
Mitsuhiro Kimura	Hosei University, Japan
Wei Li	University of Alberta, Canada
Z. Liang	Rutgers University, USA
Michael W. Lipton	IBM Corporation, USA
Xin Liu	Delft University of Technology, The Netherlands
Ming-Wei Lu	Daimler Chrysler Corporation, USA
Thomas A. Mazzuchi	The George Washington University, USA
Ann Miller	University of Missouri, USA
Satoshi Mizutani	Aichi Institute of Technology, Japan
Toshio Nakagawa	Aichi Institute of Technology, Japan
Christopher Hoang Pham	Cisco Systems Inc. & San Jose State University

Hoang Pham   Rutgers University, USA
Gordon J. Savage   University of Waterloo, Canada
Kin Wai Michael Siu   The Hong Kong Polytechnic University, Hong Kong
Young Kap Son   University of Waterloo, Canada
Kenji Takahashi   Tottori University, Japan
L.C. Tang   National University of Singapore, Singapore
Zhigang Tian   University of Alberta, Canada
Koichi Tokuno   Tottori University, Japan
Cheng Julius Wang   Daimler Chrysler Corporation, USA
Zheng Wang   Northeastern University, China
Art Wanpracha   Rutgers University, USA
Liyang Xie   Northeastern University, China
M. Xie   National University of Singapore, Singapore
Shigeru Yamada   Tottori University, Japan
Lifang Zhang   National University of Singapore, Singapore
Mingchuan Zhang   Northeastern University, China
Hao Zhang   Rutgers University, USA
Panlop Zeephongsekul   RMIT University, Australia
Ming J. Zuo   University of Alberta, Canada

Part I
System Reliability Computing

Chapter 1
Central Limit Theorem for a Family of Reliability Measures

P. Zeephongsekul

School of Mathematical and Geospatial Sciences,
RMIT University
GPO Box 2476V
Melbourne, Victoria 3001
Australia

1.1 Introduction

The collection of fundamental results, known as *Central Limit Theorems* (CLTs), occupies a distinguished position in the vast edifice which comprises modern statistics. One of their main contributions to statistical theory is that they enable probabilistic statements to be made concerning the limits of sums of random variables (and vectors) when the number of terms comprising these sums approaches infinity. One of the earliest CLT, the *deMoivre – Laplace Theorem* [1], demonstrated that the binomial distributions could be approximated by a Normal distribution when the number of independent trials of the Bernoulli experiment underlying the binomial distribution is allowed to approach infinity. Since then, the amount of research which has gone into refining and generalizing this simple CLT has been vast and comprehensive, and the reader is referred to Sect. V of the book [2] for its excellent coverage of both the classical and modern aspects of the CLTs.

Classical CLTs have concentrated on *independent and identically distributed* (iid) random variables. These were later extended to cover cases where the random variables are allowed to be heterogeneous and even dependent. In an important breakthrough, CLTs have been extended to situations where the underlying random variables take values in more general metric spaces than the traditional Euclidean spaces, especially function spaces [2, 3]. In this context, and of significance to this paper, are CLTs involving limit theorems on random sets and on space of real-valued continuous functions with compact supports [4,5] and [6]. These results, called *functional CLTs*, made possible the application of CLTs to the much richer domain of stochastic processes.

Klement [7] use functional CLTs and an extension of the celebrated embedding theorem due to Radström [8] to prove a CLT for sum of iid fuzzy random variables. Using this embedding theorem, they showed that under certain conditions, the metric of a pivotal quantity involving the sum of iid fuzzy random variables converges in distribution, *i.e.*, weakly, to the supremum norm of a Normal random element in a compact space. Although not specified in that paper, this Normal random element

Hoang Pham, *Recent Advances in Reliability and Quality in Design*
©Springer 2008

was introduced earlier by Puri and Ralescu [9] (see also [10] for some extensions) and it involves the *support functions* of compact and convex subsets [11].

In this chapter, we prove a CLT for family of *gauge measures*, a concept which was introduced in [12]. These measures have been applied in gauging the reliability of software and other systems, due to their role as an accumulator of the random fuzzy marks associated with a marked point process. As should be evident from examples given in [12], marked point processes have very wide applicability, both methodologically and practically, and have been used in modeling many physical and biological phenomena. As such, gauge measures can also be expected to yield the same utility, especially in their role as fuzzy measures of system reliability.

The organization of the chapter is as follows. Following this Introduction, in Sect. 1.2, we summarize and define all key concepts, including the notion of a marked point process, fuzzy random variables, gauge measures and the concept of normality for fuzzy random variables. (For more details concerning marked point processes, the reader is referred to references such as [13] and [14].) The concept of fuzzy random variables used in this paper was first introduced by Puri and Ralescu [15], although other related definitions of fuzzy variables exist, notably those due to Kwakernaak [16], Nahmias [17] and Krätschmer [18]. In Sect. 1.3, we prove our main result on the CLT of gauge measures and give other related results. In Sect. 1.4, we provide examples which serve to illustrate some of our results and finally, Sect. 1.5 summarizes this chapter and provides suggestions for future work in this area.

1.2 Fuzzy Sets Concepts

For the rest of the chapter, the symbol $I_A(x)$ denotes the indicator function of a set A, $\max A\,(\min A)$ is the maximal (minimal) element of A when they exist, R is the set of real numbers, $R_+ = [0, \infty)$ the non-negative half line, Z_+ the subset of non-negative integers and $\mathscr{R}^n$ the n-dimensional Euclidean space. The symbol δ_x refers to the *Dirac delta* function concentrated on x defined by $\delta_x(A) = I_A(x)$ for any set A. The *bounded* Borel subsets of R_+ will be denoted by $\mathscr{B}$. The symbols $\xrightarrow{as}$, $\xrightarrow{p}$ and $\xrightarrow{d}$ represent almost sure convergence, convergence in probability and convergence in distribution (weak convergence) respectively. Finally, the symbol $X \overset{d}{=} Y$ signifies that the random elements X and Y have the same distribution.

1.2.1 Fuzzy Sets

A fuzzy set on $\mathscr{R}^n$ is defined by means of its membership function $\mu: \mathscr{R}^n \longmapsto I = [0, 1]$ which we will assume satisfies the following properties:

(i)   there is an element $x \in \mathscr{R}^n$ such that $\mu(x) = 1$;
(ii) μ is upper semicontinuous;

(iii)   for each $0 < \alpha \leq 1$, the α-level set

$$\mu_\alpha = \{x \in \mathscr{R}^n : \mu(x) \geq \alpha\} \tag{1.1}$$

is convex;

(iv)   the support of μ, *i.e.*,

$$supp\,\mu = \overline{\bigcup_{\alpha \in I} \mu_\alpha} \tag{1.2}$$

is bounded.

Properties (ii), (iii) and (iv) imply that for all $\alpha \in (0, 1]$, μ_α is compact and convex [11]. The space of fuzzy sets which satisfy the above properties will be denoted by $\mathscr{F}(\mathscr{R}^n)$. Another important consequence of the above properties is that for any $\mu \in \mathscr{F}(\mathscr{R}^n)$,

$$\mu(x) = \sup\{\alpha \in I : x \in \mu_\alpha\} \,. \tag{1.3}$$

Let $\oplus$ and $\odot$ represent the *Minkowski sum* and *product* respectively, *i.e.*, for any two nonempty subsets A and B in $\mathscr{R}^n$ and scalar c,

$$A \oplus B = \{a + b : a \in A,\ b \in B\}$$
$$c \odot A = \{ca : a \in A\} \,,$$

The following identities hold for any μ and $v \in \mathscr{F}(\mathscr{R}^n)$ and also define the sum and scalar multiplication of fuzzy sets respectively:

$$(\mu + v)_\alpha = \mu_\alpha \oplus v_\alpha \quad \text{and} \quad (c\mu)_\alpha = c \odot \mu_\alpha \,. \tag{1.4}$$

Before we define the relevant metric on $\mathscr{F}(\mathscr{R}^n)$, we first recall that the Hausdorff metric between two compact and convex subsets A and B in $\mathscr{R}^n$ is defined by

$$d(A, B) = \max\{\rho(A, B), \rho(B, A)\}$$
$$\text{where} \quad \rho(A, B) = \max_{a \in A} \min_{b \in B} \||a - b\| \,, \quad \rho(B, A) = \max_{b \in B} \min_{a \in A} \||a - b\| \tag{1.5}$$

and where $\|.\|$ denotes the Euclidean norm in $\mathscr{R}^n$. (For arbitrary sets, max and min in (1.5) are replaced by sup and inf respectively.) We will also denote the magnitude of a compact and convex set K by

$$\|K\| = d(\{0\}, K) \,.$$

A compact and convex subset K in $\mathscr{R}^n$ can be identified by its *support function*, which is defined as

$$s(K, u) = \sup\{\langle u, x \rangle : x \in K\} \,, \quad u \in S^{n-1} \tag{1.6}$$

where $\langle .,.\rangle$ is the inner product and S^{n-1} is the $(n-1)$-dimensional unit sphere of $\mathscr{R}^n$ (refer to Fig. 1.1 for an example of support functions when $n=2$).

The following are some properties of support functions that will be required later on:

(i) $s(K_1 \oplus K_2, u) = s(K_1, u) + s(K_2, u)$,
(ii) $s(\lambda \odot K, u) = \lambda s(K, u), \quad \lambda \geq 0$,
(iii) $|s(K, u)| \leq \|K\|\|u\| = \|K\|$.

Here is a very useful relationship between the Hausdorff metric and support functions (cf. Theorem 1.8.11, [19]):

$$d(A, B) = \sup\{|s(A, u) - s(B, u)| : u \in S^{n-1}\} \tag{1.7}$$

for any compact and convex sets A and B.

We will now introduce the metric $d_\infty(\mu, \nu), \mu, \nu \in \mathscr{F}(\mathscr{R}^n)$, which will be required in this paper:

$$d_\infty(\mu, \nu) = \sup\{d(\mu_\alpha, \nu_\alpha) : \alpha \in I\} . \tag{1.8}$$

It can be shown (cf. [11], [7]), that $(\mathscr{F}(\mathscr{R}^n), d_\infty)$ is a complete metric space but not a separable one. We denote the family of Borel subsets of $\mathscr{F}(\mathscr{R}^n)$ by $\mathscr{E}$. Note that combining (1.7) and (1.8) gives

$$d_\infty(\mu, \nu) = \sup\{|s(\mu_\alpha, u) - s(\nu_\alpha, u)| : (\alpha, u) \in I \times S^{n-1}\} . \tag{1.9}$$

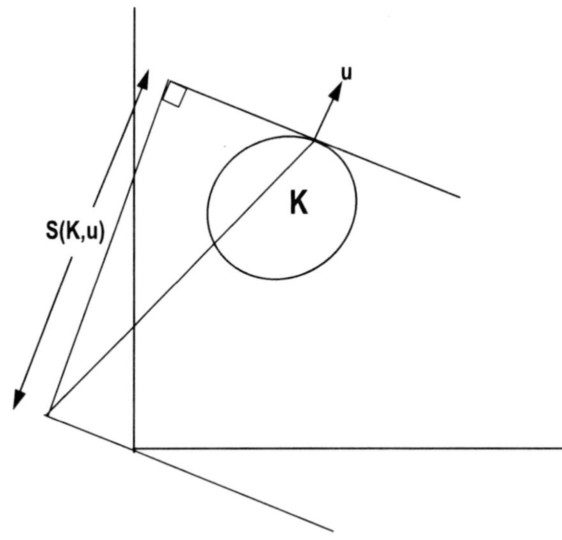

Fig. 1.1 Support function $s(K, u)$

1.2.2 Fuzzy Random Variables

A fuzzy random variable is a natural extension to the concept of random sets, in the sense that its α-level projection is a set-valued mapping from a probability space, which we will denote by $(\Omega, \mathscr{A}, P)$ hereafter. This is made precise in the following definition. A comprehensive coverage of random sets can be found, for example, in [20].

Definition 1. A fuzzy random variable (frv) X is a measurable mapping from a probability space into $(\mathscr{F}(\mathscr{R}^n), d_\infty)$ such that its corresponding α-level mapping X_α defined by

$$X_\alpha(\omega) = \{x \in \mathscr{R}^n : X(\omega)(x) \geq \alpha\} \tag{1.10}$$

is a random set for every $\alpha \in I$.

If X is a frv such that $E\|suppX\| < \infty$, Puri and Ralescu [15] proved the existence of a unique fuzzy set, the expected value of X, denoted by $E(X)$, satisfying the property

$$E(X)_\alpha = E(X_\alpha) . \tag{1.11}$$

We remark that the expected value on the right hand side of (1.11) is an *Aumann integral* [21] of the random set-valued function X_α and is defined by

$$E(X_\alpha) = \left\{ \int f \, dP : f \text{ is a selector of } X_\alpha \right\} .$$

On the left hand side of the equation, $E(X)_\alpha$ refers to the α-level of the fuzzy set $E(X)$. Thus, we can build up the fuzzy set $E(X)$ from (1.11) using (1.3).

Another useful characterization of $E(X)$ utilizes support functions where it can be shown that (Lemma 3.1, [9])

$$s(E(X)_\alpha, u) = E\left\{s(X_\alpha, u)\right\} , \quad u \in S^{n-1} . \tag{1.12}$$

We next consider $Var(X)$, the variance of a frv X. The following definition of $Var(X)$ is due to Körner [22]:

Definition 2. The variance of a frv X is given by

$$Var\{X\} = E\left\{ \int_0^1 \int_{S^{n-1}} |s(X_\alpha, u) - s(E\{X\}_\alpha, u)|^2 \, d\lambda(u) \, d\alpha \right\} \tag{1.13}$$

where λ is the Lebesgue measure on the unit sphere S^{n-1}.

By expanding the squared term in (1.13), it is easy to prove that the following identity holds for $Var(X)$:

$$Var(X) = E\|X\|_2^2 - \|E(X)\|_2^2 \tag{1.14}$$

where, for any fuzzy sets $\mu \in \mathscr{F}(\mathscr{R}^n)$,

$$\|\mu\|_2 = \sqrt{\int_0^1 \int_{S^{n-1}} |s(\mu_\alpha, u)|^2 \, d\lambda(u) \, d\alpha} < \infty. \tag{1.15}$$

Example 1. This is a simple example illustrating the calculation of $E(X)$ and $Var(X)$. It will be used several times later on to illustrate other concepts. Let $\Omega = \{\omega_1, \omega_2\}$, $\mathscr{A}$ the family of all subsets of Ω and $P(\omega_1) = P(\omega_2) = \frac{1}{2}$. Let X takes on values $X(\omega_1) = \mu_1$ and $X(\omega_2) = \mu_2$ where

$$\mu_1(x) = \begin{cases} 5x & \text{if } 0 \leq x \leq 0.2 \\ 2 - 5x & \text{if } 0.2 < x \leq 0.4 \\ 0 & \text{otherwise} \end{cases}$$

and

$$\mu_2(x) = \begin{cases} -5 + 10x & \text{if } 0.5 \leq x \leq 0.6 \\ 4 - 5x & \text{if } 0.6 < x \leq 0.8 \\ 0 & \text{otherwise}. \end{cases}$$

Since $\mu_{1\alpha} = \left[\frac{\alpha}{5}, \frac{2-\alpha}{5}\right]$ and $\mu_{2\alpha} = \left[\frac{\alpha+5}{10}, \frac{4-\alpha}{5}\right]$, $\alpha \in I$, it follows that

$$E(X_\alpha) = \left(\frac{1}{2} \odot \left[\frac{\alpha}{5}, \frac{2-\alpha}{5}\right]\right) \oplus \left(\frac{1}{2} \odot \left[\frac{\alpha+5}{10}, \frac{4-\alpha}{5}\right]\right)$$
$$= \left[\frac{3\alpha+5}{20}, \frac{3-\alpha}{5}\right].$$

Applying (1.3), it is easy to show that

$$E(X)(x) = \begin{cases} \frac{20x-5}{3} & \text{if } 0.25 \leq x \leq 0.4 \\ 3 - 5x & \text{if } 0.4 < x \leq 0.6 \\ 0 & \text{otherwise}. \end{cases}$$

(Refer to Fig. 1.2 for graphs of μ_1, μ_2 and $E(X)$. Note that $E(X)$ certainly produces an averaging effect on the two fuzzy sets.)

Next, we use identity (1.14) to obtain $Var(X)$. Since X takes values in $\mathscr{F}(\mathscr{R}^1)$, we have $n = 1$ and $S^0 = \{+1, -1\}$. Note also that on S^0, $\lambda\{1\} = \lambda\{-1\} = \frac{1}{2}$. Hence

$$E\|X\|_2^2 = \frac{1}{4}\int_0^1 \left(\frac{2-\alpha}{5}\right)^2 d\alpha + \frac{1}{4}\int_0^1 \left(\frac{4-\alpha}{5}\right)^2 d\alpha$$
$$+ \frac{1}{4}\int_0^1 \left(\frac{\alpha}{5}\right)^2 d\alpha + \frac{1}{4}\int_0^1 \left(\frac{\alpha+5}{10}\right)^2 d\alpha = 0.453$$

$$\text{and} \quad \|E(X)\|_2^2 = \frac{1}{2}\int_0^1 \left(\frac{3-\alpha}{5}\right)^2 d\alpha + \frac{1}{2}\int_0^1 \left(\frac{3\alpha+5}{20}\right)^2 d\alpha = 0.203,$$

resulting in $\quad Var(X) = 0.250$.

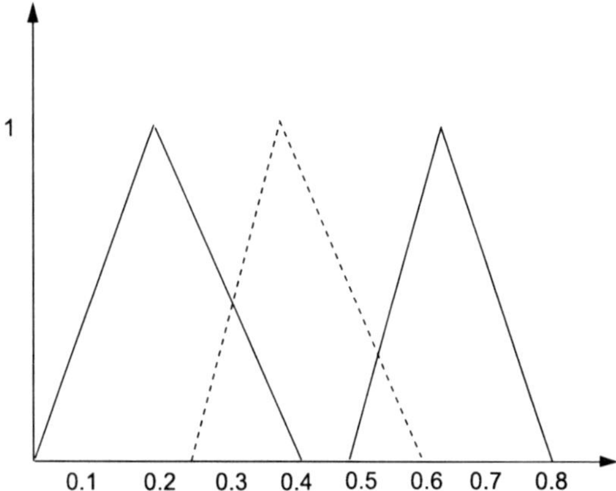

Fig. 1.2 Fuzzy sets μ_1, μ_2 and $E(X)$

1.2.3 Point Process with Random Fuzzy Marks and Corresponding Gauge Measure

Let $(\Omega, \mathscr{A}, P)$ denote a probability space and N_s the space consisting of counting measures $\omega_s = \sum_{n \geq 1} \delta_{t_n}$, where $0 = t_0 < t_1 < t_2 < \ldots < t_n < \ldots$ are the times of occurrences of an event (these times are also known as the *atoms* of the counting measure and reside in R_+). We will assume $\omega_s(B) < \infty$ for all $\omega_s \in N_s$ and $B \in \mathscr{B}$ but $\omega_s(R_+) = \infty$ almost surely. On N_s, let $\mathscr{N}_s$ be the σ-algebra generated by the sets $\{\omega_s : \omega_s(B) = j\}$ where $B \in \mathscr{B}$ and $j \in Z_+$.

A *simple point process* (spp) is a measurable mapping ξ_s from $(\Omega, \mathscr{A}, P)$ into $(N_s, \mathscr{N}_s)$. A distribution P_s on $\mathscr{N}_s$ can be defined through ξ_s by $P_s(A) = P\xi_s^{-1}(A)$, $A \in \mathscr{N}_s$. Since $(N_s, \mathscr{N}_s)$ is fixed, a s.p.p. ξ_s can be identified by its induced measure P_s. Any member $\omega_s \in \mathscr{N}_s$ is an *outcome* of P_s.

A *marked point process* (mpp) is an spp with an auxiliary variable called *mark* associated with each point. These marks are either deterministic or random. Marked point processes have been applied in diverse areas of science and engineering. Here are some examples of marked point processes.

- Quality control inspection of manufactured items: the times when the items were inspected constitute the underlying point process and the corresponding mark at each point specifies whether or not the item passed inspection.
- Queues with batch arrivals: the times of arrivals to the service facility and corresponding size of each batch represents the underlying point process and marks respectively.
- Photoelectric effect: light beams hit a photo-electric plate at various points, which constitute the underlying point process. The number of photo-electrons emitted after each hit are the marks.

- Debugging a software program: the locations in the codes where the bugs were discovered constitute the point process with corresponding marks indicating whether or not these bugs were successfully removed from the program.

(Other examples can be found in the book by Snyder and Miller [13]. A good source of materials on the theory of point and related processes is the book by Daley and Vere-Jones [14].)

This chapter considers the novel situation where each mark is an frv. To extend the concept of a spp to mpp with random fuzzy marks, we let N_m be the space consisting of counting measure $\omega_m = \sum_{n \geq 1} \delta_{(t_n, z_n)}$ where each z_n assumes values in $\mathscr{F}(\mathscr{R}^n)$. Thus, for each fixed atom t_n of a s.p.p. ω_s, the random mark $z_n : R_+ \to \mathscr{F}(\mathscr{R}^n)$ is a measurable mapping from t_n to a member of the fuzzy sets. We will assume $\omega_m(B \times \mathscr{F}(\mathscr{R}^n)) < \infty$ for all $\omega_m \in N_m$ and $B \in \mathscr{B}$ but $\omega_m(R_+ \times \mathscr{F}(\mathscr{R}^n)) = \infty$ almost surely. On N_m, let $\mathscr{N}_m$ be the σ-algebra generated by sets $\{\omega_m : \omega_m(B \times E) = j\}$ where $B \in \mathscr{B}$, $E \in \mathscr{E}$ and $j \in Z_+$. A *random fuzzy marked point process* (rfmpp) is a measurable mapping ξ_m from $(\Omega, \mathscr{A}, \mathbf{P})$ into $(N_m, \mathscr{N}_m)$ and as in the case of an spp, we will identify a rfmpp through the measure $P_m = P\xi_m^{-1}$ induced by ξ_m.

Define the distribution $Q : \mathscr{N}_s \times \mathscr{F}(\mathscr{R}^n) \longmapsto [0, 1]$ where

1. for each ω_s, $Q(\omega_s, .)$ is a probability measure on $\mathscr{E}$;
2. for each $E \in \mathscr{E}$, $Q(, E)$ is a measurable function on N_s.

$Q(\omega_s, ..)$ is the distribution of marks corresponding to the atoms of $\omega_s = \sum_{n \geq 1} \delta_{t_n} \in N_s$. We have the following representation of P_m in terms of $Q(., .)$ and P_s:

$$P_m(A) = \int_{N_s} Q(\omega_s, A_{\omega_s}) \, dP_s(\omega_s), \quad A \in \mathscr{N}_m \tag{1.16}$$

where $A_{\omega_s} = \{z_i \in \mathscr{F}(\mathscr{R}^n) : \omega_m = \sum_{n \geq 1} \delta_{(t_n, z_n)} \in A\}$ is the ω_s-section of the set A. In the rest of the chapter, where there is no possibility of confusion, we will ignore the subscripts s and m in the above notations, which distinguish simple from marked point processes respectively. This is done for the sake of typographic clarity and convenience.

We next give a formal definition of a *gauge measure* corresponding to a rfmpp.

Definition 3. For each $\omega = \sum_{n \geq 1} \delta_{(t_n, z_n)} \in N_m$, an outcome of a rfmpp, let $\omega_s = \sum_{n \geq 1} \delta_{t_n}$ be the corresponding spp and $z(t, \omega)$ the projection of ω into $\mathscr{F}(\mathscr{R}^n)$, i.e.,

$$z(t, \omega) = \begin{cases} z_n & \text{if } t = t_n \\ 0 & \text{otherwise}; \end{cases}$$

then the *fuzzy gauge measure* $G(\omega)$ is defined by the measurable function $G : \mathscr{N}_m \longmapsto \mathscr{F}(\mathscr{R}^n)$ taking values on $\mathscr{B}$ where

$$G(\omega)(B) = \int_B z(t, \omega) \, d\omega_s(t)$$
$$= \sum_{n \geq 1} z_n I_B(t_n), \tag{1.17}$$

$B \in \mathscr{B}$.

We note that a gauge measure is not a measure in the strict sense of the word, since it is a sum of frvs and is therefore an frv. However, it satisfies the countable additivity property of an ordinary measure.

Definition 4. The expected value $\eta_P(.)$ of a fuzzy gauge measure with respect to an rfmpp P is the fuzzy set defined by

$$\eta_P(B) = \int_{N_m} G(\omega)(B)\, dP(\omega)$$
$$= \int_{N_m} \int_B z(t,\omega)\, d\omega_s(t)\, dP(\omega)\,, \quad B \in \mathscr{B}\,. \tag{1.18}$$

We next display formulas for the variance and covariance of a fuzzy gauge measure which were introduced in [12].

Definition 5. The variance and covariance of a gauge measure $G(.)$ is defined as

$$Var\{G(B)\} = E_P\left\{ \int_0^1 \int_{S^{n-1}} |s((G(\omega)(B))_\alpha, u) - s((\eta_P(B))_\alpha, u)|^2\, d\lambda\, d\alpha \right\} \tag{1.19}$$

and

$$cov\{G(B_1), G(B_2)\} = E_P\left\{ \int_0^1 \int_{S^{n-1}} [s((G(\omega)(B_1))_\alpha, u) - s((\eta_P(B_1))_\alpha, u)] \right.$$
$$\left. \cdot [s((G(\omega)(B_2))_\alpha, u) - s((\eta_P(B_2))_\alpha, u)]\, d\lambda(u)\, d\alpha \right\} \tag{1.20}$$

respectively.

In the sequel, we shall suppress the dependence of $G(\omega)(B)$ on the outcome ω and P from η_P for the sake of typographical convenience and simply write $G(B)$ and $\eta(B)$.

1.2.4 Normal Fuzzy Random Variables

The concept of a Normal frv was introduced by Puri and Ralescu [9]. Since a frv generalizes the concepts of a random variable and random sets, a Normal frv also generalizes ordinary normal random variables and normal random sets [5]. The key to defining a reasonable concept of a Normal frv lies in the observation that the class of fuzzy sets that are Lipschitz with respect to the Hausdorff metric d (refer to (1.21) below) can be embedded into the Banach space $C(I \times S^{n-1})$ of continuous functions define on the compact set $I \times S^{n-1}$. These functions are precisely the support functions that were defined in (1.6). In the next proposition and elsewhere, $\|.\|_\infty$ refers to the usual sup norm in the space of continuous functions with compact support, *i.e.*, $\|f\|_\infty = \sup_{x \in \mathbf{C}} |f(x)|$ for a function f with argument x in a compact set $\mathbf{C}$.

Proposition 1. *(Theorem 3.1 [9], Theorem 6.1 [7]) Let $\mathscr{F}_L(\mathscr{R}^n)$ be the subset of $\mathscr{F}(\mathscr{R}^n)$ consisting of fuzzy sets μ which satisfies the Lipschitz condition with respect to the Hausdorff metric d, i.e., for every $\alpha, \beta \in (0,1]$, there exists a constant $M > 0$ such that*

$$d(\mu_\alpha, \mu_\beta) \leq M|\alpha - \beta|. \tag{1.21}$$

There exists a mapping

$$j \colon \mathscr{F}_L(\mathscr{R}^n) \longmapsto C(I \times S^{n-1})$$

which satisfies

(i) $\|j(\mu) - j(v)\|_\infty = d_\infty(\mu, v);$
(ii) $j(\mu + v) = j(\mu) + j(v);$
(iii)   *for any scalar $\lambda > 0$, $j(\lambda\mu) = \lambda j(\mu)$.*

This mapping $j(.)$ is given by

$$j(\mu) = s_\mu(\alpha, u)$$

$$where \quad s_\mu(\alpha, u) = \begin{cases} s(\mu_\alpha, u) & if \ \alpha > 0 \\ s(supp\,\mu, u) & if \ \alpha = 0. \end{cases} \tag{1.22}$$

From the above Proposition, $\mathscr{F}_L(\mathscr{R}^n)$ can be embedded isometrically and isomorphically into the metric space $C(I \times S^{n-1})$. Hence, it is reasonable to define a Normal frv in terms of a multivariate Normal distribution on $C(I \times S^{n-1})$ as follows:

Definition 6. A frv X in $\mathscr{F}_L(\mathscr{R}^n)$ is Normal if for every $(\alpha_i, u_i) \in I \times S^{n-1}, i = 1, 2, \ldots, n$, the vector $(s_X(\alpha_1, u_1), s_X(\alpha_2, u_2), \ldots, s_X(\alpha_n, u_n))$ has a multivariate Normal distribution.

The next basic result, due to Puri and Ralescu [9] (see also [10]), shows that a Normal fuzzy random variable has a very simple characterization.

Theorem 1. *Let a fuzzy random variable X with values in $\mathscr{F}_L(\mathscr{R}^n)$ satisfies $E\|supp X\| < \infty$. Then X is Normal if and only if $X = E(X) + \{\xi\}$ where ξ is a Normal random vector with mean 0.*

Note that $\{\xi\}$ refers to a fuzzy set whose α-level is equal to the Normal random vector ξ for all $\alpha \in (0,1]$. Thus, the level sets of a normal fuzzy random variable with expected value $E(X)$ consist of level sets of $E(X)$ translated by a random amount ξ, distributed as a normal random vector with mean 0.

In addition, $X + Y$ and λX are normal frvs if X and Y are normal frvs and λ is a real scalar quantity. To see this, let $\alpha > 0$, then from (1.4), (1.22) and using properties of support functions given in Sect. 1.2,

$$\begin{aligned} s_{X+Y}(\alpha, u) &= s((X+Y)_\alpha, u) \\ &= s(X_\alpha \oplus Y_\alpha, u) \\ &= s(X_\alpha, u) + s(Y_\alpha, u) \end{aligned}$$

$$\begin{aligned}
\text{and}\quad s_{\lambda X}(\alpha, u) &= s((\lambda X)_\alpha, u)\\
&= s(\lambda \odot X_\alpha, u)\\
&= \lambda\, s(X_\alpha, u)\,.
\end{aligned}$$

From the well known fact that sums and scalar multiples of Gaussian random variables are Gaussian, it follows that $X + Y$ and λX are normal frvs.

Example 2. We generate level sets from a normal fuzzy random variable with $E(X)$ given in Example 1 and ξ distributed as a normal random variable with variance equal to 1. Figure 1.3 gives the histograms of the left endpoints (LEFT) and the right endpoints (RIGHT) of these level sets. As expected, both of these endpoints are normally distributed.

Also, for future reference, we note that for any normal frv X, Theorem 1 implies

$$\begin{aligned}
s_X(\alpha, u) &= s_{E(X)+\xi}(\alpha, u)\\
&= s(E(X)_\alpha \oplus \xi, u)\\
&= s_{E(X)}(\alpha, u) + \langle \xi, u \rangle\,, \quad (\alpha, u) \in I \times S^{n-1}\,. \qquad (1.23)
\end{aligned}$$

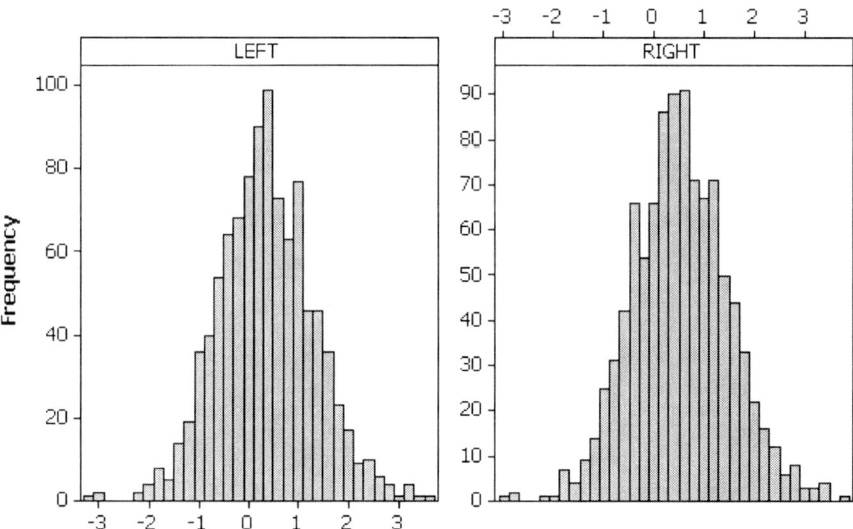

Fig. 1.3 Distributions of endpoints of level sets of a normal fuzzy random variable

1.3 A Central Limit Theorem for Gauge Measures and Related Results

1.3.1 Central Limit Theorems

A CLT for fuzzy random variables was first given by Klement *et al.* [7]. The proof of this CLT utilizes the embedding result in **Proposition 1** which maps $\mathscr{F}_L(\mathscr{R}^n)$ into $C(I \times S^{n-1})$ and then applying the general CLT for random elements in a Banach Space of continuous functions with compact supports due to Jain and Marcus [6]. The statement of this CLT follows:

Theorem 2. *Let Z_1, Z_2, $\ldots \equiv Z$ be iid fuzzy random variables with values in $\mathscr{F}_L(\mathscr{R}^n)$ satisfying*

(i) $E\|supp Z\|^2 < \infty$,

(ii) $E\left(\sup_{\alpha \neq \beta} \frac{d(Z_\alpha, Z_\beta)}{|\alpha - \beta|}\right)^2 < \infty$.

Then there exists a Normal fuzzy random variable X such that

$$\sqrt{n}\, d_\infty\left(\frac{Z_1 + Z_2 + \ldots + Z_n}{n}, E(Z)\right) \xrightarrow{d} \|X\|_\infty \tag{1.24}$$

as $n \longrightarrow \infty$.

The next theorem is our main result concerning a CLT for gauge measures of a rfmpp. We will assume here that for any outcome $\omega_m = \sum_{n \geq 1} \delta_{(t_n, z_n)}$ of a rfmmp, each z_n assumes values in $\mathscr{F}_L(\mathscr{R}^n)$ and are iid with distribution Q. The notation $N(B) = \sum_{n \geq 1} I_B(t_n), B \in \mathscr{B}$, refers to the count measure of the underlying spp. Also, we use the abbreviated notations $G(t), \eta(t)$ and $N(t)$ for $G([0,t)), \eta([0,t))$ and $N([0,t))$ respectively.

Theorem 3. *Assume that each of the iid fuzzy random variable Z of a rfmpp satisfies*

(i) $E\|supp Z\|^2 < \infty$,

(ii) $E\left(\sup_{\alpha \neq \beta} \frac{d(Z_\alpha, Z_\beta)}{|\alpha - \beta|}\right)^2 < \infty$ *and*

(iii) *the underlying spp has the following property: there exists a positive constant $c > 0$ such that*

$$\frac{N(t)}{t} \xrightarrow{p} c, \tag{1.25}$$

as $t \longrightarrow \infty$,

then

$$\frac{d_\infty(G(t), \eta(t))}{\sqrt{t}} \xrightarrow{d} \|X\|_\infty \tag{1.26}$$

where X is a Normal fuzzy random variable and $G(t)$ is the gauge measure corresponding to the rfmpp.

We note that (1.25) is not vacuous, since one of our assumptions on spp is that $N(t) \xrightarrow{as} \infty, t \longrightarrow \infty$.

A few preparatory lemmas will now be presented to help simplify the proof of the theorem.

Define $S_n = Z_1 + Z_2 + \ldots + Z_n$, it is easy to show that an equivalent expression to (1.24) is

$$\frac{d_\infty(S_n, E(S_n))}{\sqrt{n}} \xrightarrow{d} \|X\|_\infty \qquad (1.27)$$

as $n \longrightarrow \infty$. The next result expresses $d_\infty(S_n, E(S_n))$ in terms of support functions.

Lemma 1. *The following identity holds:*

$$d_\infty(S_n, E(S_n)) = \sup_{(\alpha,u) \in I \times S^{n-1}} \left| \sum_{i=1}^{n} \{s_{Z_i}(\alpha, u) - E(s_{Z_i}(\alpha, u))\} \right| \qquad (1.28)$$

where $s_Z(\alpha, u) = s(Z_\alpha, u)$ for any frv Z.

Proof. We first note that

$$s(S_{n\alpha}, u) = s(Z_{1\alpha} \oplus Z_{2\alpha} + \ldots \oplus Z_{n\alpha}, u)$$
$$= \sum_{i=1}^{n} s(Z_{i\alpha}, u) \qquad (1.29)$$

since support function preserves set addition. Also,

$$s(E(S_n)_\alpha, u) = s(E(Z_1)_\alpha \oplus E(Z_2)_\alpha \oplus \ldots \oplus E(Z_n)_\alpha, u)$$
$$= \sum_{i=1}^{n} s(E(Z_i)_\alpha, u)$$
$$= \sum_{i=1}^{n} E\{s(Z_{i\alpha}, u)\}$$

where the last equation follows from (1.12). The proof of the lemma is complete using the formula of $d_\infty(.)$ given by (1.9).

The next lemma is a well known result concerned with approximating *random sum* of iid *real* random variables by ordinary sum of iid random variables. A proof of this result is contained in the proof of Theorem 7.3.2 in Chung [23].

Lemma 2. *Let $S_n = X_1 + X_2 + \ldots + X_n$ be the sum of iid real random variables where $E(X_i) = 0$ and $Var(X_i) < \infty$. Let v_n be a sequence of random variables taking values in Z_+ such that*

$$\frac{v_n}{n} \xrightarrow{p} c$$

for some constant $c > 0$. Then the following convergence holds:

$$\frac{S_{V_n} - S_{[nc]}}{\sqrt{[nc]}} \xrightarrow{p} 0 \tag{1.30}$$

as $n \longrightarrow \infty$. The symbol $[x]$ refers to the integer part of x.

In the following lemmas and definitions preceding the proof of Theorem 3, Θ refers to a *totally bounded metric space* with metric ρ. The next two definitions can be found in Andrews [24] and Billingsley [3].

Definition 7. The *modulus of continuity* of a function $f : \Theta \longmapsto R$ is defined by

$$w(f, \delta) = \sup_{\theta \in \Theta} \sup_{\theta' \in B(\theta, \delta)} |f(\theta') - f(\theta)| \tag{1.31}$$

where $B(\theta, \delta)$ is an open sphere in Θ with center θ and radius $\delta > 0$, i.e., $B(\theta, \delta) = \{x \in \Theta : \rho(x, \theta) < \delta\}$.

Definition 8. A sequence of stochastic functions $F_n(\omega, \theta) : \Omega \times \Theta \longmapsto R$ is *stochastically equicontinuous (se)* on Θ if for every $\varepsilon > 0$, there exists $\delta > 0$ such that

$$\limsup_{n} P(w(F_n, \delta) \geq \varepsilon) < \varepsilon . \tag{1.32}$$

The next lemma is due to Andrews [24] (see also Theorem 21.9, [2]).

Lemma 3. *Let $\{F_n(\theta)\}$ be a sequence of stochastic functions with $\theta \in \Theta$, then*

$$\sup_{\theta \in \Theta} |F_n(\theta)| \xrightarrow{p} 0 \tag{1.33}$$

if and only if

(i) $F_n(\theta) \xrightarrow{p} 0$ *for all $\theta \in \Theta_0$ where Θ_0 is a dense subset of Θ,*
(ii)   *the sequence F_n is se.*

The next lemma, also due to Andrews [24] (cf. Theorem 21.12, [2]), gives conditions for random series of the form

$$F_n(\omega, \theta) = \frac{1}{n} \sum_{i=1}^{n} \{h_i(X_i(\omega), \theta) - E(h_i(X_i(\omega), \theta))\} \tag{1.34}$$

where $h_i(.) : \Omega \times \Theta \longmapsto R$ is a sequence of real-valued functions, $\{X_i(\omega)\}$ a sequence of random variables (not necessary real-valued) and $\theta \in \Theta$, to be stochastically equicontinuous.

Lemma 4. *The series $F_n(\omega, \theta)$ defined by (1.34) is se provided the following conditions hold:*

(i)   there exists a sequence of positive random variables C_i such that

$$\sup_{\theta \in \Theta} |h_i(\theta)| \leq C_i \quad \text{for all } i \tag{1.35}$$

$$\text{and} \quad \limsup_{n \to \infty} \frac{1}{n} \sum_{i=1}^{n} E\left(C_i I_{\{C_i > M\}}\right) \to 0 \quad \text{as } M \to \infty; \tag{1.36}$$

(ii)   for every $\varepsilon > 0$, there exists $\delta > 0$ such that

$$\limsup_{n \to \infty} \frac{1}{n} \sum_{i=1}^{n} P(w(h_i, \delta) > \varepsilon) < \varepsilon. \tag{1.37}$$

Proof of Theorem 3. We first observe that

$$\frac{d_\infty(G(t), \eta(t))}{\sqrt{N(t)}} = \left\{ \frac{d_\infty\left(S_{[ct]}, E\left(S_{[ct]}\right)\right)}{\sqrt{[ct]}} \right.$$
$$\left. + \frac{d_\infty(G(t), \eta(t)) - d_\infty\left(S_{[ct]}, E\left(S_{[ct]}\right)\right)}{\sqrt{[ct]}} \right\} \Big/ \sqrt{\frac{N(t)}{[ct]}}. \tag{1.38}$$

Applying Theorem 2 and noting (1.27), conditions (i) and (ii) imply

$$\frac{d_\infty\left(S_{[ct]}, E\left(S_{[ct]}\right)\right)}{\sqrt{[ct]}} \xrightarrow{d} \|X'\|_\infty$$

where X' is a Normal frv. If we can show that

$$\frac{d_\infty(G(t), \eta(t)) - d_\infty\left(S_{[ct]}, E\left(S_{[ct]}\right)\right)}{\sqrt{[ct]}} \xrightarrow{p} 0 \tag{1.39}$$

then, since condition (iii) implies

$$\sqrt{\frac{N(t)}{[ct]}} \xrightarrow{p} 1,$$

it will follow that

$$\frac{d_\infty(G(t), \eta(t))}{\sqrt{N(t)}} \xrightarrow{d} \|X'\|_\infty.$$

The proof is now complete on applying condition (iii) again, since

$$\frac{d_\infty(G(t), \eta(t))}{\sqrt{t}} = \frac{d_\infty(G(t), \eta(t))}{\sqrt{N(t)}} \cdot \sqrt{\frac{N(t)}{t}}$$

$$\xrightarrow{d} \sqrt{c}\|X'\|_\infty = \|\sqrt{c}X'\|_\infty$$

and $X = \sqrt{c}X'$ is a normal frv.

The difficulty in the proof lies in showing that (1.39) holds and this demonstration utilizes all the lemmas we have presented. We will first show that

$$S_n(\alpha, u) = \sum_{i=1}^{n} H_i(\alpha, u) \tag{1.40}$$

where

$$H_i(\alpha, u) = s_{Z_i}(\alpha, u) - E\left(s_{Z_i}(\alpha, u)\right) \tag{1.41}$$

is se. To place all the results on se random functions described previously in our setting, we let $\Theta = I \times S^{n-1}$ and ρ is the usual Euclidean metric on $I \times S^{n-1}$, i.e., $\rho((\alpha, u), (\beta, v)) = \sqrt{|\alpha - \beta|^2 + \|u - v\|^2}$. We note that (1.40) has a similar form to (1.34) where the functions $h_i(.)$ is replaced by the support function $s(.)$ and X_i by iid frvs $Z_i \equiv Z$.

The function $s_Z(\alpha, u)$ satisfies

$$|s_Z(\alpha, u) \leq \|Z_\alpha\| \|u\| = \|Z_\alpha\|$$

$$\leq \|suppZ\| \tag{1.42}$$

and condition (i) of the theorem implies

$$E\|suppZ\| < \infty . \tag{1.43}$$

Therefore,

$$E\left(\|suppZ\| I_{\|suppZ\|>M}\right) \longrightarrow 0 \quad \text{as } M \longrightarrow \infty . \tag{1.44}$$

Using a result in Klement *et al.* [7] (found in the proof of Theorem 6.1 of that paper), the function $s_Z(\alpha, u)$ is Lipschitz on $I \times S^{n-1}$ with

$$|s_Z(\alpha, u) - s_Z(\beta, v)| \leq q\|Z\|_L \rho((\alpha, u), (\beta, v)) \tag{1.45}$$

where $q > 0$ is a constant and

$$\|Z\|_L = \sup_{\alpha \neq \beta} \frac{d(Z_\alpha, Z_\beta)}{|\alpha - \beta|} + \|suppZ\| .$$

From this, conditions (i) and (ii) of the theorem imply

$$0 < E\|Z\|_L^2 < \infty .$$

Using (1.45), the modulus of continuity of s_Z satisfies

$$w(s_Z, \delta) \le q\|Z\|_L \,\delta . \tag{1.46}$$

For any $\varepsilon > 0$, let

$$\delta < \frac{\varepsilon^{3/2}}{q\sqrt{E\|Z\|_L^2}} \tag{1.47}$$

then (1.46), (1.47) and Chebyshev's inequality give

$$\limsup_{n\to\infty} \frac{1}{n}\sum_{i=1}^{n} P(w(s_{Z_i}, \delta) > \varepsilon) = P(w(s_Z, \delta) > \varepsilon)$$

$$= P\left(\|Z\|_L > \frac{\varepsilon}{\delta q}\right)$$

$$\le E\|Z\|_L^2 \times \frac{\delta^2 q^2}{\varepsilon^2}$$

$$< \varepsilon . \tag{1.48}$$

Therefore, (1.42), (1.44) and (1.48) show that conditions (1.35), (1.36) and (1.37) in Lemma 4 are satisfied with $C_i = \|supp\, Z_i\|$ respectively. Hence, the random series $S_n(\alpha, u)$ defined by (1.40) is se.

Next, referring to (1.28),

$$\frac{1}{\sqrt{[ct]}}\left|d_\infty(G(t), \eta(t)) - d_\infty\left(S_{[ct]}\right), E\left(S_{[ct]}\right)\right| \tag{1.49}$$

$$= \frac{1}{\sqrt{[ct]}}\left|\sup_{(\alpha,u)\in I\times S^{n-1}}\left|S_{N(t)}(\alpha, u)\right| - \sup_{(\alpha,u)\in I\times S^{n-1}}\left|S_{[ct]}(\alpha, u)\right|\right|$$

$$\le \frac{1}{\sqrt{[ct]}}\sup_{(\alpha,u)\in I\times S^{n-1}}\left|S_{N(t)}(\alpha, u) - S_{[ct]}(\alpha, u)\right| .$$

Since $E(H_i(\alpha, u)) = 0$, and (1.42) together with condition (i) of the theorem imply

$$E(|s_Z(\alpha, u)|^2) < \infty ,$$

we can invoke Lemma 2 to conclude that

$$\frac{S_{N(t)}(\alpha, u) - S_{[ct]}(\alpha, u)}{\sqrt{[ct]}} \xrightarrow{p} 0 . \tag{1.50}$$

Thus, condition (i) of Lemma 3 is satisfied. Finally, since the difference of two se functions and a constant multiple of an se function are also se, the term on the left hand side of (1.50) is se. Hence, condition (ii) of Lemma 3 is also satisfied and this implies that the right hand side of inequality (1.49) converges to 0 in probability and this brings the proof of Theorem 3 to its conclusion since we have shown that (1.39) holds.

1.3.2 Asymptotic Variance

The CLT of Theorem 3 raised some important questions and open problems. How is the variance of the Normal frv X in Theorem 1 related to the variance of the family of gauge measures to which it is the domain of attraction? Also, how is this variance related to $Var(G(t))$ defined in (1.19)? In an attempt to answer these questions, we first remind the reader of the following classical CLT (cf. Theorem 23.3, [2]):

Let $X_1, X_2, \ldots \equiv X$ be iid *real* random variables where $E(X) = \mu$ and $Var(X) = \sigma^2 < \infty$. Then the partial sum $S_n = X_1 + X_2 + \ldots + X_n$ satisfies

$$T_n = \frac{S_n - n\mu}{\sqrt{n}} \xrightarrow{d} N(0, \sigma^2) \tag{1.51}$$

where $N(0, \sigma^2)$ refers to the real normal random variable with mean 0 and variance σ^2.

The σ^2 in (1.51) is known as the *asymptotic variance* of T_n and

$$\lim_{n \to \infty} E\left(T_n^2\right),$$

if it exists, is called the *limiting variance* of T_n (refer to [25]).

In the classical CLT case, it is obvious that the limiting variance of T_n is equal to its asymptotic variance. However, this is not necessarily true if T_n does not have the form specified in (1.51). In general, all we can assert is that if a sequence of random variables $T_n \xrightarrow{d} Y$ and $\lim_{n \to \infty} E(T_n^2) = w^2$, then $E(Y^2) \leq w^2$ (cf. Lemma 5.1.2, [25]). The key idea here is that when equality is attained, the limiting variance can be used to infer the value of the asymptotic variance which in turn could provide the means of obtaining the variance of the Normal frv. Before proving the main result which utilizes this idea, we give some preliminary definitions. We remark that our definition of an *essential family* is new.

Definition 9. Let Y_t, $t \in R_+$, be a family of random variables such that

$$Y_t \xrightarrow{d} Y$$

for some limiting random variable Y. If

$$\lim_{t\to\infty} E\left(Y_t^2\right) = E\left(Y^2\right) \tag{1.52}$$

then the family Y_t will be said to form an *essential family*.

As previously noted, the following inequality holds in general:

$$E\left(Y^2\right) \leq \liminf_{t\to\infty} E\left(Y_t^2\right) . \tag{1.53}$$

For an *essential family* of random variables, $\lim_{t\to\infty} E(Y_t^2)$ exists and equality is attained in (1.53).

In the next definition, we define the concept of *uniform integrability* (p. 188, [2]), which is important in modern probability theory.

Definition 10. A family of random variables $Y_t, t \in R_+$ is *uniformly integrable* if it satisfies

$$\lim_{M\to\infty} \sup_t E\left(|Y_t| I_{\{|Y_t|\geq M\}}\right) = 0 .$$

Uniform integrability of $Y_t, t \in R_+$ obviously implies $E|Y_t| < \infty, \forall t \in R_+$. We also note that if either

(i) $\sup_t E|Y_t|^{1+\varepsilon} < \infty$ for some $\varepsilon > 0$, or
(ii)   there exists a random variable Y such that $E|Y| < \infty$ and

$$P(|Y_t| \geq \alpha) \leq P(|Y| \geq \alpha) , \quad \forall t \in R_+, \alpha > 0 ,$$

then the family $Y_t, t \in R_+$ is uniformly integrable (p. 32, [3]). The next Lemma (cf. (Theorem 5.4, [3]), provides one of the most important applications of uniform integrability.

Lemma 5. *If* $Y_t, t \in R_+$ *is uniformly integrable and* $Y_t \xrightarrow{d} Y$, *then* $\lim_{t\to\infty} E(Y_t) = E(Y)$.

Theorem 4. *Assume that all assumptions given in Theorem 3 hold, then the family of random variables* $\frac{d_\infty(G(t), \eta(t))}{\sqrt{t}}$, $t \in R_+$, *is an essential family if* $\frac{d_\infty^2(G(t), \eta(t))}{t}$, $t \in R_+$, *is uniformly integrable.*

Proof. Since (1.26) holds, the Continuous Mapping Theorem (Theorem 22.11, [2]) implies

$$\frac{d_\infty^2(G(t), \eta(t))}{t} \xrightarrow{d} \|X\|_\infty^2 . \tag{1.54}$$

The uniform integrability assumption and an application of Lemma 5 shows that the family $\frac{d_\infty(G(t), \eta(t))}{\sqrt{t}}$, $t \in R_+$, is essential.

Using the definition of $Var(G(t))$ given by (1.19) and also (1.9), it easy to see that

$$Var(G(t)) \leq E\left(d_\infty^2\left(G(t), \eta(t)\right)\right)$$

and therefore it follows that

$$\lim_{t\to\infty} \frac{Var(G(t))}{t} \leq \lim_{t\to\infty} \frac{E\left(d_\infty^2\left(G(t), \eta(t)\right)\right)}{t} \tag{1.55}$$

if both limits exist.

A lower bound is provided for $E\|X\|_\infty^2$ in our next result.

Corollary 1. *Assume all the assumptions in Theorem 3 hold and* $\frac{d_\infty(G(t),\eta(t))}{\sqrt{t}}, t \in R_+$, *is an essential family. In addition, let*

$$\frac{Var(N(t))}{E(N(t))} \to d \tag{1.56}$$

$$and \quad \frac{E(N(t))}{t} \to c, \tag{1.57}$$

then the following inequality holds:

$$c\left(Var(Z) + d\|E(Z)\|_2^2\right) \leq E\|X\|_\infty^2. \tag{1.58}$$

Proof. From Corollary 1 in [12], we obtain

$$\frac{Var(G(t))}{t} = \frac{E(N(t))}{t}\left(Var(Z) + \frac{Var(N(t))}{E(N(t))}\|E(Z)\|_2^2\right). \tag{1.59}$$

Since

$$\frac{E\left(d_\infty^2\left(G(t), \eta(t)\right)\right)}{t} \xrightarrow{d} E\left(\|X\|_\infty^2\right)$$

by Theorem 4, (1.55), (1.56), (1.57) and (1.59) prove (1.58).

We note that (1.58) is not vacuous, since a large class of point process, that of the renewal processes in $\mathscr{R}^1$, satisfies (1.56) and (1.57). Indeed, if the time between consecutive occurrences of the renewal process has mean μ and variance σ^2, then it is well known that

$$E(N(t)) \sim \frac{t}{\mu} \tag{1.60}$$

$$and \quad Var(N(t)) \sim \frac{t\sigma^2}{\mu^3}. \tag{1.61}$$

1.4 Further Examples and an Application

In this section, we provide more examples to demonstrate the application of some results obtained in previous sections. But first, we introduce a special type of frvs that occurs frequently in practice (see [12] and [26]).

Definition 11. Let $\Omega_1, \ldots, \Omega_m$ be a finite partition of Ω, $\mathscr{A}$ the σ-field generated by this partition and $P(\Omega_j) = p_j$, $j = 1, \ldots, m$. A frv $Z \colon \Omega \longmapsto \mathscr{F}(\mathscr{R}^n)$ is said to be finite and separably-valued if it maps each member of Ω_j to a unique fuzzy set, i.e.,

$$Z(\omega) = \mu_j, \quad \forall \omega \in \Omega_j. \tag{1.62}$$

It is clear that finite and separably-valued frv Z satisfies $E\|supp Z\| < \infty$. Also, it is easy to show for such frvs that

$$E(Z)_\alpha = \oplus_{j=1}^m (p_j \odot \mu_{j\alpha}) \tag{1.63}$$

(cf. [12]). From (1.12), we also have

$$s_{E(Z)}(\alpha, u) = \sum_{j=1}^m p_j s(\mu_{j\alpha}, u), \quad (\alpha, u) \in I \times S^{n-1}. \tag{1.64}$$

We note that Example 1 provided a simple demonstration of (1.63).

We also note that rfmpp with random fuzzy marks that are finite and separably-valued certainly satisfied condition (i) and (ii) of Theorem 3. If its underlying spp also satisfies condition (1.25), then the CLT result (1.26) holds. We also remark that condition (1.25) is satisfied by many point processes, for example, the class of renewal processes where the time between consecutive renewals has mean $\mu < \infty$ satisfies the property

$$\frac{N(t)}{t} \xrightarrow{p} \frac{1}{\mu}.$$

Since the CLT gives prominence to the random variable $\|X\|_\infty$ where X is a Normal frv, it would be interesting to display its value when X takes values in $\mathscr{F}(\mathscr{R}^1)$. Note that in $\mathscr{R}^1$, $S^0 = \{+1, -1\}$.

Proposition 2. *Let X be a Normal frv in $\mathscr{R}^1$, then*

$$\|X\|_\infty \overset{d}{=} \max\left\{\sup_{\alpha \in I} |\max(E(X)_\alpha) + \xi|, \quad \sup_{\alpha \in I} |\min(E(X)_\alpha) + \xi|\right\} \tag{1.65}$$

where ξ is a univariate Normal random variable with mean 0.

Proof. From (1.23),

$$s_X(\alpha, u) = s_{E(X)}(\alpha, u) + \langle \xi, u \rangle,$$

therefore

$$\|X\|_\infty \overset{d}{=} \sup_{(\alpha,u)\in I\times S^0} |s_X(\alpha,u)|$$

$$\overset{d}{=} \max\left\{ \sup_{\alpha\in I} |s_{E(X)}(\alpha,1)+\xi|, \quad \sup_{\alpha\in I} |s_{E(X)}(\alpha,-1)-\xi| \right\}. \tag{1.66}$$

Also,

$$s_{E(X)}(\alpha,1) = s(E(X)_\alpha,1)$$
$$= \max(E(X)_\alpha) \tag{1.67}$$
$$\text{and} \quad s_{E(X)}(\alpha,-1) = s(E(X)_\alpha,-1)$$
$$= -\min(E(X)_\alpha) \tag{1.68}$$

hence (1.67) and (1.68) into (1.66) proved (1.65).

Despite the complexity of (1.65), the calculation of the distribution of $\|X\|_\infty$ is reasonably straightforward in some cases, as our next example demonstrates.

Example 3. We revisit Example 2 where the normal frv X is defined with $E(X)$ given in Example 1 and ξ is a normal random variable with mean 0 and variance σ^2. We will derive the distribution of $\|X\|_\infty$. Since $E(X)_0 = [0.25,0.6]$ and $E(X)_1 = \{0.4\}$, it follows that $\max(E(X)_0) = 0.6, \min(E(X)_0) = 0.25$ and $\max(E(X)_1) = \min(E(X)_1) = 0.4$. Firstly, note that for any fixed ξ, $|\max(E(X)_\alpha)+\xi|$ is strictly decreasing as $\alpha \uparrow 1$. Therefore, unless

$$|\xi + \max(E(X)_0)| = |\xi+0.6| < |\xi + \max(E(X)_1)| = |\xi+0.4|$$

i.e., when $\xi < -0.5$, we will have

$$\sup_{\alpha\in I} |\max(E(X)_\alpha)+\xi| = \xi + \max(E(X)_0) = \xi + 0.6 \, ;$$

otherwise,

$$\sup_{\alpha\in I} |\max(E(X)_\alpha)+\xi| = |\xi + \max(E(X)_1)| = -\xi - 0.4 \, .$$

Next, for any fixed ξ, $|\min(E(X)_\alpha)+\xi|$ is strictly increasing as $\alpha \uparrow 1$. Hence, arguing as in the previous case, unless

$$|\xi + \min(E(X)_1)| = |\xi+0.4| < |\xi + \min(E(X)_0)| = |\xi+0.25|$$

i.e., when $\xi < -0.325$, we will have

$$\sup_{\alpha\in I} |\min(E(X)_\alpha)+\xi| = \xi + \min(E(X)_1) = \xi + 0.4 \, ;$$

otherwise,

$$\sup_{\alpha \in I} \left| \min (E(X)_\alpha) + \xi \right| = \left| \xi + \min (E(X)_0) \right| = -\xi - 0.25 \,.$$

Referring to (1.65) and from the above discussion, it is clear that

$$\|X\|_\infty = -\xi - 0.25 \quad \text{if } \xi < -0.5 \tag{1.69}$$

and

$$\|X\|_\infty = \xi + 0.6 \quad \text{if } \xi \geq -0.325 \,. \tag{1.70}$$

In case $-0.5 \leq \xi < -0.325$, it can be easily deduced that

$$\begin{aligned}
\|X\|_\infty &= \max\{-\xi - 0.25, \xi + 0.6\} \\
&= \begin{cases} -\xi - 0.25 & \text{if } -0.5 \leq \xi < -0.425 \\ \xi + 0.6 & \text{if } -0.425 \leq \xi < -0.325 \,. \end{cases}
\end{aligned} \tag{1.71}$$

Finally, (1.69), (1.70) and (1.71) allow us to conclude that

$$\|X\|_\infty = \begin{cases} -\xi - 0.25 & \text{if } \xi < -0.425 \\ \xi + 0.6 & \text{if } \xi \geq -0.425 \,. \end{cases} \tag{1.72}$$

Figure 1.4 provides a histogram from data generated using the distribution of $\|X\|_\infty$ setting $\sigma^2 = 1$.

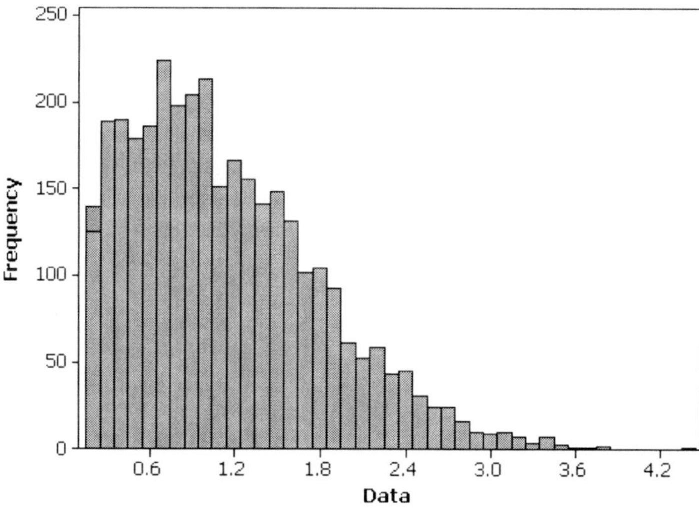

Fig. 1.4 Distribution of $\|X\|_\infty$

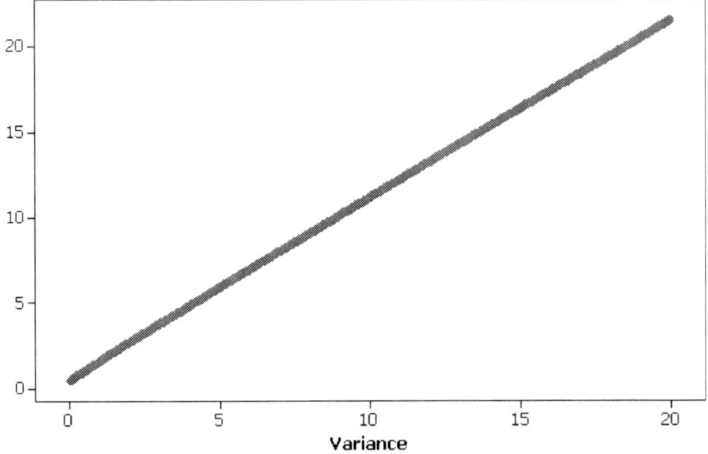

Fig. 1.5 $E(\|X\|_\infty^2)$ versus σ^2

Example 4. (Continuation) From (1.72) and defining

$$f(\xi) = \frac{1}{\sqrt{2\pi\sigma^2}}\, e^{-\frac{\xi^2}{2\sigma^2}},$$

it follows, after some simple calculations, that

$$E\left(\|X\|_\infty^2\right) = \int_{-\infty}^{-0.425} (\xi + 0.25)^2 f(\xi)\,d\xi + \int_{-0.425}^{\infty} (\xi + 0.6)^2 f(\xi)\,d\xi$$

$$= \sigma^2 + \frac{0.7\sigma}{\sqrt{2\pi}}\, e^{-\frac{0.09}{\sigma^2}} + 0.36 - 0.297\, \Phi\left(-\frac{0.425}{\sigma}\right) \tag{1.73}$$

where $\Phi(x)$ refers to the cumulative distribution function of the standard normal random variable. In Fig. 1.5, we plot $E(\|X\|_\infty^2)$ versus different values of σ^2 using (1.73). What is not in doubt is the fact that the asymptotic variance would increase with increasing σ^2. However, the graph displays a perfect straight line which implies that one is able to predict the asymptotic variance perfectly given the variance of the Normal frv and *vice versa*. This is somewhat unexpected and whether it will remain true in general requires further investigation.

1.4.1 An Application

We will now apply our Central Limit Theorem to a problem in software reliability. As discussed earlier in Sect. 1.2.3 and with expanded examples in [12], one application of a marked point process is in the debugging of the software program during testing, an integral part in the improvement of software reliability. During testing, detected faults, revealed by failures in the operation of the program, are removed from the program. The failures that occurred during this period of testing can be modeled as a stochastic point process $(t_n)_{n \geq 1}$ in R_+. When a failure occurs, a debugging effort takes place immediately to fix the faults that caused the failure. Instead of regarding the fault removal process as crisp, *i.e.*, successful with a complete removal (binary digit 1) or unsuccessful with the faults remaining in the program (binary digit 0), it is often more realistic to regard the result of each debugging effort as a fuzzy random variable Z, *i.e.*, as a graded effort with value in the closed interval $[0, 1]$. This approach is very reasonable, given that most debugging efforts are to a greater or lesser extent a subjective exercise, hence a fuzzy set approach is more appropriate than a *crisp* one.

If Z_i is the result of applying fuzzy debugging to the *ith* failure t_i, the cumulative effect of the debugging effort by time t is equal to the sum of $N(t) \equiv N[0, t)$ fuzzy random variables

$$G(t) = \sum_{n \geq 1} Z_n I_{[0,t)}(t_n)$$

$$= Z_1 + Z_2 + \ldots + Z_{N(t)} \tag{1.74}$$

(refer to (1.17)). We note that $G(t)$ is the *fuzzy analogue* of a *software reliability growth curve* which have been modeled extensively in Software Reliability Growth Models (SRGMs) literature (see, *e.g.*, [27] and [28]). However, the nomenclature "*growth curve*" is not appropriate here, as $G(t)$ is a fuzzy random variable, hence the name *gauge measure* is more suitable. Note also that the mean of this gauge measure (1.18) is the fuzzy analogue of the mean value function which is used to model most so called *Fault-Count SRGMs*. Its variance (1.19) can be used as a measure of consistency in the fuzzy debugging process.

Assume that each debugging effort results in a fuzzy random variable Z, which takes on fuzzy set μ_1 and μ_2 defined in Example 1, with probability q and $1 - q$ respectively, $q \in [0, 1]$. We could interpret μ_1 as portraying an effort which is *less than perfect* (*LP*) effort and μ_2 as portraying an *adequate* (*A*) effort. Additional grades of debugging effort can also be included, but two should suffice to illustrate an application of the main result. We also assume that the fuzzy marks z_i are iid and independent of the underlying point process describing failures.

A key question here is how to estimate the proportion of *LP* effort at time t, *i.e.*, $\frac{M}{N(t)}$ where $N(t)$ is the total number of failures observed and M is the number of failures out of $N(t)$ failures which were judged not to be perfectly debugged, *i.e.*, showing a propensity for μ_1 rather than μ_2. Since debugging is assumed here to be a fuzzy phenomena, M is not really an objectively observable random variable but rather a fuzzy random variable. However, we will show nevertheless that a confidence interval can be placed on M using the result of Theorem 3.

Proceeding as in Example 1, it is straightforward to show that

$$E(Z)(z) = \begin{cases} \frac{10z-5(1-q)}{1+q} & \text{if } \frac{1-q}{2} \le z \le \frac{3-2q}{5} \\ 2(2-q)-5z & \text{if } \frac{3-2q}{5} < z \le \frac{2(2-q)}{5} \\ 0 & \text{otherwise .} \end{cases} \tag{1.75}$$

Note that in $\mathscr{R}^1$, $S^0 = \{+1, -1\}$ and the measure $\lambda(.)$ allocates equal weight of $\frac{1}{2}$ to each member of S^0. From formulas for $\mu_{1\alpha}$ and $\mu_{2\alpha}$ given in Example 1, it follows that

$$s(Z_\alpha, 1) = \begin{cases} \frac{2-\alpha}{5} & \text{with probability } q \\ \frac{4-\alpha}{5} & \text{with probability } 1-q \end{cases} \tag{1.76}$$

and

$$s(Z_\alpha, -1) = \begin{cases} -\frac{\alpha}{5} & \text{with probability } q \\ -\frac{(\alpha+5)}{5} & \text{with probability } 1-q . \end{cases} \tag{1.77}$$

Therefore, (1.29), (1.76) and (1.77) give

$$s(G(t)_\alpha, 1) = \frac{(4-\alpha)}{5}N(t) - \frac{2}{5}M \tag{1.78}$$

and

$$s(G(t)_\alpha, -1) = -\frac{(\alpha+5)}{10}N(t) + \frac{(5-\alpha)}{10}M \tag{1.79}$$

respectively with probability $^{N(t)}C_M q^M (1-q)^{N(t)-M}$. Taking expected values of (1.78) and (1.79) with respect to M and $N(t)$ then give

$$s(\eta(t)_\alpha, 1) = \left[\frac{(4-\alpha)}{5} - \frac{2}{5}q\right] E(N(t)) \tag{1.80}$$

and

$$s(\eta_\alpha, -1) = -\left[\frac{(\alpha+5)}{10} - \frac{(5-\alpha)}{10}q\right] E(N(t)) \tag{1.81}$$

respectively. Using (1.78), (1.79), (1.80) and equation (4) in [10], it follows after some calculations that

$$d_\infty(G(t), \eta(t)) = \left|\frac{4}{5}(N(t) - E(N(t)) - \frac{2}{5}(M - qE(N(t)))\right| . \tag{1.82}$$

Next, assume that the three conditions leading to convergence (1.26) hold. (Conditions (i) and (ii) hold due to Z being finite and separably valued. A general sufficient condition for (1.25) is for $N(t)$ to be a renewal process.) Then for large t, (1.26) holds and we have approximately that

$$P\left(\sqrt{t}\,\zeta_{\beta/2} \leq \left|\frac{4}{5}(N(t) - E(N(t))) - \frac{2}{5}(M - qE(N(t)))\right| \leq \sqrt{t}\,\zeta_{1-\beta/2}\right) = 1 - \beta$$

(1.83)

where $0 < \beta < 1$ and $P(\|X\|_\infty \leq \zeta_\beta) = \beta$. Solving the inequality in (1.83), an approximate $(1 - \beta) \times 100\%$ confidence interval for M is $I_1 \cup I_2$ where

$$I_1 = \left[qE(N(t)) + 2(N(t) - E(N(t))) - \frac{5}{2}\sqrt{t}\,\zeta_{1-\beta/2}, qE(N(t))\right.$$
$$\left. + 2(N(t) - E(N(t))) - \frac{5}{2}\sqrt{t}\,\zeta_{\beta/2}\right]$$

and

$$I_2 = \left[qE(N(t)) + 2(N(t) - E(N(t))) + \frac{5}{2}\sqrt{t}\,\zeta_{\beta/2}, qE(N(t))\right.$$
$$\left. + 2(N(t) - E(N(t))) + \frac{5}{2}\sqrt{t}\,\zeta_{1-\beta/2}\right].$$

1.5 Conclusion

In this chapter, we presented and proved a Central Limit Theorem for a reliability measure which is based on point processes with random fuzzy marks. From our previous work [12], it is seen that these measures, called gauge measures, have many practical applications, especially in assessing the reliability of systems subject to systematic and random factors and where there is a degree of subjectivity involved in their assessment. The presentation and proof of the CLT were preceded by a brief discussion of the concepts of fuzzy random variables, stochastic point processes and Normal fuzzy random variables. Several examples and an application illustrating the results were also given.

The work considered in this chapter should open up many areas of application related to large sample statistical inference on gauge measures. This will be extremely useful to reliability practitioners. In future work, we hope to extend the CLT of this paper, which essentially looks at the independent and identically distributed case, to the more general case where the frvs are dependent or heterogeneous. This would extend its range of applicability tremendously although one would anticipate that success in this venture could be fraught with some difficulties.

References

1. Renyi A (1970) Foundations of probability. Holden-Day, San Francisco
2. Davidson J (1994) Stochastic limit theory: an introduction for econometricians. Oxford University Press, Oxford
3. Billingsley P (1968) Convergence of probability measures. Wiley, New York
4. Cressie N (1979) A central limit theorem for random sets. Z. Wahrscheinlichkeitstheorie verw. Gebiete 49:37–47
5. Weil W (1982) An application of the Central Limit for Banach-space-valued random variables to the theory or random sets. Z Wahrscheinlichkeitstheorie verw. Gebiete 60:203–208
6. Jain NC, Marcus MB (1975) Central limit theorem for C(S)-valued random variables. Journal of Functional Analysis 19:216–231
7. Klement EP, Puri ML, Ralescu DA (1986) Limit theorems for fuzzy random variables. Proc R Soc Lond A 407:171–182
8. Radström H (1952) An embedding theorem for spaces of convex sets. Proc Am Math Soc 3:165–169
9. Puri ML, Ralescu DA (1985) The concept of normality for fuzzy random variables. Annals of Probability 13:1373–1379
10. Feng Y (2000) Gaussian fuzzy random variables. Fuzzy Sets and Systems 111:325–330
11. Diamond P, Kloeden P (1994) Metric spaces of fuzzy sets. World Scientific, Singapore
12. Zeephongsekul P (2006) On a measure of reliability based on point processes with random fuzzy marks. International Journal of Reliability, Quality and Safety Engineering 13:237–255
13. Snyder DL, Miller MI (1991) Random point processes in time and space, 2nd edn. Springer-Verlag, New York
14. Daley DJ, Vere-Jones D (1988) An introduction to the theory of point processes. Springer-Verlag, New York
15. Puri ML, Ralescu DA (1978) Fuzzy random variables. J Math Anal Appl 64:409–422
16. Kwakernaak H (1978) Fuzzy random variables I. Definitions and theorems. Information Sciences 15:1–29
17. Nahmias S (1978) Fuzzy variables. Fuzzy Sets and Systems 1:97–101
18. Krätschmer V (2001) A unified approach to fuzzy random variables. Fuzzy Sets and Systems 123:1–9
19. Schneider R (1993) *Convex bodies: the Brunn–Minkowski theory.* Cambridge University Press, Cambridge, UK
20. Matheron G (1975) Random sets and integral geometry. Wiley, New York
21. Aumann RJ (1965) Integrals of set-valued functions. J Math Anal Appl 12:1–22
22. Körner R (1997) On the variance of fuzzy random variables. Fuzzy Sets and Systems 92:83–93
23. Chung KL (1968) A course in probability theory. Harcourt, Brace & World Inc., New York
24. Andrews DWK (1992) Generic uniform convergence. Econometric Theory 8:241–257
25. Lehmann EL (1983) Theory of Point Estimation. Springer-Verlag, New York
26. Zeephongsekul P (2001) On the variability of fuzzy debugging. Fuzzy Sets and Systems 123:29–38
27. Pham H (2000) Software reliability. Springer-Verlag, Singapore
28. Xie M (1991) *Software reliability modelling. World Scientific, Singapore*

Chapter 2
Modeling and Reliability Evaluation
of Multi-state k-out-of-n Systems

Zhigang Tian, Wei Li, Ming J. Zuo

Department of Mechanical Engineering,
University of Alberta, Canada

2.1 Introduction

The k-out-of-n system structure is a very popular type of redundancy in fault tolerant systems, with wide applications in both industrial and military systems. Examples of such fault tolerant systems include the multi-engine system in an airplane, the multi-display system in a cockpit, and the multi-transmitter system in a communication system [1].

2.1.1 Binary k-out-of-n Systems

Definition 1. A n-component system is called a binary k-out-of-n:G system if it is working whenever at least k components are working. A n-component system is called a binary k-out-of-n:F system if it is failed whenever at least k components are failed.

Efficient reliability evaluation algorithms for binary k-out-of-n systems with independent components have been provided by Barlow and Heidtmann [2] and Rushdi [3], as follows:

$$R(n,k) = p_n \cdot R(n-1,k-1) + q_n \cdot R(n-1,k), \tag{2.1}$$

where $R(n,k)$ is the recursive function, representing the reliability of a k-out-of-n:G system. p_n is the reliability of component n, and $q_n = 1 - p_n$. The boundary conditions are:

$$R(n,0) = 1,$$
$$R(n,k) = 0, \quad \text{for} \quad 0 < n < k. \tag{2.2}$$

Wu and Chen generalized the binary k-out-of-n system models to the binary weighted k-out-of-n models [4].

Hoang Pham, *Recent Advances in Reliability and Quality in Design*
©Springer 2008

Definition 2. In a binary weighted k-out-of-n system, component i carries a weight of w_i, $w_i > 0$ for $i = 1, 2, \ldots, n$. The total weight of all components is w, $w = \sum_{i=1}^{n} w_i$. The system works if and only if the total weight of working components is at least k, a pre-specified value.

The "weight" here means the contribution of a component. In the binary weighted system, the component with higher contribution to the system has higher "weight." The component with lower contribution to the system has lower "weight." For example, a jet plane usually has several engines. The engine with higher drive has higher "weight." The engine with lower drive has lower "weight."

A recursive equation is provided by Wu and Chen [4]. We use $R(i, j)$ to represent the probability that a system with j components can output a total weight of at least i. Then, $R(k, n)$ is the reliability of the weighted k-out-of-n:G system. The following recursive equation can be used for reliability evaluation of such systems.

$$R(i, j) = p_j R(i - u_j, j - 1) + q_j R(i, j - 1) \,, \tag{2.3}$$

which requires the following boundary conditions:

$$R(i, j) = 1 \,, \quad \text{for} \quad i \leq 0 \,, \quad j \geq 0 \,, \tag{2.4}$$
$$R(i, 0) = 0 \,, \quad \text{for} \quad i > 0 \,. \tag{2.5}$$

An important variation of binary weighted k-out-of-n systems is weighted voting systems [5,6].

2.1.2 Multi-state Systems

Many practical components and systems have more than two different performance levels. For example, a power generator in a power station can work at full capacity, which is its nominal capacity, say 10 mW, when there are no failures at all [7]. Certain types of failures can cause the generator to fail completely, while other failures will lead to the generator working at a reduced capacity say 4 mW. On the system level, let's consider a power generating system consisting of several power generators. The abilities of the system to meet high power load demand, normal power load demand and lower power load demand can be regarded as different system states. Another example of multi-state components is an oil transmission pipeline. The pipeline is used to transmit oil from the source to spots A, B and C aligned in order along the pipeline. We say that the pipeline is in state 0 when it cannot transmit oil to any of the spots; it is in state 1 if the oil can reach spot A; it is in state 2 if the oil can reach up to spot B, *i.e.*, spot A and B; it is in state 3 if the oil can reach up to spot C. A component with multiple failure modes can also be considered to be a multi-state component [8]. A component like this has a working state and several failure states that might have different impacts on the system level. Since Barlow and Wu [9] presented the first multi-state coherent system structure, many system structures, such as series-parallel structure, k-out-of-n structure and network structure, have been extended from the binary cases to the multi-state cases by allowing the components and the systems to take more than two possible states. With multi-state models, we can deal with reliability issues of engineering systems more accurately, and would be able to answer more questions.

2.1.3 Overview of Multi-state k-out-of-n System Modeling and Evaluation

El-Neweihi *et al.* [10] defined the first multi-state k-out-of-n system model, where the system state was defined as the state of the k-th best component. In another word, at any state j, for the system to be in state j or above, there should be at least k components in state j or above. That is, the k value is the same with respect to all states. Boedigheimer and Kapur [11] defined the multi-state k-out-of-n model from the perspective of lower and upper boundary points, and their definition is in fact consistent with that by El-Neweihi *et al.*

Huang *et al.* [12, 13] proposed the generalized multi-state k-out-of-n:G system model, where there can be different k values with respect to different states. Zuo and Tian developed an efficient recursive algorithm for reliability evaluation of generalized multi-state k-out-of-n systems with identically and independently distributed (iid) components [14] and independent components [15]. A reliability bounding approach is also developed for this model. However, Huang's model of multi-state k-out-of-n systems [12] suffers from the fact that few practical applications can fit into this model.

There might be other applications of Huang's model yet to be identified. Tian *et al.* [16] proposed another multi-state k-out-of-n model so that more practical applications can fit into it.

By allowing the components to have multiple states, Li and Zuo [17] proposed the multi-state weighted k-out-of-n system model by extending the binary weighted k-out-of-n system model to multi-state context. Obviously, the multi-state weighted-k-out-of-n system model has more flexibilities in modeling systems involving weighted-k-out-of-n structure.

In this chapter, we do not consider the case of consecutive multi-state k-out-of-n systems [18, 19].

Assumptions:

- The state space of each component and the system is $\{0, 1, 2, \ldots, M\}$.
- The state of the system is completely determined by the states of the components.

Notation:

n: The number of components of a system
M: The maximum state level of a multi-state system and its components

x_i: state of component i, $x_i = j$ if component i is in state j, $0 \leq j \leq M$, $1 \leq i \leq n$

$\mathbf{x}$:   an n-dimensional vector representing the states of all components, $\mathbf{x} = (x_1, x_2, \ldots, x_n)$

$\phi(\mathbf{x})$:   state of the system, $0 \leq \phi(\mathbf{x}) \leq M$

k_j: the k value with respect to level j of a generalized multi-state k-out-of-n system

$\mathbf{v}_j$:   a minimal cut vector to level j of an increasing multi-state k-out-of-n:F system

$P_{s,j}$: $\Pr(\phi(\mathbf{x}) \geq j)$

$Q_{s,j}$: $\Pr(\phi(\mathbf{x}) < j)$, i.e., $1 - P_{s,j}$

$r_{s,j}$: $\Pr(\phi(\mathbf{x}) = j)$

$P(\bullet)$:   the recursive function for k-out-of-n:G systems

$Q(\bullet)$:   the recursive function for k-out-of-n:F systems

$\mathbf{k}$:   the $\mathbf{k}$ vector of a multi-state k-out-of-n system, $\mathbf{k} = (k_1, k_2, \ldots, k_M)$

$\mathbf{P}$:   the component state distribution matrix for a nominal multi-state k-out-of-n system

$p_{n,j}$:   the probability of component n in state j

$\mathbf{k}^j$:   the generated $\mathbf{k}$ vector when component n is in state j

$\mathbf{P}^j$:   the generated $\mathbf{P}$ matrix when component n is in state j

2.2 Multi-state *k*-out-of-*n* System Models

2.2.1 Multi-state k-out-of-n:G System Model by Huang et al.

This model is the first model that allows different k values with respect to different states.

Definition 3 (Huang *et al.* [12]). An n-component system is called a generalized multi-state k-out-of-n:G system if $\phi(\mathbf{x}) \geq j$ ($1 \leq j \leq M$) whenever there exists an integer value l ($j \leq l \leq M$) such that at least k_l components are in state l or above.

- When $k_1 \leq k_2 \leq \cdots \leq k_M$, the system is called an increasing multi-state k-out-of-n:G system.
- When $k_1 > k_2 > \cdots > k_M$, the system is called a decreasing multi-state k-out-of-n:G system.
- When k_j is a constant, i.e., $k_1 = k_2 = \cdots = k_M = k$, the structure of the system is the same for all system state levels. This reduces to the definition of the simple multi-state k-out-of-n:G system studied by El-Neweihi [10] and Boedigheimer and Kapur [11]. Such systems are called constant multi-state k-out-of-n:G systems.

Zuo and Tian [14] proposed the definition of the generalized multi-state k-out-of-n:F system, which is the image of the corresponding multi-state k-out-of-n:G system:

Definition 4. An n-component system is called a generalized multi-state k-out-of-n:F system if $\phi(\mathbf{x}) < j$ $(1 \leq j \leq M)$ whenever the states of at least k_l components are below l for all l such that $j \leq l \leq M$.

We provide an example to illustrate the modeling of an engineering system as a decreasing multi-state k-out-of-n:G model [14].

Example 1. Consider a power station with three generators. Each generator is treated as a component and there are 3 components in this system. Each generator may be in three possible states, 0, 1, and 2. When a generator is in state 2, it is capable of generating 10 mW in power output; in state 1, 2 mW; and in state 0, 0 mW. The total power output of the system is equal to the sum of the power output from all three generators. The system may also be in three different states: 0, 1, and 2. When the total output is greater than or equal to 10 mW, the system is considered to be in state 2; otherwise but greater than or equal to 4 mW, in state 1; otherwise, in state 0. Based on these descriptions, the system can be considered to be a decreasing multi-state k-out-of-n:G system with the following parameters:

$$n = 3, \quad M = 2, \quad k_1 = 2, \quad k_2 = 1.$$

Using the terminology in Definition 1, we can describe this model as follows: The system is in state 2 whenever at least 1 component is in state 2; in state 1 or above whenever either at least 1 component is in state 2 or at least 2 components are in state 1 or above; and in state 0 otherwise.

2.2.2 Multi-state k-out-of-n System Model by Tian et al.

The multi-state k-out-of-n:G system model by Tian *et al.* [16] is presented as follows:

Definition 5. An n-component system is called a multi-state k-out-of-n:G system if $\phi(\mathbf{x}) \geq j$ $(1 \leq j \leq M)$ whenever at least k_l components are in state l or above for all l such that $1 \leq l \leq j$.

Intuitively, to be in state j or above, the system has to meet all the requirements on the number of components at states from 1 to j. In Huang's model of multi-state k-out-of-n:G system in Definition 3, however, the system is in state j or above if any of the requirements on the number of components at states from j to M can be met.

A special case of the proposed multi-state k-out-of-n:G system model given in Definition 5 is defined as follows:

Definition 6. A multi-state k-out-of-n:G system is called a decreasing multi-state k-out-of-n:G system if $k_1 > k_2 > \cdots > k_M$.

As will be shown later, the reliability evaluation of the general case of multi-state k-out-of-n:G system given in Definition 5 can be handled through a decreasing multi-state k-out-of-n:G system given in Definition 6.

The multi-state k-out-of-n:F system model is also defined as follows [16]:

Definition 7. An n-component system is called a multi-state k-out-of-n:F system if $\phi(\mathbf{x}) < j$ ($1 \leq j \leq M$) whenever there exists an integer value l ($1 \leq l \leq j$) such that at least k_l components are in states below l.

There is an equivalent multi-state k-out-of-n:G system with respect to each multi-state k-out-of-n:F system, and *vice versa*. As to be discussed later, the minimal path vectors of the multi-state k-out-of-n:G have special patterns, which enables us to develop efficient reliability evaluation algorithms for it. A multi-state k-out-of-n:F system can be evaluated via its equivalent multi-state k-out-of-n:G system.

Multiple states are interpreted as multiple levels of capacity [16]. To fit into the multi-state k-out-of-n:G system model, an engineering system should have the following characteristics: (1) the system is supposed to meet multiple types of demands. (2) a component has different levels of capacity, where higher level of capacity means that the component can contribute to meet additional types of demands on the system level. (3) each type of demand on the system level requires at least a certain number of components to contribute to meet the demand. In the following, we present an example of this category of applications of the multi-state k-out-of-n:G system model.

Example 2. Consider an oil supply system [16], as shown in Fig. 2.1. Oil is delivered from the oil source to three stations through four oil pipelines. A pipeline is considered to be a multi-state component (thus $n = 4$). Due to the possible failures in different parts of a pipeline and due to the pumping performance of the oil source, a pipeline might be in four possible states:

- state 0: oil cannot reach any stations.
- state 1: oil can reach up to station 1.
- state 2: oil can reach up to station 2.
- state 3: oil can reach up to station 3.

Each station has different demands on oil:

- station 1: requires at least four pipelines working to meet its demand.
- station 2: requires at least two pipelines working to meet its demand.
- station 3: requires at least three pipelines working to meet its demand.

On the system level, the oil supply system has four states:

- system state 0: it cannot meet the oil demand of any of the stations.
- system state 1: it can meet the oil demand of station 1 only.
- system state 2: it can meet the oil demands of station 1 and station 2 only.
- system state 3: it can meet the oil demands of station 1, 2 and 3.

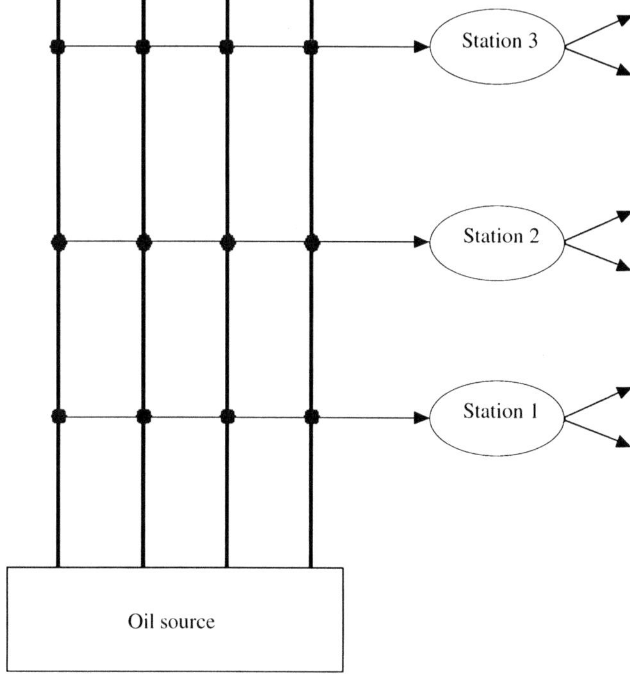

Fig. 2.1 An oil supply system

In practice, we might be only interested in the reliability of the oil supply system in terms of meeting the demands of all the stations, *i.e.*, the probability of the system in state 3. Based on the descriptions above, this oil supply system can be regarded as a multi-state k-out-of-n:G system given in Definition 5 with $n = 4$, and $k_1 = 4$, $k_2 = 2$, $k_3 = 3$.

Similar applications can be found in power supply systems and telecommunication systems.

2.2.3 Multi-state Weighted k-out-of-n System Model

Suppose there are n components in a system, each component may be in $M + 1$ states: $\{0, 1, 2, \ldots, M\}$, component i, when in state j, has a weight value of w_{ij}.

In Sect. 2.1.1, we have described the binary weighted k-out-of-n:G system model. In a binary weighted k-out-of-n:G system, the weight of the system is the sum of the weights of working components. When a component fails, its contribution to system weight is 0.

In a multi-state context, a component may be in different states. When it is in a different state, it may have a different contribution to the system. When it fails completely, its contribution to the system is zero. In the multi-state weighted k-out-of-n:G model to be defined in this section, a certain component in a certain state has a certain contribution to the system's performance. The formal definition of Model I of the multi-state weighted k-out-of-n:G system is given as follows.

Definition 8. The system is in state j or above if the total weight of all components is equal to or greater than k_j, a pre-specified value [17].

Let ϕ be the structure function of the system representing the state of the system and W the total weight of all components. Then, this definition means $\Pr\{\phi \geq j\} = \Pr\{W \geq k_j\}$. Since state 0 is the worst state of the system, we have $\Pr\{\phi \geq 0\} = 1$.

Huang *et al.* [12] proposed the general multi-state k-out-of-n system model in 2000. In their definition, for the system state to be not lower than a given value j, the number of components whose states are not lower than j must be at least k_j, a pre-specified value. In this definition, the components whose states are below j do not make any contribution for the system to be in state j or above. Based on this idea, we define a new model, Model II, of the multi-state weighted k-out-of-n:G system. In the proposed definition, for the system to be in state j or above, the sum of the weights of the components that are in state j or above must be not less than k_j, a pre-specified value. The difference between Model I and Model II presented in this section is whether the components whose states are below j are making any contributions for the system to be in state j or above. The formal definition of Model II is given below.

Definition 9. The system is in state j or above if the total weight of the components in state j or above is equal to or greater than k_j, a pre-specified value [17].

Let ϕ be the structure function of the system and W_j be the sum of the utilities of the components whose states are j or above. We then have $\Pr\{\phi \geq j\} = \Pr\{W_j \geq k_j\}$.

Applications of multi-state weighted k-out-of-n system can be found in aircraft, telecommunication networks, traffic systems, satellites, electric generation and distribution systems, mining and distribution systems, space shuttles and computer systems. Three practical examples are given below.

Example 3. Consider a conveyor belt set, where there are several conveyors working together in parallel. During the life of the conveyor, it will deteriorate from being perfect to failing completely. When it deteriorates, it cannot work under the full load situation any more. That means corresponding to different states, it has different weights. When we require that in the set, the total drive of all the conveyors to be at least 100 kW to transmit coals, it is a multi-state Model I weighted k-out-of-n system. Because of the wear-out, some of the conveyors cannot work under full load. In this situation, if we require that in the set, the total drive generated by the conveyors which still can work under full-load to be 10,0kw drive (because the coal block is big), it is a multi-state Model II weighted k-out-of-n system.

Example 4. A modern aircraft usually has several engines. We consider all the engines together as a system and every engine as a component. Every engine may work in different states such as working, partially working, or failed, and corresponding to different states it can supply different thrust. The thrust forces are the performance utilities corresponding to the component's different states. The total thrust of the air plane is the sum of the thrust forces of all the engines, no matter they fail, fully work or partially work. This is an example of multi-state Model I weighted *k*-out-of-*n* system.

Example 5. Suppose that there are five bridges on a river. Different bridges have different allowed maximum load levels for vehicles. When the bridge becomes older and older, it will deteriorate more and more. With the increase of the deterioration level, the allowed maximum load level of the bridge may be reduced. That is, the bridge can be in different states and corresponding to each state there is a specified maximum load level for vehicles as the performance measure. In this example, the multi-state weighted system is the bridge system with five bridges. Each bridge is a multi-state weighted component. The performance of the component and the system are measured by the allowed maximum load levels for vehicles. When a large truck motorcade wants to go across the river, some bridges may not be used because their specified maximum load levels for the vehicles are less than the weight of the big truck. This is an example of Model II multi-state weighted *k*-out-of-*n* systems. In this situation, we only consider the bridges that can bear the big trucks. This point is the difference between Example 5 and Example 4. In Example 4, all the engines can make contribution, so they are all considered in the system.

2.3 Reliability Evaluation of Multi-state *k*-out-of-*n* Systems

The reliability evaluations of multi-state *k*-out-of-*n* systems using recursive algorithms will be presented in the section. Examples will be used to illustrate the correctness and efficiency of the algorithms.

First, let us look at the reliability evaluation problem of multi-state *k*-out-of-*n* systems. A multi-state *k*-out-of-*n* system has multiple states. The reliability evaluation problem of a multi-state *k*-out-of-*n* system is calculating the probabilities of the system in different possible states. $r_{s,j}$ is used to represent the probability of the system in state *j*. Thus, the problem is to calculate $r_{s,j}$ for all *j*.

$r_{s,j}$ can be calculated through $P_{s,j}$, the probability of the system in state *j* or above. It can also be calculated through $Q_{s,j}$, the probability of the system in states below *j*. In some cases, certain $P_{s,j}$ or $Q_{s,j}$ is all that we are interested in, and we do not have to calculate all the $r_{s,j}$ values.

2.3.1 Fundamental Elements of Recursive Algorithms

Recursive algorithms have been efficient ways for system reliability evaluations [1]. As to be shown later, the efficient evaluation algorithms for multi-state k-out-of-n systems are all recursive algorithms [14, 16, 17].

There are three fundamental elements in a recursive algorithm. We would like to use the recursive algorithm for binary k-out-of-n systems, (2.1) and (2.2), as an example to illustrate these elements.

2.3.1.1 Recursive Function

Recursive function is the key function that calls itself in the recursive algorithm. It has some parameters, which change during the course of the recursive algorithm.

In the recursive algorithm for binary k-out-of-n systems, $R(n, k)$ is the recursive function, in which n and k are the parameters.

2.3.1.2 Updating Algorithm

The updating algorithm decides how the recursive function calls itself. In other words, the updating algorithm decides the relationship between a recursive function with certain parameters and recursive functions with different parameters, which are typically less complex.

In the recursive algorithm for binary k-out-of-n systems, (2.1) shows the updating algorithm. The recursive function $R(n, k)$ can be calculated via calculating two recursive functions $R(n - 1, k)$ and $R(n - 1, k - 1)$, which have smaller parameters and thus easier to calculate.

2.3.1.3 Boundary Conditions

When one of the boundary conditions is met, the value of recursive function is a certain value, or can be determined in a specific and simple way.

In the recursive algorithm for binary k-out-of-n systems, (2.2) shows the boundary conditions. There are two boundary conditions in this case. The first one is when $k = 0$, the value of the recursive function $R(n, k)$ is 1. The other boundary condition is when $0 < n < k$, the value of the recursive function $R(n, k)$ is 0.

2.3.2 Reliability Evaluation of the Multi-state k-out-of-n Model Defined by Huang et al.

For any multi-state k-out-of-n:G system defined by Huang *et al.* [12], there is an equivalent multi-state k-out-of-n:F system defined by Zuo and Tian [14]. The reliability evaluation of a multi-state k-out-of-n:G system can be made via its equivalent multi-state k-out-of-n:F system. In this section, we will present the recursive evaluation algorithms for multi-state k-out-of-n:F systems with iid components [14] and with independent components [15].

Before presenting the algorithms, the definition of nominal increasing multi-state k-out-of-n:F system proposed in [14] should be introduced:

Definition 10. [14] A nominal increasing multi-state k-out-of-n:F system is the same as an increasing multi-state k-out-of-n:F system, except that the probability of a component in all possible states may be less than 1.

$Q_{s,j}$ of a general multi-state k-out-of-n:F system can be calculated through a generated increasing multi-state k-out-of-n:F system [14]. Thus, the following discussion is focused on the increasing case.

When the components are iid the recursive algorithm proposed in [14] is used for the performance evaluation of multi-state k-out-of-n systems. Specifically, we need to calculate $Q_{s,j}$, probability of the system in states below j, to any state j of the system.

The **Recursive Function** is denoted by $Q(m, N, \mathbf{k}, \mathbf{p})$, which represents the recursive function used in the recursive algorithm, where m is the number of possible states of the nominal increasing multi-state k-out-of-n:F system minus 1, N is the number of components of the nominal increasing multi-state k-out-of-n:F system, $\mathbf{k}$ is the $\mathbf{k}$ vector of the nominal increasing multi-state k-out-of-n:F system where $\mathbf{k} = (k_1, k_2, \ldots, k_m)$, and $\mathbf{p}$ is the probability vector of the nominal increasing multi-state k-out-of-n:F system where $\mathbf{p} = (p_0, p_1, \ldots, p_m)$.

The recursive function $Q(m, N, \mathbf{k}, \mathbf{p})$ is designed to represent the probability for the system to be in state "0" of a nominal increasing multi-state k-out-of-n:F system with $m + 1$ possible states, totally N components, vector $\mathbf{k}$, and probability vector $\mathbf{p}$. Thus, to calculate the $Q_{s,j}$ value of any multi-state k-out-of-n:F system, we can first transform it into a nominal increasing multi-state k-out-of-n:F system to state j and do the calculation using the recursive algorithm.

The **Updating Algorithm** of the recursive algorithms is as follows:

$$Q(m, N, \mathbf{k}, \mathbf{p}) = \sum_{i=k_1}^{k_2-1} \binom{N}{i} p_0^i \cdot Q(m-1, N-i, \dot{\mathbf{k}}, \dot{\mathbf{p}}) + Q(m-1, N, \ddot{\mathbf{k}}, \ddot{\mathbf{p}}) \qquad (2.6)$$

where $\dot{\mathbf{k}} = (k_2 - i, k_3 - i, \ldots, k_m - i)$, $\dot{\mathbf{p}} = (p_1, p_2, \ldots, p_m)$, $\ddot{\mathbf{k}} = (k_2, k_3, \ldots, k_m)$, $\ddot{\mathbf{p}} = (p_0, p_1 + p_2, p_3, \ldots, p_m)$.

The **Boundary Condition** for the recursive algorithm is

$$Q(1, N, \mathbf{k}, \mathbf{p}) = \sum_{i=k_1}^{N} \binom{N}{i} p_0^i \cdot p_1^{N-i} \qquad (2.7)$$

From (2.6) and (2.7), we can see that this algorithm is actually recursive on the parameter m, not on the number of components N. Therefore, we can apply the recursive algorithm to large systems including a large number of components, without leading to exponential growth of computation time.

We will use one example to illustrate and verify the recursive algorithm for increasing multi-state k-out-of-n:F systems proposed in the previous section.

Example 6. We consider an increasing multi-state k-out-of-n:F system with 10 iid components and 4 possible states, *i.e.*, $n = 10$ and $M = 3$. The $\mathbf{k}$ vector is $\mathbf{k} = (k_1, k_2, k_3) = (3, 6, 8)$. The state distribution of components is $\mathbf{p} = (0.1, 0.3, 0.4, 0.2)$.

To calculate $Q_{s,3}$, we get a nominal increasing multi-state k-out-of-n:F system to state 3, with $m = 1$, $N = 10$, $\mathbf{k} = (k_3) = 8$, and $\mathbf{p} = (p_0 + p_1 + p_2, p_3) = (0.8, 0.2)$. Using the recursive algorithm, the value $Q_{s,3}$ is equal to $Q(m, N, \mathbf{k}, \mathbf{p})$, which is 0.6778.

To calculate $Q_{s,2}$, we get a nominal increasing multi-state k-out-of-n:F system to state 2, with $m = 2$, $N = 10$, $\mathbf{k} = (k_2, k_3) = (6, 8)$, and $\mathbf{p} = (p_0 + p_1, p_2, p_3) = (0.4, 0.4, 0.2)$. Using the recursive algorithm, the value $Q_{s,2}$ is equal to $Q(m, N, \mathbf{k}, \mathbf{p})$, which is 0.1523.

To calculate $Q_{s,1}$, we get a nominal increasing multi-state k-out-of-n:F system to state 1, with $m = 3$, $N = 10$, $\mathbf{k} = (k_1, k_2, k_3) = (3, 6, 8)$, and $\mathbf{p} = (p_0, p_1, p_2, p_3) = (0.1, 0.3, 0.4, 0.2)$. Using the recursive algorithm, the value $Q_{s,1}$ is equal to $Q(m, N, \mathbf{k}, \mathbf{p})$, which is 0.0308.

We can get the probability of the system at each individual state

$$r_{s,0} = 0.0308\,, \quad r_{s,1} = 0.1214\,, \quad r_{s,2} = 0.5255\,, \quad r_{s,0} = 0.3222\,.$$

We also used the enumerating method to calculate the state distribution of this increasing multi-state k-out-of-n:F system. The results agree with those we get using the proposed recursive algorithm, which verifies the correctness of the recursive algorithm.

When the components are independent, the algorithm proposed by Tian *et al.* can be used for the evaluation of multi-state k-out-of-n:F system [15], that is, calculating the probability of a nominal increasing multi-state k-out-of-n:F systems with independent components in nominal state "0."

The **Recursive Function** we are using in this recursive algorithm is denoted by $Q(n, \mathbf{k}, \mathbf{P})$, where

- n: the number of components of the nominal increasing multi-state k-out-of-n:F system,
- $\mathbf{k}$: the $\mathbf{k}$ vector of the nominal increasing multi-state k-out-of-n:F system, $\mathbf{k} = (k_1, k_2, \ldots, k_M)$, where M is the number of possible states of the nominal increasing multi-state k-out-of-n:F system minus 1,
- $\mathbf{P}$: the components state distribution matrix for the nominal increasing multi-state k-out-of-n:F system,

$$\mathbf{P} = \begin{pmatrix} p_{1,0} & p_{1,1} & \cdots & p_{1,M} \\ p_{2,0} & p_{2,1} & \cdots & p_{2,M} \\ \vdots & \vdots & \vdots & \vdots \\ p_{n,0} & p_{n,1} & \cdots & p_{n,M} \end{pmatrix}$$

The recursive function $Q(n, \mathbf{k}, \mathbf{P})$ is designed to represent the probability in nominal state "0" of a nominal increasing multi-state k-out-of-n:F system with n independent components, vector $\mathbf{k}$, and probability matrix $\mathbf{P}$. To evaluate $Q_{s,j}$, the states below j are combined into nominal state "0" of the generated nominal increasing multi-state k-out-of-n:F system with vector $\mathbf{k}$ and probability matrix $\mathbf{P}$, and thus we have $Q_{s,j} = Q(n, \mathbf{k}, \mathbf{P})$.

The **Updating Algorithm** is as follows:

$$Q(n, \mathbf{k}, \mathbf{P}) = \sum_{j=0}^{M} p_{n,j} \cdot Q(n-1, \mathbf{k}^{j}, \mathbf{P}^{j}) \tag{2.8}$$

The basic idea of this algorithm is similar to that of the recursive algorithm for binary multi-state k-out-of-n systems: we enumerate the cases where component n is in different possible states, and thus evaluate a system with n components via evaluating several systems with $n-1$ components. For each certain j on the right hand side of 2.8), we need to reorganize $\mathbf{k}$ and $\mathbf{P}$ to generate $\mathbf{k}^{j}$ and $\mathbf{P}^{j}$.

First for $j \neq M$,

$$k_{l}^{j} = k_{l} - 1, \quad \text{for} \quad l \geq j+1$$
$$k_{l}^{j} = k_{l}, \quad \text{for} \quad l < j+1. \tag{2.9}$$

The idea is that if component n is in state j, the required number of components for any state above j should be decreased by 1. For $j = M$, we have $k_{l}^{j} = k_{l}$ for $1 \leq l \leq M$. $\mathbf{P}^{j}$ is obtained by deleting the nth row from matrix $\mathbf{P}$, and thus $\mathbf{P}^{j}$ is a $n-1$ by $M+1$ matrix.

There is a special case when generating $\mathbf{k}^{j}$ and $\mathbf{P}^{j}$, under which a certain state will be "absorbed" by the adjacent lower state, and we should make further transformations on $\mathbf{k}^{j}$ and $\mathbf{P}^{j}$. This special case is the case when $k_{h}^{j} = k_{h+1}^{j}$ for a state h ($0 \leq h \leq M-1$). In this case, state $h+1$ is absorbed by state h. k_{h+1}^{j} is deleted from $\mathbf{k}^{j}$; $p_{i,h}^{j} = p_{i,h}^{j} + p_{i,h+1}^{j}$ for $1 \leq i \leq n-1$, and then the column $h+1$ is deleted from $\mathbf{P}^{j}$. Thus, the number of possible states is decreased from $M+1$ to M. This is in fact a major reason that the computation time using this algorithm will not increase exponentially with the increase of n. By generating $\mathbf{k}^{j}$ and $\mathbf{P}^{j}$ in this way, $\mathbf{k}^{j}$ will always be a strictly increasing vector.

The **Boundary Conditions** for the recursive algorithm are as follows:

Boundary condition 1: $k_M > n$. In this case, $Q(n, \mathbf{k}, \mathbf{P}) = 0$.

Boundary condition 2: $M = 1$. In this case, the increasing multi-state k-out-of-n:F system is reduced to a binary k-out-of-n:F system with independent components. The recursive algorithm in (2.1) and (2.2) for binary k-out-of-n systems can thus be used to evaluate $Q(n, \mathbf{k}, \mathbf{P})$.

Example 7. In this example, we consider an increasing multi-state k-out-of-n:F system with independent components. The system under consideration has five independent components and five possible states. The $\mathbf{k}$ vector is $\mathbf{k} = (k_1, k_2, k_3, k_4) = (1, 2, 3, 4)$. The state distribution matrix is:

$$P = \begin{pmatrix} 0.1 & 0.2 & 0.1 & 0.4 & 0.2 \\ 0.2 & 0.1 & 0.1 & 0.3 & 0.3 \\ 0.1 & 0.2 & 0.1 & 0.2 & 0.4 \\ 0.4 & 0.2 & 0.1 & 0.1 & 0.2 \\ 0.1 & 0.1 & 0.2 & 0.2 & 0.4 \end{pmatrix}$$

Using the recursive algorithm, we obtained

$$\mathbf{Q}_s = (0.2526, 0.2939, 0.3350, 0.5261)$$
$$\mathbf{r}_s = (0.2526, 0.0413, 0.0412, 0.1910, 0.4739) \,.$$

We also used the enumerating method to evaluate this system, and the results agree with the results by the recursive algorithm, which illustrates the correctness of the recursive algorithm in case of independent components.

2.3.3 Reliability Evaluation of the Multi-state *k-out-of-n* Model Defined by Tian et al.

Similar to the nominal increasing multi-state k-out-of-n:F system model defined by Zuo and Tian [14], here we define the nominal decreasing multi-state k-out-of-n:G system model.

Definition 11. A nominal decreasing multi-state k-out-of-n:G system is the same as an decreasing multi-state k-out-of-n:G system, except that the probability of a component in all possible states may be less than 1.

A nominal decreasing multi-state k-out-of-n:G system is a result of only considering part of the component states and/or combining several adjacent states together. The nominal decreasing multi-state k-out-of-n:G system model plays an important role in the evaluation of multi-state k-out-of-n:G systems. The $P_{s,j}$ for any state j of a general multi-state k-out-of-n:G system can be evaluated by transforming this general multi-state k-out-of-n:G system into a nominal decreasing multi-state k-out-of-n:G system, and calculating the probability of this nominal decreasing multi-state k-out-of-n:G system in the highest nominal state.

To evaluate the state distribution of a decreasing multi-state k-out-of-n:G system, we need to calculate $P_{s,1}, P_{s,2}, \ldots, P_{s,M}$. $P_{s,j}$ is the probability that the system is in state j or above.

When the components are iid the evaluation of multi-state k-out-of-n:G systems can be done using the following algorithm. Based on the form of minimal path vector for decreasing multi-state k-out-of-n:G systems [16], to any state j, we have

$$P_{s,j} = \Pr(\mathbf{x} \geq (\underbrace{j, j, \ldots, j}_{k_j}, \underbrace{j-1, \ldots, j-1}_{k_{j-1}}, \ldots, \underbrace{1, \ldots, 1}_{k_1}, 0, \ldots, 0)^\star) \qquad (2.10)$$

$$\underbrace{}_{n}$$

Zuo and Tian [14] developed an efficient algorithm for the reliability evaluation of multi-state k-out-of-n:F systems under Huang's definition [12], based on minimal cut vectors. In their algorithm, the probability of the system in states below j, *i.e.*, $Q_{s,j}$, is equal to the probability that there exists a minimal cut vector of the system so that each component state is not bigger than the corresponding element of the minimal path vector:

$$Q_{s,j} = \Pr(\mathbf{x} \leq (\underbrace{j-1, j-1, \ldots, j-1}_{k_j}, \underbrace{j, \ldots, j}_{k_{j+1}}, \ldots, \underbrace{M-1, \ldots, M-1}_{k_M}, M, \ldots, M)^\star)$$

$$\underbrace{}_{n}$$

$$(2.11)$$

The probability calculations of $P_{s,j}$ in (2.10) and $Q_{s,j}$ in (2.11), we would say, are mathematically the same. Thus, we would be able to reformat $P_{s,j}$ and use the algorithm by Zuo and Tian [14] to perform the probability calculation. Actually, $P_{s,j}$ in (2.10) can be expressed in the form in (2.11) by reversing the order of the component states, *i.e.*, letting $p_j = p_{M-j}$. Under the reversed order of component states, $P_{s,j}$ can be expressed as

$$P_{s,j} = \Pr(\mathbf{x} \leq$$
$$(\underbrace{M-j, \ldots, M-j}_{k_j}, \underbrace{M-j+1, \ldots, M-j+1}_{k_{j-1}}, \ldots, \underbrace{M-1, \ldots, M-1}_{k_1}, M, \ldots, M)^\star)$$

$$\underbrace{}_{n}$$

$$(2.12)$$

The probability in (2.12) has the same form as that in (2.11), and thus can be calculated using the algorithm by Zuo and Tian [14].

Zuo and Tian's algorithm [14], which is simple and elegant, has been shown to be very efficient. In their efficiency investigation example of an increasing multi-state k-out-of-n:F system with $k_1 = 1$, $k_2 = 2$, $k_3 = 3$, $k_4 = 4$ and $k_5 = 5$, the reliability evaluation time only increases approximately linearly with the increase of the number of components [14].

When the components are independent, the evaluation of multi-state k-out-of-n:G systems can be performed using the following algorithm. The recursive algorithm to be presented in this section is very similar to the algorithm in Sect. 2.2.3. The difference is that the algorithm in this section focuses on $P_{s,j}$, while the algorithm in Sect. 2.2.3 focuses on $Q_{s,j}$. And of course they are for different models.

The $P_{s,j}$ for any state j of a multi-state k-out-of-n:G system can be evaluated by transforming this multi-state k-out-of-n:G system into a nominal decreasing multi-state k-out-of-n:G system, and calculating the probability of this nominal decreasing multi-state k-out-of-n:G system in the highest nominal state. Therefore, in the following part, we will focus on evaluating the probability of a nominal decreasing multi-state k-out-of-n:G systems with independent components in the highest nominal state.

The **Recursive Function** we are using in this recursive algorithm is denoted by $P(n, \mathbf{k}, \mathbf{P})$, where

- n: the number of components of the nominal decreasing multi-state k-out-of-n:G system,
- $\mathbf{k}$: the $\mathbf{k}$ vector of the nominal decreasing multi-state k-out-of-n:G system, $\mathbf{k} = (k_1, k_2, \ldots, k_M)$, where M is the number of possible states of the nominal decreasing multi-state k-out-of-n:G system minus 1,
- $\mathbf{P}$: the component state distribution matrix for the nominal decreasing multi-state k-out-of-n:G system,

$$\mathbf{P} = \begin{pmatrix} p_{1,0} & p_{1,1} & \cdots & p_{1,M} \\ p_{2,0} & p_{2,1} & \cdots & p_{2,M} \\ \vdots & \vdots & \vdots & \vdots \\ p_{n,0} & p_{n,1} & \cdots & p_{n,M} \end{pmatrix}$$

The recursive function $P(n, \mathbf{k}, \mathbf{P})$ is designed to represent the probability in the highest nominal state of a nominal decreasing multi-state k-out-of-n:G system with n independent components, vector $\mathbf{k}$, and probability matrix $\mathbf{P}$. To evaluate $P_{s,j}$ in (2.10), the states j and s above are combined into the highest nominal state of the generated nominal decreasing multi-state k-out-of-n:G system with vector $\mathbf{k}$ and probability matrix $\mathbf{P}$. Thus we have $P_{s,j} = P(n, \mathbf{k}, \mathbf{P})$.

The **Updating Algorithm** is as follows:

$$P(n, \mathbf{k}, \mathbf{P}) = \sum_{j=0}^{M} p_{n,j} \cdot P(n-1, \mathbf{k}^j, \mathbf{P}^j) \tag{2.13}$$

The basic idea of this algorithm is similar to that of the recursive algorithm for binary multi-state k-out-of-n systems: we enumerate the cases where component n

is in different possible states, and thus evaluate a system with n components via evaluating several systems with $n - 1$ components. For each certain j on the right hand side of (2.13), we need to reorganize $\mathbf{k}$ and $\mathbf{P}$ to generate $\mathbf{k}^j$ and $\mathbf{P}^j$.

First for $0 < j < M$,

$$k_l^j = k_l, \quad \text{for} \quad l > j$$
$$k_l^j = k_l - 1, \quad \text{for} \quad l \le j. \tag{2.14}$$

The idea is that if component n is in state j, the required number of components for any state equal to or below j should be decreased by 1. For $j = M$, we have $k_l^j = k_l - 1$ for $1 \le l \le M$. For $j = 0$, we have $k_l^j = k_l$ for $1 \le l \le M$. $\mathbf{P}^j$ is obtained by deleting the nth row from matrix $\mathbf{P}$, and thus $\mathbf{P}^j$ is a $n - 1$ by $M + 1$ matrix.

There is a special case when generating $\mathbf{k}^j$ and $\mathbf{P}^j$, under which a certain state will be "absorbed" by the adjacent upper state, and we should make further transformations on $\mathbf{k}^j$ and $\mathbf{P}^j$. This special case is the case when $k_h^j = k_{h-1}^j$ for a state h ($1 \le h \le M$). In this case, state $h - 1$ is absorbed by state h. k_{h-1}^j is deleted from $\mathbf{k}^j$; $p_{i,h}^j = p_{i,h}^j + p_{i,h-1}^j$ for $1 \le i \le n - 1$, and then the column $h - 1$ is deleted from $\mathbf{P}^j$. Thus, the number of possible states is decreased from $M + 1$ to M. This is in fact a major reason that the computation time using this algorithm will not increase exponentially with the increase of n. By generating $\mathbf{k}^j$ and $\mathbf{P}^j$ in this way, $\mathbf{k}^j$ will always be a strictly increasing vector.

The **Boundary Conditions** for the recursive algorithm are as follows:

Boundary condition 1: $k_1 > n$. In this case, $P(n, \mathbf{k}, \mathbf{P}) = 0$.

Boundary condition 2: $M = 1$. In this case, the decreasing multi-state k-out-of-n:G system is reduced to a binary k-out-of-n:G system with independent components. The recursive algorithms by Barlow and Heidtmann [2] and Rushdi [3] can be used for the reliability evaluation.

Example 8. In this example, we consider an decreasing multi-state k-out-of-n:G system with independent components. The system under consideration has five independent components and five possible states. The $\mathbf{k}$ vector is $\mathbf{k} = (k_1, k_2, k_3, k_4) = (4, 3, 2, 1)$. The state distribution matrix is

$$\mathbf{P} = \begin{pmatrix} 0.2 & 0.4 & 0.1 & 0.2 & 0.1 \\ 0.3 & 0.3 & 0.1 & 0.1 & 0.2 \\ 0.4 & 0.2 & 0.1 & 0.2 & 0.1 \\ 0.2 & 0.1 & 0.1 & 0.2 & 0.4 \\ 0.4 & 0.2 & 0.2 & 0.1 & 0.1 \end{pmatrix}$$

Using the recursive algorithm, we obtained

$$\mathbf{P}_s = (0.5261, 0.3350, 0.2939, 0.2526)$$
$$\mathbf{r}_s = (0.4739, 0.1910, 0.0412, 0.0413, 0.2526).$$

We also used the enumerating method to evaluate this system, and the results agree with the results by the recursive algorithm, which illustrates the correctness of the recursive algorithm in case of independent components.

2.3.4 Reliability Evaluation of Multi-state Weighted k-out-of-n Systems

The reliability evaluation recursive method for the binary weighted k-out-of-n system already exists. In this section, the reliability evaluation recursive algorithms for the multi-state weighted k-out-of-n models are brought forward.

Levitin *et al.* [20] developed the Universal Generating Function (UGF) approach to evaluate multi-state systems [21–27], which can be used to deal with a wide range of multi-state systems. This technique allows one to find the entire system performance distribution based on the performance distributions of its elements by using a rapid algebraic procedure. Although UGF is a universal technique, we need to develop specific operators for different kinds of systems based on their specific logical structures. Since the binary weighted k-out-of-n system is a special case of the multi-state weighted-k-out-of-n system, first we provide the specific operators in the UGF reliability evaluation method for the binary weighted k-out-of-n system. This process can also help us understand the UGF method for the multi-state weighted-k-out-of-n system.

In the binary weighted k-out-of-n system, UGF for the components is:

$$U_i(z) = (1 - p_i)z^{0 \times u_i} + p_i z^{u_i} \tag{2.15}$$

To obtain the UGF of the system based on the individual UGF of the components, the following composition operator Ω can be used:

$$U_s(z) = \Omega(U_1(z), U_2(z), \ldots, U_n(z)) \tag{2.16}$$

In the above equation:

$$\Omega(U_1(z) \cdots U_k(z), U_{k+1}(z) \cdots U_n(z)) = \Omega(U_1(z) \cdots U_{k+1}(z), U_k(z) \cdots U_n(z)) \tag{2.17}$$

$$\Omega(U_1(z) \cdots U_k(z), U_{k+1}(z) \cdots U_n(z)) = \Omega(\Omega(U_1(z) \cdots U_k(z)), \Omega(U_{k+1}(z) \cdots U_n(z))) \tag{2.18}$$

$$\Omega(U_1(z), U_2(z)) = \Omega\left[\sum_{j=1}^{J} p_{1j}z^{g_{1j}} \cdot \sum_{l=1}^{L} p_{2l}z^{g_{2l}} \right] = \sum_{j=1}^{J}\sum_{l=1}^{L} p_{1j}p_{2l}z^{(g_{1j}+g_{2l})} \tag{2.19}$$

Having binary weighted k-out-of-n output performance distribution (OPD) in the above form, one can obtain the system reliability for an arbitrary k using the following operator δ_A:

$$R_s(k) = \delta_A(U_s(z), k) = \delta_A\left(\sum_{k=1}^{K} p_k z^{G_k}, k \right) = \sum_{k=1}^{K} p_k \alpha(G_k - k) \tag{2.20}$$

The function $\alpha(x)$ in the above equation means:

$$\alpha(x) = \begin{cases} 1, x \geq 0 \\ 0, x < 0 \end{cases}$$

To present a recursive algorithm for reliability evaluation of the defined multi-state weighted k-out-of-n systems, we first define the following notations:

n: the number of components in the system
M: the highest possible state of each component
$w_{i,j}$: the weight of component i when it is in state j
$p_{i,j}$: $\Pr\{\text{Component } i \text{ is in state } j\}$
$q_{i,j}$: $\Pr\{\text{Component } i \text{ is in a state below } j\}$, $q_{i,j} = \sum_{l=0}^{j-1} p_{i,l}$.
k_j: the minimum total weight required to ensure that the system is in state j or above.

When Model I is considered, let $R^I_j(k_j, n)$, the **Recursive Function**, denote the probability for the system to be in state j or above based on the defined Model I.

The **Updating Algorithm** for evaluation of the state distribution of the system (Model I) is as follows:

$$R^I_j(k_j, i) = \sum_{r=0}^{r=M} p_{i,r} \cdot R^I_j(k_j - w_{i,r}, i-1) \,. \tag{2.21}$$

The **Boundary Conditions** for this recursive equation are:

$$R^I_j(k, 0) = 0 \,, \quad \text{when} \quad 0 < k \le k_j \,,$$
$$R^I_j(k, i) = 1 \,, \quad \text{when} \quad i \ge 0 \quad \text{and} \quad k \le 0 \,.$$

We would also like to investigate the use of the UGF approach for reliability evaluation of multi-state weighted k-out-of-n systems. The UGF of each multi-state component is now given by:

$$U_i(z) = p_{i,0} z^{w_{i,0}} + p_{i,1} z^{w_{i,1}} + \cdots p_{i,M} z^{w_{i,M}} \,. \tag{2.22}$$

To obtain the system UGF using the component UGFs, we still use the same operator Ω and δ_A as in the UGF method for binary weighted k-out-of-n systems.

When Model II is considered, let $R^{II}_j(k, n)$, the **Recursive Function**, denote the probability for the n component system to have a sum of useful weights of at least k when one is evaluating the probability for the system to be in state j or above.

The following **Updating Algorithm** is proposed for evaluation of the performance distribution of the system based on Definition 9.

$$R^{II}_j(k_j, n) = q_{n,j} \cdot R^{II}_j(k_j, n-1) + \sum_{r=j}^{r=M} p_{n,r} \cdot R^{II}_j(k_j - w_{n,r}, n-1) \,. \tag{2.23}$$

The **Boundary Conditions** for (2.23) are:

$$R^{II}_j(j, 0) = 0 \,, \quad \text{for} \quad j = 1, 2, 3, \cdots, k_j \,.$$
$$R^{II}_j(k, i) = 1 \,, \quad \text{for} \quad k \le 0 \quad \text{and} \quad i = 0, 1, 2, \cdots, n \,.$$

Furthermore, extending the UGF from Model I to Model II, we only need to change the UGF for the individual component to the following form:

$$U_i(z) = q_{i,j}z^0 + p_{i,j}z^{w_{i,j}} + \cdots p_{i,k}z^{w_{i,k}} + \cdots + p_{i,M}z^{w_{i,M}} . \tag{2.24}$$

Example 9. Consider a multi-state weighted k-out-of-n:G system with three components. Every component has three possible states: 0, 1, 2. Tables 2.1 and 2.2 give the reliability distribution and the weight distribution of all the components.

In this example, $n = 3$, $M = 2$, $k_1 = 5$, and $k_2 = 10$. We can use (2.21) to calculate the reliability of the system.

$$R_1^l(5,3) = \sum_{r=0}^{2} p_{3,r} \cdot R_1^l(5 - w_{3,r}, 3 - 1)$$
$$= p_{3,0} \cdot R_1^l(5 - 1, 2) + p_{3,1} \cdot R_1^l(5 - 3, 2) + p_{3,2} \cdot R_1^l(5 - 5, 2)$$
$$= p_{3,0} \cdot R_1^l(4, 2) + p_{3,1} \cdot R_1^l(2, 2) + p_{3,2} \cdot R_1^l(0, 2)$$
$$= p_{3,0} \cdot R_1^l(4, 2) + p_{3,1} + p_{3,2} ,$$

$$R_1^l(4,2) = \sum_{r=0}^{2} p_{2,r} \cdot R_1^l(4 - w_{2,r}, 2 - 1)$$
$$= p_{2,0} \cdot R_1^l(4 - 1, 1) + p_{2,1} \cdot R_1^l(4 - 3, 1) + p_{2,2} \cdot R_1^l(4 - 4, 1)$$
$$= p_{2,0} \cdot R_1^l(3, 1) + p_{2,1} \cdot R_1^l(1, 1) + p_{2,2} \cdot R_1^l(0, 1)$$
$$= p_{2,0} \cdot p_{1,2} + p_{2,1} + p_{2,2}$$
$$= 0.4 * 0.7 + 0.2 + 0.4 = 0.88 ,$$

$$R_1^l(5,3) = p_{3,0} \cdot R_1^l(4, 2) + p_{3,1} + p_{3,2} = p_{3,0} \cdot 0.88 + p_{3,1} + p_{3,2}$$
$$= 0.3 * 0.88 + 0.5 + 0.2 = 0.964 ,$$

$$R_2^l(10,3) = \sum_{r=0}^{2} p_{3,r} \cdot R_2^l(10 - w_{3,r}, 3 - 1)$$
$$= p_{3,0} \cdot R_2^l(10 - 1, 2) + p_{3,1} \cdot R_2^l(10 - 3, 2) + p_{3,2} \cdot R_2^l(10 - 5, 2)$$
$$= p_{3,0} \cdot R_2^l(9, 2) + p_{3,1} \cdot R_2^l(7, 2) + p_{3,2} \cdot R_2^l(5, 2) ,$$

Table 2.1 Reliability distribution of the components, $p_{i,j}$

	$j=0$	$j=1$	$j=2$
$i=1$	0.1	0.2	0.7
$i=2$	0.4	0.2	0.4
$i=3$	0.3	0.5	0.2

Table 2.2 Weight distribution of the components, $w_{i,j}$

	$j=0$	$j=1$	$j=2$
$i=1$	1	2	3
$i=2$	1	3	4
$i=3$	1	3	5

$$R_2^I(5,2) = \sum_{r=0}^{2} p_{2,r} \cdot R_2^I(5 - w_{2,r}, 2 - 1)$$

$$= p_{2,0} \cdot R_2^I(5 - 1, 1) + p_{2,1} \cdot R_2^I(5 - 3, 1) + p_{2,2} \cdot R_2^I(5 - 4, 1)$$

$$= p_{2,0} \cdot R_2^I(4, 1) + p_{2,1} \cdot R_2^I(2, 1) + p_{2,2} \cdot R_2^I(1, 1)$$

$$= p_{2,0} \cdot 0 + p_{2,1} \cdot (p_{1,1} + p_{1,2}) + p_{2,2}$$

$$= 0.2 \cdot (0.2 + 0.7) + 0.4 = 0.58 ,$$

$$R_2^I(7,2) = p_{1,2} \cdot p_{2,2} = 0.7 * 0.4 = 0.28 ,$$

$$R_2^I(9,2) = 0 ,$$

$$R_2^I(10,3) = p_{3,1} \cdot 0.28 + p_{3,2} \cdot 0.58 = 0.5 \cdot 0.28 + 0.2 \cdot 0.58 = 0.256 .$$

Example 10. The multi-state weighted system studied in Example 9 is considered here. The UGF for the three components are as follow:

$$U_1(z) = 0.1z^1 + 0.2z^2 + 0.7z^3 ,$$

$$U_2(z) = 0.4z^1 + 0.2z^3 + 0.4z^4 ,$$

$$U_3(z) = 0.3z^1 + 0.5z^3 + 0.2z^5 .$$

Based on the individual UGFs of the components, we can obtain the system UGF by using operator Ω as follows:

$$U_s(z) = \Omega(U_1(z), U_2(z), U_3(z))$$

$$= \Omega((0.1z^1 + 0.2z^2 + 0.7z^3), (0.4z^1 + 0.2z^3 + 0.4z^4), (0.3z^1 + 0.5z^3 + 0.2z^5))$$

$$= 0.012z^3 + 0.02z^5 + 0.008z^7 + 0.006z^5 + 0.01z^7 + 0.004z^9 + 0.012z^6$$

$$+ 0.02z^8 + 0.008z^{10} + 0.024z^4 + 0.04z^6 + 0.016z^8 + 0.012z^6 + 0.02z^8$$

$$+ 0.008z^{10} + 0.024z^7 + 0.04z^9 + 0.016z^{11} + 0.084z^5 + 0.14z^7 + 0.056z^9$$

$$+ 0.042z^7 + 0.07z^9 + 0.028z^{11} + 0.084z^8 + 0.14z^{10} + 0.056z^{12} .$$

The equation given above is the output performance distribution of the multi-state weighted k-out-of-n:G system Model I. From this equation, we can obtain the system state distribution for the case of $k_1 = 5$ and $k_2 = 10$ using the operator δ_A, as shown in (2.20).

$$R_1^I(5,3) = \delta_A (U_s(z), 5)$$

$$= 0.02 + 0.008 + 0.006 + 0.01 + 0.004 + 0.012 + 0.02 + 0.008 + 0.04$$

$$+ 0.016 + 0.012 + 0.02 + 0.008 + 0.024 + 0.04 + 0.016 + 0.084 + 0.14$$

$$+ 0.056 + 0.042 + 0.07 + 0.028 + 0.084 + 0.14 + 0.056 = 0.964 ,$$

$$R_2^I(10,3) = \delta_A(U_s(z), 10) = 0.008 + 0.008 + 0.016 + 0.028 + 0.14 + 0.056$$

$$= 0.256 .$$

So we get the same result as in Example 9.

In summary, the state distribution of the system is as follows:

$$\Pr(\phi \geq 0) = 1 \,,$$
$$\Pr(\phi \geq 1) = 0.964 \,,$$
$$\Pr(\phi \geq 2) = 0.256 \,,$$
$$\Pr(\phi = 2) = 0.256 \,,$$
$$\Pr(\phi = 1) = 0.964 - 0.256 = 0.708 \,,$$
$$\Pr(\phi = 0) = 1 - 0.964 = 0.036 \,.$$

Example 11. We consider the same set of components used in Example 9. However, the system state is determined based on Definition 9. Thus, we have $n = 3$, $M = 2$, $k_1 = 5$, and $k_2 = 10$. Equation (2.23) is used below to calculate the state distribution of the system.

$$R_1^{II}(5,3) = q_{3,1} \cdot R_1^{II}(5,2) + \sum_{r=1}^{2} p_{3,r} \cdot R_1^{II}(5 - w_{3,r}, 3 - 1)$$
$$= 0.3 \cdot R_1^{II}(5,2) + 0.5 \cdot R_1^{II}(2,2) + 0.2 \,,$$

$$R_1^{II}(5,2) = q_{2,1} \cdot R_1^{II}(5,1) + \sum_{r=1}^{2} p_{2,r} \cdot R_1^{II}(5 - w_{2,r}, 2 - 1) = 0.54 \,,$$

$$R_1^{II}(2,2) = q_{2,1} \cdot R_1^{II}(2,1) + \sum_{r=1}^{2} p_{2,r} \cdot R_1^{II}(2 - w_{2,r}, 2 - 1) = 0.96 \,,$$

$$R_1^{II}(5,3) = 0.3 \cdot R_1^{II}(5,2) + 0.5 \cdot R_1^{II}(2,2) + 0.2$$
$$= 0.3 \cdot 0.54 + 0.5 \cdot 0.96 + 0.2 = 0.842 \,,$$

$$R_2^{II}(10,3) = q_{3,2} \cdot R_2^{II}(10,2) + \sum_{r=2}^{2} p_{3,r} \cdot R_2^{II}(10 - w_{3,r}, 3 - 1)$$
$$= q_{3,2} \cdot R_2^{II}(10,2) + p_{3,2} \cdot R_2^{II}(5,2) \,,$$
$$R_2^{II}(10,2) = 0 \,,$$

$$R_2^{II}(5,2) = q_{2,2} \cdot R_2^{II}(5,1) + \sum_{r=2}^{2} p_{2,r} \cdot R_2^{II}(5 - w_{2,r}, 2 - 1) = 0.28 \,,$$

$$R_2^{II}(10,3) = q_{3,2} \cdot R_2^{II}(10,2) + p_{3,2} \cdot R_2^{II}(5,2) = p_{3,2} \cdot 0.28 = 0.056 \,.$$

Example 12. We still consider the multi-state weighted system in Example 11. When we calculate the probability that the system is in state 1 or above, the UGF for the three components should be written as follows:

$$U_1(z) = 0.1z^0 + 0.2z^2 + 0.7z^3 \,,$$
$$U_2(z) = 0.4z^0 + 0.2z^3 + 0.4z^4 \,,$$
$$U_3(z) = 0.3z^0 + 0.5z^3 + 0.2z^5 \,.$$

Based on the individual UGF of the components, we can obtain the system UGF by using operator Ω:

$$
\begin{aligned}
U_s(z) &= \Omega(U_1(z), U_2(z), U_3(z)) \\
&= 0.012z^0 + 0.024z^2 + 0.11z^3 + 0.012z^4 + 0.06z^5 + 0.216z^6 + 0.12z^7 \\
&\quad + 0.08z^8 + 0.118z^9 + 0.148z^{10} + 0.044z^{11} + 0.056z^{12}\,.
\end{aligned}
$$

The above form is the multi-state weighted k-out-of-n Model I OPD. We can obtain the system reliability for the arbitrary $k_1 = 5$ based on this form using the operator δ_A, as shown in (2.20):

$$
R_1^{I\!I}(5,3) = \delta_A(U_s(z), 5) = 0.842\,.
$$

When we calculate the probability that the system is in state 2 or above, the UGF for the three components should be written as follows:

$$
\begin{aligned}
U_1(z) &= 0.3z^0 + 0.7z^3\,, \\
U_2(z) &= 0.6z^0 + 0.4z^4\,, \\
U_3(z) &= 0.8z^0 + 0.2z^5\,.
\end{aligned}
$$

Based on the individual UGF of the components, we can obtain the system UGF by using operator Ω:

$$
\begin{aligned}
U_s(z) &= \Omega(U_1(z), U_2(z), U_3(z)) \\
&= 0.144z^0 + 0.336z^3 + 0.096z^4 + 0.036z^5 \\
&\quad + 0.224z^7 + 0.084z^8 + 0.024z^9 + 0.056z^{12}\,.
\end{aligned}
$$

The above form is the multi-state weighted k-out-of-n Model II OPD. We can obtain the system reliability for the arbitrary $k_2 = 10$ based on this form using the operator δ_A, as shown in (2.20):

$$
R_2^{I\!I}(10,3) = \delta_A(U_s(z), 10) = 0.056\,.
$$

So we get the same result as in Example 11.

In summary, we have the state distribution of the system as follows:

$$
\begin{aligned}
\Pr(\phi \geq 2) &= R_2^{I\!I}(10,3) = 0.056\,, \\
\Pr(\phi \geq 1) &= R_1^{I\!I}(5,3) = 0.842\,, \\
\Pr(\phi \geq 0) &= 1\,, \\
\Pr(\phi = 2) &= 0.056\,, \\
\Pr(\phi = 1) &= 0.842 - 0.056 = 0.786\,, \\
\Pr(\phi = 0) &= 1 - 0.842 = 0.158\,.
\end{aligned}
$$

Examples 9 and 10 are for multi-state weighted k-out-of-n systems Model I. Examples 11 and 12 are for multi-state weighted k-out-of-n systems Model II. They use the same set of components. The only difference between these two classes of examples is which components' weights are used in determining the state of the system. The state distributions obtained from them are given in Table 2.3.

Table 2.3 Comparison of system state distributions of Model I and Model II

	Model I	Model II
$\Pr(\phi \geq 2)$:	0.256	0.056
$\Pr(\phi \geq 1)$:	0.964	0.842
$\Pr(\phi \geq 0)$:	1.000	1.000

From Table 2.3, it is apparent that the probability for the Model II system to be not less than a specific state is less than that for the Model I system. Model I may be applied whenever the contribution of every component is useful, no matter how bad or good a component is. Model II is applicable when components in bad states cannot make any contribution for operation in high system states.

2.4 Conclusions

The k-out-of-n structure is a very popular redundancy structure, and has been widely used in many systems. Meanwhile, in many real applications, components and systems might have multiple states. "Multiple states" of a component or system might be interpreted as multiple levels of performance, multiple functions, multiple levels of benefits, or multiple failure modes. The multi-state k-out-of-n system models reported in this chapter enable us to model multiple levels of k-out-of-n requirements of the system. Recursive algorithms have been developed for efficient performance evaluations of the multi-state k-out-of-n system models.

Future work includes performing case studies of practical systems with multiple levels of k-out-of-n requirements, and developing more general models for modeling more general and complex situations.

Acknowledgements This research work was supported by the Natural Sciences and Engineering Research Council of Canada.

References

1. Kuo W, Zuo MJ (2003) Optimal Reliability Modeling: Principles and Applications. Wiley, New York, pp 452–503
2. Barlow RE, Heidtmann KD (1984) Computing k-out-of-n system reliability. IEEE Transactions on Reliability 33:322–323

3. Rushdi AM (1986) Utilization of symmetric switching functions in the computation of k-out-of-n system reliability. Microelectronics and Reliability 26:973–987
4. Wu JS, Chen RJ (1994) An algorithm for computing the reliability of a weighted-k-out-of-n system. IEEE Transactions on Reliability 43:327–328
5. Nordmann L, Pham H (1999) Weighted voting systems. IEEE Transactions on Reliability 48:42–49
6. Nordmann L, Pham H (1997) Reliability of decision making in human-organizations. IEEE Transactions on Systems Man and Cybernetics Part A-Systems and Humans 27:543–549
7. Lisnianski A, Levitin G (2003) Multi-state system reliability: assessment, optimization and applications. World Scientific, Singapore
8. Elsayed EA (1996) Reliability Engineering. Addison Wesley Longman, New York
9. Barlow RE, Wu AS (1978) Coherent systems with multi-state components. Mathematics of Operations Research 3:275–281
10. El-Neweihi E, Proschan F, Sethuraman J (1978) Multi-state coherent system. Journal of Applied Probability 15:675–688
11. Boedigheimer RA, Kapur KC (1994) Customer-driven reliability models for multi-state coherent systems. IEEE Transactions on Reliability 43:46–50
12. Huang J, Zuo MJ, Wu YH (2000) Generalized multi-state k-out-of-n:G systems. IEEE Transactions on Reliability 49:105–111
13. Zuo MJ, Huang J, Kuo W (2003) Multi-state k-out-of-n systems. In: Pham H (ed) Handbook of Reliability Engineering. Springer, London, pp 3–15
14. Zuo MJ, Tian Z (2006) Performance evaluation for generalized multi-state k-out-of-n systems. IEEE Transactions on Reliability 55:319–327
15. Tian Z, Zuo MJ, Yam RCM (2005) Performance evaluation of generalized multi-state k-out-of-n systems with independent components. Proceedings of the Fourth International Conference on Quality and Reliability, Beijing, China, August 9–11, pp 515–520
16. Tian Z, Zuo MJ, Yam RCM. The multi-state k-out-of-n systems and their performance evaluation. IIE Transactions, Submitted June 2006 (under review)
17. Li W, Zuo MJ (2007) Reliability evaluation of multi-state weighted k-out-of-n systems. Reliability Engineering and System Safety 92:15–22
18. Huang J, Zuo MJ, Fang Z (2003) Multi-state consecutive-k-out-of-n systems. IIE Transactions 35:527–534
19. Yamamoto H, Zuo MJ, Akiba T, Tian Z (2006) Recursive formulas for the reliability of multi-state consecutive-k-out-of-n:G systems. IEEE Transactions on Reliability 55:98–104
20. Levitin G (2005) Universal Generating Function in Reliability Analysis and Optimization. Springer-Verlag, London
21. Levitin G, Lisnianski A, Elmakis D (1997) Structure optimization of power system with different redundant elements. Electric Power Systems Research 43:19–27
22. Levitin G, Lisnianski A, Ben-Haim H, Elmakis D (1998) Redundancy optimization for series-parallel multi-state systems. IEEE Transactions on Reliability 47:165–172
23. Levitin G, Lisnianski A (1998) Structure optimization of power system with bridge topology. Electric Power Systems Research 45:201–208
24. Lisnianski A, Levitin G, Ben Haim H (2000) Structure optimization of multi-state system with time redundancy. Reliability Engineering and System Safety 67:103–112
25. Levitin G, Lisnianski A (2001) Reliability optimization for weighted voting system. Reliability Engineering and System Safety 71:131–138
26. Levitin G, Lisnianski A (2001) Structure optimization of multi-state system with two failure modes. Reliability Engineering and System Safety 72:75–89
27. Levitin G (2002) Optimal series-parallel topology of multi-state system with two failure modes. Reliability Engineering and System Safety 77:93–107

Further Reading

Beaulieu NC (1991) On the generalized multinomial distribution, optimal multinomial detectors, and generalized weighted partial decision detectors. IEEE Transactions on Communications 39:193–194

Chen Y, Yang QY (2005) Reliability of two-stage weighted-k-out-of-n systems with components in common. IEEE Transactions on Reliability 54:431–440

Natvig B (1982) Two suggestions of how to define a multi-state coherent system. Applied Probability 14:391–402

Tian Z, Richard RCM, Zuo MJ, Huang H (2007) Reliability bounds for multi-state k-out-of-n systems. IEEE Transactions on Reliability 92:137–148

Chapter 3
On Weighted Least Squares Estimation for the Parameters of Weibull Distribution

L.F. Zhang, M. Xie, L.C. Tang

Department of Industrial and Systems Engineering,
National University of Singapore

3.1 Introduction

The two-parameter Weibull distribution is one of the most widely used life distributions in reliability studies. It has shown to be satisfactory in modeling the phenomena of fatigue and life of many devices such as ball bearings, electric bulbs, capacitors, transistors, motors and automotive radiators. In recent years, a number of modifications of the traditional Weibull distribution have been proposed and applied to model complex failure data sets. For references see, *e.g.*, [1–4].

Many methods have been proposed for the estimation of Weibull parameters. The maximum likelihood estimation (MLE) method and the least squares estimation (LSE) method are frequently used. MLE is considered to have many good statistical properties and is usually preferred by researchers. The LSE method, especially when used with the Weibull probability plot (WPP), is very convenient, and hence is preferred by practitioners. While MLE has been intensively studied in the literature, LSE is less discussed and has a good potential. Some early research has been carried out to find the best linear unbiased estimators (BLUE) [5] and best linear invariant estimators (BLIE) [6]. However, these methods are computationally intensive and require reference to tables proposed by the authors. Berger and Lawrence [7] examined the nonlinear regression technique and compared it with the conventional LSE via Monte Carlo simulations. They concluded that the non-linear regression technique performs similar to, if not worse than, the LSE. WLSE and robust regression techniques have also been examined by some other researchers. This chapter focuses on WLSE techniques. For robust regression techniques, see *e.g.*, [8].

One problem with LSE is that it treats each data point equally under the assumption that the variance of the error term is a constant. It is common that the first few data points in a Weibull sample are more scattered than the remaining part of the sample. For this reason, LSE has a low efficiency and appropriate weights to different data points can be used to improve it. WLSE can maximize the efficiency of parameter estimation by giving each data point its proper amount of influence over the parameter estimates.

Several authors have examined WLSE techniques for estimating the Weibull parameters and proposed different methods for calculating weights. White [9] gave a numerical example of WLSE for estimating the Weibull parameters. However, the calculation of weights of White's method is very complicated. In addition, the author gave few comments on the WLSE technique. Bergman [10] and Faucher and Tyson [11] gave a closer look at the WLSE techniques for Weibull parameter estimation, and they each proposed a simple formula for calculating weights. Both of them pointed out that WLSE has better performance over LSE on parameter estimation. In recent years, Hung [12] and Lu *et al.* [13] re-examined WLSE techniques and they also proposed different formulae for calculating weights. Lu *et al.* [13] also compared their WLSE technique with the three existing techniques of Bergman [10], Faucher and Tyson [11] and Hung [12] for estimating the two Weibull parameters. In the following, these methods are simply denoted by "Lu," "Bergman," "F&T" and "Hung" respectively. With Monte Carlo simulations, Lu *et al.* [13] concluded that Bergman (as well as Hung, which is basically the same) in most cases generates larger mean square errors (MSE) than the others and methods of F&T and Lu perform similar. All four methods, as well as the method of White [9], follow the routine to calculate weights by the reciprocal of the variance of the dependent variable values, *i.e.*, $Y_{(i)} = \ln\left[-\ln(1 - F_i)\right]$. However, the variances used in Bergman, F&T, Hung and Lu are only approximate values obtained either from the theory of error propagation, *e.g.*, Bergman, Hung and Lu or the interval length between two percentiles, *e.g.*, F&T. It is likely that errors are introduced by using such approximations. Meanwhile, we noted that the exact values of the variances can be determined through analytical means. Therefore, more appropriate weights can be obtained.

In this chapter, we present the formulae to calculate more appropriate weights to be used in WLSE for estimating the Weibull parameters. A simple approximation formula is also proposed through numerical method which can be used when the sample size is within 20 to simplify the calculation for weights. Through Monte Carlo simulations, the proposed WLSE methods are compared with some existing methods. Simulation results show that the proposed procedure is slightly better than the existing WLSE methods. The relative efficiency of LSE over the proposed WLSE method is only about $70-80\%$.

Notations and Abbreviations

α, β	scale parameter and shape parameter
$\hat{\alpha}, \hat{\beta}$	estimators of α and β
A, B	coefficients of a simple linear regression model $\hat{Y} = A + BX$
n	sample size
i	order number of observations in a sample
t_i	i^{th} observation (time to failure or time to censor) in a sample
$t_{(i)}$	order statistic
F_i	$F_i = F(t_{(i)})$, failure probability of the ith failure in a complete sample

$\hat{F}_i$	estimator of F_i
r	number of failures in a censored sample
j	order number of failure in a censored sample, $j = 1, 2, \cdots, r$
l	order number of censor in a censored sample, $l = 1, 2, \cdots, n - r$
I_j	original order number (failure and censors merged) of the jth failure in a censored sample
t_{fj}	j^{th} failure in a censored sample (order statistic)
t_{sl}	l^{th} censor in a censored sample (order statistic)
F_{fj}	$F_{fj} = F(t_{fj})$, failure probability of the jth failure in a censored sample
$\hat{F}_{fj}$	estimator of F_{fj} for a censored sample
$\hat{R}_i, \hat{R}_{fj}$	estimator of reliability, $\hat{R}_i = 1 - \hat{F}_i$, $\hat{R}_{fj} = 1 - \hat{F}_{fj}$
m_{fj}	modified failure order number of the jth failure in a censored sample
Δ_j	increment for calculating m_{fj}
w_i	weights
$Y_{(i)}, Z_{(i)}$	order statistics of Y, Z
$Var(\cdot)$	variance of $Y_{(i)}$ and $Z_{(i)}$
$S_{(\cdot)}$	standard deviation
SS, SS'	sum of squares in the objective functions of LSE and WLSE
w_app	approximated w_i
b_U	unbiased estimator of β
$w_{m_{fj}}$	weights for the m_{fj}^{th} failure in a censored sample
cdf	cumulative distribution function
pdf	probability density function
WPP	Weibull Probability Plot
LSE	Least Squares Estimation
WLSE	Weighted Least Squares Estimation
MLE	Maximum Likelihood Estimation
MSE	Mean Square Error
MFON	Modified failure order number

3.2 Basic Concepts in Lifetime Data Analysis

Lifetime data measure the life of products. Lifetime data can be collected from life testing experiments or from field. There are different types of lifetime data: complete data, singly censored data (type I or type II), multiply censored data, etc. Censored units are called censors or suspensions, and their failure times are known only to be beyond their present running times (*i.e.*, the censoring times). If all units are started on test together and all censors have a common running time, the data are singly censored. Such data are classified into time censored or type I censored if the test is stopped at a predetermined time, and failure censored or type II censored if the test is stopped when a predetermined number of failures occur. If units begin their services at different times and thus when the test stops before all units are failed censoring times and failure times are intermixed, the data are said to be multiply censored data.

The singly censored data can be treated as a special case of the multiply censored data. However, they are usually examined separately in lifetime data analysis. For more detailed classification of data type, see *e.g.*, Nelson [14].

Statistical analysis of lifetime data can be based on parametric, non-parametric or semi-parametric models [15]. Parametric analysis is most frequently used. This involves fitting a statistical distribution to lifetime data from a representative sample to model the life of the product, estimate the model parameters via the sample data, and finally use the model to estimate life characteristics such as reliability. The commonly used statistical distributions, also known as life distributions, include Exponential, Weibull, Extreme-Value, Gamma and Lognormal.

Obviously, it is important to have an efficient method for estimating the model parameters. Commonly used parameter estimation methods for life distributions include graphical estimation, method of moment and method of maximum likelihood.

Graphical estimation frequently refers to probability plot, while sometimes hazard plot is used instead. Special probability papers are designed based on the linearization of the cumulative density function (cdf) of life distributions so that a straight line relationship is exhibited on these probability papers. The straight line can be fitted by eye or by least squares regression technique. After the line is fitted, parameter estimates can be usually obtained via the slope and scale of the fitted line. Probability plotting is frequently used for estimating the parameters of Exponential, Weibull, Extreme-Value, and Lognormal distributions.

Method of moment solves parameter estimates by equating population and sample moments. It can be efficient when the number of parameters is low and when there is no censor in the sample. This method works well for the Gamma distribution. However, it is not efficient for the Weibull distribution because the population moments do not have simple forms. One way to overcome this problem is to make use of the relationship between Weibull and Extreme-Value distribution and deduce the Weibull parameter estimates from the moment estimates of the Extreme-Value distribution parameters.

Maximum likelihood estimation finds parameter estimates by maximizing the likelihood function so that the parameter values obtained are most consistent with the sample data. It is preferred by statisticians due to its good statistical prospective. MLE can be used for all the life distributions and it generates reasonable parameter estimates in most cases. Under some circumstances, however, it generates inconsistent results or the estimators do not exist. Although MLE usually involves complicated calculation which can be hardly solved manually, many statistical software packages have programs for calculating MLE of parameters of various distributions.

Selection of parameter estimation methods should not be arbitrary. In fact, different types of lifetime data (complete or censored, large data set or small data set, etc.) may need different estimation methods. For example, for the Weibull distribution, there is no method always outperforms the others in view of the properties of the estimators including bias and MSE. In addition, other factors may affect the selection of parameter estimation methods such as the computation simplicity.

3.3 Common Estimation Methods for Weibull Distribution

The cdf of the two-parameter Weibull distribution is expressed by

$$F(t) = 1 - \exp\left[-\left(\frac{t}{\alpha}\right)^{\beta}\right] \tag{3.1}$$

The scale parameter α is frequently called characteristic life, since it is the time at which 63.2% of the population failed regardless of the value of β. The shape parameter β is of great importance to the Weibull distribution because it determines the shape of the probability density function (pdf), characterizes the failure rate trend (increasing, constant or decreasing) an indicate some failure modes such as initial, random and wear-out [2].

Let $t_1, t_2, \cdots, t_i, \cdots, t_n$ $(i = 1, 2, \cdots, n)$ denote a random sample from a two-parameter Weibull distribution, and $t_{(3.1)} \le t_{(3.2)} \le \cdots \le t_{(i)} \le \cdots \le t_{(n)}$ the order statistics. This sample can be a complete sample where all the observations are failures, or it can be a censored sample where some of the observations are failures and the others are censors. For a censored sample, let $t_{f1} \le t_{f2} \le \cdots \le t_{fj} \le \cdots \le t_{fr}$ $(j = 1, 2, \cdots, r)$ denotes r ordered failure times and $t_{s1} \le t_{s2} \le \cdots \le t_{sl} \le \cdots \le t_{s(n-r)}$ $(l = 1, 2, \cdots, n - r)$ denotes the remaining ordered censoring times. This censored sample is singly censored if $t_{s1} = t_{s2} = \cdots = t_{sl} = \cdots = t_{s(n-r)}$, or multiply censored if censoring times are different and intermixed with failure times.

The objective of parameter estimation is to estimate α and β using the sample data. In the following, the theoretical background and estimating equations are presented for three most widely used Weibull parameter estimation methods: WPP, LSE and MLE.

3.3.1 Weibull Probability Plot

By taking natural logarithm twice on both sides of (3.1), *i.e.*, the Weibull cdf, it can be linearized as

$$\ln\left[-\ln(1 - F(t))\right] = \beta \ln t - \beta \ln \alpha \tag{3.2}$$

The Weibull probability paper is specially designed based on (3.2) so that the plot of t vs. $F(t)$ is a straight line if the Weibull distribution fits. The values of $F(t)$, commonly called Y-axis plotting positions, depend on the unknown parameters α and β, and hence can only be estimated. Several estimators for $F(t)$ have been proposed to be applied to complete data and censored data, respectively. For complete data, the Bernard estimator [16] is most widely used. For censored data, the method of Herd–Johnson [17, 18] and Johnson's modified method [18] are widely used. Their formulas are given below.

$$\text{\textit{Bernard estimator}} \qquad \hat{F}_i = \frac{i - 0.3}{n + 0.4} \tag{3.3}$$

$$\text{\textit{Herd–Johnson estimator}} \qquad \begin{cases} \hat{R}_{fj} = \left(\dfrac{n + 1 - I_j}{n + 2 - I_j} \right) \cdot \hat{R}_{f(j-1)} \\[2ex] \hat{F}_{fj} = 1 - \hat{R}_{fj} \end{cases} \tag{3.4}$$

$$\text{\textit{Johnson's modified estimator}} \qquad \begin{cases} \Delta_j = \dfrac{n + 1 - m_{f(j-1)}}{n - I_j + 2} \\[2ex] m_{fj} = m_{f(j-1)} + \Delta_j \\[2ex] \hat{F}_{fj} = \dfrac{m_{fj} - 0.3}{n + 0.4} \end{cases} \tag{3.5}$$

A widely used procedure of WPP is to plot $t_{(i)}$ along the horizontal axis $\hat{F}_i$ and along the vertical axis. In the case of censored data, t_{fj} versus $\hat{F}_{fj}$ is plotted. After the data points are placed on the Weibull probability paper, a straight line can be fitted to the points by eye. However, move objective estimates can be obtained by fitting the straight line with the least squares regression technique. After the straight line is fitted, the shape parameter is estimated by the slope of the regression line and the scale parameter is estimated by the ratio of the intercept and slope.

WPP is widely used by engineers. It is generally available in statistical software packages that involve reliability data analysis. The plot can be easily produced with a 'best fit' linear regression line generated by the least squares technique. Figure 3.1 is an example of computer generated WPP.

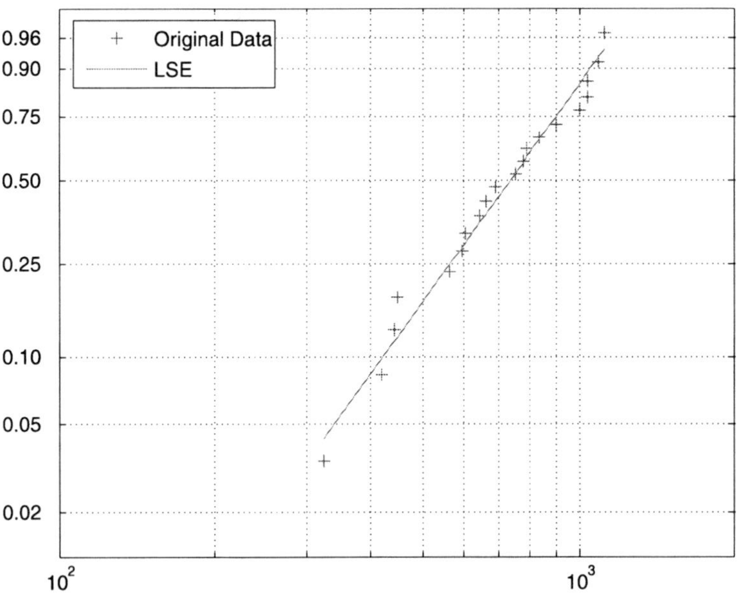

Fig. 3.1 An example of computer generated WPP

3.3.2 Least Squares Estimation Method

LSE uses the least squares regression to estimate the two parameters based on the linearized Weibull cdf in (3.2).

Setting $X = \ln t$ and $Y = \ln\left[-\ln(1 - F(t))\right]$, (3.2) turns to a simple equation, *i.e.*

$$Y = \beta X - \beta \ln \alpha \tag{3.6}$$

Thus, the estimation of α and β can be transferred to the estimation of the regression coefficients for a simple linear regression model of the form $\hat{Y} = A + BX$ (or $Y = A + BX + e$), where $A = -\beta \ln \alpha$, $B = \beta$ and e is the error term.

The objective function of LSE is

$$\min SS = \sum_{i=l}^{r} \left[y_i - (\beta x_i - \beta \ln \alpha)\right]^2 \tag{3.7}$$

For complete data, $r = n$.

By taking partial derivatives of with SS with regard to α and β respectively, and setting the results to 0, the LS analysis of (3.6) generates following estimating equation for β and α.

$$\hat{\beta} = \frac{r \sum\limits_{i=1}^{r} x_i y_i - \sum\limits_{i=1}^{r} x_i \cdot \sum\limits_{i=1}^{r} y_i}{r \sum\limits_{i=1}^{r} x_i^2 - \left(\sum\limits_{i=1}^{r} x_i\right)^2}$$
$$\hat{\alpha} = \exp\left(-\frac{\sum\limits_{i=1}^{r} y_i - \hat{\beta} \sum\limits_{i=1}^{r} x_i}{r\hat{\beta}}\right) \tag{3.8}$$

For complete data, $r = n$.

x_i and y_i in (3.8) are the values X and Y for a specific sample. For a complete sample, they can be obtained by $x_i = \ln(t_{(i)})$ and $y_i = \ln\left[-\ln(1 - \hat{F}_i)\right]$, where $\hat{F}_i$ can be calculated by (3.3), for example. For a censored sample, they can be obtained by $x_i = \ln(t_{fj})$ and $y_i = \ln\left[-\ln(1 - \hat{F}_{fj})\right]$, where $\hat{F}_{fj}$ may be calculated by (3.4) or (3.5). For both cases, values of X and Y come from order statistics.

The calculation of $\hat{F}_{fj}$ is very important to LSE for censored data because only failure times are used in the estimating equation of LSE, and the influence of censors are reflected by the estimation of $\hat{F}_{fj}$ or y_i. For some discussions of the estimation of $\hat{F}_{fj}$, see *e.g.*, [19, 20] and [21]. However, their methods have not been popularized and should be used with caution.

3.3.3 Maximum Likelihood Estimation Method

MLE is one of the most widely used tools for statistical inference. Cohen (1965) introduced the maximum likelihood equations for estimating the two Weibull parameters from complete samples, type I and type II singly censored samples and multiply censored samples. According to Cohen [22], the estimating equation of MLE for complete Weibull samples is given by

$$\frac{\sum_{i=1}^{n} t_i^{\hat{\beta}} \ln t_i}{\sum_{i=1}^{n} t_i^{\hat{\beta}}} - \frac{1}{\hat{\beta}} = \frac{1}{n} \sum_{i=1}^{n} \ln t_i$$

$$\hat{\alpha} = \left(\sum_{i=1}^{n} t_i^{\hat{\beta}} \Big/ n \right)^{1/\hat{\beta}} \tag{3.9}$$

The estimating equation of MLE for singly censored samples, either type I or type II censored, is given by

$$\frac{\sum_{i=1}^{r} t_i^{\hat{\beta}} \ln t_i}{\sum_{i=1}^{r} t_i^{\hat{\beta}}} - \frac{1}{\hat{\beta}} = \frac{1}{r} \sum_{i=1}^{r} \ln t_i \tag{3.10}$$

$$\hat{\alpha} = \left[\left(\sum_{i=1}^{r} t_i^{\hat{\beta}} \right) \Big/ r \right]^{1/\hat{\beta}}$$

For multiply censored samples, the estimating equation is

$$\frac{\sum_{j=1}^{r} t_{fj}^{\hat{\beta}} \ln t_{fj} + \sum_{l=1}^{n-r} t_{sl}^{\hat{\beta}} \ln t_{sl}}{\sum_{j=1}^{r} t_{fj}^{\hat{\beta}} + \sum_{l=1}^{n-r} t_{sl}^{\hat{\beta}}} - \frac{1}{\hat{\beta}} = \frac{1}{r} \sum_{j=1}^{r} \ln t_{fj} \tag{3.11}$$

$$\hat{\alpha} = \left[\left(\sum_{j=1}^{r} t_{fj}^{\hat{\beta}} + \sum_{l=1}^{n-r} t_{sl}^{\hat{\beta}} \right) \Big/ r \right]^{1/\hat{\beta}}$$

The Newton–Raphson method can be used to solve the estimating equations. Although the calculation is complicated, nowadays many statistical software packages provide programs for calculating the ML estimators of various distributions. Electronic spreadsheet such as Excel can also solve the estimating equations of MLE.

3.3.4 Comparisons of the Methods

Different parameter estimation methods may result in widely differing estimates; therefore, it is important to have objective criteria to instruct the selection of one estimation method over other alternatives. Common criteria include unbiasedness, minimum variance or MSE, consistency, sufficiency and simplicity.

Among the three basic estimation methods previously described, WPP is the simplest method and it can serve as a simple tool for model validation and outlier identification. MLE is considered to have a good statistical perspective since the estimators are asymptotically unbiased, asymptotically efficient and consistent. MLE has been intensively examined in the literature. Compared with MLE, LSE has some advantages: first, it has a closed form solution that can be easily calculated; second, it can be easily incorporated into WPP to offer a graphical presentation; third, LSE can be refined by other sophisticated linear regression estimation techniques such as WLSE methods and robust regression estimation methods to deal with harsh data conditions such as small sample size, high censoring level or outliers. In addition, researchers have shown that LSE may perform significantly better than MLE in view of the bias of the shape parameter estimator for both complete data and censored data [23–25].

The LSE method, especially when used with WPP, is very convenient and hence is preferred by practitioners. As previously stated, LSE is basically the least squares estimation of the regression coefficients for a simple linear regression model. Based on this, more sophisticated linear regression techniques such as WLS techniques can also be applied to estimate the two Weibull parameters.

3.4 Weighted Least Squares Estimation Methods and Related Work

In LSE, each data point is equally treated as if each of them provides same amount of precise information. This is, however, hardly the truth. By doing so, LSE actually has a low efficiency. The idea of WLSE is to give each data point its proper amount of influence by assigning each data point a weight, w_i. It can often be used to maximize the efficiency of parameter estimation.

3.4.1 Estimating Equation of WLSE

The objective function of WLSE is

$$\min SS' = w_i \sum_{i=1}^{r} [y_i - (\beta x_i - \beta \ln \alpha)]^2 \tag{3.12}$$

For complete data, $r = n$.

By taking partial derivatives of SS' with regard to α and β respectively, and setting the results to 0, the estimating equation for α and β of the WLSE method is

$$\hat{\beta} = \frac{\sum_{i=1}^{r} w_i \cdot \sum_{i=1}^{r} w_i x_i y_i - \sum_{i=1}^{r} w_i x_i \cdot \sum_{i=1}^{r} w_i y_i}{\sum_{i=1}^{r} w_i \cdot \sum_{i=1}^{r} w_i x_i^2 - \left(\sum_{i=1}^{r} w_i x_i\right)^2}$$

$$\hat{\alpha} = \exp\left(-\frac{\sum_{i=1}^{r} w_i y_i - \hat{\beta} \sum_{i=1}^{r} w_i x_i}{\hat{\beta} \sum_{i=1}^{r} w_i}\right) \tag{3.13}$$

Equation (3.13) can be applied to both complete data and censored data. For complete data, $r = n$. Values of x_i and y_i in (3.13) can be obtained in the same way as in LSE. The only problem is then to determine w_i. As a special case, when $w_i = 1$ for all data points, the WLS estimators reduce to the LS estimators.

3.4.2 Calculation of Weights and Assumptions

Conventionally, $X = \ln t$ is treated as the independent variable and $Y = \ln[-\ln(1 - F(t))]$ as the dependent variable. This matches with WPP, which plots t along the X-axis and $F(t)$ along the Y-axis.

Assume that there is no measurement error on the failure time t so that the uncertainty of error all comes from the uncertainty of Y.

As a common practice, weights can be calculated through the reciprocal of the variance of Y values. For the Weibull case, since Y values come from order statistics, w_i is calculated by

$$w_i = 1/Var(Y_{(i)}) \tag{3.14}$$

Since values of the dependent variable Y are not measured but estimated, $Var(Y_{(i)})$ can only be obtained through analytical means.

3.4.3 Related Work

Some research has been carried out to examine WLSE for the Weibull distribution. White [9] defined a log-Weibull order statistic and deduced the formula for calculating the variance of this statistic. Some of the results are tabulated in the paper. Based on the result, he gave a numerical example of WLSE for estimating the Weibull parameters. However, the calculation of weights of White's method is complicated. In addition, there is no detailed comparison between White's WLSE and LSE. Later,

Bergman [10], Faucher and Tyson [11], Hung [12] and Lu *et al.* [13] each proposed a simple formula for calculating weights. Their methods are briefly discussed in the following.

Bergman [10]

Bergman applied the theory of error propagation on the relationship $Y_{(i)} = \ln\left[-\ln(1 - F_i)\right]$ and obtained

$$S_{Y_{(i)}} \approx \frac{dY_i}{dF_i} \cdot S_{F_i} = -S_{F_i} \cdot \left[(1 - \hat{F}_i) \ln\left(1 - \hat{F}_i\right)\right]^{-1} \tag{3.15}$$

By assuming standard deviation S_{F_i} is a constant, according to (3.15), $S_{Y_{(i)}}$ is proportional to $\left[(1 - \hat{F}_i) \ln\left(1 - \hat{F}_i\right)\right]$, and Bergman determined the formula for weights as

$$w_i = \left[(1 - \hat{F}_i) \ln(1 - \hat{F}_i)\right]^2 \tag{3.16}$$

Here the author used $i/(n + 1)$ and $(i - 0.5)/n$ to calculate $\hat{F}_i$. A simulation experiment was conducted to compare his WLSE with LSE for each of the plotting positions. It was concluded that the WLSE with $\hat{F}_i = i/(n + 1)$ gave somewhat smaller bias than LSE.

Hung [12]

Hung proposed a formula for calculating weights in a way very similar to that of Bergman [10]. His formula of weights is given by

$$w_i = \frac{\left[(1 - \hat{F}_i) \ln\left(1 - \hat{F}_i\right)\right]^2}{\sum\limits_{i=1}^{n} \left[(1 - \hat{F}_i) \ln\left(1 - \hat{F}_i\right)\right]^2} \tag{3.17}$$

Since $\sum\limits_{i=1}^{n} \left[(1 - \hat{F}_i) \ln\left(1 - \hat{F}_i\right)\right]^2$ results in a constant independent of i, Hung's formula is basically same as Bergman's formula.

Hung suggested $\hat{F}_i$ to be calculated by the method of Drapella and Kosznik [26]. The formula is

$$\begin{cases} y_i = \dfrac{n!}{(i - 1)!(n - i)!} \sum\limits_{v=0}^{i-1} (-1)^v \binom{i-1}{v} \times \dfrac{-0.5774 - \ln(n - i + v + 1)}{n - i + v + 1} \\ \hat{F}_i = 1 - \exp(-\exp(y_i)) \end{cases} \tag{3.18}$$

Hung's simulation results showed that his WLSE procedure always provides smaller MSE than LSE does.

For both methods of Bergman [10] and Hung [12], there is an assumption that the uncertainty of F_i is constant. We believe that this is not a good assumption because F_i is an order statistic and its uncertainty depends on the order number i.

Faucher and Tyson [11]

This method uses the concept of percentiles. We know the pth percentile of F_i can be calculated from the following equation.

$$\sum_{k=1}^{i} \binom{n}{k-1} F_i^{k-1} (1 - F_i)^{n+1-k} = 1 - p \tag{3.19}$$

For example, when $p = 0.5$ we can solve for the exact median rank of F_i. The authors estimated the uncertainty of F_i through the difference of two percentiles, *i.e.*, the 20th percentile and the 80th percentile of F_i, denoted by F_i' and F_i'' respectively. Then with the relationship $Y_{(i)} = \ln\left[-\ln(1 - F_i)\right]$, they concluded that the uncertainty of $Y_{(i)}$ is proportional to the difference of $\ln\left[-\ln(1 - F_i')\right] - \ln\left[-\ln(1 - F_i'')\right]$, and their formula for weights was determined as

$$w_i = \frac{1}{\left(\ln\left[-\ln\left(1 - F_i'\right)\right] - \ln\left[-\ln\left(1 - F_i''\right)\right]\right)^2} \tag{3.20}$$

Their selection of the two percentiles, however, is somewhat subjective.

Based on the values calculated by (3.20), they also proposed a simple approximation formula as

$$w_i = 3.3\hat{F}_i - 27.5\left[1 - (1 - \hat{F}_i)^{0.025}\right] \tag{3.21}$$

They suggested $(i - 0.3)/(n + 0.4)$ to calculate $\hat{F}_i$.

Their Monte Carlo experiment compared the proposed WLSE procedure with LSE in view of bias and standard deviation of the estimators. The results showed that the WLSE procedure significantly reduces the standard deviation of the estimators.

Lu et al. [13]

They defined an intermediate variable $C = -\ln(1 - F)$. From the Weibull cdf, C follows the standard exponential distribution.

Applying the theory of error propagation, they obtained

$$Var\left(Y_{(i)}\right) = Var\left(\ln C_{(i)}\right) \approx \frac{Var\left(C_{(i)}\right)}{\left[E\left(C_{(i)}\right)\right]^2} = \sum_{j=1}^{i} \frac{1}{(n-j+1)^2} \Bigg/ \left[\sum_{j=1}^{i} \frac{1}{(n-j+1)}\right]^2 \tag{3.22}$$

Therefore, their formula for weights is

$$w_i = \left[\sum_{j=1}^{i} \frac{1}{(n-i+j)} \right]^2 \bigg/ \sum_{j=1}^{i} \frac{1}{(n-i+j)^2} \tag{3.23}$$

The advantage of this method is that it avoids calculating $\hat{F}_i$. Authors of this paper gave an overview of the existing WLSE techniques and conducted an intensive Monte Carlo simulation experiment to compare different methods. They concluded that Bergman's method (as well as Hung's, which is basically the same) in most cases generates larger MSE than the others. F&T [11] and their method show a similar level of performance.

In summary, the four WLSE methods presented above all have the advantage of simplicity. However, the variances used in all the four methods are only approximated values obtained either from the theory of error propagation, *e.g.*, Bergman [10], Hung [12] and Lu *et al.* [13] or the interval length between two percentiles, *e.g.*, Faucher and Tyson [11]. It is likely that errors are introduced by using such approximations. In the next section, an improved method for calculating weights is presented.

3.5 An Improved Method for Calculating Weights

3.5.1 Calculation for 'Best' Weights

'Best' weights can be obtained through the reciprocal of the exact values of $Var(Y_{(i)})$. Here the word 'best' is quoted because the weights are truly best only under the assumption that there is no measurement error on t. However, this assumption is often violated in real life. Following shows the procedure for deducing exact $Var(Y_{(i)})$.

Introducing a new random variable Z defined as

$$Z = \ln\left[(t/\alpha)^\beta\right] = \beta \ln t - \beta \ln \alpha \tag{3.24}$$

where t follows the two-parameter Weibull distribution.

Based on the Weibull cdf, the cdf of Z can be determined as

$$F(z) = P(Z \le z) = P\left(\ln\left[(t/\alpha)^\beta\right] \le z\right) = P(t \le \alpha \cdot e^{z/\beta})$$

$$= 1 - \exp\left[-\left(\frac{\alpha \cdot e^{z/\beta}}{\alpha}\right)^\beta\right] = 1 - \exp(-e^z) \tag{3.25}$$

Thus Z follows a parameter-free distribution.

Comparing (3.24) with (3.2), *i.e.*, the linearized Weibull cdf, we have

$$Z = \ln\left[-\ln(1 - F(t))\right] \tag{3.26}$$

We have also defined $Y = \ln\left[-\ln(1 - F(t))\right]$; therefore, obviously the values of $y_i = \ln\left[-\ln(1 - \hat{F}_i)\right]$ can be looked on as the values taken on by the order statistic of Z. Therefore, $Var(Y_{(i)})$ is same as $Var(Z_{(i)})$.

Since Z follows a parameter-free distribution, the analysis is much easier. From (3.25), the cdf of $Z_{(i)}$ can be determined as

$$
F\left(z_{(i)}\right) = i\binom{n}{i} \int_{-\infty}^{z} F^{i-1}(z)(1 - F(z))^{n-i} f(z)\,dz
$$

$$
= i\binom{n}{i} \int_{-\infty}^{z} (1 - \exp(-e^{z}))^{i-1} (\exp(-e^{z}))^{n-i} d(1 - \exp(-e^{z})) \quad (3.27)
$$

The mean and variance of $Z_{(i)}$ can be deduced from its cdf. The results are

$$
E\left(Z_{(i)}\right) = i\binom{n}{i} \cdot \sum_{k=0}^{i-1}\left\{(-1)^{k}\binom{i-1}{k} \cdot \frac{-c - \ln(n - i + k + 1)}{n - i + k + 1}\right\} \quad (3.28)
$$

$$
E\left(Z_{(i)}^{2}\right) = 1.978112 + i\binom{n}{i}\sum_{k=0}^{i-1} \times
$$

$$
\left\{(-1)^{k}\binom{i-1}{k} \cdot \frac{2c\ln(n - i + k + 1) + \ln^{2}(n - i + k + 1)}{n - i + k + 1}\right\} \quad (3.29)
$$

$$
Var\left(Z_{(i)}\right) = E\left(Z_{(i)}^{2}\right) - E^{2}\left(Z_{(i)}\right) \quad (3.30)
$$

where $c = 0.577216$ is the Euler's constant. The Appendix gives the detailed derivation of (3.28) and (3.29). Our (3.28) is same as the formula of Drapella and Kosznik [26], see (3.18).

Thus the weights used in WLSE can be calculated as

$$
w_i = 1/Var\left(Z_{(i)}\right) \quad (3.31)
$$

Table 3.1 lists the values of weights directly calculated by (3.28)–(3.30) for selected sample sizes. From Table 3.1, we can see that for $n = 5, 6$, the weights are increasing with the order number i, and the weights for the last two data points are much higher than those for the first two data points. From $n = 7$ onwards, however, the largest weights are not given to the last data point but a little bit earlier. The weights for the end part of the sample are still much larger than those for the beginning part of the sample. This indicates that we should not treat each data point in a sample equally.

Since the weight for each observation is given relative to the weights for other observations, they can be normalized in some way. For example, the normalized weights can be obtained by dividing the weight for each observation by the mean weight over the whole sample. In this way, the sum of the normalized weights equals to the sample size.

Table 3.1 Weights calculated by (3.28)–(3.30) for selected sample sizes (the largest weights in each column are highlighted)

$i\backslash n$	5	6	7	8	10	12	14	16	18	20
1	0.2675	0.2269	0.1970	0.1741	0.1414	0.1190	0.1028	0.0904	0.0807	0.0729
2	0.6779	0.5761	0.5009	0.4431	0.3600	0.3032	0.2619	0.2305	0.2058	0.1859
3	1.0838	0.9286	0.8108	0.7190	0.5857	0.4939	0.4269	0.3759	0.3357	0.3033
4	1.4263	1.2538	1.1071	0.9875	0.8091	0.6841	0.5921	0.5218	0.4663	0.4215
5	**1.5446**	1.5013	1.3673	1.2364	1.0250	0.8708	0.7556	0.6668	0.5964	0.5393
6		**1.5133**	**1.5416**	1.4440	1.2266	1.0509	0.9155	0.8097	0.7252	0.6564
7			1.4754	**1.5605**	1.4023	1.2202	1.0699	0.9494	0.8520	0.7721
8				1.4353	1.5306	1.3719	1.2158	1.0843	0.9758	0.8859
9					**1.5628**	1.4946	1.3489	1.2125	1.0957	0.9971
10					1.3564	**1.5667**	1.4626	1.3310	1.2100	1.1049
11						1.5410	1.5452	1.4355	1.3167	1.2080
12						1.2836	**1.5758**	1.5193	1.4129	1.3050
13							1.5089	**1.5709**	1.4942	1.3938
14							1.2181	1.5700	1.5539	1.4715
15								1.4726	**1.5810**	1.5337
16								1.1595	1.5554	1.5738
17									1.4351	**1.5811**
18									1.1071	1.5359
19										1.3979
20										1.0599

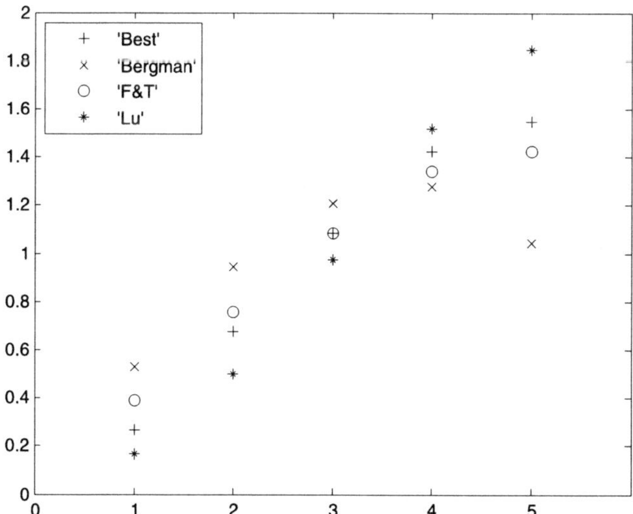

Fig. 3.2 Comparison of normalized weights calculated by different methods at $n = 5$

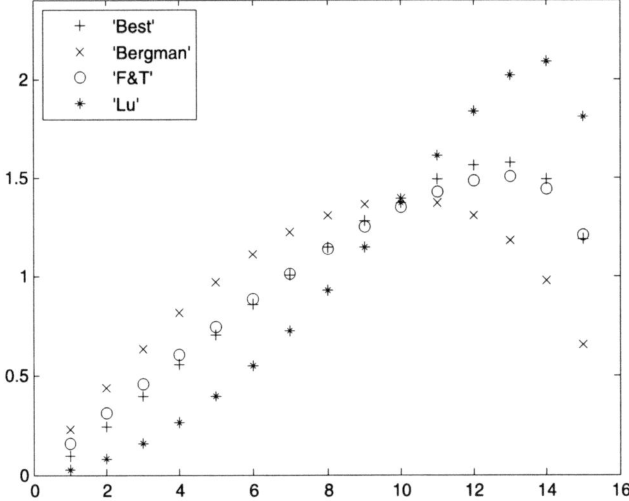

Fig. 3.3 Comparison of normalized weights calculated by different methods at $n = 15$

By normalizing the weights, we can compare the best weights calculated by (3.28)–(3.30) with other existing methods for calculating weights. For example, we compared the best weights with weights calculated by Bergman [(3.16)], F&T [(3.21)] and Lu [(3.23)] for a sample of size 5 and a sample of size 15 respectively. Figures 3.2 and 3.3 show the different values of weights calculated by different methods at each data point in the two samples. It can be observed that: 1) weights of F&T are very close to the best weights; 2) compared to the best weights, Bergman and Lu present reversed trends. Bergman underestimates the last few points and overestimates the remaining points, while Lu overestimates the last few points and underestimates the remaining points; 3) Lu gives very large weights to the last few data points and very small weights to the beginning two data points. Bergman gives much smaller weights to the last few points than the other methods.

3.5.2 An Approximation for 'Best' Weights for Small and Complete Samples

The above method for calculating weights is not convenient without the aid of a computer program. Also note that when the sample size becomes large, say $n \geq 30$, the binomial coefficients in (3.28)–(3.29) will become extremely large, making it hard to generate accurate results for the weights. Combined with the fact that the conventional LSE procedure performs not very well mainly for small samples, say $n \leq 20$, therefore, we explores WLSE for only small samples.

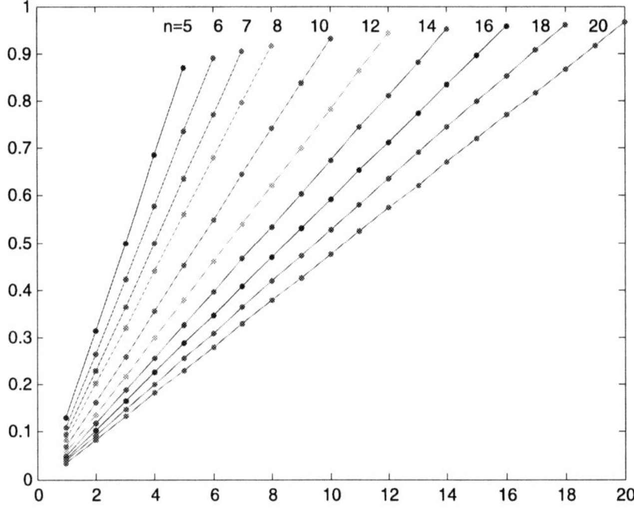

Fig. 3.4 Plot of $\hat{F}_i = (i - 0.3)/(n + 0.4)$ at selected sample sizes

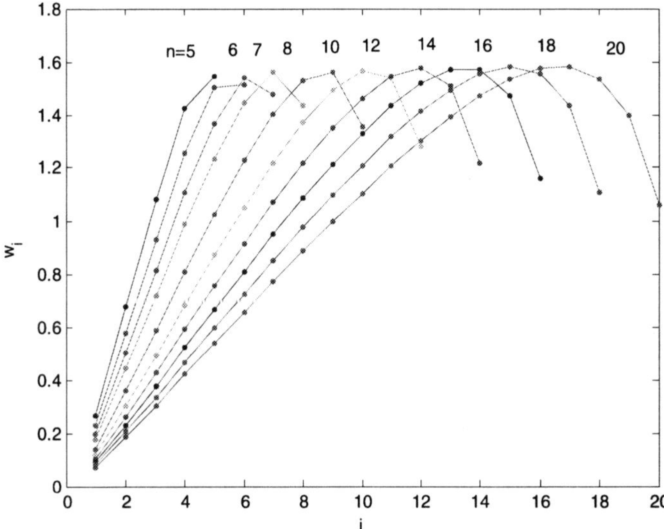

Fig. 3.5 Plot of weights at selected sample sizes. Weights are calculated by (3.28)–(3.30)

To simplify the calculation, some approximation formula can be proposed through numerical analysis. It is possible to directly model w_i as the function of order number i and sample size n, however, it is easier to model it as the function of $\hat{F}_i$. Figure 3.4 is the plot of $\hat{F}_i$, calculated by the Bernard estimator $\hat{F}_i = (i - 0.3)/(n + 0.4)$, at selected sample sizes. Figure 3.5 plots the weights directly calculated by (3.28)–(3.30) for the same sample sizes. It is obvious that the two figures show similar patterns.

The proposed approximation model, which specifies the relationship between w_i and $\hat{F}_i$, is a polynomial model expressed by

$$w\_app(i) = p_0 + p_1\hat{F}_i + p_2\hat{F}_i^2 + p_3\hat{F}_i^3 + p_4\hat{F}_i^4 \tag{3.32}$$

where $w\_app(i)$ denotes the approximated value of w_i, and p_0, p_1, p_2, p_3, p_4 are the model parameters to be determined.

The model parameters can be determined by the non-linear curve fitting technique. The objective function is

$$\min \sum_{i=1}^{n} \left[w_i - \left(p_0 + p_1\hat{F}_i + p_2\hat{F}_i^2 + p_3\hat{F}_i^3 + p_4\hat{F}_i^4 \right) \right]^2 \tag{3.33}$$

To solve this function, multidimensional unconstrained non-linear minimization, *i.e.*, the Nelder–Mead method [27], was used. The calculation was executed in MATLAB 7 and the built-in function *fminsearch* was used.

The best values of w_i, calculated by (3.28)–(3.30), and $\hat{F}_i$, calculated by $\hat{F}_i = (i - 0.3)/(n + 0.4)$, for samples of sizes 2–20 were used in (3.33) to determine five model parameters. The results are

$$p_0 = -0.076, p_1 = 3.610, p_2 = -6.867, p_3 = 13.54, p_4 = -9.231 \tag{3.34}$$

Thus, the approximation formula for calculating weights is

$$w\_app(i) = -0.076 + 3.610\hat{F}_i - 6.867\hat{F}_i^2 + 13.54\hat{F}_i^3 - 9.231\hat{F}_i^4 \tag{3.35}$$

3.5.3 Application Procedure

The application procedure of the proposed WLSE method for estimating the Weibull parameters in the case of small, complete samples is summarized as follows:

Step 1: Rank the failure times from smallest to largest and calculate the Y-axis plotting positions by $\hat{F}_i = (i - 0.3)/(n + 0.4)$;
Step 2: Plot the failure times t_i against $\hat{F}_i$ on WPP. If the Weibull distribution fits, the data points should appear to be on a straight line;
Step 3: Calculate the weights using (3.35);
Step 4: Obtain the estimates for α and β using (3.13).

Nowadays, many statistical software packages and electrical spreadsheet provide the WLS programs and users just need to provide the weights. With a simple formula for weights, WLSE can be carried out as convenient as LSE. In addition, WLSE can also be incorporated into WPP. Some software, MATLAB 7, for example, has the program for generating WPP. By simply replacing the code of LS fitting with WLS fitting, the WPP with a straight line generated by WLS can be obtained.

3.5.4 Numerical Example

Following is a randomly generated Weibull sample with $\alpha = 1, \beta = 2$. Five methods including the proposed one, 'Bergman' (3.16), 'F&T' (3.21), 'Lu' [3.23], and ordinary LSE were applied to this sample and the estimates for and were calculated as can be seen in Table 3.2. Figure 3.6 is the WPP with straight lines generated by each method.

Generated failure time (from the smallest to the largest):

0.2153, 0.6394, 0.7607, 0.8112, 1.0024, 1.2612, 1.3418, 1.4468, 1.5011, 1.8998.

From Fig. 3.6 and Table 3.2, it can be concluded that the LSE line is greatly affected by the first point and considerably underestimates β. The proposed method provides the best estimate for β. F&T performs very close to the proposed one. Bergman underestimates β and Lu overestimates β. All methods overestimate α but the difference in $\hat{\alpha}$ is smaller than that in $\hat{\beta}$.

This example shows that the five methods can generate quite different estimation results. In real cases, we do not know the true parameter values so it is difficult to judge which method performs best. Therefore, we conducted a Monte Carlo simulation experiment to find out which method tends to generate the best estimates. The experiment is presented in the next section.

3.5.5 Monte Carlo Study

Monte Carlo experiments were conducted to examine the proposed procedure on estimating the Weibull parameters for small, complete data sets. The following techniques were compared in the experiment (for all methods, $\hat{F}_i$ are calculated by $\hat{F}_i = (i - 0.3)/(n + 0.4)$, *i.e.*, the Bernard estimator):

1. Exact W: The weights are calculated by (3.28)–(3.30);
2. App. W: The proposed procedure with weights calculated by (3.35);
3. F&T: The weights are calculated by (3.21);
4. Lu: The weights are calculated by (3.23);
5. LSE: The ordinary LSE method.

Bergman's method is not considered because Lu *et al.* [13] have shown that it is significantly inferior to the other methods.

Table 3.2 Estimation results for the numerical example

	Proposed	'Bergman'	'F&T'	'Lu'	LSE
$\hat{\alpha}$	1.2526	1.2774	1.2547	1.2465	1.2863
$\hat{\beta}$	2.0639	1.9350	2.0318	2.2034	1.7221

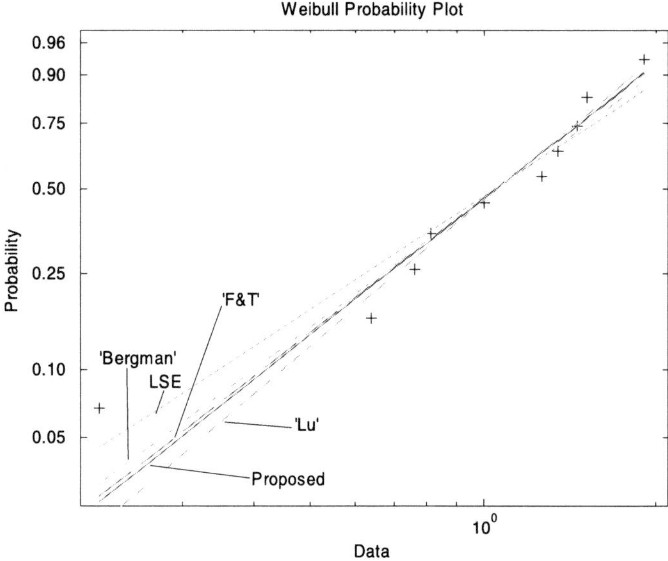

Fig. 3.6 WPP of the numerical example

Weibull samples of different sizes were randomly generated with selected values of α and β, and then for each sample generated, estimates of α and β were obtained by the above techniques. By repeating this process for a large number of times, the average (or mean) values, standard deviations and MSEs of the estimates were calculated as the comparison criteria. The simulation conditions are: sample sizes $n = 5, 6, 8, 10, 15, 18, 20$, $\alpha = 1$, $\beta = 0.5, 1, 2$ and the repetition number $M = 20,000$.

It should be noted that the weights of all WLSE techniques examined are independent of the values of α and β. Therefore, the two pivotal functions, $\hat{\beta}/\beta$ and $\hat{\beta}\ln(\hat{\alpha}/\alpha)$ for the LS estimated α and β also hold for the WLS estimated Weibull parameters. The advantage of using pivotal functions is that the distribution of them can be derived from the normalized Weibull distribution (*i.e.*, $\alpha = \beta = 1$), so that simulations can be greatly reduced. In this experiment, the true value of β was fixed at 1 to assess the estimators of β, and to assess the estimators of α, three true values of β were used because $\hat{\alpha}/\alpha$ is not a pivotal function. α is just a scale parameter, so we fix its true value to 1 in the experiment.

The simulation results are shown in Tables 3.3–3.6. The following conclusions can be observed.

Estimation of β (Table 3.3)

1. In view of both standard deviation and MSE of $\hat{\beta}$, WLSE methods are significantly better than LSE. The ratio of MSE of App. W and LSE is about 70% at $n = 20$. Among the WLSE methods examined, Exact W and App. W always

generate smallest standard deviation and MSE, followed by the F&T method. Lu has slightly larger standard deviation and MSE than the other three WLSE methods.

2. In view of the bias, all WLSE methods perform similarly and they only outperform LSE at $n = 5$. In most cases, the bias of $\hat{\beta}$ of the LSE is much smaller, about $2 - 3\%$ less than that of the other methods.
3. App. W performs very close to Exact W.

Estimation of α (Tables 3.4–3.6)

1. In view of the bias, WLSE methods outperform LSE in nearly all cases. The bias of Lu is always smallest among WLSE methods, followed by that of the Exact W and App. W. The bias of $\hat{\alpha}$ of Lu is $5 - 10\%$ less than that of the LSE, and the bias of $\hat{\alpha}$ of Exact W and App. W are $3 - 5\%$ less than that of the LSE.
2. In view of both standard deviation and MSE of $\hat{\alpha}$, WLSE methods always outperform LSE. Among the WLSE methods examined, Lu always generates smallest standard deviation and MSE, followed by the Exact W, App. W and F&T. At $\beta = 0.5$, the ratio of MSE of Lu and LSE is about 70%.

Table 3.3 Mean, standard deviation and MSE of $\hat{\beta}$ obtained by different methods. The values in brackets are MSE

Method	5	6	8	10	12	15	18	20
							n	
Exact W	1.006 ±0.563	0.972 ±0.447	0.948 ±0.335	0.946 ±0.286	0.945 ±0.254	0.950 ±0.222	0.953 ±0.198	0.958 ±0.188
	(0.317)	(0.201)	(0.115)	(0.085)	(0.068)	(0.051)	(0.041)	(0.037)
App. W	1.006 ±0.562	0.971 ±0.447	0.947 ±0.335	0.946 ±0.286	0.946 ±0.254	0.950 ±0.222	0.955 ±0.198	0.960 ±0.189
	(0.316)	(0.201)	(0.115)	(0.085)	(0.068)	(0.051)	(0.041)	(0.037)
Bergman	1.006 ±0.564	0.971 ±0.452	0.947 ±0.343	0.944 ±0.295	0.943 ±0.263	0.948 ±0.232	0.951 ±0.208	0.955 ±0.197
	(0.318)	(0.205)	(0.121)	(0.093)	(0.073)	(0.056)	(0.046)	(0.041)
F&T	1.007 ±0.562	0.972 ±0.448	0.948 ±0.336	0.946 ±0.287	0.945 ±0.255	0.950 ±0.223	0.953 ±0.199	0.957 ±0.189
	(0.316)	(0.201)	(0.116)	(0.085)	(0.068)	(0.052)	(0.042)	(0.037)
Lu	1.005 ±0.567	0.971 ±0.451	0.948 ±0.339	0.948 ±0.292	0.949 ±0.261	0.953 ±0.228	0.956 ±0.204	0.961 ±0.195
	(0.321)	(0.204)	(0.118)	(0.089)	(0.070)	(0.054)	(0.044)	(0.040)
LSE	1.046 ±0.592	1.009 ±0.481	0.978 ±0.370	0.968 ±0.319	0.962 ±0.287	0.961 ±0.255	0.959 ±0.230	0.961 ±0.220
	(0.353)	(0.231)	(0.137)	(0.103)	(0.084)	(0.067)	(0.057)	(0.051)

Table 3.4 Mean, standard deviation and MSE of $\hat{\alpha}$ obtained by different methods at $\beta = 0.5$. The values in brackets are MSE

	Method	5	6	8	10	12	15	18	20
								n	
	Exact W	1.383 ± 1.344	1.316 ±1.154	1.219 ±0.921	1.191 ±0.807	1.154 ±0.721	1.129 ±0.624	1.103 ±0.557	1.093 ±0.523
		(1.954)	(1.431)	(0.896)	(0.687)	(0.544)	(0.406)	(0.320)	(0.282)
	App. W	1.395 ±1.356	1.325 ±1.162	1.224 ±0.925	1.194 ±0.808	1.155 ±0.722	1.127 ±0.623	1.101 ±0.556	1.091 ±0.522
		(1.996)	(1.455)	(0.906)	(0.691)	(0.545)	(0.405)	(0.319)	(0.281)
	Bergman	1.502 ±1.485	1.425 ±1.268	1.310 ±1.008	1.269 ±0.873	1.221 ±0.779	1.182 ±0.664	1.147 ±0.589	1.133 ±0.552
$\beta=0.5$		(2.456)	(1.787)	(1.111)	(0.835)	(0.655)	(0.474)	(0.369)	(0.322)
	F&T	1.405 ±1.365	1.334 ±1.169	1.233 ±0.931	1.202 ±0.813	1.163 ±0.727	1.135 ±0.627	1.108 ±0.559	1.098 ±0.525
		(2.026)	(1.478)	(0.921)	(0.703)	(0.555)	(0.412)	(0.324)	(0.286)
	Lu	1.333 ±1.302	1.264 ±1.115	1.167 ±0.888	1.142 ±0.780	1.109 ±0.700	1.088 ±0.608	1.066 ±0.544	1.059 ±0.513
		(1.806)	(1.313)	(0.817)	(0.629)	(0.501)	(0.377)	(0.300)	(0.266)
	LSE	1.528 ±1.495	1.454 ±1.286	1.342 ±1.032	1.304 ±0.902	1.256 ±0.812	1.216 ±0.689	1.181 ±0.612	1.167 ±0.573
		(2.512)	(1.861)	(1.183)	(0.906)	(0.725)	(0.521)	(0.408)	(0.356)

Table 3.5 Mean, standard deviation and MSE of $\hat{\alpha}$ obtained by different methods at $\beta = 1$. The values in brackets are MSE

	Method	5	6	8	10	12	15	18	20
					n				
$\beta=1$	Exact W	1.067 ±0.497 (0.251)	1.055 ±0.447 (0.203)	1.040 ±0.384 (0.149)	1.035 ±0.344 (0.120)	1.027 ±0.311 (0.097)	1.022 ±0.278 (0.078)	1.020 ±0.256 (0.066)	1.016 ±0.240 (0.058)
	App. W	1.072 ±0.499 (0.254)	1.059 ±0.449 (0.205)	1.042 ±0.385 (0.150)	1.036 ±0.344 (0.120)	1.027 ±0.311 (0.097)	1.021 ±0.278 (0.078)	1.019 ±0.255 (0.066)	1.015 ±0.240 (0.058)
	Bergman	1.110 ±0.521 (0.284)	1.096 ±0.470 (0.230)	1.076 ±0.403 (0.168)	1.066 ±0.359 (0.134)	1.054 ±0.324 (0.108)	1.045 ±0.289 (0.085)	1.040 ±0.265 (0.072)	1.034 ±0.248 (0.063)
	F&T	1.076 ±0.500 (0.256)	1.063 ±0.450 (0.207)	1.046 ±0.386 (0.151)	1.039 ±0.346 (0.121)	1.030 ±0.312 (0.098)	1.025 ±0.279 (0.079)	1.023 ±0.256 (0.066)	1.018 ±0.241 (0.058)
	Lu	1.047 ±0.489 (0.242)	1.033 ±0.440 (0.195)	1.017 ±0.378 (0.144)	1.012 ±0.339 (0.115)	1.006 ±0.307 (0.094)	1.003 ±0.276 (0.076)	1.003 ±0.254 (0.064)	1.000 ±0.239 (0.057)
	LSE	1.122 ±0.523 (0.288)	1.109 ±0.472 (0.235)	1.090 ±0.407 (0.173)	1.081 ±0.363 (0.138)	1.069 ±0.328 (0.112)	1.060 ±0.293 (0.090)	1.055 ±0.269 (0.075)	1.049 ±0.252 (0.066)

Table 3.6 Mean, standard deviation and MSE of $\hat{\alpha}$ obtained by different methods at $\beta = 2$. The values in brackets are MSE

	Method	5	6	8	10	12	15	18	20
					n				
$\beta=2$	Exact W	1.006 ±0.234 (0.055)	1.004 ±0.214 (0.046)	1.003 ±0.186 (0.035)	1.003 ±0.168 (0.028)	1.003 ±0.153 (0.024)	1.002 ±0.136 (0.019)	1.002 ±0.125 (0.016)	1.002 ±0.119 (0.016)
	App. W	1.008 ±0.235 (0.055)	1.006 ±0.215 (0.046)	1.004 ±0.186 (0.035)	1.003 ±0.168 (0.028)	1.004 ±0.153 (0.024)	1.001 ±0.136 (0.019)	1.002 ±0.125 (0.016)	1.002 ±0.119 (0.016)
	Bergman	1.026 ±0.241 (0.059)	1.023 ±0.220 (0.049)	1.020 ±0.191 (0.037)	1.018 ±0.172 (0.030)	1.017 ±0.158 (0.025)	1.012 ±0.140 (0.020)	1.012 ±0.128 (0.017)	1.012 ±0.123 (0.017)
	F&T	1.010 ±0.235 (0.055)	1.008 ±0.215 (0.046)	1.006 ±0.187 (0.035)	1.005 ±0.168 (0.028)	1.005 ±0.154 (0.024)	1.003 ±0.136 (0.019)	1.004 ±0.126 (0.016)	1.004 ±0.120 (0.016)
	Lu	0.996 ±0.233 (0.054)	0.993 ±0.213 (0.045)	0.991 ±0.185 (0.034)	0.992 ±0.167 (0.028)	0.993 ±0.153 (0.023)	0.992 ±0.136 (0.019)	0.994 ±0.126 (0.016)	0.996 ±0.120 (0.016)
	LSE	1.032 ±0.240 (0.059)	1.030 ±0.220 (0.049)	1.027 ±0.191 (0.037)	1.025 ±0.172 (0.030)	1.024 ±0.158 (0.026)	1.020 ±0.140 (0.020)	1.019 ±0.129 (0.017)	1.018 ±0.123 (0.017)

3. The standard deviation and MSE of $\hat{\alpha}$ of all methods are decreasing with the increase of β.
4. App. W performs very close to Exact W.

3.6 Discussions

Recent developments on parameter estimation for the Weibull distribution have focused on small samples and censored samples because they are common in field data and the traditional MLE and LSE have been found that do not perform very well under such conditions. The estimators are found to be biased and have low efficiency [23–25]. As is well known, WLS is an efficient method that makes good use of small data sets and it has the ability to handle a sample where data points inside are of varying quality.

The key of WLSE is to calculate the weights. Based on the theoretical deduction of the variance of $Y_{(i)}$, the formulas for calculating the best values of weights are provided, *i.e.*, (3.28)–(3.30). However, when the sample size is large, say , the binomial coefficients in (3.28), (3.29) will become extremely large so that it is hard to generate accurate results for the weights. For example, MATLAB 7 generates negative values for weights at which is obviously wrong. A possible solution for this is to calculate weights through (3.36), (3.37) when the sample size is large. As can be seen in the Appendix, (3.36), (3.37) are the intermediate results in deriving (3.28), (3.29). The Simpson rule [28] might be applied to calculate the integrals in (3.36), (3.37).

$$E\left(z_{(i)}\right) = i\binom{n}{i}\int_0^{+\infty} \ln v \cdot (e^v - 1)^{i-1} e^{-nv}\,dv \tag{3.36}$$

$$E\left(z_{(i)}^2\right) = i\binom{n}{i}\int_0^{+\infty} \ln^2 v \cdot (e^v - 1)^{i-1} \cdot e^{-nv}\,dv \tag{3.37}$$

The simple formula for calculating weights proposed in this chapter is limited to small, complete samples. The proposed WLSE helps to improve the efficiency of parameter estimation, which is justified by Monte Carlo experiment. However, it can also be seen from the experiment results that WLSE methods present large bias in dealing with small samples. This can be improved by using bias correction methods. For the shape parameter estimator of the proposed WLSE method, the plot of bias vs n presents a hyperbolic appearance, therefore, an unbiasing formula is proposed as in (3.38).

$$b_U = b \cdot \left(0.986 + \frac{1.521}{n} - \frac{8.339}{n^2} + \frac{3.527}{n^3} + \frac{6.345}{n^4}\right) \tag{3.38}$$

This equation can be added in the end of the procedure described in Sect. 5.3.

Censored data are commonly encountered in reliability data analysis and it adds difficulty to parameter estimation. For a censored sample, LSE uses only failure data points to conduct regression and the influence of censored items is reflected through the estimation of $\hat{F}_{fj}$, more basically speaking, through the modified failure order numbers (MFON) of each failed data point, denoted by m_{fj}. Several methods have been proposed for calculating MFON for multiply censored data, recently; see, e.g., [19–21]. The Herd–Johnson method and Johnson's modified method [17, 18] are the most widely used methods for calculating $\hat{F}_{fj}$ for multiply censored data. Their formulas are given in 3.4, 3.5. To apply WLSE to multiply censored data, a possible procedure is expressed as follows.

Step 1:   Calculate m_{fj} and $\hat{F}_{fj}$ for each failure data point using (3.5);
Step 2:   Calculate the weight for each failure data point based on its MFON. The weights are calculated through linear interpolation.

$$w_{m_j} = w(Int_j) + (m_j - Int_j)\left[w(Int_j + 1) - w(Int_j)\right] \tag{3.39}$$

where $Int_j = int[m_{fj}]$, that is the integral part of m_{fj}. For small samples, $w(Int_j)$ can be calculated by (3.35) and for large samples, $w(I_j)$ can be calculated by (3.21) or (3.23).
Step 3:   Obtain the estimates for α and β using (3.13).

3.7 Conclusions

This chapter proposes a simple formula for calculating the weights to be used in WLSE for estimating the two Weibull parameters in the case of small, complete samples of size $n \leq 20$. Compared with the existing WLSE methods for the Weibull distribution, the proposed method has a better statistical foundation because it is based on the theoretical deduction of the exact variance of the ordered Y values. The Monte Carlo experiment shows that the proposed method is slightly better than the others and significantly better than LSE in view of standard deviation and MSE of the two parameter estimators. However, it is noteworthy that there is no measurement error and no outliers in the generated Weibull samples in the Monte Carlo experiment. When there are outliers in a sample, WLSE might be worse than LSE. Therefore, WLSE should be used with caution.

WLSE can also be applied to censored data. A procedure is suggested which requires the calculation of modified failure order number. Then the weights for each failure data point can be calculated by linear interpolation.

References

1. Lai CD, Xie M, Murthy DNP (2003) A modified Weibull distribution. IEEE Transactions on Reliability 52:33–37
2. Murthy DNP, Xie M, Jiang RY (2003) Weibull Models. Wiley, New Jersey
3. Murthy DNP, Bulmer M, Ecclestion JA (2004) Weibull model selection for reliability modeling. Reliability Engineering & System Safety 86:257–267
4. Xie M, Tang Y, Goh TN (2002) A modified Weibull extension with bathtub-shaped failure rate function. Reliability Engineering & System Safety 76:279–285
5. White JS (1964) Least-squares unbiased censored linear estimation for the log-Weibull (extreme value). Journal of Industrial Math Soc 14:21–60
6. Mann NR (1967) Tables for obtaining the best linear invariant estimates of parameters of the Weibull distribution. Technometrics 9:629–645
7. Berger RW, Lawrence K (1974) Estimating Weibull parameters by linear and non-linear regression. Technometrics 16:617–619
8. Lawson C, Keats JB, Montgomery DC (1997) Comparison of robust and least-squares regression in computer-generated probability plots. IEEE Transactions on Reliability 46:108–115
9. White JS (1969) The moments of log-Weibull order statistics. Technometrics 11:65–72
10. Bergman B (1986) Estimation of Weibull parameters using a weight function. Journal of Materials Science Letters 5:611–614
11. Faucher B, Tyson WR (1988) On the determination of Weibull parameters. Journal of Materials Science Letters 7:1199–1203
12. Hung WL (2001) Weighted least squares estimation of the shape parameter of the Weibull distribution. Quality and Reliability Engineering International 17:467–469
13. Lu HL, Chen CH, Wu JW (2004) A note on weighted least-squares estimation of the shape parameter of the Weibull distribution. Quality and Reliability Engineering International 20:579–586
14. Nelson WB (2004) Applied Life Data Analysis. Wiley, New York
15. Pham H (2003) Handbook of Reliability Engineering. Springer, UK
16. Bernard A, Bosi-Levenbach EC (1953) The plotting of observations on probability paper. Statistica Neerlandica 7:163–173

17. Herd GR (1960) Estimation of Reliability from Incomplete Data. Proceedings of the 6th International Symposium on Reliability and Quality Control
18. Johnson LG (1964) The Statistical Treatment of Fatigue Experiments. Elsevier, New York
19. Campean IF (2000) Exponential age sensitive method for failure rank estimation. Quality and Reliability Engineering International 16:291–300
20. Hastings NAJ, Bartlett HJG (1997) Estimating the failure order-number from reliability data with suspended items. IEEE Transactions on Reliability 46:266–268
21. Wang WD (2004) Refined rank regression method with censors. Quality and Reliability Engineering International 20:667–678
22. Cohen AC (1965) Maximum likelihood estimation in the Weibull distribution based on complete and on censored samples. Technometrics 7:579–588
23. Montanari GC, Mazzanti G, Cacciari M, Fothergill JC (1997a) In search of convenient techniques for reducing bias in the estimation of Weibull parameter for uncensored tests. IEEE Transactions on Dielectrics and Electrical Insulation 4:306–313
24. Montanari GC, Mazzanti G, Cacciari M, Fothergill JC (1997b) Optimum estimators for the Weibull distribution of censored data: singly-censored tests. IEEE Transactions on Dielectrics and Electrical Insulation 4:462–469
25. Montanari GC, Mazzanti G, Cacciari M, Fothergill JC (1998) Optimum estimators for the Weibull distribution from censored test data: progressively-censored tests. IEEE Transactions on Dielectrics and Electrical Insulation 5:157–164
26. Drapella A, Kosznik S (1999) An alternative rule for placement of empirical points on Weibull probability paper. Quality and Reliability Engineering International 15:57–59
27. Nelder JA, Mead R (1965) A simplex method for function minimization. Computer Journal 7: 308–313
28. Thisted AT, Thisted RA (1988) Elements of Statistical Computing. CRC Press, Fla

Further Reading

Ross R (1994) Graphical methods for plotting and evaluating Weibull distributed data. IEEE Transactions on Dielectrics and Electrical Insulation 1:247–253

Appendix

Derivation of (3.28) and (3.29)

The cdf of $Z_{(i)}$ is given by (3.27), *i.e.*

$$F\left(z_{(i)}\right) = i\binom{n}{i}\int_{-\infty}^{z} F^{i-1}(z)(1-F(z))^{n-i}f(z)\,\mathrm{d}z$$

$$= i\binom{n}{i}\int_{-\infty}^{z}\left(1-\mathrm{e}^{-\mathrm{e}^{z}}\right)^{i-1}\left(\mathrm{e}^{-\mathrm{e}^{z}}\right)^{n-i}\mathrm{d}\left(1-\mathrm{e}^{-\mathrm{e}^{z}}\right)$$

and its pdf is

$$f\left(z_{(i)}\right) = i\binom{n}{i} F^{i-1}(z)(1 - F(z))^{n-i} f(z)$$

$$= i\binom{n}{i}\left(1 - e^{-e^{z}}\right)^{i-1}\left(e^{-e^{z}}\right)^{n-i} e^{-e^{z}} e^{z}$$

The mean of $Z_{(i)}$, by definition, can be obtained by

$$E\left(Z_{(i)}\right) = \int_{-\infty}^{\infty} z f\left(z_{(i)}\right) \, dz = i\binom{n}{i}\int_{-\infty}^{+\infty} z\left(1 - e^{-e^{z}}\right)^{i-1}\left(e^{-e^{z}}\right)^{n-i} e^{-e^{z}} e^{z} \, dz$$

Setting $v = e^{z}$, so that $z = \ln v$, $dz = dv/v$, and the above equation becomes

$$E\left(Z_{(i)}\right) = i\binom{n}{i}\int_{0}^{+\infty} \ln v \cdot (1 - e^{-v})^{i-1} e^{-(n-i+1)v} \, dv$$

$$= i\binom{n}{i}\int_{0}^{+\infty} \ln v \cdot (e^{v} - 1)^{i-1} e^{-nv} \, dv$$

Using the binominal theorem, $i.e.$

$$(x + a)^{n} = \sum_{k=0}^{n}\binom{n}{k} x^{k} a^{n-k}$$

we have

$$(e^{v} - 1)^{i-1} = \sum_{k=0}^{i-1}\binom{i-1}{k}(-1)^{k} e^{v(i-1-k)}$$

Thus

$$E\left(Z_{(i)}\right) = i\binom{n}{i}\sum_{k=0}^{i-1}(-1)^{k}\binom{i-1}{k}\int_{0}^{+\infty} \ln v \cdot e^{(n-i+k+1)v} \, dv$$

Let $T = (n - i + k + 1)t$, after replacing, we have

$$E\left(Z_{(i)}\right) = i\binom{n}{i}\sum_{k=0}^{i-1}(-1)^{k}\binom{i-1}{k}\int_{0}^{+\infty}[\ln T - \ln(n - i + k + 1)] \cdot \frac{e^{-T}}{n - i + k + 1} \, dT$$

$$= i\binom{n}{i}\sum_{k=0}^{i-1}(-1)^{k}\binom{i-1}{k}\frac{1}{n - i + k + 1}\left[\int_{0}^{+\infty} \ln T \cdot e^{-T} \, dT - \right.$$

$$\left. \ln(n - i + k + 1)\int_{0}^{+\infty} e^{-T} \, dT\right]$$

Since

$$\int_0^{+\infty} \ln T \cdot e^{-T}\, dT = -c = -0.577216$$

where c is the Euler's constant,

$$\int_0^{+\infty} e^{-T}\, dT = 1$$

and

$$i \binom{n}{i} \sum_{k=0}^{i-1} (-1)^k \binom{i-1}{k} \frac{1}{n-i+k+1} = 1$$

Finally we have

$$E\left(Z_{(i)}\right) = i \binom{n}{i} \cdot \sum_{k=0}^{i-1} \left\{ (-1)^k \binom{i-1}{k} \cdot \frac{-c - \ln(n-i+k+1)}{n-i+k+1} \right\}$$

which is (3.28).

Similarly, we can obtain

$$E(Z_{(i)}^2) = i \binom{n}{i} \int_{-\infty}^{+\infty} z^2 (1 - e^{-e^z})^{i-1} (e^{-e^z})^{n-i} e^{-e^z} e^z\, dz$$

$$= i \binom{n}{i} \int_0^{+\infty} \ln^2 v \cdot (e^v - 1)^{i-1} \cdot e^{-nv}\, dv$$

$$= i \binom{n}{i} \sum_{k=0}^{i-1} (-1)^k \binom{i-1}{k} \int_0^{+\infty} \ln^2 v \cdot e^{-(n-i+k+1)v}\, dv$$

$$= i \binom{n}{i} \sum_{k=0}^{i-1} (-1)^k \binom{i-1}{k} \int_0^{+\infty} [\ln T - \ln(n-i+k+1)]^2 \cdot \frac{e^{-T}}{n-i+k+1}\, dT$$

$$= i \binom{n}{i} \sum_{k=0}^{i-1} (-1)^k \binom{i-1}{k} \frac{1}{n-i+k+1} \left[\int_0^{+\infty} \ln^2 T \cdot e^{-T}\, dT \right.$$

$$\left. + 2c \ln(n-i+k+1) + \ln^2(n-i+k+1) \right]$$

Since

$$\int_0^{+\infty} \ln^2 T \cdot e^{-T}\, dT = 1.978112$$

Finally we have,

$$E\left(Z_{(i)}^2\right) = 1.978112 + i \binom{n}{i} \sum_{k=0}^{i-1}$$
$$\left\{ (-1)^k \binom{i-1}{k} \cdot \frac{2c\ln(n-i+k+1) + \ln^2(n-i+k+1)}{n-i+k+1} \right\}$$

which is (3.29).

Chapter 4
Periodic and Sequential Imperfect Preventive Maintenance Policies for Cumulative Damage Models

Toshio Nakagawa, Satoshi Mizutani

Department of Marketing and Information Systems,
Aichi Institute of Technology,
1247 Yachigusa, Yakusa-cho, Toyota 470-0392, Japan

4.1 Introduction

Many serious accidents have happened recently and caused heavy damage as systems have become large-scale and complex. There is considerable anxiety that big earthquakes in the near future might happen in Japan and might destroy large high buildings and old plants such as chemical and power plants, and inflict serious damage on wide areas. Furthermore, public infrastructure in most advanced nations is becoming old. Maintenance policies for such industrial systems and public infrastructure should be established scientifically and practically according to their occasions and circumstances [1].

The maintenance models and associated optimization problems were summarized in [1–4]. The standard preventive maintenance (PM) of an operating unit is performed at periodic times based on its age or operating time [1]. Most models have assumed that the unit after PM becomes like new, but this assumption might not be true. The unit after PM usually might be only younger, and its improvement would depend on the resources available for PM. PM models in which the age or failure rate of the unit after PM reduces in proportion to that before PM [5, 6] are called the imperfect PM. Similar imperfect repair models were considered and studied in [7, 8]. Some chapters [9–11] of recently published books have summarized many results of imperfect PM and repair. The PM of large complex systems such as computers, radar, aeroplanes and plants should be performed frequently as the unit ages. A sequential PM policy where the PM is performed at fixed intervals T_k has been proposed in [12, 13]. In some actual situations, however, the PM only seems imperfect in the sense that it does not make the unit new [14].

As one example of maintenance models, we can consider the cumulative damage model where the total damage is additive. Such reliability models and their optimal maintenance policies have been discussed extensively in [15]. We can apply the cumulative damage model to a gas turbine engine of a cogeneration system [16]: A gas turbine engine is generally used as the power source of a cogeneration system because its size is small, its exhaust gas emission is clean, and both its noise and vibration level are low. The turbine engine suffers mechanical damage when it is turned on and operated. Therefore, the engine has to be overhauled when it has exceeded the number of cumulative turn-on cycles or the total operating time. The damage models have also been applied to crack growth models [17–19], welded joints [20], floating structures [21], reinforced concrete structures [22], and plastic automotive components [23]. Such stochastic models of fatigue damage of materials in engineering systems have been described in detail [24–26].

We take up the preventive maintenance (PM) of cumulative damage models of a unit where the total damage is additive [15]. First, the PM is performed at periodic times kT ($k = 1, 2, \ldots$) and an amount of damage incurred for each PM interval $((k-1)T, kT]$ has an identical distribution $G(x)$ for a specified $T > 0$. The total damage after kth PM becomes aZ_k ($0 < a < 1$) when it was Z_k before PM, $i.e.$, the kth PM reduces the total damage Z_k to a Z_k. Suppose that the unit fails when the total damage has exceeded a failure level K at some PM time (Fig. 4.1). Then, we adopt two replacement policies when the unit is replaced at time nT and when the total damage has exceeded a managerial level Z ($0 \le Z \le K$) before failure. An example is shown when $G(x) = 1 - e^{-\mu x}$.

Next, we take up the cumulative damage model with minimal repair at failure [27]: The unit is subject to shocks that occur in a Poisson process. At each shock, the unit suffers random damage that is additive, and fails with probability $p(x)$ when the total damage is x. If the unit fails, then it undergoes only minimal repair. We apply a sequential PM policy to this model where each PM is imperfect: The PM is performed at the intervals of sequential times T_k ($k = 1, 2, \ldots, n$),

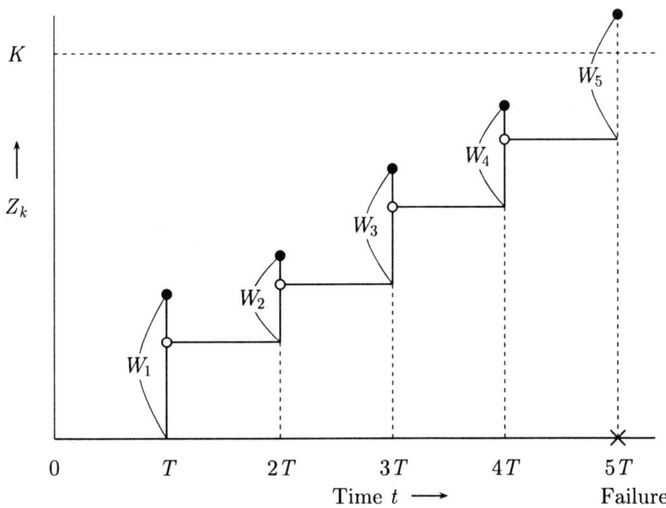

Fig. 4.1 Process of periodic imperfect PM for a cumulative damage model

and the unit is replaced at time $T_1 + T_2 + \cdots + T_n$. The amount of damage after the kth PM becomes $a_k Z_k$ when it was Z_k before PM, *i.e.*, the kth PM reduces the total damage Z_k to $a_k Z_k$. For the above PM model, we obtain the expected cost per unit of time until replacement, and discuss optimal PM times T_k^* ($k = 1, 2, \ldots, n$). Furthermore, suppose that the unit has to be operating for a finite interval $(0, S]$. Then, putting that $\sum_{k=1}^{n} T_k = S$, we compute an optimal number n^* and optimal times T_k^* ($k = 1, 2, \ldots, n^* - 1$) that minimize the expected cost until replacement.

4.2 Periodic PM

Consider the periodic PM policy where the PM is performed at periodic times kT ($k = 1, 2, \ldots$) for a specified $T > 0$. We call an interval from the $(k-1)$th PM to the kth PM *periodic k*. It is assumed that W_k is an amount of damage caused in periodic k and has an identical distribution $G(x) \equiv \Pr\{W_k \leq x\}$, where $W_0 \equiv 0$. The total damage is additive and is known only immediately before the PM.

We introduce an improvement factor in PM where the kth PM reduces $100(1-a)\%$ ($0 < a \leq 1$) of the total damage. Letting Z_k be the total damage at periodic k, *i.e.*, just before the kth PM, the kth PM reduces it to aZ_k. Then, we have the relation

$$Z_k = \sum_{j=1}^{k} a^{k-j} W_j \quad (k = 1, 2, \ldots) . \tag{4.1}$$

Noting that $G_j(x) \equiv \Pr\{a^{k-j} W_j \leq x\} = G(x/a^{k-j})$, the total damage has a distribution $G^{(k)}(x) \equiv \Pr\{Z_k \leq x\} = G_1(x) * \cdots * G_k(x)$, where the asterisk denotes the Stieltjes convolution, *i.e.*, $a(t) * b(t) \equiv \int_0^t b(t-u)\, da(u)$. In particular, when $a = 1$, *i.e.*, the PM is completely useless, $G^{(k)}(x)$ represents the k-fold Stieltjes convolution of $G(x)$ with itself.

When $0 < a < 1$ and $G(x) = 1 - e^{-\mu x}$,

$$G_j(x) = 1 - e^{-\mu x/a^{k-j}} \quad (j = 1, 2, \ldots, k) .$$

Thus, we have

$$G^{(1)}(x) = G_1(x) = 1 - e^{-\mu x} ,$$

$$G^{(2)}(x) = \int_0^x G^{(1)}(x-y)\, dG_1(y) = \frac{1}{1-a}(1 - e^{-\mu x}) + \frac{1}{1-a^{-1}}(1 - e^{-\mu x/a}) ,$$

$$G^{(3)}(x) = \int_0^x G^{(2)}(x-y)\, dG_1(y)$$

$$= \frac{1}{(1-a)(1-a^2)}(1 - e^{-\mu x}) + \frac{1}{(1-a^{-1})(1-a)}(1 - e^{-\mu x/a})$$

$$+ \frac{1}{(1-a^{-2})(1-a^{-1})}(1 - e^{-\mu x/a^2}) ,$$

and generally,

$$G^{(k)}(x) = \int_0^x G^{(k-1)}(x-y)\,dG_1(y)$$

$$= \sum_{i=1}^{k} \left[\frac{1}{\Pi_{\substack{j=1 \\ j \neq i}}^{k}(1 - a^{j-i})} \right] \left(1 - e^{-\mu x/a^{i-1}} \right) \quad (k = 1, 2, \ldots)\,. \tag{4.2}$$

where $\Pi_{\substack{j=1 \\ j \neq i}}^{1} \equiv 1$. This is easily proved by using the mathematical induction.

Suppose that the unit fails when the total damage has exceeded a failure level K and its failure is detected at time kT $(k = 1, 2, \ldots)$ before PM. Then, we consider the following two replacement policies [15]: First, the unit is replaced at time nT or at failure, whichever occurs first. Because the probability that the unit is replaced before failure at time nT is $G^{(n)}(K)$, and the mean time to replacement is

$$nTG^{(n)}(K) + \sum_{k=0}^{n-1}(k+1)T\left[G^{(k)}(K) - G^{(k+1)}(K)\right] = T\sum_{k=0}^{n-1}G^{(k)}(K)\,,$$

the expected cost rate is [1]

$$C_1(n) = \frac{c_K - (c_K - c_n)G^{(n)}(K)}{T\sum_{k=0}^{n-1}G^{(k)}(K)} \quad (n = 1, 2, \ldots)\,, \tag{4.3}$$

where c_K is the replacement cost at failure and c_n $(<c_K)$ is the replacement cost at time nT.

We find an optimal number n^* that minimizes $C_1(n)$. From the inequality $C_1(n+1) - C_1(n) \geq 0$, we have

$$Q(n+1)\sum_{k=0}^{n-1}G^{(k)}(K) + G^{(n)}(K) \geq \frac{c_K}{c_K - c_n} \quad (n = 1, 2, \ldots)\,, \tag{4.4}$$

where $Q(n) \equiv [G^{(n-1)}(K) - G^{(n)}(K)]/G^{(n-1)}(K)$ and $G^{(0)}(x) \equiv 1$ for $x \geq 0$. It is clearly seen that if $Q(n)$ is strictly increasing, then the left-hand side of (4.4) is also strictly increasing. In this case, if there exists some n such that (4.4), an optimal n^* is given by a unique minimum that satisfies (4.4).

Therefore, substituting (4.2) into (4.4), we can compute an optimal n^* and the resulting cost $C_1(n^*)$ in (4.3) for specified μK, a and c_K/c_n. Table 4.1 presents optimum n^* and the expected cost rate $TC_1(n^*)/c_n$ for a and μK when $c_K/c_n = 5$. For example, when $a = 0.9$ and $\mu K = 5$, $i.e.$, the unit fails at 5 periods on average without doing PM, it should be replaced before failure at 4 periods. This indicates that optimal n^* increase with μK and decrease with a because the unit becomes hard to fail as μK becomes large and a becomes small.

Secondly, the unit is replaced before failure when the total damage has exceeded a failure level Z $(0 \leq Z \leq K)$ at time kT. Because the probability that the unit is

Table 4.1 Optimal number n^* and expected cost rate $TC_1(n^*)/c_n$ when $c_K/c_n = 5$

μK	$a = 0.8$		$a = 0.85$		$a = 0.9$		$a = 0.95$	
	n^*	$TC_1(n^*)/c_n$	n^*	$TC_1(n^*)/c_n$	n^*	$TC_1(n)/c_n$	n^*	$TC_1(n^*)/c_n$
1	3	2.311	2	2.341	2	2.368	2	2.393
2	3	1.272	2	1.316	2	1.347	2	1.378
3	4	0.760	3	0.819	3	0.872	2	0.898
4	∞	0.191	4	0.546	3	0.595	3	0.634
5	∞	0.089	6	0.377	4	0.433	3	0.480
6	∞	0.050	∞	0.128	5	0.326	4	0.370
7	∞	0.033	∞	0.072	6	0.253	5	0.298
8	∞	0.026	∞	0.045	8	0.197	6	0.246
9	∞	0.023	∞	0.032	11	0.154	7	0.206
10	∞	0.021	∞	0.026	∞	0.078	8	0.175

replaced at failure is $\sum_{k=0}^{\infty} \int_0^Z [1 - G_{k+1}(K - x)] \, dG^{(k)}(x)$, and the mean time to replacement is

$$\sum_{k=0}^{\infty} (k + 1)T \left\{ \int_0^Z [1 - G_{k+1}(K - x)] \, dG^{(k)}(x) \right.$$
$$\left. + \int_0^Z [G_{k+1}(K - x) - G_{k+1}(Z - x)] \, dG^{(k)}(x) \right\} = T \sum_{k=0}^{\infty} G^{(k)}(Z) ,$$

the expected cost rate is [1]

$$C_2(Z) = \frac{c_Z + (c_K - c_Z) \sum_{k=0}^{\infty} \int_0^Z [1 - G_{k+1}(K - x)] \, dG^{(k)}(x)}{T \sum_{k=0}^{\infty} G^{(k)}(Z)} , \tag{4.5}$$

where c_K is the replacement cost at failure and c_Z $(<c_K)$ is the replacement cost at total damage Z.

It is difficult to envisage an optimal Z^* that minimizes $C_2(Z)$ analytically. In particular, when $a = 1$, the expected cost rate is [15]

$$\widetilde{C}_2(Z) = \frac{c_Z + (c_K - c_Z) \left[\overline{G}(K) + \int_0^Z \overline{G}(K - x) \, dM_G(x) \right]}{T[1 + M_G(Z)]} , \tag{4.6}$$

where $\overline{G}(x) \equiv 1 - G(x)$ and $M_G(x) \equiv \sum_{k=1}^{\infty} G^{(k)}(x)$ represents the renewal function of $G(x)$ in $[0, t]$. If $M_G(K) > c_Z/(c_K - c_Z)$, then an optimal Z^* $(0 < Z^* < K)$ to minimize $\widetilde{C}_2(Z)$ is given by a finite and unique solution that satisfies

$$\int_{K-Z}^K [1 + M_G(K - x)] \, dG(x) = \frac{c_Z}{c_K - c_Z} , \tag{4.7}$$

and the resulting cost rate is

$$T\widetilde{C}_2(Z^*) = (c_K - c_Z)\overline{G}(K - Z^*) . \tag{4.8}$$

4.3 Sequential PM

Consider a sequential PM policy for the unit where the PM is performed at fixed intervals T_k $(k = 1, 2, \ldots, n-1)$ and the replacement is made at $T_1 + T_2 + \cdots + T_n$ [13]. We call an interval from the $(k-1)$th PM to the kth PM *period k*.

Suppose that shocks occur in a Poisson process with rate λ. Random variables N_k $(k = 1, 2, \ldots, n)$ denote the number of shocks in period k, *i.e.*, $\Pr\{N_k = j\} = [(\lambda T_k)^j / j!] \exp(-\lambda T_k)$ $(j = 0, 1, 2, \ldots)$. Further, we denote W_{kj} the amount of damage caused by the jth shock in period k where $W_{k0} \equiv 0$. It is assumed that W_{kj} is non-negative, independent and identically distributed, and has an identical distribution $\Pr\{W_{kj} \le x\} \equiv G(x)$ for all k and j. The total damage is additive, and $G^{(j)}(x)$ $(j = 1, 2, \ldots)$ is the j-fold Stieltjes convolution of $G(x)$ with itself and $G^{(0)}(x) \equiv 1$ for all $x \ge 0$. Then, it follows that

$$\Pr\{W_{k1} + W_{k2} + \cdots + W_{kj} \le x\} = G^{(j)}(x) \quad (j = 0, 1, 2, \ldots). \tag{4.9}$$

When the total damage becomes x at a shock, the unit fails with probability $p(x)$, which increases with x from 0 to 1. If the unit fails between PMs, it undergoes only minimal repair, and hence, the total damage remains unchanged by any minimal repair. It is assumed that all times required for any PM and minimal repair are negligible.

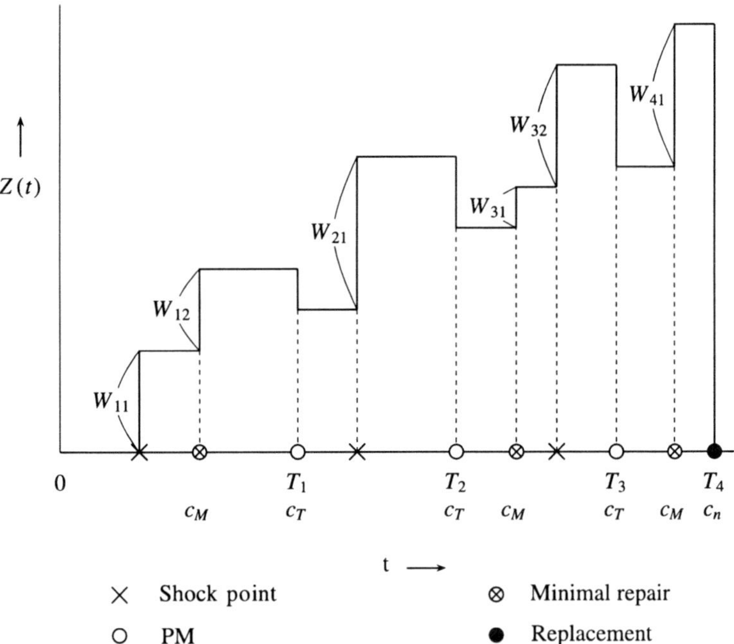

Fig. 4.2 Process for imperfect PM

Next, we introduce an improvement factor in PM where the kth PM reduces $100(1 - a_k)\%$ $(0 \le a_k \le 1)$ of the total damage. Letting Z_k be the total damage at the end of period k, *i.e.*, just before the kth PM, the kth PM reduces it to $a_k Z_k$. Since the total damage during period k is additive and is not removed by minimal repair, we have the relation

$$Z_k = a_{k-1} Z_{k-1} + \sum_{j=1}^{N_k} W_{kj} \quad (k = 1, 2, \ldots, n), \tag{4.10}$$

where $Z_0 \equiv 0$ and $\sum_{j=1}^{0} \equiv 0$.

Let c_T be the cost of each PM, c_n be the cost of replacement with $c_n > c_T$, and c_M be the cost of minimal repair. Then, from the assumption that the unit fails with probability $p(\cdot)$ only at shocks, the total cost in period k is

$$\widetilde{C}(k) = c_T + c_M \sum_{j=1}^{N_k} p(a_{k-1} Z_{k-1} + W_{k1} + W_{k2} + \cdots + W_{kj}) \quad (k = 1, 2, \ldots, n-1),$$

$$\tag{4.11}$$

$$\widetilde{C}(n) = c_n + c_M \sum_{j=1}^{N_n} p(a_{n-1} Z_{n-1} + W_{n1} + W_{n2} + \cdots + W_{nj}). \tag{4.12}$$

Further, we assume that $p(x)$ is exponential, *i.e.*, $p(x) = 1 - e^{-\theta x}$ for $\theta > 0$. Letting $G^*(\theta)$ be the Laplace–Stieltjes transform of $G(x)$, *i.e.*, $G^*(\theta) \equiv \int_0^\infty e^{-\theta x} \, dG(x)$,

$$E\left\{\exp\left[-\theta(W_{k1} + W_{k2} + \cdots + W_{kj})\right]\right\} = \int_0^\infty e^{-\theta x} \, dG^{(j)}(x) = [G^*(\theta)]^j. \tag{4.13}$$

Using the law of total probability in (4.11), the expected cost in period k is

$$E\{\widetilde{C}(k)\} = c_T + c_M E\left\{ \sum_{j=1}^{N_k} p\left(a_{k-1} Z_{k-1} + W_{k1} + W_{k2} + \cdots + W_{kj}\right) \right\}$$

$$= c_T + c_M \sum_{i=1}^{\infty} \sum_{j=1}^{i} E\left\{ 1 - \exp\left[-\theta\left(a_{k-1} Z_{k-1} + W_{k1} + W_{k2} + \cdots + W_{kj}\right)\right] \right\}$$

$$\Pr\{N_k = i\}.$$

Let $B_k^*(\theta) \equiv E\{\exp(-\theta Z_k)\}$. Then, since Z_{k-1} and W_{kj} are independent of each other, we have, from (4.13),

$$E\left\{ 1 - \exp[-\theta(a_{k-1} Z_{k-1} + W_{k1} + \cdots + W_{kj})] \right\} = 1 - B_{k-1}^*(\theta a_{k-1})[G^*(\theta)]^j.$$

Thus, from the assumption that N_k has a Poisson distribution with rate λ,

$$
\begin{aligned}
E\{\widetilde{C}(k)\} &= c_T + c_M \sum_{i=1}^{\infty} \frac{(\lambda T_k)^i}{i!} \, e^{-\lambda T_k} \sum_{j=1}^{i} \left\{ 1 - B_{k-1}^*(\theta a_{k-1})[G^*(\theta)]^j \right\} \\
&= c_T + c_M \left\{ \lambda T_k - \frac{G^*(\theta)}{1 - G^*(\theta)} B_{k-1}^*(\theta a_{k-1}) \left[1 - e^{-\lambda[1-G^*(\theta)]T_k} \right] \right\} \\
&\quad (k = 1, 2, \ldots, n-1) .
\end{aligned}
\tag{4.14}
$$

Similarly, the expected cost in period n is

$$
E\{\widetilde{C}(n)\} = c_n + c_M \left\{ \lambda T_n - \frac{G^*(\theta)}{1 - G^*(\theta)} B_{n-1}^*(\theta a_{n-1}) \left[1 - e^{-\lambda[1-G^*(\theta)]T_n} \right] \right\} .
\tag{4.15}
$$

Letting $A_r^k \equiv \Pi_{j=r}^k a_j$ for $r \le k$ and $\equiv 0$ for $r > k$, we have, from (4.10),

$$
a_{k-1} Z_{k-1} = \sum_{r=1}^{k-1} A_r^{k-1} \sum_{j=1}^{N_r} W_{rj} .
$$

Thus, recalling that W_{ij} are independent and have an identical distribution $G(x)$, we have [12]

$$
\begin{aligned}
B_{k-1}(\theta a_{k-1}) &= E\left\{ e^{-\theta a_{k-1} Z_{k-1}} \right\} \\
&= E\left\{ \exp\left[-\theta \sum_{r=1}^{k-1} A_r^{k-1} \sum_{j=1}^{N_r} W_{rj} \right] \right\} .
\end{aligned}
$$

Since

$$
\begin{aligned}
E\left\{ \exp\left[-\theta A_r^{k-1} \sum_{j=1}^{N_r} W_{rj} \right] \right\} &= \sum_{i=0}^{\infty} \Pr\{N_r = i\} E\left\{ \exp\left[-\theta A_r^{k-1} \sum_{j=1}^{i} W_{rj} \right] \right\} \\
&= \sum_{i=0}^{\infty} \frac{(\lambda T_r)^i}{i!} \, e^{-\lambda T_r} \left[G^*\left(\theta A_r^{k-1} \right) \right]^i \\
&= \exp\left\{ -\lambda T_r \left[1 - G^*\left(\theta A_r^{k-1} \right) \right] \right\} ,
\end{aligned}
$$

we have consequently,

$$
B_{k-1}(\theta a_{k-1}) = \exp\left\{ -\sum_{j=1}^{k-1} \lambda T_j \left[1 - G^*\left(\theta A_j^{k-1} \right) \right] \right\} .
\tag{4.16}
$$

Substituting (4.16) into (4.14) and (4.15) respectively, the expected cost in period k is

$$E\{\widetilde{C}(k)\} = c_T + c_M \left(\lambda T_k - \frac{G^*(\theta)}{1 - G^*(\theta)} \exp\left\{ -\sum_{j=1}^{k-1} \lambda T_j \left[1 - G^* \left(\theta A_j^{k-1}\right)\right] \right\} \right. $$
$$\left. \left\{1 - e^{-\lambda T_k[1 - G^*(\theta)]}\right\} \right) \quad (k = 1, 2, \ldots, n-1), \tag{4.17}$$

and

$$E\{\widetilde{C}(n)\} = c_n + c_M \left(\lambda T_n - \frac{G^*(\theta)}{1 - G^*(\theta)} \exp\left\{ -\sum_{j=1}^{n-1} \lambda T_j \left[1 - G^* \left(\theta A_j^{n-1}\right)\right] \right\} \right. $$
$$\left. \left\{1 - e^{-\lambda T_n[1 - G^*(\theta)]}\right\} \right). \tag{4.18}$$

Therefore, the expected cost rate until replacement is

$$\mathbf{C}(T_n) = \frac{\sum_{k=1}^{n-1} E\{\widetilde{C}(k)\} + E\{\widetilde{C}(n)\}}{\sum_{k=1}^{n} T_k}$$
$$= \frac{(n-1)c_T + c_n - c_M \left[\{G^*(\theta)/[1 - G^*(\theta)]\} \sum_{k=1}^{n} \exp\left\{ -\sum_{j=1}^{k-1} \lambda T_j \left[1 - G^* \left(\theta A_j^{k-1}\right)\right] \right\} \left\{1 - e^{-\lambda T_k[1 - G^*(\theta)]}\right\} \right]}{\sum_{k=1}^{n} T_k}$$
$$+ \lambda c_M \quad (n = 1, 2, \ldots). \tag{4.19}$$

We derive optimal times T_k^* which minimize $\mathbf{C}(T_n)$ in (4.19) when $a_k \equiv a$ and $G(x) = 1 - e^{-\mu x}$. In this case, (4.19) is rewritten

$$\widetilde{\mathbf{C}}_1(T_n) \equiv c_M - \frac{\mathbf{C}(T_n)}{\lambda}$$
$$= \frac{c_M(\mu/\theta) \sum_{k=1}^{n} \exp\left\{ -\sum_{j=1}^{k-1} \lambda T_j \left[\theta a^{k-j}/\left(\theta a^{k-j} + \mu\right)\right] \right\} \left\{1 - e^{-\lambda T_k \frac{\theta}{\theta + \mu}}\right\} - (n-1)c_T - c_n}{\lambda \sum_{k=1}^{n} T_k}$$
$$(n = 1, 2, \ldots). \tag{4.20}$$

We discuss optimal T_k^* numerically which maximize $\widetilde{\mathbf{C}}_1(T_n)$. When $n = 1$,

$$\widetilde{C}_1(T_1) = \frac{c_M(\mu/\theta)\left[1 - e^{-\lambda T_1 \frac{\theta}{\theta + \mu}}\right] - c_n}{\lambda T_1}. \tag{4.21}$$

Thus, we have easily

$$\widetilde{C}_1(0) \equiv \lim_{T_1 \to 0} \widetilde{C}_1(T_1) = c_M \frac{\mu}{\theta + \mu} ,$$

$$\widetilde{C}_1(\infty) \equiv \lim_{T_1 \to \infty} \widetilde{C}_1(T_1) = 0 .$$

Differentiating $\widetilde{C}_1(T_1)$ with respect to T_1 and putting it equal to zero, we have

$$\frac{1}{\theta} \left[1 - e^{-\lambda T_1 \frac{\theta}{\theta + \mu}} \right] - \frac{\lambda T_1}{\theta + \mu} e^{-\lambda T_1 \frac{\theta}{\theta + \mu}} = \frac{c_n}{\mu c_M} , \tag{4.22}$$

the left-hand side of which is strictly increasing from 0 to $1/\theta$. Therefore, if $c_M > (\theta/\mu)c_n$ then there exists a finite and unique T_1^* ($0 < T_1^* < \infty$) which satisfies (4.22). Conversely, if $c_M \le (\theta/\mu)c_n$ then $T_1^* = 0$. An optimal time is approximately

$$\widetilde{T}_1 = \frac{\theta + \mu}{\lambda} \sqrt{\frac{c_n}{\mu \theta c_M}} . \tag{4.23}$$

Furthermore, differentiating $\widetilde{\mathbf{C}}_1(T_n)$ in (4.20) with respect to T_k and putting it equal to zero, we have

$$\frac{\theta}{\theta + \mu} \exp \left\{ - \sum_{j=1}^{k} \lambda T_j \left[\frac{\theta a^{k-j}}{\theta a^{k-j} + \mu} \right] \right\}$$

$$- \sum_{i=k+1}^{n} \frac{\theta a^{i-k}}{\theta a^{i-k} + \mu} \exp \left\{ - \sum_{j=1}^{i-1} \lambda T_j \left[\frac{\theta a^{i-j}}{\theta a^{i-j} + \mu} \right] \right\} \left\{ 1 - e^{-\lambda T_i \frac{\theta}{\theta + \mu}} \right\}$$

$$= \frac{\widetilde{\mathbf{C}}_1(T_n)}{(\mu/\theta)c_M} \quad (k = 1, 2, \ldots, n-1) , \tag{4.24}$$

$$\frac{\theta}{\theta + \mu} \exp \left\{ - \sum_{j=1}^{n} \lambda T_j \left[\frac{\theta a^{n-j}}{\theta a^{n-j} + \mu} \right] \right\} = \frac{\widetilde{\mathbf{C}}_1(T_n)}{(\mu/\theta)c_M} . \tag{4.25}$$

Thus, we may solve numerically the simultaneous (4.24) and (4.25), and obtain the expected cost $\widetilde{\mathbf{C}}_1(T_n)$. Next, comparing $\widetilde{\mathbf{C}}_1(T_n)$ for all $n \ge 1$, we can get the optimal number n^* and T_k^* ($k = 1, 2, \ldots, n^*$).

4.4 PM for a Finite Interval

A unit has to be operating for a finite interval $(0, S]$, and be replaced at a specified time S. The other assumptions are the same as those of the previous model, except that $T_1 + T_2 + \cdots + T_n = S$. Then, we consider the optimal policy that maximizes the expected cost

$$\widetilde{C}_2(T_n) = \frac{\mu c_M}{\theta} \sum_{k=1}^{n} \exp\left\{ -\sum_{j=1}^{k-1} \lambda T_j \left[\frac{\theta a^{k-j}}{\theta a^{k-j} + \mu} \right] \right\}$$
$$\left\{ 1 - e^{-\lambda T_k \frac{\theta}{\theta+\mu}} \right\} - (n-1)c_T - c_n \quad (n = 1, 2, \ldots). \tag{4.26}$$

For example, when $n = 1$,

$$\widetilde{C}_2(S) = \frac{\mu c_M}{\theta} \left\{ 1 - e^{-\lambda S \frac{\theta}{\theta+\mu}} \right\} - c_n . \tag{4.27}$$

When $n = 2$,

$$\widetilde{C}_2(T_1) = \frac{\mu c_M}{\theta} \left\{ 1 - e^{-\lambda T_1 \frac{\theta}{\theta+\mu}} + e^{-\lambda T_1 \frac{\theta a}{\theta a+\mu}} \left[1 - e^{-\lambda(S-T_1)\frac{\theta}{\theta+\mu}} \right] \right\} - c_T - c_n . \tag{4.28}$$

Differentiating $\widetilde{C}_2(T_1)$ with respect to T_1 and putting it equal to zero,

$$\frac{\theta}{\theta+\mu} \left[e^{-\lambda T_1 \left(\frac{\theta}{\theta+\mu} - \frac{\theta a}{\theta a+\mu} \right)} - e^{-\lambda(S-T_1)\frac{\theta}{\theta+\mu}} \right] - \frac{\theta a}{\theta a+\mu} \left[1 - e^{-\lambda(S-T_1)\frac{\theta}{\theta+\mu}} \right] = 0 . \tag{4.29}$$

Letting $Q(T)$ be the left-hand side of (4.29),

$$Q(0) = \left(\frac{\theta}{\theta+\mu} - \frac{\theta a}{\theta a+\mu} \right) \left(1 - e^{-\lambda S \frac{\theta}{\theta+\mu}} \right) > 0 ,$$
$$Q(S) = \frac{-\theta}{\theta+\mu} \left[1 - e^{-\lambda S \left(\frac{\theta}{\theta+\mu} - \frac{\theta a}{\theta a+\mu} \right)} \right] < 0 ,$$
$$Q'(T) = -\left(\frac{\theta}{\theta+\mu} \right) \left(\frac{\theta}{\theta+\mu} - \frac{\theta a}{\theta a+\mu} \right)$$
$$\left[e^{-\lambda T_1 \left(\frac{\theta}{\theta+\mu} - \frac{\theta a}{\theta a+\mu} \right)} + e^{-\lambda(S-T_1)\frac{\theta}{\theta+\mu}} \right] < 0 .$$

Thus, there exists an optimal T_1^* $(0 < T_1^* < S)$ which satisfies (4.29).
    When $n = 3$,

$$\widetilde{C}_2(T_1, T_2) = \frac{\mu c_M}{\theta} \left\{ 1 - e^{-\lambda T_1 \frac{\theta}{\theta+\mu}} + e^{-\lambda T_1 \frac{\theta a}{\theta a+\mu}} \left[1 - e^{-\lambda T_2 \frac{\theta}{\theta+\mu}} \right] \right.$$
$$\left. + e^{-\lambda T_1 \frac{\theta a^2}{\theta a^2+\mu} - \lambda T_2 \frac{\theta a}{\theta a+\mu}} \left[1 - e^{-\lambda(S-T_1-T_2)\frac{\theta}{\theta+\mu}} \right] \right\} - 2c_T - c_n . \tag{4.30}$$

Differentiating $\widetilde{C}_2(T_1, T_2)$ with respect to T_1 and T_2 and putting them equal to zero, we have respectively,

$$\frac{\theta}{\theta + \mu}\left[e^{-\lambda T_1 \frac{\theta}{\theta+\mu}} - e^{-\lambda T_1 \frac{\theta a^2}{\theta a^2+\mu} -\lambda T_2 \frac{\theta a}{\theta a+\mu} -\lambda(S-T_1-T_2)\frac{\theta}{\theta+\mu}} \right]$$

$$-\frac{\theta a}{\theta a + \mu} e^{-\lambda T_1 \frac{\theta a}{\theta a+\mu}} \left[1 - e^{-\lambda T_2 \frac{\theta}{\theta+\mu}} \right]$$

$$-\frac{\theta a^2}{\theta a^2 + \mu} e^{-\lambda T_1 \frac{\theta a^2}{\theta a^2+\mu} -\lambda T_2 \frac{\theta a}{\theta a+\mu}} \left[1 - e^{-\lambda(S-T_1-T_2)\frac{\theta}{\theta+\mu}} \right] = 0 , \tag{4.31}$$

$$\frac{\theta}{\theta + \mu}\left[e^{-\lambda T_1 \frac{\theta a}{\theta a+\mu} -\lambda T_2 \frac{\theta}{\theta+\mu}} - e^{-\lambda T_1 \frac{\theta a^2}{\theta a^2+\mu} -\lambda T_2 \frac{\theta a}{\theta a+\mu} -\lambda(S-T_1-T_2)\frac{\theta}{\theta+\mu}} \right]$$

$$-\frac{\theta a}{\theta a + \mu} e^{-\lambda T_1 \frac{\theta a^2}{\theta a^2+\mu} -\lambda T_2 \frac{\theta a}{\theta a+\mu}} \left[1 - e^{-\lambda(S-T_1-T_2)\frac{\theta}{\theta+\mu}} \right] = 0 . \tag{4.32}$$

In general, differentiating $\widetilde{C}_2(T_n)$ with respect to T_k $(k = 1, 2, \ldots, n-1)$ $(n \geq 2)$ and putting them equal to zero, we have

$$\frac{\theta}{\theta + \mu}\left\{ \exp\left[-\sum_{j=1}^{k} \lambda T_j \frac{\theta a^{k-j}}{\theta a^{k-j} + \mu} \right] - \exp\left[-\sum_{j=1}^{n} \lambda T_j \frac{\theta a^{n-j}}{\theta a^{n-j} + \mu} \right] \right\}$$

$$-\sum_{i=k+1}^{n} \frac{\theta a^{i-k}}{\theta a^{i-k} + \mu} \exp\left[-\sum_{j=1}^{i-1} \lambda T_j \frac{\theta a^{i-j}}{\theta a^{i-j} + \mu} \right]$$

$$\left\{ 1 - e^{-\lambda T_i \frac{\theta}{\theta+\mu}} \right\} = 0 \quad (k = 1, 2, \ldots, n-1) , \tag{4.33}$$

where it should be noted that $T_n = S - T_1 - T_2 - \cdots - T_{n-1}$.

Therefore, we may solve the simultaneous (4.33), and obtain the expected cost $\widetilde{C}_2(T_n)$ in (4.26). Next, comparing $\widetilde{C}_2(T_n)$ for all $n \geq 1$, we can get the optimal number n^* and times T_k^* $(k = 1, 2, \ldots, n^* - 1)$ for a specified S.

Table 4.2 presents T_k and $\widetilde{C}_2(T_n)$ when $a = 0.5$, $\mu/\theta = 10$, $c_n/c_M = 5$, $c_T/c_M = 1.0$ and $S = 40$.

Comparing $\widetilde{C}_2(T_1, T_2, \ldots, T_{n-1})$ for $n = 1, 2, \ldots, 10$, the expected cost $\widetilde{C}_2(\cdot)$ is maximum, i.e., $C_2(\cdot)$ in (4.26) is minimum at $n^* = 8$. In this case, the optimum PM number is $n^* = 8$ and optimum PM times are 7.80, 11.46, 15.18, 18.91, 22.64, 26.35, 29.99, 40. This indicates the interesting result that the last PM time interval is the largest and the first one is the second. At first they increase, then remain constant for some period, and then decrease for large n, that is, the PM time intervals follows a upside-down bathtub curve [28] for $2 \leq k \leq n-1$. Figure 4.3 shows the PM interval times T_k $(k = 1, 2, \ldots, 10)$, and depicts roughly a standard bathtub curve.

Table 4.2 $a = 0.5$, $\mu/\theta = 10$, $c_n/c_M = 5$, $c_T/c_M = 1.0$, $S = 40$

	$n=1$	$n=2$	$n=3$	$n=4$	$n=5$	$n=6$	$n=7$	$n=8$	$n=9$	$n=10$
λT_1	40.00	13.17	12.41	11.37	10.32	9.36	8.52	7.80	7.17	6.63
λT_2		26.83	5.60	5.27	4.82	4.38	3.99	3.66	3.37	3.11
λT_3			21.99	5.23	4.87	4.45	4.06	3.72	3.42	3.17
λT_4				18.22	4.78	4.45	4.07	3.73	3.44	3.18
λT_5					15.22	4.35	4.06	3.73	3.44	3.18
λT_6						13.01	3.97	3.71	3.44	3.18
λT_7							11.33	3.64	3.42	3.18
λT_8								10.01	3.35	3.16
λT_9									8.96	3.10
λT_{10}										8.10
$\dfrac{\tilde{C}_2(T_n)}{c_M}$	4.74	5.86	6.87	7.70	8.34	8.78	9.05	9.17	9.16	9.03

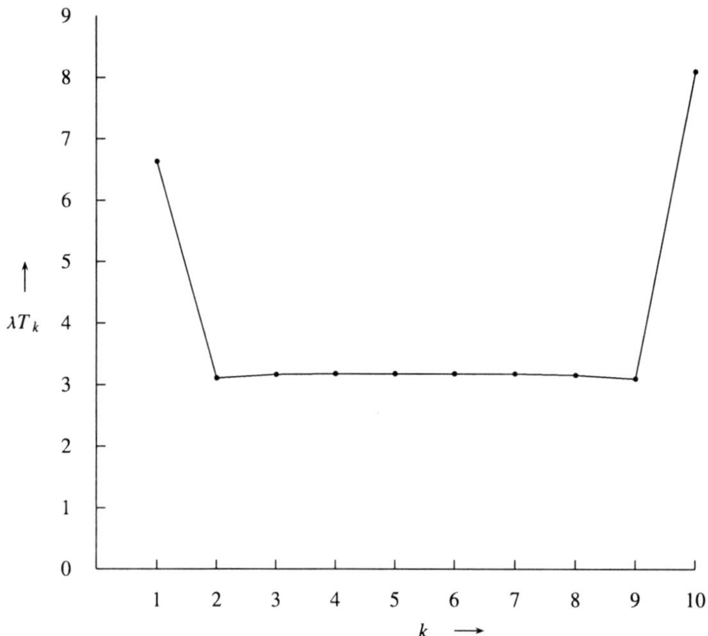

Fig. 4.3 Optimum PM time interval T_k for $k = 1, 2, \ldots, 10$

4.5 Conclusions

We have applied imperfect PM policies to a cumulative damage model where the total damage is additive. First, we have considered two replacement policies for the periodic PM model where a unit is replaced before failure at time nT and when the total damage has exceeded a managerial level Z. The optimal n^* that minimizes the expected cost rate is computed numerically when the amount of damage incurred for each periodic interval is distributed exponentially. Next, we have studied the sequential PM policy for a cumulative damage model with minimal repair at failures in two cases where the unit has to be operating for infinite and finite time intervals. Two expected costs until replacement are derived analytically when shocks occur in a Poisson process. The optimal PM times T_k^* and number n^* are computed numerically. It is of great interest that optimum PM time intervals follow two types of bathtub curves. It might be necessary to inquire into why PM times describe the two different curves.

References

1. Nakagawa T (2005) Maintenance Theory of Reliability. Springer, London
2. Barlow RE, Proschan F (1965) Mathematical Theory of Reliability. Wiley, New York
3. Osaki S (ed) (2002) Stochastic Models in Reliability and Maintenance. Springer, Berlin
4. Pham H (ed) (2003) Handbook of Reliability Engineering. Springer, London
5. Nakagawa T (1979) Optimal policies when preventive maintenance is imperfect. IEEE Trans Reliability R-28:331–332
6. Nakagawa T, Yasui K (1987) Optimum policies for a system with imperfect maintenance. IEEE Transactions on Reliability R-36:631–633
7. Brown M, Proschan F (1983) Imperfect repair. Journal of Applied Probability 20:851–859
8. Li HJ, Shaked M (2003) Imperfect repair models with preventive maintenance. Journal of Applied Probability 40:1043–1059
9. Nakagawa T (2000) Imperfect Preventive Maintenance Models. In: Ben-Daya M, Duffuaa SO, Raouf A (eds) Maintenance, modeling and optimization. Kluwer Academic, Boston, pp 201–214
10. Hang H, Pham H (2003) Optimal Imperfect Maintenance Models. In: Pham H (ed) Handbook of Reliability Engineering. Springer, London, pp 397–414
11. Nakagawa T (2002) Imperfect preventive maintenance models. In: Osaki S (ed) Stochastic Models in Reliability and Maintenance. Springer, Berlin, pp 125–143
12. Nguyen DC, Murthy DNP (1981) Optimal preventive maintenance policies for repairable systems. Operations Research 29:1181–1194
13. Nakagawa T (1986) Periodic and sequential preventive maintenance policies. Journal of Applied Probability 23:536–542
14. Nakagawa T (1988) Sequential imperfect preventive maintenance policies. IEEE Transactions on Reliability 37:295–298
15. Nakagawa T (2006) Shock and Damage Models in Reliability Theory. Springer, London
16. Kodo I, Nakagawa T (2006) Maintenance of a Cumulative Damage Model and Its Application to Gas Turbine Engine of Cogeneration System. In: Pham H (ed) Reliability modeling, analysis and optimization. World Scientific Publications, Singapore
17. Sobczyk K, Trebick J (1989) Modelling of random fatigue by cumulative jump process. Engineering Fracture Mechanics 34:477–493
18. Scarf PA, Wang W, Laycok PJ (1996) A stochastic model of crack growth under periodic inspections. Reliability Engineering and System Safety 51:331–339
19. Hopp WJ, Kuo YL (1998) An optimal structured policy for maintenance of partially observable aircraft engine components. Navel Research Logistics 45:335–352
20. Lukić M, Cremona C (2001) Probabilistic optimization of welded joints maintenance versus fatigue and fracture. Reliability Engineering and System Safety 72:253–264
21. Garbotov Y, Soares CG (2001) Cost and reliability based strategies for fatigue maintenance planning of floating structures. Reliability and System Safety 73:293–301

22. Petryna YS, Pfanner D, Shangenberg F, Krätzig WB (2002) Reliability of reinforced concrete structures under fatigue. Reliability and System Safety 77:253–261
23. Campean IF, Rosala GF, Grove DM, Henshall E (2005) Life modelling of a plastic automotive component. Proceedings of the Annual Reliability and Maintainability Symposium, pp 319–325
24. Sobczyk K (1987) Stochastic models for fatigue damage of materials. Advanced Applied Probability 19:652–673
25. Sobczyk K, Spencer Jr BF (1992) Random Fatigue: From Data to Theory. Academic Press, New York
26. Dasgupta A, Pecht M (1991) Material failure mechanisms and damage models. IEEE Transactions on Reliability 40:531–536
27. Kijima M, Nakagawa T (1992) Replacement policies of a shock model with imperfect preventive maintenance. European Journal of Operational Research 57:100–110
28. Mie J (1995) Bathtub failure rate and upside-down bathtub mean residual life. IEEE Transactions on Reliability 44:388–391

Chapter 5
Some Alternative Approaches to System Reliability Modeling

Jerzy K. Filus[1], Lidia Z. Filus[2]

[1] Department of Mathematics and Computer Science,
Oakton Community College, Des Plaines, Il 60016, USA
[2] Department of Mathematics, Northeastern Illinois University,
Chicago, IL 60625, USA

5.1 Introduction

This work contains a review and development of the ideas concerning some alternative methods in stochastic modeling of reliability (see [1]) that we have worked on over the last decade. Part of the presented material is published for the first time. Another part which has already appeared or will appear in the coming months (see, for example [2–5]), now is presented in a newer, "simpler" way, as our teaching experience on that has grown during many professional discussions.

There are three basic features that may characterize the results presented in this work.

5.1.1 New Kinds of Stochastic Dependences

The core of the theory is an analytical description and application of new kinds of stochastic dependences. The dependencies are mostly considered in association with modeling various types of stochastically dependent lifetimes in reliability theory framework. Two distinct methods of the underlying multivariate pdf constructions are developed. The first method, called the parameter replacement method, somehow resembles the conditioning methods developed, between others, in [6–9]. The two approaches, however, are essentially different and the classes of new multivariate pdfs, obtained in the cited papers, and those presented here are disjoint. A comparison of the two approaches is given in Sect. 5.4.5 (also see [5]).

The second method we use is called the (pseudoaffine) transformations method. That in many cases appears to be equivalent to the first, since some classes of the stochastic models obtained by either of the two methods are identical.

Special importance of the transformation method, as it may serve as a tool in statistical testing of the obtained models, is explained in Sect. 5.5.2.

5.1.2 Joint Probability Distributions

The new classes of the obtained multivariate pdfs are applied in reliability theory as stochastic models of parallel multicomponent systems, in the form of joint probability distributions of the random vectors of the component lifetimes. For references to this kind of model, see, for example [2–5, 10–14].

The models we constructed turn out to be, in a sense, "complementary" to the classical Freund [15] model as well as to Marshall and Olkin [16].

According to the Freund model (the bivariate case), the time period of the physical "impact" of, say, component e_1 (or rather lack of it) on the other component e_2, takes place after the e_1 component failure, and both components work independently before the first failure in the system. In the case of the Marshall and Olkin model, both components work independently all the time, but a single time instant in which either or both of them may fail. In case of the phenomena we model, unlike with the Freund pattern, component e_1 has a constant (continuous) impact on e_2 during the common activity period, *i.e.*, before the first failure occurs, only. For more on that, see Sect. 5.4.

5.1.3 Determination of pdfs

In applications of the presented models it is often possible to define various common patterns for sequences of n-variate pdfs as $n = 1, 2, \ldots$ grows. In many cases that may be considered "important", the complexity of the so determined pdfs in consecutive formulae may grow relatively slowly with the dimension n. That weak sensitivity of computational complexity of the pdfs on the growth of the dimension opens the way for extensions of the invol.ved random vectors to discrete time stochastic processes. Our second main reliability application of the introduced stochastic dependences relies on the use of the constructed stochastic processes as models for reliability of (single) systems with repair. In that case, the terms of the stochastic processes are considered to be stochastically dependent times of the system functioning between its failures, where each failure is followed by a repair.

The common pattern for the stochastic dependences makes them relatively easy to describe. As a result we often are able to relax significantly the common Markovianity assumption of the models, without stepping into excessive computational complexity that up to now relentlessly accompanied such attempts. This, potentially, allows us to incorporate more information concerning the system's past. That, in turn, may possibly increase the accuracy of the predictions associated with such the models.

5.1.4 Application of Stochastic Dependences

At this point, we should add that the kind of stochastic dependences we define, as well as the obtained particular models, are of a significantly wider generality than reliability applications alone.

Because of that, in some parts of the paper we occasionally mention some other, close to our main reliability stream subjects, particularly those associated with extensions of the multivariate normal pdfs toward the pseudonormal. For the last see [17–19]. See also a short notice in [20] on pages 217, 218.

At the end of the paper we give several simple bivariate examples of analytical calculations. One of them (Sect. 5.10.4) is devoted to an asymmetric bivariate Gaussians pdf. extension.

5.2 A New Bivariate Probability Densities Construction

The method of pdf construction presented in this section and in Sect. 5.3 is called the "parameter replacement method". It is distinct from the alternative "triangular transformation method" described later in this paper.

We start our considerations with a description of a device for a two-stage procedure in a process of stochastic modeling the reliability of a parallel system that is composed of two components.

5.2.1 Modeling of Component Lifetime

At the first stage of the procedure, it is assumed that the lifetime of each of the two system components e_1, e_2 is tested independently (each separately from the other) in idealized laboratory conditions. Therefore, at this stage, the component lifetimes, represented by the usual non-negative r. variables T_1, T_2, are, by nature, stochastically independent. Here it is assumed that, as a result of a testing procedure, one obtains estimated pdfs $f_1(t_1; \theta_1)$, $f_2(t_2; \theta_2)$ of the lifetimes T_1, T_2 respectively, where for $j = 1, 2$, θ_j denotes a scalar or a vector parameter of each the pdf $f_j(t_j; \theta_j)$ of T_j.

At the second stage new components e_1^*, e_2^* are installed into the real system. The components e_1^*, e_2^* are assumed to be statistically identical to the previous components e_1, e_2 in their states before the laboratory conditions testing procedure started. For that reason we will denote them by the same symbols e_1, e_2 as before. We assume that in the system some physical phenomena associated with the operation of one component may contribute to the failure mechanism(s) of other component(s). Therefore, an additional "stress" may be put on the remaining component(s). (Here the symbol '(s)' indicates that the above phenomena may also be considered in systems containing more than two components.)

In this situation, additional changes in the physical structure of e_2, such as microdamage, may occur. These may accelerate (or sometimes delay) the process leading to failure.

Note that now the lifetimes, say X_1, X_2 of the components, when they work in the system, differ from the lifetimes T_1, T_2 in laboratory conditions because in the new conditions they become statistically shorter or longer than at the original (laboratory) stage. The new lifetimes X_1, X_2, which, in general, should remain in some association with the T_1, T_2, are considered to be stochastically dependent.

Our task is to find the joint probability density of the r. vector (X_1, X_2) as the system reliability stochastic model.

In order to be able to describe the underlying stochastic dependences that constitutes this model, some further assumptions are made:

1. In the system we consider, there is no physical influence of component e_2 on e_1, essential for the failure mechanism of e_1. Obviously, there is always mutual stochastic dependence between the lifetimes X_1, X_2 of the components.
2. The random event $X_1 < X_2$ occurs with probability approximately equal to 1.

To quantify in stochastic terms the, often vague or poorly recognized, physical influence of component e_1 on e_2, we will attempt to reflect the changes in the physical structure of e_2 by corresponding changes in the failure rate of component e_2. If component e_1 fails before e_2 fails, the original failure rate, say $\lambda_2(x_2, \theta_2)$, of component e_2 (which corresponds to the original pdf $f_2(x_2, \theta_2)$ of the laboratory conditions lifetime T_2) is subjected to some change in value. This change of value of the failure rate $\lambda_2(x_2, \theta_2)$ here is assumed to be reflected by a change in its (scalar or a vector) parameter θ_2.

In turn, the magnitude of change in parameter θ_2 obviously depends on the (random) time X_1 over which the stress caused by the influence of e_1 on e_2 was endured by e_2. If then the random event $X_1 = x_1$ happens, the new value θ_2^* of the parameter θ_2 of the failure rate $\lambda_2(x_2, \theta_2)$ clearly depends on that time x_1. Therefore, it seems reasonable to treat the new parameter θ_2^* of the failure rate, say $\lambda_2(x_2, \theta_2^*)$ of the lifetime X_2 of e_2, when in the system, as a (continuous) function of time x_1. This postulate justifies the functional notation $\theta_2^* = \theta_2(x_1)$. As a consequence of the above, the failure rate $\lambda_2(x_2, \theta_2)$, corresponding to lifetime T_2, turns into a "slightly" different failure rate, say $\lambda_2(x_2, \theta_2(x_1))$, which corresponds to lifetime X_2, given that $X_1 = x_1$ happened. In such a way, we have obtained the (proposed) conditional failure rate $\lambda_2(x_2|x_1)$ of lifetime X_2, given $X_1 = x_1$, as defined by the formula:

$$\lambda_2(x_2|x_1) = \lambda_2(x_2, \theta_2(x_1)) \tag{5.1}$$

where $\theta_2(x_1)$ is a suitably chosen continuous function of the "influence time" x_1.

At this point it should be noted that the probability density $f_2(x_2, \theta_2)$ of T_2 is subjected to "the same" transformation as the failure rate $\lambda_2(x_2, \theta_2)$ of T_2 is. In both, parameter θ_2 is transformed into a different one: $\theta_2^* = \theta_2(x_1)$. Thus, parallel to the definition of the conditional failure rate (5.1), the formula:

$$g_2(x_2|x_1) = f_2(x_2, \theta_2(x_1)) \tag{5.2}$$

determines the conditional pdf of the random variable X_2, given $X_1 = x_1$. As a consequence of the assumption made above (component e_2 has no physical influence on e_1), we obtain $X_1 = T_1$, in distribution, so that the marginal pdf $g_1(x_1)$ of lifetime X_1 is given to be $f_1(x_1)$. One therefore obtains the joint pdf $g(x_1, x_2)$ of the random vector (X_1, X_2) in the usual form of an arithmetic product:

$$g(x_1, x_2) = g_2(x_2|x_1)g_1(x_1) . \tag{5.3}$$

The above pattern of construction can be illustrated by the following example.

Example 1. Suppose that the lifetimes of the components e_1 and e_2 in laboratory conditions are the following independent and exponentially distributed random variables T_1, T_2 with $f_k(t_k; \theta_k) = (1/\theta_k)\exp[-t_k/\theta_k]$ as the pdf of $T_k(k = 1, 2)$. If we replace parameter θ_2 in the above pdf $f_2(x_2; \theta_2)$, by a "slightly" different parameter function $\theta_2^* = \theta_2(x_1)$, then pdf $f_2(x_2; \theta_2)$ of T_2 transforms into the conditional pdf $g_2(x_2|x_1) = f_2(x_2; \theta_2(x_1))$ of X_2, given $X_1 = x_1$. Assuming for the marginal pdf $g_1(x_1)$ of the elementary events $X_1 = x_1$, that $g_1(x_1) = f_1(x_1)$, one obtains a wide class of bivariate exponential pdfs:

$$g(x_1, x_2) = g_1(x_1)g_2(x_2|x_1) = (\theta_1)^{-1}\exp[-x_1/(\theta_1)](\theta_2(x_1))^{-1}\exp[-x_2/(\theta_2(x_1))] , \tag{5.4}$$

where, in particular, one may specify:

$$\theta_2(x_1) = \theta_2(1 + Ax_1^r) , \tag{5.5}$$

(with A, r being positive real "parameters of the parameter function").

For analytical calculations associated with the model given by (5.4) and (5.5), see Sect. 5.10.2. Another analytically interesting model can be obtained by setting:

$$\theta_2(x_1) = \theta_2 \exp\left[A'x_1^{r'}\right] , \tag{5.6}$$

with positive parameters A' and r'.

Note that both factors $g_1(x_1)$ and $g_2(x_2|x_1)$ of $g(x_1, x_2)$ in (5.4) are exponentials, which justifies the pdf's name "exponential". However, the marginal pdf $g_2(x_2)$ of X_2 is, in general, not an exponential.

It is well known that application of normal pdfs to model lifetimes is very limited. Nevertheless, in many areas of reliability theory the normal pdfs are used as models of system or system component (random) strength. A typical example of such application is the strength of bundles of fibers investigated, for example, in [21].

Consider the following extension of the multivariate normal pdfs.

Example 2. Let the random variables T_1, T_2 be independent, having normal pdfs $f_1(x_1) = N(\mu_1, \sigma_1)$ and $f_2(x_2) = N(\mu_2, \sigma_2)$ respectively. Then the corresponding bivariate "pseudonormal" pdf $g(x_1, x_2)$ of the related random vector (X_1, X_2) is given by the product formula:

$$g(x_1, x_2) = g_1(x_1) \cdot g_2(x_2|x_1) \, ,$$

with the invariant pdf $g_1(x_1) = f_1(x_1) = N(\mu_1, \sigma_1)$, and with conditional pdf

$$g_2(x_2|x_1) = N(\mu_2(x_1), \sigma_2(x_1)) \, ,$$

where the replaced parameters of $f_2(x_2)$ are the following (arbitrary) continuous "parameter functions"

$$\mu_2 = \mu_2(x_1), \ \sigma_2 = \sigma_2(x_1) \text{ of the random event } X_1 = x_1 \, .$$

More explicitly, one obtains the following class of bivariate pseudonormal pdfs:

$$g(x_1, x_2) = \left[\sigma_1 \sqrt{(2\pi)}\right]^{-1} \exp\left[-(x_1 - \mu_1)^2/2\sigma_1^2\right] \left[\sigma_2(x_1) \sqrt{(2\pi)}\right]^{-1}$$
$$\times \exp\left[-(x_2 - \mu_2(x_1))^2/2[\sigma_2(x_1)]^2\right] \, ,$$

where $\mu_2(x_1) = E[X_2|x_1]$ is the (in general, non-linear) *regression function,* and $[\sigma_2(x_1)]^2 = Var[X_2|x_1]$ is the conditional variance, which may also be chosen to be an arbitrary function continuous in x_1.

In particular, one may consider the (pseudonormal) nonlinear regression function of the form:

$$E[X_2|x_1] = \mu_2(x_1) = \mu_2 + a(x_1 - \mu_1) + A(x_1 - \mu_1)^n \, ,$$

with a and A being arbitrary real numbers, and $n = 2, 3, \dots$.

For the coefficient "A" ("small", in a comparison to "a"), the term $A(x_1 - \mu_1)^n$ may be considered as a non-linear "correction" to the regular Gaussian (linear) regression function. The purpose of that correction is mainly to enhance the accuracy in various modeling situations. Particularly interesting seems to be the "quadratic case" $n = 2$ that by its nature is non-symmetric, contrary to the regular bivariate normal. At this point, note the potential usefulness of the quadratic correction in situations where empirical data show significant asymmetric tendency, while "the best" (normal) model available is the symmetric one. For specific analytical calculations associated with this model see Sect. 10.4.

5.2.2 Choice of Subclass of Continuous Functions

The above construction of the class of bivariate and pseudonormal pdfs is "general" in the sense that, when modeling real life phenomena, the pattern it defines contains a number of possibilities, *i.e.*, particular cases one has to chose from. Each such case is determined by a choice of a particular class of continuous parameter functions $\theta_2(x_1)$ that may, possibly, be "the best", with respect to a potential fit to given empirical data.

At this stage of the modeling procedure (in a reliability or in some other possible framework), the task is to make a proper choice of a subclass of the class of continuous functions. In some instances, the choice may possibly be dictated by human experience or theoretical knowledge, but even so, the final process of discrimination among the 'last' few singled out possibilities [such as by examples (5.5) and (5.6) in the exponential case] must be based on a given empirical information. For example, suppose that some real life phenomenon (such as the reliability of a two-component system considered above) is modeled by use of scheme (5.4) for the bivariate exponential pdfs $g(x_1, x_2)$. Also suppose that after "theoretical considerations" the two (or possibly more) "final candidates" for the proper stochastic model are determined by the classes of parameter functions (5.5) and (5.6).

Now, given a random sample from the random vector (X_1, X_2) we estimate, in some way (*e.g.*, using the such as maximum likelihood method), the parameters A and r of parameter function (5.5) and then the parameters A' and r' of (5.6), while for both the common parameter "θ_2" is assumed to be estimated earlier in the first (laboratory conditions) stage. All what remains now is to compare the data fit of candidates (5.5) and (5.6) by the use of proper statistical tests. One of them could be, for example, the probability ratio test. On the other hand, in situations similar to the one described in Example 2, often the discriminating tests may easily be reduced to parametric ones. In cases like that a comparison of the fit to the (same) data, between a usual bivariate normal model versus an alternative quadratic pseudonormal in the simplest case, reduces to parametric testing of the zero hypothesis $A = 0$ (in favor of the normal model) versus the alternative hypothesis $A = A^*$ (in favor of the pseudonormal), where A^* is the earlier obtained nonzero estimate of the parameter A. In fact, it will be more realistic to consider the more complex zero hypothesis: $a = a^*, A = 0$ versus the alternative $a' = a^*, A = A^*$ (with $a \neq a'$ and $A^* \neq 0$). Here, notice that, in general, the linear coefficient a in the ordinary linear regression function of the normal model is different from the coefficient a' of the 'linear part' of the quadratic regression in the pseudonormal.

Naturally, the method of modeling sketched above, requires further development. This is an open problem for future research.

Not that in reality, the parameter replacement method presented above can be regarded as the process of randomizing some deterministic parameters (scalars or vectors)), say θ_2, by considering them as continuous functions $\theta_2()$ of a random variable, say X_1, having a known pdf. To avoid misunderstandings at this point, we stress that this kind of parameter randomization does not rely on the usual simple compounding widely used in the theory and applications of probability. One can easily realize that the random structure of the introduced parameter functions and the underlying bivariate (as well as similar, later introduced, multivariate) pdfs is significantly more complex and conceptually different. For a quick comparison realize the following two facts: 1) in the case of ordinary compounding, the class of all continuous parameter functions $\theta_2(X_1)$ is reduced to only one function, namely the identity $\theta_2(X_1) = X_1$; and 2) in the vast majority of frameworks where simple compounding is applied, the random variable X_1 is not regarded as another marginal of the "same" random vector, say (X_1, X_2), where X_2 is the r. variable under consideration. In other words, the interpretation of the usual "random parameter" X_1 is more specific in the framework considered here.

In the literature, the classical analog to the so described stochastic dependences is the bivariate normal, where the parameter function "replacing" the expectation parameter $E[X_2] = \mu_2$ is the conditional expectation, say $E[X_2|X_1 = x_1] = \mu_2(x_1)$. The latter apparently is (only) a linear function of x_1. Extending this linear (regression) functions class to the class of the arbitrary continuous parallels extending the class of normals to some of the pseudonormals.

For a comparison with other works in the literature, see [6–9], which use a similar type of the parameter randomization, see Sect. 5.4.5.

5.3 Multivariate Extensions of the Bivariate Models

The bivariate stochastic model constructed in Sect 5.1 can be extended to an arbitrary (reasonable) set of the r. variables $\{X_1, X_2, \ldots, X_{n-1}, X_n\}$, $n = 3, 4, \ldots$ Now consider, a similar to the case $n = 2$, parallel system whose components are e_1, $e_2, \ldots, e_n$. Let $f_1(t_1; \theta_1)$, $f_2(t_2; \theta_2), \ldots, f_n(t_n; \theta_n)$ be the pdfs of the independent component lifetimes $T_1, T_2, \ldots, T_n$ respectively, tested in laboratory conditions.

The stochastically dependent lifetimes of the components, when working in the system, will be denoted by $X_1, X_2, \ldots, X_n$.

Similar to the bivariate case, we impose the following two basic conditions:

1. for each $j = 2, 3, \ldots, n$, the components $e_1, e_2, \ldots, e_{j-1}$ may have physical impact on component e_j, while the components $e_{j+1}, \ldots, e_n$ do not. Consequently, no component has ever a physical influence on e_1.
2. the probability of the random event $X_1 < X_2 < \ldots < X_n$ is approximately equal to 1.

Note that the reason for imposing the above pretty strong restrictions, such as the order $X_1 < X_2 < \ldots < X_n$ for the n random variables, together with the anti-symmetry requirement specified by the condition (1), is not necessarily dictated by properties of the modeled realities (such as the multicomponent system). In our opinion, the real cause of that "imperfection" lies in a limited capability of the method (for example, the parameter replacement method) that works well under the anti-symmetry assumption only. In particular, huge difficulties arise, whenever the model construction starts with an attempt to find a stochastic description in form of a joint probability density.

Fortunately, other stochastic description of the phenomena of the similar, but in turn mutual interactions between the system components turned out to be possible, when using slightly different means. However, in this case a somehow different idea governs the models construction. For example, as a starting point of those constructions, the most convenient form of the stochastic model to work with, is the joint survival function rather than the joint probability density so far used in this paper.

Opposite to the models build in this work, the "mutual interaction models" are pretty sensitive for increment of the dimension. Still, most of the corresponding trivariate survival functions are analytically treatable. Some results on that can be found in [22].

A more developed version of this set of the problems now is in preparation.

Note also that even for the models considered in this work, condition (5.2) alone can be dropped in most of other than the system reliability, applications. This especially is the case when a time interpretation is given to the subscripts $j = 1, 2, \ldots$, in all the X'_j. Thus, the condition (5.2) is dropped in all the applications such as those considered in Sect. 5.8.

The stochastic dependencies between the lifetimes $X_1, X_2, \ldots, X_n$ of the components basically are of the same nature as in the bivariate case. For each $j = 2, 3, \ldots, n$, the way the components $e_1, e_2, \ldots, e_{j-1}$ (when working in the system) physically and stochastically affect component e_j is quite the same as the way component e_1 influences component e_2 in the previously described bivariate model. Stochastically, their total influence on the failure rate $\lambda_j(t_j; \theta_j)$ of component e_j's lifetime T_j, in laboratory conditions, is reflected by a change in the value of the (scalar or vector) parameter θ_j. As in the bivariate case, the magnitude of that change is assumed to depend on the (random) times $X_1, \ldots, X_{j-1}$ the stresses put by components $e_1, e_2, \ldots, e_{j-1}$ on e_j lasted. Consequently, the parameter θ_j of the failure rate $\lambda_j(t_j; \theta_j)$ of the (original) lifetime T_j is "replaced" by parameter $\theta_j^* = \theta_j(x_1, \ldots, x_{j-1})$, of the ("in system") lifetime X_j, which is continuously dependent on the times $x_1, \ldots, x_{j-1}$, as the random events $(X_1 = x_1, \ldots, X_{j-1} = x_{j-1})$ happen. The continuous functions $\theta_j^* = \theta_j(x_1, \ldots, x_{j-1})$ are called "parameter functions".

Note that, as in the bivariate case, these parameter functions (whenever known) determine the following conditional failure rate for each component e_j, when in the system:

$$\lambda_j(x_j | x_1, \ldots, x_{j-1}) = \lambda_j(x_j ;\quad \theta_j(x_1, \ldots, x_{j-1})) . \tag{5.7}$$

In parallel, as a result of the same transformation of the parameters θ_j into the parameter functions $\theta_j(x_1, \ldots, x_{j-1})$, the pdfs $f_j(t_j; \theta_j)$ of T_j are transformed into the conditional pdfs $g_j(x_j | x_1, \ldots, x_{j-1})$ of the lifetimes X_j, given the random events $(X_1 = x_1, \ldots, X_{j-1} = x_{j-1})$.

The conditional pdfs are defined by the following sequence of identities:

$$g_j(x_j | x_1, \ldots, x_{j-1}) = f_j(x_j; \theta_j(x_1, \ldots, x_{j-1})), j = 2, 3, \ldots, n . \tag{5.8}$$

Formula (5.8) provides a full description of the stochastic dependencies among the component lifetimes $X_1, \ldots, X_n$, when in the system.

It should also be realized that the mathematical descriptions (5.7) and (5.8) of the phenomena are equivalent to each other.

The general pattern for the joint pdfs of the random vectors $(X_1, \ldots, X_n)$ construction, for $n = 1, 2, 3, \ldots$, may be thought of as a result of the following recurrence procedure. This procedure contains the following steps:

1. For $n = 1$, the "original" marginal pdf $g_1(x_1)$ of X_1 is given by the assumption that $g_1(x_1) = f_1(x_1)$, where $f_1(x_1)$ is an "arbitrary" pdf of T_1, given in advance.
2. For $n = 2$, the bivariate pdf $g(x_1, x_2)$ of the r. vector (X_1, X_2) one obtains by the method described in Sect. 5.1. (Recall that it was obtained in the factored form

$$g(x_1, x_2) = g_1(x_1)g(x_2|x_1)) \, .$$

3. For $n \geq 3$, if, for some $j = 3, 4, \ldots, n$, the joint pdf $g^{j-1}(x_1, \ldots, x_{j-1})$ of the random vector $(X_1, \ldots, X_{j-1})$ is obtained as the result of the $(j-1)^{\text{th}}$ iteration, then the jth iteration yields the jth joint pdf $g(x_1, \ldots, x_j)$ of the r. vector $(X_1, \ldots, X_j)$ as the product

$$g^j(x_1, \ldots, x_j) = g^{j-1}(x_1, \ldots, x_{j-1})g_j(x_j|x_1, \ldots, x_{j-1}) \, , \qquad (5.9)$$

where the conditional pdf $g_j(x_j|x_1, \ldots, x_{j-1})$ is given by formula (5.8), for a given continuous parameter function $\theta_j(x_1, \ldots, x_{j-1})$.

The process stops when $j = n$.

As the final result one obtains the sought after joint pdf $g(x_1, \ldots, x_n)$ of the component lifetimes $X_1, \ldots, X_n$.

From the above recurrence procedure description, it follows that for each $n = 2, 3, \ldots$ the obtained joint pdf $g(x_1, \ldots, x_n)$ can always be represented as the product of exactly n factors:

$$g(x_1, \ldots, x_n) = g_1(x_1)g_2(x_2|x_1)g_3(x_3|x_1, x_2)\ldots g_n(x_n|x_1, \ldots, x_{n-1}) \, . \qquad (5.10)$$

Example 3. To illustrate the general n-variate case, assume that the r. variables $T_1, \ldots, T_n$ are independent, each having, for $i = 1, \ldots, n$, the following Weibull pdf:

$$f_i(t_i) = (\gamma_i/\beta_i)(t_i - \alpha_i)^{\gamma(i)-1} \exp\left[-(t_i - \alpha_i)^{\gamma(i)}/\beta_i\right] \, , \quad t_i > \alpha_i = 0 \, , \text{ elsewhere} \, .$$

The two conventions: $f_i(t_i) = W_i(\alpha_i; \beta_i, \gamma(i))$ and $\gamma_i = \gamma(i)$ will be adopted.

Applying the pattern of the parameter replacement method (in a simple procedure shown below) one obtains the class of Weibullian joint pdfs $g(x_1, \ldots, x_n)$ of the r. vectors $(X_1, \ldots, X_n)$ in the product form (5.10). According to the procedure, first we set $g_1(x_1) = f_1(x_1) = W_1(\alpha_1, \beta_1, \gamma(1))$. Next, for each $j = 2, \ldots, n$, the conditional pdf $g_j(x_j|x_1, \ldots, x_{j-1})$ of X_j becomes a factor in product (5.10). It is noteworthy that every such factor remains Weibullian, each one with respect to x_j alone. The conditional pdf is given by the following specification of the (three parameter Weibullian) parameter function:

$$g_j(x_j|x_1, \ldots, x_j) = W_j(\alpha_j(x_1, \ldots, x_{j-1}); \beta_j(x_1, \ldots, x_{j-1}); \gamma_j(x_1, \ldots, x_{j-1})) \, .$$
$$(5.11)$$

"Theoretically" the continuous parameter functions $\alpha_j(x_1,\ldots,x_{j-1})$, $\beta_j(x_1,\ldots,x_{j-1})$, and $\gamma_j(x_1,\ldots,x_{j-1})$ can be declared in a basically, arbitrary way. They represent the Weibullian parameters of the shift, the scale, and the shape respectively. Each of them continuously depends on occurring random events $(X_1,\ldots,X_{j-1}) = (x_1,\ldots,x_{j-1})$. Once used as stochastic models (of system reliability, for example), the "arbitrariness" of the parameter functions must obviously be replaced by a rather careful choice of possibly "the best" of them. For such a choice use of statistical methods of estimation and discrimination is necessary.

5.4 A Comparison with Freund, Marshall and Olkin, and some Other Models

The system reliability models proposed above are in a specific relation with the two now deemed classical reliability models by on the one hand Freund [15], and on the other hand by Marshall and Olkin [16]. The nature of this association of the models is neither generalization nor specification, but rather all three types of system component life dependencies, in a sense, appear to be each other "complementary". This relationship between the three models is the topic of this section.

5.4.1 The Freund Model

The physical motivation for the Freund (bivariate) model generally relies on the following phenomena (see [15]). Suppose two components e_1, e_2 form a parallel system whose task is to share a common load. One can imagine the components being two engines of a plane, or two high vol.tage electric power lines, or two traffic lines, *etc.* Until failure of either of the two components (each of them assumed to have a constant hazard rate) the components are assumed to work independently in both a physical and a stochastic sense, *i.e.*, none of them has an influence on the reliability of the other.

Also, contrary to the Marshall and Olkin model, a failure of any component does not cause (nor is it accompanied by) an immediate failure of the other with a positive probability. Once one of the components fails the common load has to be carried by the surviving component. An increment of the load or other stress, causes a change in its hazard rate to another (constant) value. Notice that in the Freund model the "phenomenon of dependence" occurs after failure.

5.4.2 The Marshall and Olkin Models

On the other hand, the "dependencies" present in the Marshall and Olkin models only occur instantaneously at the precise time(s) of failure(s). They rely on the so-called multiple failures, caused by fatal shocks when more than one component can fail at a given time instant. In this model it is assumed that between (or after) consecutive failures the components work independently each having a constant failure (hazard) rate.

5.4.3 Classification of Stochastic Dependency Models

Taking under consideration the two component parallel systems, as in the models cited above, and adding to them the models proposed in this work, one may classify the phenomena of stochastic dependencies between the lifetimes by setting each of the models into one of three basic mutually excluding types. The following three conditions may serve as criteria for the classification:

1. the "Freund type": one component "affects" the other after first failure.
2. the "Marshall–Olkin type": the "dependence takes place" at the very moment of the first failure.
3. the "third type": the phenomenon of physical interactions between the components, reflected by a stochastic influence of one of them on the lifetime of the other only takes place before the first failure. The work conditions that the surviving component experiences after the failure of the other are "normal" (here the "laboratory conditions").

One can expect that in real life situations a combination of the three types of the component dependences may be present.

5.4.4 Physical Impacts Outside the System

Quite a different type of stochastic dependence that is not subjected to the above classification comprises cases where a source of corresponding physical impacts lies "outside" the system, in particular environmental conditions. See Lindley and Singpurwalla [13] for further details. In that case, instead of mutual physical impacts of the components considered in this paper, a common physical impact on all the components from outside environmental factors implies the (mutual) stochastic dependences between the component lifetimes. This type of dependence is proposed to be named 4) the "Lindley–Singpurwalla type".

5.4.5 "Third Type" Stochastic Dependence Models

Other examples of the "third type" stochastic dependence models are due to the well known group of the authors who construct the multivariate pdfs based on similar methods of conditioning we present here, see [6–9], and other related papers. For more references see [6]. The type of conditioning through randomizing the parameter, used in those papers, somehow resembles the "method of parameter replacement" we apply here. The basic definitions of the conditional pdfs used in the papers we cite are very similar to the ones given by (5.2) or (5.8) and elsewhere in this text. On the other hand, those methods essentially differ from the ones described in our work. First, to the best of our knowledge, none of the obtained multivariate pdfs has the form given by (5.3) or (5.9) or (5.10), and therefore the class of all the obtained models seems to be disjoint from the ones presented in our work. This is because of the two special rules we adopt in this work. According to our knowledge they are not applied in the cited works:

(a)   The predetermined order in conditioning (see (5.10)), with exactly n-1 conditional pdfs chosen to be specified, is imposed.
(b)   These n-1 conditional pdfs are always completed by exactly one (initial) marginal pdf ($g_1(x_1)$ in (5.10)).

It is noteworthy that, using the method of parameter replacement, the resulting n-variate pdfs are always uniquely characterized and constructed in the product form of (5.10), which is very simple. For other remarks on this subject see [5].

To describe the conditioning methods used in the above cited works more specifically, we restrict ourselves to the bivariate constructions pattern. In those cases the authors assume they are given both conditional pdfs $g_2(x_2|x_1)$ and $g_1(x_1|x_2)$ which basically are defined in the same manner as we have used in (5.2). Usually both the conditional pdfs are assumed to belong to one of the classes such as the Gaussians, Weibullian, exponential, or another major class. The construction often relies on the fact that the functional equation

$$g_2(x_2|x_1)g_1(x_1) = g_1(x_1|x_2)g_2(x_2)$$

always holds for appropriate marginal densities $g_1(x_1)$ and $g_2(x_2)$, which are aimed to be found.

So the task is to solve the above equation or analyze its properties. One disadvantage of this is that the two unknown functions must be determined from one equation and therefore the solutions are usually not unique.

When using the parameter replacement method considered here, only one of the two conditional pdfs is predetermined (the "order"). The rule we impose can be described as follow. The initial pdf $g_1(x_1)$ is only matched with the conditional pdf $g_2(x_2|x_1)$, and $g_2(x_2)$ only matches $g_1(x_1|x_2)$. In the framework we consider it is essential that no other combination of the four pdfs is admitted. Consequently, in each of the above cases the bivariate $g(x_1, x_2)$ is immediately uniquely determined.

The order restriction in the general n-dimensional case becomes very natural when the ordering factor has a "time" interpretation. One of the outcomes for that ordering procedure is an extension of the construction toward stochastic processes. For that, see Sect. 5.6.

5.5 The Transformation Method for the pdfs Construction

5.5.1 Direct Transformations of Random Vectors

In a variety of important cases the same pdfs $g(x_1, \ldots, x_n)$ of the r. vectors $(X_1, \ldots, X_n)$ as the ones obtained by the method of parameter replacement, can also be obtained by some direct transformations of the random vectors, *i.e.*,

$$(T_1, \ldots, T_n) \rightarrow (X_1, \ldots, X_n),$$

where the r. variables $T_1, \ldots, T_n$ may be assumed to be independent Weibull or gamma, or normal.

These transformations are mostly given by means of some new extension of the class of affine transformations $R^n \rightarrow R^n$.

This class of (easily reversible) transformations we call pseudoaffine, and is determined by the following scheme:

$$\begin{aligned}
X_1 &= \varphi_0 T_1 + \psi_0 \\
X_2 &= \varphi_1(X_1)T_2 + \psi_1(X_1) \\
X_n &= \varphi_{n-1}(X_1, \ldots, X_{n-1})T_n + \psi_{n-1}(X_1, \ldots, X_{n-1}),
\end{aligned} \qquad (5.12)$$

where the arbitrary continuous functions $\varphi_0, \varphi_1(), \ldots, \varphi_{n-1}()$, and $\psi_0, \psi_1(), \ldots, \psi_{n-1}()$ play the same role in the pdfs $g(x_1, \ldots, x_n)$ of the resulting random vectors $(X_1, \ldots, X_n)$ syntax structure as the parameter functions of the pdfs obtained by the previous method (*i.e.*, the method of parameters replacement). Notice that the Jacobians of the transformations inverse to (5.12) take on the remarkably simple form of an arithmetic product:

$$[\varphi_0]^{-1} \cdot [\varphi_1(x_1)]^{-1} \ldots [\varphi_{n-1}(x_1, \ldots, x_{n-1})]^{-1}.$$

This makes possible easy calculation of the pdf $g(x_1, \ldots, x_n)$.

A more general class of new transformations is the class of pseudopower transformations:

$$\begin{aligned}
X_1 &= \varphi_0 \cdot (T_1)^{a(0)} + \theta_0 \\
X_2 &= \varphi_1(X_1) \cdot (T_2)^{a(1)} + \theta_1(X_1) \\
X_n &= \varphi_{n-1}(X_1, \ldots, X_{n-1}) \cdot (T_n)^{a(n-1)} + \theta_{n-1}(X_1, \ldots, X_{n-1}),
\end{aligned} \qquad (5.13)$$

where the exponents $a(0), \ldots, a(n-1)$ may depend on values of the r. variables $X_1, \ldots, X_{n-1}$ in such a way that for $j = 1, \ldots, n-1, a(j) = a(j)(X_1, \ldots, X_j)$.

Applying transformations (5.13) to a set of independent Weibullian r. variables $T_1, \ldots, T_n$, one obtains a wide class of n-variate Weibullian pdfs of $(X_1, \ldots, X_j)$.

This class is significantly wider than the one obtained by means of pseudoaffines only.

The common property, essential for both classes of transformations, is the fact that the Jacobi matrices of all the mappings are triangular. This property is the defining property for a much wider class of the so-called "triangular mappings". That class of transformations properly includes the class of all the pseudopowers. Because of the simple product form of the Jacobians the triangular mappings, which possibly are not the pseudopowers, can also be useful in probabilistic investigations, similar to those presented in this work.

5.5.2 On the Role of the Pseudoaffine and Pseudopower Transformations in Statistical Analysis and Sampling

The use of pseudoaffine or pseudopower transformations is not mandatory for the pdfs construction, as the simpler alternative method of parameter replacement exists. However, the transformation method is not only about mathematical elegance. From an investigation of the algebraic structure of the class of the pseudoaffines, as well as of the pseudopowers, it was established that:

(a) Every n-variate Weibullian, exponential, or pseudonormal pdf, obtained by the transformations method, can be uniquely determined by a unique pseudoaffine transformation as the pdf of an output random vector $(X_1, \ldots, X_n)$. That r. vector is considered to be the image (under the unique transformation) of the "fixed" random vector $(T_1, \ldots, T_n)$, where the independent r. variables $T_1, \ldots, T_n$ are distributed according to the standard exponential or the standard normal pdfs respectively.

(b) The class of the above constructed Weibullian, pseudoexponential, or pseudonormal pdfs, and the corresponding class of the defining pseudoaffines, are equivalent in the sense of existence of a one to one relationship between the transformations and the pdfs. Moreover, the two equivalent classes have the same (in the sense of algebraic isomorphism) algebraic structure of a group.

The algebraic properties of the classes facilitate statistical analysis of those pdfs that are obtainable by the use of the transformations. Suppose, for example a data in the form of a random sample from a random vector $(X_1, \ldots, X_n)$ has been obtained. A possible task may be to compare two alternative pseudonormal (a normal versus a non-normal pseudonormal for example) models for that data with respect to the best accuracy criterion. This task can be formulated as a non-parametric statistical hypothesis. Denote the hypothetical normal pdf of the r. vector $(X_1, \ldots, X_n)$ by $f(x_1, \ldots, x_n)$ and the alternative pseudonormal model for the same data by $g(x_1, \ldots, x_n)$. Denote the corresponding defining pseudoaffine transformations by Φ_f and Φ_g respectively. Applying to each particular observation $(x_1, \ldots, x_n)_j$ (where $j = 1, 2, \ldots, k$; and k is size of the sample) of the random vector $(X_1, \ldots, X_n)$ the inverses Φ_f^{-1}, Φ_g^{-1} of the defining transformations, one obtains two corresponding "observations":

$$\{(t_1, \ldots, t_n)\}_{j,f} \text{ and } \{(t_1, \ldots, t_n)\}_{j,g} \text{ of the same (input) r. vector } (T_1, \ldots, T_n).$$

Since the random variables $T_1, \ldots, T_n$ are independent and distributed as standard normal, the inference on possibly complicated n-variate normal and pseudonormal pdfs reduces to the simple procedure of testing (comparing) the consistency of two different sets of one transformed data with the same standard normal pdf. The independence of the r. variables $T_1, \ldots, T_n$ allows us to split the testing procedure into n independent simpler procedures.

Another benefit of the transformation approach is the ease of the sampling procedures for values of the r. vectors $(X_1, \ldots, X_n)$ having a pseudonormal (or pseudo-exponential) pdf $g(x_1, \ldots, x_n)$ that corresponds to the transformation Φ_g. The procedures are simply reduced to n independent samplings from the standard normals $T_1, \ldots, T_n$ (or standard exponentials respectively) and then applying transformation Φ_g to the obtained values $\{(t_1, \ldots, t_n)\}_{j,g}$ $j = 1, 2, \ldots, k$.

5.6 Extension of the Random Vector Models to Stochastic Processes

5.6.1 Discrete Time Interpretation

In this work a discrete time interpretation is given to the subscript $j (1 \leq j \leq n, n \geq 2)$ which labels the jth row in (5.1). Letting $n \to \infty$ in formula (5.1) one obtains the following extension of the pattern of the pseudoaffine transformations that will be called infinite pseudoaffines:

$$X_1 = \varphi_0 T_1 + \psi_0 \tag{5.12*}$$
$$X_j = \varphi_{j-1}(X_1, \ldots, X_{j-1})T_j + \psi_{j-1}(X_1, \ldots, X_{j-1})$$
$$\text{where } j = 2, 3, \ldots$$

The 'time' j in (5.12*) is assumed to take on all positive integer values, while all the transformations satisfying (5.12*) are thought of as $R^\infty \to R^\infty$ transformations, with R^∞ here understood as the class of all sequences of real numbers (possibly endowed with the Frechet metric). The basic pattern of the constructions mainly relies on transforming through (5.12*) some stochastic processes $\{T_1, T_2, \ldots\}$, which usually is chosen to belong to certain important, well known, classes of processes (for example normal). As a result, in each such case, a new class of the corresponding st. processes $\{X_1, X_2, \ldots\}$ is determined. The classes so obtained mostly are new as, according to our best knowledge, the transformations here used are new, at least in such probabilistic framework that we consider in this paper. (For the validity of that construction pattern realize that every so obtained stochastic processes $\{X_1, X_2, \ldots\}$ can be identified with the (unique) sequence $\{h^{(n)}(x_1, \ldots, x_n)\}_{n=2,3,\ldots}$ of the well defined joint pdfs of the random vectors $(X_1, \ldots, X_n)$, each being the output of the first n rows in (5.12*). The consistency of the underlying pdfs is obvious.) The various analytical properties of the processes often turn out to be very interesting, and seem to promise to be useful in applications, especially in reliability theory.

In some cases, in particular in the Weibullian case, the classes of the so obtained st. processes can be fruitfully extended even more if the infinite pseudoaffine (5.12*) is replaced by the following, more general, infinite pseudopower ($R^\infty \to R^\infty$) transformations scheme. (For a discussion of the corresponding finite pseudopower transformations class, in a probabilistic setting, see [5].)

$$X_1 = \varphi_0 T_1^{\alpha(0)} + \psi_0$$

$$X_j = \varphi_{j-1}(X_1, \ldots, X_{j-1}) T_j^{\alpha(j-1)(X_1,\ldots,X_{j-1})} + \psi_{j-1}(X_1, \ldots, X_{j-1}) \qquad (5.13*)$$

where, for $j = 2, 3, \ldots$ the symbols $\alpha(j-1)(X_1, \ldots, X_{j-1})$ will be called "exponent parameter functions" with the r.v.s $X_1, \ldots, X_{j-1}$ as their arguments, while the exponent parameter function $\alpha(0)$ is considered to be a nonzero real constant. Note that if for all $j = 1, 2, \ldots$ the conditions $\alpha(0) = \alpha(1)(X1) = \cdots = \alpha(j-1)(X_1, \ldots, X_{j-1}) = 1$ hold, then (5.13*) reduces to the pseudoaffine (5.12*).

5.6.2 Stochastic Processes Memory

With the analytical tools now at our disposal, possibly some light can be shed on the problem of the stochastic processes memory. First of all, observe that every stochastic process $\{X_j\}_{j=1,2,\ldots}$, that is constructed by either of the two methods (or just by any single transformation) one may consider as defined by the pattern for a unique sequence $\{g_j(x_j|x_1, \ldots, x_{j-1})\}_{j=2,3,\ldots}$ of the conditional pdfs, completed by an initial pdf $g_1(x_1)$ of X_1. In accordance with (5.10), these data uniquely determine the sequence $\{h^{(n)}(x_1, \ldots, x_n)\}_{n=2,3,\ldots}$ of consistent joint pdfs of all the r. vectors $(X_1, \ldots, X_n), n = 2, 3, \ldots$, and consequently the whole stochastic process $\{X_j\}_{j=1,2,\ldots}$ is uniquely determined. An associated fact of significant importance is that such a pattern of the constructions, at least theoretically, guarantees the possibility of containing all the past memory $(X_1, \ldots, X_{j-1}) = (x_1, \ldots, x_{j-1})$ of the stochastic process at any given time instant j, in a compact and usually simple analytical formula for the conditional pdf $g_j(x_j|x_1, \ldots, x_{j-1})$. The stochastic processes which possesses that property we will call "long memory stochastic process".

Example 4. Consider the following class of Weibull long memory stochastic processes, determined by (5.11). Recall that, according to our consideration in Sect. 5.3, for every time moment $j = 2, 3, \ldots$ the conditional pdf of X_j is a Weibull pdf in x_j, given by the following general pattern:

$$g_j(x_j|x_1, \ldots, x_j) = W_j(\alpha_j(x_1, \ldots, x_{j-1}); \quad \beta_j(x_1, \ldots, x_{j-1}); \quad \gamma_j(x_1, \ldots, x_{j-1})),$$

completed by a univariate Weibull pdf $g_1(x_1)$ of the initial term X_1. For all $j \geq 2$, we set $\alpha_j(x_1, \ldots, x_{j-1}) = 0$ for the shift parameter. For the scale parameter we assume that

$$\beta_j(x_1,\ldots,x_{j-1}) = \lambda_j \left(1 + a_{1,k(1)}x_1^\rho + a_{2,k(2)}x_2^\rho \ldots, + a_{j-1,k(j-1)}x_{j-1}^\rho\right), \quad (5.14)$$

$(\lambda_j, \rho; a_{1,k(1)}, a_{2,k(2)}, \ldots, a_{j-1,k(j-1)}$ are constant real "parameters of the parameter function"), and for the shape parameter $\gamma_j(x_1,\ldots,x_{j-1}) = \gamma_j$ is a non-zero constant (possibly varying with j). Also, we set the Weibull pdf $g_1(x_1) = W_1(0;\beta_1;\gamma_1)$, with β_1 and γ_1 being non-zero constants. All that defines the Weibullian st. process $\{X_j\}_{j=1,2,\ldots}$ that will be used as a proposed model for the reliability of systems with repair in Sect. 5.8.

Note that in the particular model in Example 4, the "backward" stochastic dependencies (*i.e.*, the "correlations") of the times $X_1,\ldots,X_j$ (between the consequtive failures or the corresponding repairs), as $j = 2,3,\ldots$ are determined by the corresponding sequences $\{a_{1,k(1)}, a_{2,k(2)}, \ldots, a_{j-1,k(j-1)}\}_{j=2,3,\ldots}$ of the coefficients of the scale parameter functions $\beta_j(x_1,\ldots,x_{j-1})$. For all the times j, the r. variables $X_1,\ldots,X_j$ become independent, if an only if all the above a coefficients are 0. Thus, in this relatively simple case, the long memory of the considered stochastic process is determined by the time-variable j, sequences of the "a" parameters. Difficulties in the model applications that may occur, are associated with a growing number of the parameters ("a" in our particular example) to be statistically estimated as time "j" grows. For a "high" value of j the associated errors in the coefficients estimation may cumulate to a level that would rather be more harmful than beneficial for the accuracy in modeling the given realities. As usual, to some extend and whenever possible, the cure lies in increasing an amount of data such as sample sizes.

Independently of that, two other possible means to overcome the difficulties rely on some modification of the models that are similar to the model in Example 4. Both modifications, developed below, are based on a simple observation that, in many practical cases, as the time distance from the present, say j to the events that took place in the past grows, the influence of those remote events (such as past damage to the system) on the system present state decay. So the events, here represented by the "a" coefficients of the considered particular model, far enough in time, may simply be neglected.

First modification (a memory restriction) of the models is based on use of the common notion of "k-Markovianity" (known in literature under the slightly longer name "k^{th}-order Markov chains", see for example [23]) with some limited values of the number $k = 1, 2, \ldots$. This subject will be developed in Sect. 5.7.

The second modification is based on concept of "forgetting factors", and will be developed in Sect. 5.8.3, in a direct association with maintenance problems.

5.7 Application of k-Markovian Stochastic Processes

The k-Markovianity ($k = 0, 1, 2, \ldots$) property of stochastic processes is a natural generalization of the ordinary ($k = 1$) Markovianity. At this place it seems to be instructive as to introduce that concept within framework of the pseudoaffine transformations. Let us start with an analysis of the simplest case *i.e.*, with the ordinary Markovian stochastic processes associated with the models we construct.

5.7.1 Finite Dimensional Pseudoaffine Transformations

Before that let us return for a moment to the finite dimensional pseudoaffine transformations given by (5.12). For $j = 3, 4, \ldots, n$ consider any r. vector $(X_1, \ldots, X_j)$ and the related conditional pdf $g_j(x_j | x_1, \ldots, x_{j-1})$ of the r. variables X_j, given the values $X_1 = x_1, \ldots, X_{j-1} = x_{j-1}$. Recall the well known definition of the Markov property for n-variate pdfs.

Definition 1. The n-variate r. vector $(X_1, \ldots, X_n)$ with pdf $g(x_1, \ldots, x_n)$, $n \geq 3$, is said to have the Markov property if for every $j = 3, 4, \ldots, n$, the following equality holds:

$$g_j(x_j | x_1, \ldots, x_{j-2}, x_{j-1}) = g_j(x_j | x_{j-1}) . \tag{5.15}$$

In Definition 1, it is understood that the conditional pdfs (5.15) do not depend on the values $x_1, \ldots, x_{j-2}$ of the r. variables $X_1, \ldots, X_{j-2}$, while they may depend on x_{j-1}. The notion of Markov (or k-Markov) property often occurs in association with the n-variate Gaussian pdfs. Here, we refer to it in order to signalize the possibility of constructing a wide class of "nice" Markov (not necessarily Gaussian) stochastic processes with discrete time. To start with their construction consider (5.12*) along with the pseudopowers (5.13*). An interesting fact is that, even if not all, most of the coming definitions and assertions, spelled out in association with the pseudoaffines, are also valid for the pseudopower transformations defined by (5.13*).

5.7.2 Markovian Pseudonormal Processes

Consider an infinite sequence of output r. variables $X_1, X_2, \ldots$ given by (5.12*) or (5.13*), and the associated sequence of, in general, independent and identically distributed input r. variables $T_1, T_2, \ldots$.

We say the sequence $\{X_1, X_2, \ldots\}$ satisfies the Markov property if and only if (5.15) holds for all corresponding r. vectors $(X_1, \ldots, X_n)$, for $n = 3, 4, \ldots$. The so defined random sequences $\{X_1, X_2, \ldots\}$ form a large collection of classes of discrete time Markov stochastic processes. One of these classes, with potential applications in reliability modeling (of, for example random strength of various sorts of materials, or possibly of time of repair) is the class of Markovian pseudonormal processes, determined by the previous considerations in this paper. It is a commonly known subclass of the class of the Markovian normal processes.

In what follows, we will show a natural correspondence between the k-Markovian stochastic processes ($k = 1, 2, \ldots$) and the defining them "k-Markovian transformations".

Definition 2. Any transformation given by (5.12) or (5.12*) for which all the parameter functions satisfy:

$$\varphi_{j-1}(x_1, \ldots, x_{j-1}) = \varphi_{j-1}(x_{j-1}), \; \psi_{j-1}(x_1, \ldots, x_{j-1}) = \psi_{j-1}(x_{j-1}), \qquad (5.16)$$

(*i.e.*, for each $j = 2, \ldots, n$ the parameter functions do not depend on $x_1, \ldots, x_{j-2}$) will be called (finite or infinite, according the cases $n < \infty$ and $n = \infty$ respectively) "Markovian pseudoaffine".

Notice that the ordinary non-singular linear or the affine transformations are "Markovian" as well. In association with Proposition 1 below, it is purposeful to single out a subfamily of the family of transformations determined by Definition 2. The next definition does this.

Definition 2*. Finite and infinite Markovian pseudoaffine transformations will be called strictly Markovian iff not all of the associated parameter functions in (5.16), as well as the functions $\varphi_1 x_1)$, $\psi_1 x_1)$ are reduced to constants.

The following fairly obvious proposition is given without proof.

Proposition 1. *Suppose a given stochastic process $\{X_j\}$ is obtained from a stochastic process $\{T_j\}$ by use of (5.12*). For the process $\{X_j\}$ to be Markovian, it is sufficient that the following two conditions are simultaneously satisfied: 1) the considered transformation (5.12*) is strictly Markovian, and 2) all the r. variables T_1, $T_2, \ldots$ are mutually independent.*

From Proposition 1 one obtains the following:

Corollary 1. *Suppose that a transformation satisfying (5.12*) or (5.13*) which brings stochastic process $\{T_j\}$ into stochastic process $\{X_j\}$, $j = 1, 2, \ldots$ is strictly Markovian. Then, if the process $\{T_j\}$ is Markovian (the trivial case when all the r. variables T_1, $T_2, \ldots$ are mutually independent is excluded), then the process $\{X_j\}$ is not Markovian but it is 2-Markovian= i.e., for every $j = 3, 4, \ldots$ all the underlying conditional pdfs satisfy:*

$$g_j(x_j | x_1, \ldots, x_{j-3}, x_{j-2}, x_{j-1}) = g_j(x_j | x_{j-2}, x_{j-1}), \qquad (5.17)$$

in the sense that each of the conditional pdfs present in (5.17) does not depend on the values $x_1, \ldots, x_{j-3}$, but only on x_{j-2}, x_{j-1}.

5.7.3 k-Markovianity for Pseudoaffine Transformations

Now we define the corresponding notion of k-Markovianity for the pseudoaffine transformations.

Definition 3. Let $n = 3, 4, \ldots$. Consider any pseudoaffine transformation subject to scheme (5.1) or (5.1*), all of whose parameter functions satisfy:

$$\varphi_{j-1}(x_1, \ldots, x_{j-1}) = \varphi_{j-1}(x_{j-k}, x_{j-k+1}, \ldots, x_{j-2}, x_{j-1}),$$
$$\psi_{j-1}(x_1, \ldots, x_{j-1}) = \psi_{j-1}(x_{j-k}, x_{j-k+1}, \ldots, x_{j-2}, x_{j-1}), \qquad (5.18)$$

(*i.e.* for each $j = 2, \ldots, n$, and fixed $k = 0, 1, 2, \ldots, j-1$, the above functions do not depend on the values $x_1, \ldots, x_{j-k-1}$).

All members of the class of such transformations will be called (finite or infinite according to the cases $n < \infty$ and $n = \infty$ respectively) "k-th order Markovian pseudoaffine transformations".

Note that a pseudoaffine transformation is Markovian, in the sense of Definition 2, if it is first ($k = 1$) order Markovian, in the sense of Definition 3.

As a convention, we assume that if $k = 0$ the parameter functions in (5.18) are considered to be constant, and therefore the resulting 0-th order pseudoaffine transformations are simply the well known regular affine.

We will need the following counterpart for Definition 2*.

Definition 3*. Finite and infinite k-th order Markovian pseudoaffine transformations ($k = 1, 2, \ldots, j; j = 2, 3, \ldots$) will be called "strictly k-Markovian pseudoaffines" if all the associated parameter functions in (5.18) essentially depend on the variables $x_{j-k}, x_{j-k+1}, \ldots, x_{j-2}, x_{j-1}$. For differentiable parameter functions in (5.18), which is a rather typical case, the last conditions simply reduce to the usual ones:

$$\partial \varphi_{j-1}/\partial x_m \neq 0, \quad \text{and} \quad \partial \psi_{j-1}/\partial x_m \neq 0, \ m = j-k, j-k+1, \ldots, j-1.$$

For the sake of completion, we formulate the following definition of the k-Markovian stochastic processes.

Definition 4. Given a stochastic process $\{X_j\}_{j=1,2,\ldots}$ defined by an initial pdf $g_1(x_1)$ of X_1, and by a sequence $\{g_j(x_j|x_1, \ldots, x_{j-1})\}_{j=2,3,\ldots}$ of all the conditional pdfs of the r. variables X_j, given $X_1 = x_1, \ldots, X_{j-1} = x_{j-1}$. Let k be a fixed positive integer. The stochastic process $\{X_j\}$ is said to be a strictly "k-Markovian process" if and only if for every $j > k$, the conditional pdf $g_j(x_j|\ldots)$ of the r. variable X_j, given a past $X_1, \ldots, X_{j-1}$ ($j = 2, 3, \ldots$), satisfies the following condition:

$$g_j(x_j|x_1, \ldots, x_{j-1}) = g_j(x_j|x_{j-k}, x_{j-k+1}, \ldots, x_{j-2}, x_{j-1}). \qquad (5.19)$$

This is to be understood that the conditional density $g_j(x_j|x_1, \ldots, x_{j-1})$ essentially depends on the "information" $x_{j-k}, x_{j-k+1}, \ldots, x_{j-2}, x_{j-1}$ in sense of Definition 3*, and that these pdfs do not depend on the values $x_1, \ldots, x_{j-k-1}$. Moreover, if all the r. variables $X_1, X_2, \ldots$ are mutually independent then the process will be considered "zero-Markovian" or "of Markovianity zero".

In the particular framework we consider, the name "memory size" for the nonnegative integer 'k' in Definition 4 seems to be appropriate.

Using the notion of k-Markovian transformation, as defined above, we can formulate the following generalization of Proposition 1.

Proposition 2. *For each $k = 0, 1, 2, \ldots$ any strictly k-Markovian stochastic process can be obtained by using as input a 0-Markovian process $\{T_j\}$ and applying to it a strictly k-Markovian pseudoaffine transformation.*

The proof only requires simple standard calculations.

Proposition 2 is especially (but not only) addressed to such stochastic processes $\{T_j\}$ for which all the (independent) terms $T_1, T_2, \ldots$ are distributed according to Weibullian, or gamma, or, in particular, exponential pdfs. Also, whenever appropriate, we consider normal pdfs. The essential motivations for applying the notion of k-Markovian stochastic processes, in our setting, are the following:

1. Apparently, for $k = 2, 3, \ldots$, the k-Markovian models are capable of incorporating more information on the process past than are the regular Markovians. This fact may become a vital in situations when applying two st. processes as alternatives in modeling the same reality.

    Often, in situations such as modeling the systems with repair (or, mathematically identical, stochastic description of various kinds of medical treatment), having at a disposal a longer past history of system performance may significantly enhance our knowledge of the 'present'.

    (The present "state of an interest" here is understood as estimated probability distributions, or expected values, of some important random quantities. As an example of such a quantity may serve the parameter considered in the following, time X_j until the next failure of a system after j runs followed by $(j-1)$ repairs. Another mathematically equivalent example is an estimated conditional density (such as the investigated throughout $g_j(x_j|x_1, \ldots, x_{j-1})$) or at least a conditional expectation of the time X_j to, say, the next heart attack, given an illness history.) The k-Markovian stochastic models are then expected to be more accurate, in the sense of a better fit to statistical data. Also 'longer' future predictions on probability distributions of r. variables, say, $X_{j+1}, X_{j+2}, \ldots$ made 'at a given time j', in many instances are expected to be more precise than those, based on the ordinary Markovian models.

2. The usually associated with such attempts increment in computational complexity is not that dramatic for the models like that. As it turns out, most of the important calculations associated with the k-Markovian (for $k \geq 2$) stochastic processes, can be handled, also in an analytical way.

3. Possibility of a direct application of the k-Markovian st. processes as models, for reliability of systems with repair arises.

5.8 Maintenance Models

5.8.1 Reliability and Maintenance of Systems

The well known problem in the theory of reliability and maintenance of systems such as, for example, cars, machinery, electronic equipment *etc.*, with repair (see for example [23]), can briefly be depicted as follows. At each failure a system is to be repaired, and then it renews its performance until the next failure.

The times $X_1, X_2, \ldots$ between consecutive failures usually are described by a discrete time ('j') stochastic process $\{X_j\}$, where the non-negative random variables X_j ($j = 1, 2, \ldots$) here are only assumed to have probability densities (pdfs). Consider the system failure after a jth working period. One of the associated basic problems of a critical importance is to predict the (random) time X_{j+1} until the next failure, i.e., to estimate its probability distribution or at least some of the distribution parameters like the expectation. If any stochastic dependences between the times $X_1, X_2, \ldots$ are taken under consideration, the stochastic processes, used as models in such situations, are actually always assumed to be at most Markovian in a narrow or in a wider sense. In practice this is reflected by a common restriction on information of the system repair history that one may reasonably incorporate into the model without enormous increase of the complexity of underlying computations. That history, even when available, must be reduced to the system state at the last failure only, if the stochastic models are to be used at all. More precisely, the only information on the system past (in the sense of lifetime), that can efficiently be applied in planning the future repair policy at the jth ($j = 2, 3, \ldots$) repair, is the total time of the system performance up to the most recent failure. According to our best knowledge, no subdivision of that total lifetime into the times between past repairs was ever successfully incorporated into an analytical stochastic model. On the other hand, researchers are aware that the reduction of system repair history, to the state of the system at a recent failure only, may affect, possibly badly, precision of the required forecasts. Among other possibilities, such essential phenomena like those, when the lifetimes X_j show any kind of stochastic decreasing tendency (decreasing expectations, for example) or a periodic behavior, must, by necessity, be ignored. The problem lies in lack of computationally efficient enough stochastic non-Markovian models, say, $\{X_j\}$ for the repair processes, considered to be stochastically determined by their entire (or a significant parts of) work-repair histories. The only exception seems to be represented by the class of the non-Markovian normal st. processes, *i.e.*, the processes based on proper (non-Markovian) multivariate normal distributions. Unfortunately, the normal distributions are very seldom satisfactory as models for system lifetimes. Attempts in research (the main source of this information is based on private communications) to build proper non-Markovian models, which are not normal (first of all the Weibullian), faced overwhelming computational complexity barriers. This work provides an attempt to construct non-Markovian models, which are satisfactory simple with respect to the underlying computations. The most essential factor that makes the constructions possible is obtaining an associated with the constructed st. process $\{X_j\}$, uniform functional pattern for the corresponding sequences of the conditional pdfs:

$$\{g_j(x_j|x_1, \ldots, x_{j-1})\}, \quad j = 2, 3, \ldots$$

for all the r. variables X_j, with $j \geq 2$, given their 'pasts': $X_1 = x_1, \ldots, X_{j-1} = x_{j-1}$.

These sequences are chosen in such a way, that every j^{th} pdf, with respect to the corresponding variable x_j, belongs to the same class of the pdfs. Two examples where all the pdfs $g_j(x_j|x_1,\ldots,x_{j-1})$ are Weibullian with respect to x_j, are given as applications in the repair models that follow, (see (5.20) and (5.21)).

Before entering the actual maintenance considerations, we would like to mention an interesting relationship between the two stochastic processes involved. The (main part of) actual model is a sequence of the (dependent) random variables X_1, $X_2,\ldots,X_j,\ldots$ which represents the times between consecutive failures of a (one-unit) system when it is repaired. This sequence may be considered as obtained, (by use of either of the two methods described in this paper) from the sequence of the (independent) r. variables $T_1, T_2,\ldots,T_j,\ldots$. Note that the T sequence (which, in particular, may be considered to be a common 'renewal process') models "the same" system's times between failures under the policy that after each failure, the system is replaced by an "identical" new one. In turn, the X sequence models the same system under the repair policy. The relation between the two stochastic processes, which mathematically one is obtained from the other, corresponds to the two different methods of the system exploitation. However, on the other hand, it also relates (for each $n = 2, 3,\ldots$), any two r. vectors $(T_1, T_2,\ldots,T_n)$, $(X_1, X_2,\ldots,X_n)$, viewed as the components $e_1, e_2,\ldots,e_n$ lifetimes. In this case, the first r. vector applies to the laboratory conditions, while the second describes lifetimes of the same components, when in the system (see Sects. 5.2 and 5.3). The above may shed some additional light on the nature of the pseudoaffine (triangular) transformations.

5.8.2 Aging Systems Repaired at Each Failure

Consider the model of an aging system repaired at each failure. The main part of the system maintenance model is a stochastic process, belonging to the class of the Weibullian processes $\{X_1, X_2,\ldots\}$, where X_1 is the time to the first failure, and for $j = 2, 3,\ldots X_j$ is the time of the system performance between the $(j-1)^{\text{th}}$ and j^{th} failure. Suppose there is available a predetermined list L (formally represented by a finite set) of possible symptoms and/or causes of all the failures that may occur. Thus any particular j^{th} failure of the system ($j = 1, 2,\ldots$) may be associated with exactly one non-empty subset $f_j^{(k)}$ of L, $k = 1,\ldots,q$, which may be regarded as the "k^{th} kind of the jth failure" which is proposed to be called the "failure type". The set $f_j^{(k)}$ should be understood as an elementary random event that has the associated probability $p_j^{(k)}$. (In the formal language, the space of elementary events is, in this case, a family of subsets of the (finite) set of failure symptoms L).

In turn, assume that with each kind of the jth failure $f_j^{(k)}$ there is an associated another finite set $R_j^{(k)} = \left\{ r_j^{(k,1)}, r_j^{(k,2)},\ldots,r_j^{(k,u)} \right\}$ of all kinds of repair available. (In particular, the set $R_j^{(k)}$ may be constant over j and/or k.)

Finally, with every kind of the repair $r_j^{(k,l)}$, $l = 1,\ldots,u$; (following the corresponding failure $f_j^{(k)}$) that belongs to the set $R_j^{(k)}$ a cost $c_j^{(k,l)}$ is associated.

On the other hand, the repairs are also assumed to differ from each other by a quality, so that a more expensive repair of the $f_{j-1}^{(k)}$ failure may result in a longer average time to the next failure and/or in a smaller probabilities of some "unfortunate" events that the next failure, say $f_j^{(k)}$, will be a bad one, possibly fatal or the one requiring a very expensive repair in the next stage(s).

For more clarity it is assumed that the state of the system at any time is known. Also, the time lengths of the repairs and waiting times, *etc.*, are not included in this particular setting.

Now let the stochastically dependent times of system functioning between the $(j-1)^{\mathrm{th}}$ and j^{th} failure be modeled by the components X_j, $j = 1,2,\ldots$ of some Weibullian stochastic process, as defined before. Suppose that, for some $j = 2,3,\ldots$, the $(j-1)^{\mathrm{th}}$ failure occurred, and that, for $m = 1,\ldots,j-2$, all the elementary r. events $X_m = x_m$ have happened (and, possibly, are recorded by a given service station). By the maintenance history of the system, at the given time $(j-1)$, we mean a given value $f_{j-1}^{(k)}$ of the present kind of failure together with recorded values of the sequence $\{H_m\}_{j-1}$, $m = 1,\ldots,j-2$, where $H_m = x_m$; $\left(f_m^{(k)}, r_m^{(k,l)}\right)$, $l = 1,2,\ldots,u_m$.

The first question that arises is: suppose the maintenance history at the $(j-1)^{\mathrm{th}}$ failure is given. What will be the pdf (or just the expectation) of the time X_j "from now on" to the next failure, if the $r_{j-1}^{(k,l)\mathrm{th}}$ $(l = 1,\ldots,u_{j-1})$ kind of the repair would be chosen?

The second question may be stated as follows: given the same maintenance history as in the previous case and the same current kind of failure $f_{j-1}^{(k)}$, suppose the $r_{j-1}^{(k,l)}$ kind of repair is chosen. What are the probabilities $p_j^{(k,l)}$ of the corresponding kinds of the failures $f_j^{(k)}$ that may occur at the (next) time j, for $k = 1,\ldots,q$; $l = 1,\ldots,u_{j-1}$.

This question suggests that in order to make the procedure available for practical use, a huge amount of a statistical work should be done.

We concentrate on a partial answer to the first question only.

To make an efficient (and rationally based) prediction, as for the pdf of the "future" time X_j of the system operating until a next failure, one of the Weibull conditional pdfs $g(x_j|x_1,\ldots,x_{j-1})$, should be chosen (and then testified by means of statistical methods based on past data of similar events) as a model. In particular, one may possibly determine (using statistical methods) and then apply, one of the members of the following class of Weibull conditional pdfs:

$$g_j(x_j|x_1,\ldots,x_{j-1})$$
$$= \left[\lambda_j\left(1 + a_{1,k(1)}x_1^\rho + a_{2,k(2)}x_2^\rho + \cdots + a_{j-1,k(j-1)}x_{j-1}^\rho\right)\right]x_j^{\gamma-1}$$
$$\exp\left\{-\left[\lambda_j\left(1 + a_{1,k(1)}x_1^\rho + a_{2,k(2)}x_2^\rho + \cdots + a_{j-1,k(j-1)}x_{j-1}^\rho\right)\right]x_j^\gamma\right\} \qquad (5.20)$$

where all the coefficients are assumed be positive.

On the other hand, the above conditional pdfs apparently depend on the "parameters $a_{i,k()}$, $i = 1, \ldots, j-1$, of the (scale) parameter function":

$$\lambda_j \left(1 + a_{1,k(1)} x_1^\rho + a_{2,k(2)} x_2^\rho + \cdots + a_{j-1,k(j-1)} x_{j-1}^\rho \right),$$

while the parameters $\{a_{i,k}\}$ themselves could be assumed to be dependent on the maintenance history of the system up to the last failure. The last statement can be expressed symbolically by "$a_k = a_k(\{H_m\}_{j-1})$". Note that there is always a finite number of possible values for each a_k as determined by the maintenance history.

The unavoidable question, concerning the "best" policy for choices of the repairs, after given types of failures, should lead to a (general) framework for a particular set of optimization problems that might be formulated as part of an anticipated larger theory.

The underlying idea behind such a potential theory is based on an observation that there may emerge a need as to 'balance' the system efficiency (in the sense of maximizing the total length of the times X_1, X_2, ...) against the total cost of the repairs, so that the possible rewards could possibly be defined in terms of underlying objective functions such as 'maximum expected profit', to be obtained from the system's exploitation. Other proposed models are also given in the form of the conditional (Weibullian in x_j) pdfs:

$$g_j(x_j|x_1, \ldots, x_{j-1})$$
$$= \lambda_j \exp\left[b_{1,k(1)} x_1^\rho + b_{2,k(2)} x_2^\rho + \cdots + b_{j-1,k(j-1)} x_{j-1}^\rho \right] x_j^{\gamma-1}$$
$$\exp\left[-\lambda_j \exp\left[b_{1,k(1)} x_1^\rho + b_{2,k(2)} x_2^\rho + \cdots + b_{j-1,k(j-1)} x_{j-1}^\rho \right] x_j^\gamma \right], \qquad (5.21)$$

where the coefficients $b_{i,k(i)}$ ($i = 1, \ldots, j-1$) are arbitrary (possibly negative).

Somewhat simplified versions of the (5.20) and (5.21) are obtained if only the conditional expectations of the lifetimes are of an interest. Then, for $j = 2, 3, \ldots$ we have the following wide range of (continuous only) regression functions:

$$E\left[X_j|x_1, \ldots, x_{j-1}\right]$$
$$= \left[\lambda_j \left(1 + a_{1,k(1)} x_1^\rho + a_{2,k(2)} x_2^\rho + \cdots + a_{j-1,k(j-1)} x_{j-1}^\rho \right) \right]^{-1/\gamma} \Gamma(1 + 1/\gamma),$$
$$(5.20^*)$$

or

$$E\left[X_j|x_1, \ldots, x_{j-1}\right]$$
$$= \left\{ \lambda_j \exp\left[b_{1,k(1)} x_1^\rho + b_{2,k(2)} x_2^\rho + \cdots + b_{j-1,k(j-1)} x_{j-1}^\rho \right] \right\}^{-1/\gamma} \Gamma(1 + 1/\gamma).$$
$$(5.21^*)$$

The conditional expectations in (5.20^*), (5.21^*) correspond to the pdfs in (5.20) and (5.21) respectively.

Both cases simplify to the exponential when $\gamma = 1$.

Also consider the important case $\gamma = 2$.

Other than Weibullian models, especially the gamma, may also be considered.

5.8.3 "Forgetting Factors" Method

In reference to our considerations in Sect. 5.6.2 on possible difficulties associated with in growing number of the coefficients present in the Weibullian models (5.20) and (5.21) as time j grows, the following method of "forgetting factors," that in a sense reduces unnecessary (long but negligible) stochastic process memory, will briefly be described in this section. For convenience recall the model (5.20):

$$g_j(x_j|x_1,\ldots,x_{j-1})$$
$$= \left[\lambda\left(1 + a_{1,k(1)}x_1^\beta + a_{2,k(2)}x_2^\beta + \cdots + a_{j-1,k(j-1)}x_{j-1}^\beta\right)\right]x_j^{\gamma-1}$$
$$\exp\left\{-\left[\lambda\left(1 + a_1x_1^\beta + a_2x_2^\beta + \cdots + a_{j-1}x_{j-1}^\beta\right)\right]x_j^\gamma\right\},$$

Introduce the corresponding time distances; representing the "memory durations" (at the moment x_{j-1} of $(j-1)^{\text{th}}$ failure) to be equal $y_k = x_{j-1} - x_k$, for $k = 1, 2, \ldots, j-1$; as $j = 3, 4, \ldots$.

So that y_k is the (non-negative) time distance between the "presence" x_{j-1} and the time of the kth failure occurred.

For each $k = 1, 2, \ldots, j-2$, define the kth "forgetting factor" $\varphi(y_k)$ to be a continuous (or only piecewise-continuous) function of the argument y_k that is assumed to possess the following properties:

1) $\varphi(0) = 1$,
2) $\varphi(y_k)$ is non-increasing.

Additionally we also may assume that

3) $\varphi(y_k) \to 0$, as $y_k \to \infty$.

As examples, we propose the following three "candidates" for the forgetting factors:

A) $\varphi(y_k) = (1 + c\log(1 + y_k))^{-1}$
B) $\varphi(y_k) = (1 + cy_k^u)^{-1}$, and especially:
C) $\varphi(y_k) = \exp\left[-cy_k^u\right]$,

where the coefficients c, u are any positive real constants.

In case C) especially, aim portant particular case arises when assuming $u = 1$.

On the other hand combinations of the above three cases may be useful too.

Now we are in a position to define the "time forgetting" Weibull stochastic process $\{X_j\}$ as the one defined by the initial Weibull pdf $g_1(x_1)$ of X_1, the conditional pdf $g_2(x_2|x_1)$ of X_2 given $X_1 = x_1$, and by all the remaining terms (as $j = 3, 4, \ldots$)

of the conditional pdfs $g_j(x_j|x_1,\ldots,x_{j-1})$ of X_j, given the whole past $x_1,\ldots,x_{j-1}$ by the following formula as derived from (5.20) by incorporating to it the forgetting phenomenon:

$$
\begin{aligned}
&g_j(x_j|x_1,\ldots,x_{j-1}) \\
&= \Big[\lambda\Big(1 + \varphi(y_1)a_{1,k(1)}x_1^\beta + \varphi(y_2)a_{2,k(2)}x_2^\beta + \cdots + \varphi(y_{j-2})a_{j-2,k(j-2)}x_{j-2}^\beta \\
&\quad + \varphi(y_{j-1})a_{j-1,k(j-1)}x_{j-1}^\beta\Big)\Big]x_j^{\gamma-1}\exp\Big\{-\Big[\lambda\Big(1 + \varphi(y_1)a_{1,k(1)}x_1^\beta \\
&\quad + \varphi(y_2)a_{2,k(2)}x_2^\beta + \cdots + \varphi(y_{j-2})a_{j-2,k(j-2)}x_{j-2}^\beta \\
&\quad + \varphi(y_{j-1})a_{j-1,k(j-1)}x_{j-1}^\beta\Big)\Big]x_j^\gamma\Big\}\,.
\end{aligned}
\tag{5.22}
$$

As, for example, the time y_m grows the associated forgetting coefficient $\varphi(y_m)$ $a_{m,k(m)}$ decays. The method relies on ignoring such coefficients whenever they become smaller than some number given in advance.

The same method can be applied for the model (5.21) or other similar.

5.9 Additional Remarks

5.9.1 Extention of Class of Pseudoaffine Transformations

Using the method of pdfs parameter replacement, one obtains a significantly wider generalization than that by extending the class of the pseudoaffine transformations to the class of the pseudopower.

According to this simple method it is enough just to declare the factors of the pdfs $g(x_1,\ldots,x_n)$ given by (5.1) to be obtained by replacing (for $j=2,\ldots,n$) the parameters $\alpha_j,\beta_j,\gamma_j,\delta_j$ of the generalized gamma pdfs $f_j(t_j)$ by arbitrary parameter functions $\alpha_j^*(\#),\beta_j^*(\#),\gamma_j^*(\#),\delta_j^*(\#)$ respectively. (The symbol # here denotes the string of arguments $x_1,\ldots,x_{j-1}$.)

Recall that for $j=1$ all these parameter functions are assumed to be constant.

Using the pdfs factorization () one obtains the class of the 4-parameter generalized gamma pdfs that can be described as follows:

$$
g_1(x_1) = gG\left(x_1;\alpha_1^*,\beta_1^*,\gamma_1^*,\delta_1^*\right),
$$

and for $j=2,\ldots,n$

$$
\begin{aligned}
&g_j(x_j|x_1,\ldots,x_{j-1}) \\
&= gG\left(x_j;\alpha_j^*(\#),\beta_j^*(\#),\gamma_j^*(\#),\delta_j^*(\#)\right).
\end{aligned}
$$

Setting in the latter two equalities (for $j=1,\ldots,n$), $\delta_j=\delta_j^*(\#)\equiv 1$, one obtains, as proper subclass, the class of 3-parameter "generalized Weibullian" pdfs. Setting additionally $\gamma_j=\gamma_j^*(\#)\equiv 1$ one obtains the class of 2-parameter exponential pdfs.

The last two classes of the pdfs are much wider than the classes of the "ordinary pseudoweibullians" or of the "ordinary pseudoexponentials" respectively, that were constructed using the transformations method.

5.9.2 *Extension of n-variate pdf Classes*

One can obtain wider extension of the n-variate pdfs classes considered in this work if the currently held assumption that "the pdfs $f_1(t_1), \ldots, f_n(t_n)$ of the input r. variables $T_1, \ldots, T_n$ belong to the same class of pdfs" is relaxed.

This relaxation, while at first may seem to be unjustified, turns out to be natural when modeling the reliability of the multicomponent systems. Clearly, it may easily happen that the system components $e_1, \ldots, e_n$ are each of a different nature, and may therefore be subject to different mechanisms of failure.

5.9.3 *Extended Applications of Stochastic Dependences*

The pattern for the described above stochastic dependences through a transformation of parameters into parameter functions seems to be proper for applications in a much wider area of the real world phenomena, than the reliability problems only.

One may realize that in stochastic models in form of the joint pdfs of the r. vectors (X_1, X_2), that we considered throughout, the meaning of the two random variables need not to be restricted to the cases we investigated. Thus the meaning of both the r. variables X_1, X_2, virtually, may be arbitrary and yet the realities described by each of them might be of different natures. Nevertheless, the essence of the stochastic dependences of an X_2 from an X_1, as we have described it may remain the same. Therefore, it could be said, that, to a measure "universal," model of the stochastic dependence may work for "any kind" of the real world phenomena.

In this perhaps "universal" pattern, a random quantity X_1, may be considered as an explanatory variable for "any kind" of an associated (in any possible way) random quantity X_2 of interest. Even if not always this would be the case, it nevertheless may be worth to testify statistically that possibility as a zero hypothesis.

5.9.4 *Multivariate Analysis*

This bivariate pattern has its natural extension to the multivariate analysis. The description of common nature of all the considered in this work as multivariate stochastic dependences that may occur in statistical practice can be summarized as follows.

A set of (any kind of) random variables $X_1, \ldots, X_{j-1}$, "influences stochastically" some other random variable (or, possibly, also a random vector) X_j, $(j = 2, 3, \ldots)$

(that has a pdf or a cdf $f_j(x_j; \theta)$, and θ is a scalar or a vector parameter(s),) by having an "impact" on the parameter(s) θ, that results in turning its value into another value θ $(x_1, \ldots, x_{j-1})$. Here the arguments $x_1, \ldots, x_{j-1}$ of the foregoing parameter function are particular values of the explanatory random variables $X_1, \ldots, X_{j-1}$.

For a kind of a representative illustration of the previous statements, notice that the common regression function in any multivariate normal distribution model satisfies the formulated above rule as a specific case, with $\theta = \mu_j$ being replaceable for any continuous (only) regression function $\mu_j(X_1, \ldots, X_{j-1})$. Needless to say, that the variance, of possibly that same normal r. variable X_j, may also be replaced by some other continuous function, say, $\sigma^2(X_1, \ldots, X_{j-1})$.

5.10 Some Analytic Examples

5.10.1 Pseudolinear Transformations

Consider the following class of the pseudolinear transformations R^2 R^2, which is a special case (as $n = 2$) of (5.12):

$$X_1 = aT_1$$
$$X_2 = \varphi(X_1)T_2 \quad \text{a.s.} , \tag{5.23}$$

where T_1, T_2 are independent r. variables, and for $k = 1, 2$, T_k has a following one parameter exponential pdf $f_k(t_k) = (1/\theta_k)\exp[-t_k/\theta_k]$. Moreover, the symbol "a" denotes a positive real number, and $\varphi(x_1)$ is assumed to be any positive continuous function, in general satisfying $\varphi(0) = 1$. Simple calculations yield the following formula for the bivariate pseudoexponential pdf of the r. vector (X_1, X_2):

$$g(x_1, x_2) = g_1(x_1)\{g_2(x_2|x_1)\} \tag{5.24}$$
$$= (a\theta_1)^{-1}\exp[-x_1/(a\theta_1)]\left\{(\theta_2\varphi(x_1))^{-1}\exp[-x_2/(\theta_2\varphi(x_1))]\right\} .$$

Notice that $E[X_1] = a\theta_1$, and $E[X_2|x_1] = \theta_2\varphi(x_1)$, where $E[X_1]$ is the expected value of X_1, and $E[X_2|x_1]$ is the conditional expectation of X_2 given $X_1 = x_1$.

Therefore the expectation of X_2 is given by $E[X_2] = \theta_2 E_1[\varphi(X_1)]$, where the expression $E_1[]$ denotes the expectation with respect to the pdf $g_1(x_1)$.

Recall that $g_1(x_1) = (1/a)f_1(x_1/a)$, where $f_1(t_1)$ is the exponential pdf of T_1 as given above. Also one easily obtains the following second moments of X_1, and X_2:

$$E\left[X_1^2\right] = 2a^2\theta_1^2 ,$$
$$E\left[X_2^2|x_1\right] = 2\theta_2^2[\varphi(x_1)]^2 ,$$
$$E\left[X_2^2\right] = 2\theta_2^2 E_1\left[(\varphi(X_1))^2\right] , \text{ and the variance of } X_2 \text{ is:}$$
$$Var(X_2) = \theta_2^2\left\{2E_1\left[(\varphi(X_1))^2\right] - (E_1[\varphi(X_1)])^2\right\} .$$

For the covariance of the r. variables X_1, X_2 one obtains:

$$\mathrm{Cov}(X_1, X_2) = \theta_2 \left\{ E_1[X_1 \varphi(X_1)] - \theta_1 E_1[\varphi(X_1)] \right\} .$$

5.10.2 Further Analytic Calculations

More analytic calculations can be performed if some specific classes of the parameter functions $\varphi(x_1)$ are employed in (5.23) and (5.24). One of the possible 'nice' parameter function may take, for example, the form $\varphi(x_1) = \cosh(cx_1)$, where c is a non-zero real number.

Note that in the latter case the condition $\varphi(0) = 1$ is satisfied. Recall that for the system reliability applications framework this condition is substantial. Other models that also seem to be proper in the reliability problems setting, one obtains by employing the class of power functions $\varphi(x_1) = 1 + Ax_1^r$, where A, r are arbitrary positive real numbers.

Now formula (5.24) takes on a more specific form:

$$g(x_1, x_2) = (a\,\theta_1)^{-1} \exp[-x_1/(a\,\theta_1)] \left\{ (\theta_2(1 + Ax_1^r))^{-1} \exp[-x_2/\theta_2(1 + Ax_1^r)] \right\} .$$

$$(5.25)$$

To simplify further considerations assume $a = 1$.

The following set of results, concerning the class of the pdfs given by (5.25), can easily be obtained by using elementary calculations, so that we only list some of them.

If the r. vector (X_1, X_2) has the joint pdf given by (5.25), one obtains the following results:

$$E\left[X_2^k | x_1\right] = k!\theta_2^k (1 + Ax_1^r)^k , \quad \text{for } k = 1, 2, \ldots$$

Using the binomial expansion one obtains:

$$E\left[X_2^k | x_1\right] = k!\theta_2^k \sum_{j=0}^{k} \left(C_j^k A^j x_1^{r(k-j)} \right) , \tag{5.26}$$

where $C_j^k = k!/j!(k - j)!$, and therefore for the k-th moment of X_2 we have:

$$E\left[X_2^k\right] = k!\theta_2^k \sum_{j=0}^{k} \left[C_j^k A^j \Gamma(r(k - j) + 1)\theta_1^{r(k-j)} \right] . \tag{5.27}$$

It is easily seen that for the r. variable X_2 considered in (5.26), and (5.27) all the moments exist and, for every positive integer k of a reasonable magnitude, can easily be obtained. In particular, for $k = 1$ one obtains the expectations:

$$E[X_2 | x_1] = \theta_2(1 + Ax_1^r) , \quad \text{and} \quad E[X_2] = \theta_2(1 + AE_1[X_1^r]) . \tag{5.28}$$

Using the familiar formula

$$E[X_1^r] = \theta_1^r \Gamma(r+1) , \tag{5.29}$$

which is valid for any real parameter $r > 0$, one obtains other version of (5.28), *i.e.*,

$$E[X_2] = \theta_2(1 + A\theta_1^r \Gamma(r+1)) . \tag{5.28*}$$

For the variance one obtains:

$$Var(X_2) = \theta_2^2 \left[1 + 2\Gamma(r+1)A\theta_1^r + \gamma A^2 \theta_1^{2r} \right] ,$$

where $\gamma = 2\Gamma(2r+1) - (\Gamma(r+1))^2$.

The covariance of X_1, X_2 is given by the simple formula:

$$\mathrm{Cov}(X_1, X_2) = A\theta_1 \theta_2 r E[X_1^r] .$$

Using (5.29), one obtains the following alternative formula:

$$\mathrm{Cov}(X_1, X_2) = rA\Gamma(r+1)\theta_1^{r+1}\theta_2 , \quad \text{for any real } r > 0 .$$

5.10.3 Simplification of Calculations

In the previous example we had chosen the class of the parameter functions $\varphi(x_1) = 1 + Ax_1^r$ as the one satisfying the requirement $\varphi(0) = 1$. If, however, for theoretical benefits such as the generality, this requirement is omitted, then the choice of parameter functions having the form $\varphi(x_1) = Ax_1^r$ makes the above formulae and the underlying analytical calculations significantly simpler, so more nice stochastic properties for this model can be derived. We do not intend to explore this case extensively. Notice only the two especially interesting facts:

1. Setting in (5.24) $a = 1$, and $\varphi(x_1) = Ax_1^r$ one obtains that

$$Pr(X_2 = 0 | X_1 = 0) = 1 .$$

2. Formula (5.27) simplifies dramatically and one obtains:

$$E[X_2^\alpha] = \Gamma(\alpha+1)\Gamma(\alpha r+1)(A\theta_2)^\alpha \theta_1^{\alpha r} , \tag{5.27*}$$

where in place of $k = 1, 2, \ldots$ present in (5.27) the symbol α in (5.27*) denotes any positive real number. Placing an integer k back into (5.27*) one obtains a remarkably simple formula for the kth moment of X_2. Thus for $k = 1, 2, \ldots$, we have

$$E\left[X_2^k\right] = k!(kr)!(A\theta_2)^k . \tag{5.27**}$$

5.10.4 An Example of a Non-symmetric Pseudonormal Class of pdfs

The most representative class of bivariate non-symmetrical pseudonormal pdfs, with respect to the simplicity and possible applications, seems to be the one defined by the following class of pseudolinear transformations:

$$X_1 = T_1 \,,$$
$$X_2 = T_2 + AX_1^{2k} \,,$$

where k is a positive integer, and the r. vector (T_1, T_2) is an arbitrary normal with the correlation coefficient ρ. The conditional density $h_2(x_2|x_1)$ for the resulting r. vector (X_1, X_2) satisfies:

$$h_2(x_2|x_1) = \left(1/\sigma_2\left(1-\rho^2\right)2\pi\right)\exp\left[-\left(x_2 - ax_1 - Ax_1^{2k}\right)^2 \Big/ 2\left(\sigma_2^2(1-\rho^2)\right)\right] \,.$$

The expectations $E[X_1] = 0$, and $E[T_2] = 0$, but, in this case, the lack of symmetry causes $E[X_2] \neq 0$. We have then the regression function $E[X_2|x_1] = ax_1 + Ax_1^{2k}$, and upon ordinary integrating one obtains:

$$E[X_2] = A j(2k-1)\sigma_1^{2k} \,.$$

(Recall that $j(w) = (1)(3)\ldots(w)$ for any odd integer w.)

Notice also the fact that $E[X_2] \to 0$ as $A \to 0$ so the parameter 'A' may serve as a measure of the asymmetry of the joint pdf.

The variances:

$$Var(X_1) = \sigma_1^2 \,, \quad \text{and}$$
$$Var(X_2) = s_2^2 = \sigma_2^2 + \gamma(k)A^2\sigma_1^{4k} \,, \tag{5.30}$$

where $\gamma(k) = j(4k-1) - (j(2k-1))^2$.

Higher central moments

We are able to find the third and fourth central moments of X_2 in a reasonable form. After elementary but pretty tedious calculations we obtain for $k = 1, 2, \ldots$ the following asymmetry coefficient of the marginal X_2's pdf:

$$v(k) = E\left[(X_2 - m_2)^3\right] \Big/ (Var(X_2))^{3/2}$$
$$= \alpha(k)A^3\sigma_1^{6k} + \beta(k)a^2A\sigma_1^{2k} + \sigma_1^{(2k+2)} \Big/ \left(\sigma_2^2 + \gamma(k)A^2\sigma_1^{4k}\right)^{3/2} \,,$$

where the expectation $m_2 = E[X_2]$ was given above, and

$$\alpha(k) = j(6k-1) - 3j(2k-1)j(4k-1) + (j(2k-1))^3 \,,$$
$$\beta(k) = 3j(2k-1)[2k(2k+1)-1] \,,$$

with $\gamma(k)$ as given in above.

Note that the asymmetry of the density of X_2 tends to zero as $A \to 0$ and has a common sign with A.

For the particular case $k = 1$ one obtains

$$v(1) = \left(7A^3\sigma_1^6 + 15a^2A\sigma_1^4\right) / \left(\sigma_2^2 + 2A^2\sigma_1^4\right)^{3/2} .$$

The fourth central moment is given in the following way.

Let $m_2 = E[X_2]$ be defined as above.

Then one obtains the formula:

$$E\left[(X_2 - m_2)^4\right] = 3\sigma_2^4 + A^2\sigma_1^{4k}\left\{\alpha^*(k)A^2\sigma_1^{4k} + \beta^*(k)\sigma_2^2 + \gamma^*(k)a^2\sigma_1^2\right\} ,$$

where

$$\alpha^*(k) = 6(j(2k-1))^2 j(4k-1) - 4j(2k-1)j(6k-1) - 3(j(2k-1))^4 + j(8k-1),$$

$$\beta^*(k) = 6\left[j(4k-1) - (j(2k-1))^2\right] ,$$

$$\gamma^*(k) = 6\left[(4k(4k+1) - 1)j(4k-1) + 2(j(2k-1))^2 - 2j(2k-1)j(2k+1)\right] .$$

For the quadratic case, *i.e.*, with $k = 1$, we obtain

$$E[(X_2 - m_2)^4] = 3\sigma_2^4 + A^2\sigma_1^4[60A^2\sigma_1^4 + 12\sigma_2^2 + 318a^2\sigma_1^2] .$$

The calculation of the ordinary moments $E[X_2^r]$, $r = 3, 4, \ldots$ is essentially easier than that of the central ones.

Nevertheless, they are represented by relatively long strings of the writing terms that make them hard to read. For this reason we do not present them here.

Correlation coefficients

Again, let c denote a correlation coefficient of the variables X_1, X_2. After some calculations we have obtained the covariance $E[X_1 X_2]$, which turned out to be equal just as_1^2.

Also we obtained the simple formula for the correlation coefficient as follows

$$c = as_1/s_2 ,$$

where $s_1 = \sigma_1$, and s_2 is given by (5.30).

Notice too, that in the considered case we have other unexpected result: $E[X_1 X_2] = E[T_1 T_2]$, while $c \neq \rho$ as $s_2 \neq \sigma_2$.

References

1. Barlow RE, Proschan F (1975) Statistical Theory of Reliability and Life Testing. Holt, Rinehart and Winston, New York
2. Filus, JK, Filus LZ (2005) On two new methods for constructing multivariate probability distributions with system reliability motivations. Applied Stochastic Models and Data Analysis Conference, 'ASMDA 2005', Brest, France, May 17–20, Conference Proceedings, pp. 1232–1240

3. Filus, JK, Filus LZ (2006) On some new classes of multivariate probability distributions. Pakistan Journal of Statistics 22:21–42
4. Filus, JK, Filus LZ (2006) On methods for construction new multivariate probability distributions in system reliability framework. 12th International Conference on Reliability and Quality in Design, ISSAT 2006, Chicago, USA, Conference Proceedings, pp. 245–252
5. Filus, JK, Filus LZ (2007) On new multivariate probability distributions and stochastic processes with system reliability and maintenance applications. Methodology and Computing in Applied Probability June issue
6. Arnold BC, Castillo E, Sarabia JM (1992) Conditionally specified distributions, lecture notes in statistics – 73, Springer-Verlag, New York
7. Arnold BC, Strauss DJ (1988) Bivariate distributions with exponential conditionals. Journal of the American Statistical Association 83:522–527
8. Arnold BC, Strauss D (1991) Bivariate distributions with conditionals in prescribed exponential families. Journal of the Royal Statistical Society Series B 53:365–375
9. Castillo E, Galambos J (1990) Bivariate distributions with Weibull conditionals. Analysis Mathematica 16:3–9
10. Filus JK (1991) On a type of dependencies between Weibull lifetimes of system components. Reliability Engineering and System Safety 31:267–280
11. Hayakawa Y (1994) The construction of new bivariate exponential distributions from a bayesian perspective. Journal of the American Statistical Association 89:1044–1049
12. Heinrich G, Jensen U (1995) Parameter estimation for a bivariate life-distribution in reliability with multivariate extensions. Metrika 42:49–65
13. Lindley DV, Singpurwalla ND (1986) Multivariate distributions for the life lengths of components of a system sharing a common environment. Journal of Applied Probability 23:418–431
14. Lu J (1989) Weibull extensions of the Freund and Marshall–Olkin bivariate exponential models. IEEE Transactions on Reliability 38:615–619
15. Freund JE (1961) A bivariate extension of the exponential distribution. Journal of the American Statistical Association 56:971–977
16. Marshall AW, Olkin I (1967) A generalized bivariate exponential distribution. Journal of Applied Probability 4:291–303
17. Filus JK, Filus LZ (2000) A class of generalized multivariate normal densities. Pakistan Journal of Statistics 16:11–32
18. Filus JK, Filus LZ (2001) On some bivariate pseudonormal densities. Pakistan Journal of Statistics 17:1–19
19. Filus, JK, Filus LZ (2006) On pseudonormal extension of the class of multivariate normal probability distributions. International Conference Statistical Methods for Biomedical and Technical Systems, BIOSTAT 2006, Limassol, Cyprus, Conference Proceedings, pp. 79–86
20. Kotz S, Balakrishnan N, Johnson NL (2000) Continuous multivariate distributions, vol. 1, 2nd edn. Wiley New York, Chichester, Weinheim, Brisbane, Singapore, Toronto
21. Harlow DG. (1977) Probabilistic models for the tensile strength of composite materials. PhD Thesis, Cornell University
22. Filus, JK, Filus LZ (2007) A stochastic model of reliability of systems with continuously dependent components. Proceedings of the International Conference ASMDA 2007
23. Wang H (2002) A survey of maintenance policies of deteriorating system. Journal of Operational Research 139:469–489

Further Reading

Wang YH (1992) Approximated kth-order two-state Markov chains. Journal of Applied Probability 29:861–868

Chapter 6
The Optimal Burn-in: State of the Art and New Advances for Cost Function Formulation

Xin Liu[1], Thomas A. Mazzuchi[2]

[1]Delft Institute of Applied Mathematics, Delft University of Technology,
PO Box 5031, 2600GA Delft, The Netherlands
[2]Department of Engineering Management and Systems Engineering,
The George Washington University, USA

6.1 Introduction

Burn-in is a quality screening technique used to induce early failures that would be costly if experienced by the customer. As a method to screen out the earlier failures of the products, burn-in testing has been widely used in electronic manufacturing as well as many other areas such as the military and aerospace industries since the 1950s. Burn-in has proven to be a very effective quality control procedure which can improve products' quality, enhance their reliability for operational life, and bring both profit and goodwill to the manufacturers.

The motivation of doing the burn-in test can be simply expressed by Sultan (1986): "for electronic products with an initially high failure rate, burn-in is used to reduced the warranty cost ...The effect of burn-in is to reduce the infant mortality region so that the finished product will operate in the region near the constant failure rate." Jensen and Petersen (1982) provide more insightful information by noting that "... if all parts were made properly in the first place, then burn-in, or indeed any other form of post-production screening, should not be necessary. For quite a large number of electronic components or mechanical devices with a long history of production this will certainly be true. However, in a time of rapidly changing technologies and production methods, coupled with an increasing awareness of reliability, most manufacturers will be forced to instigate some sort of reliability screening to minimize the number of early failures in the field."

Various aspects of burn-in testing have been investigated by many researchers over the last 40 years or so. Most interest centers on the topic of how long and under what conditions should the burn-in process be conducted in order to get the most profit or the highest reliability. The "best" burn-in time is often calculated based on certain specific optimization criteria given by the manufacturers and the decision makers according to their specific situation and interest. The "best" burn-in time derived from these criteria is usually called the *optimal burn-in time*. In this chapter, the objective is two-fold. Firstly, the authors will give a dedicated and thorough state-of-art on many aspects of burn-in testing, such as the nature of the test (static vs. dynamic burn-in), the treatment of failures during burn-in, different warranty policies, the various distribution assumptions used, the inference method applied (Bayesian vs. classical models), the optimality criteria (cost vs. reliability), *etc.* Among these aspects, special attention will be given to the cost function review and analysis. Secondly, based on the literature review work of the optimal burn-in cost models, the authors found that although a variety of cost models for burn-in have been constructed for different situations over the past years, most of them are only valid for the case that the component failures during burn-in are followed immediately by a failure treatment. This so-called "in-time failure treatment" (IFT) is not always applied, and there are some practical situations where the failure treatments can be only implemented at the end of burn-in whenever the failure happens. The latter situation, as referred to as "after burn-in failure treatment" (AFT) has not received much attention in the literature. Therefore, a new cost optimization model will be proposed especially for it by introducing the concept of "per-item-output." Properties and theorems about this new model are given, followed by a numerical example. Application of this new model to the industry policy-making strategy is finally discussed.

6.2 State-of-art on Optimal Burn-in Research

The burn-in process has many facets and each has been the subject of competing research. First of all, the conditions under which the manufacture must function need to be addressed. These include such decisions as the number of items to be delivered to the customer, the form of the warrantee policy, and the level of assembly at which the burn-in is to be conducted. Next, the manner in which the burn-in test is conducted needs to be considered. Important aspects here are whether to conduct the test at normal or accelerated conditions, whether to scrap or repair failed items during testing, whether it is possible to stop the testing when a failure occurs or to only observe failed items at the end of the test, and whether the determination of successful burn-in can be made sequentially or only at a fixed time. Modeling assumptions such as the form of the distribution (single or mixture) used for product lifetimes and the inference procedure (classical or Bayesian) if any used for determining the distribution parameters. Finally, the criteria (cost or reliability or a mixture) under which optimality is defined is also a main issue for research.

6.2.1 Failure Time Model

6.2.1.1 Failure Rate Assumptions

Assume that product under test has lifetime T, with density function $f(t)$, cumulative distribution function (cdf) $F(t)$ and survival function $R(t) = 1 - F(t)$. The

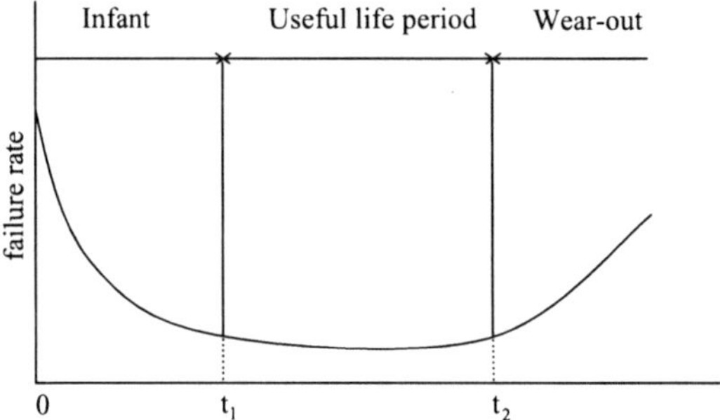

Fig. 6.1 Bathtub failure rate curve

failure rate (FR) function $r(t)$ is then given by $r(t) = f(t)/R(t)$. It is assumed that $f(t)$ is continuous and thus the FR is well defined for all $t \geq 0$.

Most electronic, electro-mechanical, and mechanical product lifetime distributions follow the so-called "bathtub failure rate" depicted in Fig. 6.1. The bathtub failure rate consists of three periods: in the first period the FR decreases until a certain time point, then remains at an (almost) constant level, and finally increases. The initial decreasing FR is due to manufacturing defects and the final increasing FR is due to aging. A formal definition of the bathtub failure rate is presented in Mi (1997).

Definition 5. (Mi (1997)): A failure-rate function $r(t)$ is said to have a bathtub shape if there exist $0 \leq t_1 \leq t_2 \leq \infty$, such that

$$r(t) \begin{cases} & \text{strictly decreases, if} \quad 0 \leq t \leq t_1 \\ \text{is} & \text{a constant, say } r_0, \text{ if} \quad t_1 \leq t \leq t_2 \\ & \text{strictly increases, if} \quad t_2 \leq t \end{cases}$$

The points t_1 and t_2 are called the change points of $r(t)$. The time interval $[0, t_1]$ is called the infant mortality period; the interval $[t_1, t_2]$, where $r(t)$ is flat and attains its minimum value is called the normal operating life or useful life; and the interval $[t_2, \infty)$ is called the wear-out period. Actually, in real applications, the second period is not always an exact constant but approximately a constant level. Furthermore, some parts may shrink to a point. For example, if $t_1 = t_2 = 0$, then the FR is strictly increasing. If $t_1 = t_2 = \infty$, then the FR is strictly decreasing. In general, if $t_1 = t_2$, then the constant interval degenerates to a single point. These situations are regarded as special cases for the bathtub failure rate.

Among these three parts, special attention has been paid to the first part in the burn-in problems since infant mortality failure critically affects both the manufacturer's reputation and cost. Prevention of these failures is actually the reason for doing burn-in testing. Researchers have different ideas as to why this initial period exhibits a decreasing FR. Jensen and Petersen (1982) state that this part comes from the failures in the weak subpopulation of a bimodal lifetime distribution, while another

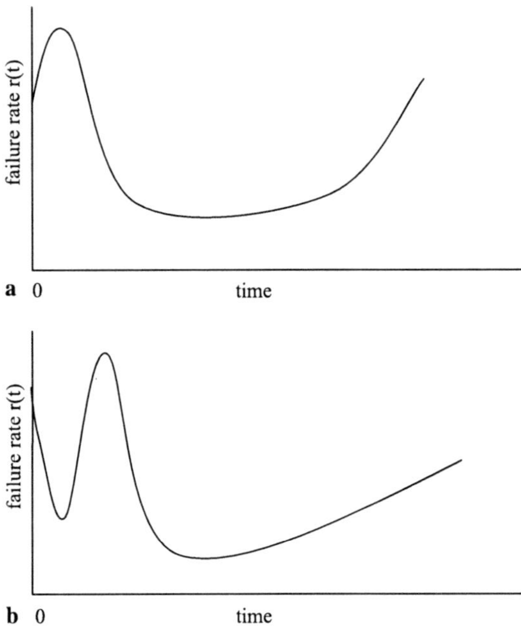

Fig. 6.2 a S-shape failure rate curve; **b** W-shape failure rate curve

common hypothesis noted by Kuo and Kuo (1983) shows that it originates from a skew part of a single population. Yun, Lee and Ferreira (2002) conclude that when this period exists, burn-in procedure must be taken into consideration to reduce the manufacturing cost.

Modifications of the bathtub failure rate have also been proposed by other researchers. For example, the so called *S-shape model* (Yuan and Shih, 1991) or *generalized bathtub curve* (Kuo and Kuo, 1983) presented in Fig. 6.2a has been proposed. This curve is derived from well separated mixed Weibull distributions. Jensen and Petersen (1982) state that this shape is more typical for products without screenings for early failures. In addition, the "W-shape" failure rate curve of Fig. 6.2b was proposed by Su and Wu (2001). This curve is derived from a case study of the production of a switch mode rectifier with the application of the back-propagation neural network method. The failure rate is estimated using the model. This failure rate curve is identical to the S-shaped curve, except for having an additional decreasing part at the start of the lifetime. Some mixed Weibull distributions have this kind of failure rate shape.

A *Unimodal (hump) failure rate* assumption is used by Chang (2000). This is presented in Fig. 6.3a. This is actually a special case of S-shape failure rate which includes three periods: at first it strictly increases and reaches the highest value, then strictly decreases and finally is approximately a constant. So it is just a S-shape type without the wear-out phase. The point γ (see Fig. 6.4) at which the failure rate reaches the highest value is called the *critical time*. Based on the single lifetime

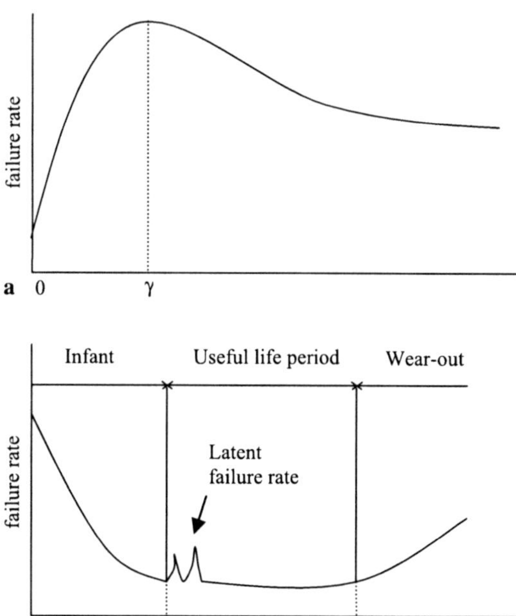

Fig. 6.3 a Unimodal (hump) failure rate curve-special S-shape; **b** Latent failure related bathtub curve

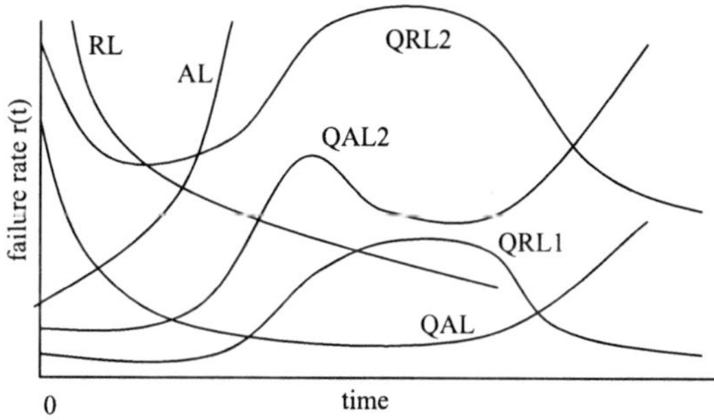

Fig. 6.4 Types of failure rate from reliability laws point of view

distribution point of view, the author says that an exact bathtub shape is not easy to obtain.

Another special case for bathtub failure rate curve is presented by Yan and English (1997). It integrates the *latent failure*, which is another typical failure pattern that has been identified recently by some practitioners and researchers in the electronics industry. The latent defects occur during the product useful life, *i.e.*, the second part of the bathtub curve. They cannot be detected by inspection or functional testing. It is only when the (possibly accelerated) operating environment is applied over time that the latent defects are exposed and converted to patent defects.

Therefore, the defects can be detected and so there will be a jump in the failure rate curve as the latent failures are exposed. The failure rate shape used in this paper just shows the characteristic of this phenomenon: there is a piece of oscillating curve at the beginning of the useful period (Fig. 6.3b).

No matter bathtub shape, S-shape or even W-shape failure rate, they all have the similar characteristics: at the beginning of the lifetime, they all start with a high risk and unstable failure rate period which is the infant mortality phase, and then decrease rapidly into a steady-state period which is at low risk and more-or-less constant.

Baskin (2002) further classifies the types of failure rate from the reliability laws point of view. He lists six types of failure rate as follows:

AL   *the aging law of reliability (increasing failure rate)*
RL   *the rejuvenating law of reliability (decreasing failure rate)*
QAL1   *the quasiaging law of reliability of first type (bathtub-shaped)*
QAL2   *the quasiaging law of reliability of second type*
QRL1   *the quasi-rejuvenating law of reliability of first type (lognormal type)*
QRL2   *the quasi-rejuvenating law of reliability of second type.*

These failure rate curves are shown in Fig. 6.4. The author also states that *all* known distribution functions enter into one of these reliability laws depending on parameter values. From Fig. 6.4, it can be observed that each of these can be classified into the failure rate types we discussed before, *i.e.*, bathtub, S-shape or W-shape.

6.2.1.2 Distribution Type (Model Selection)

The choice for the lifetime distribution type of the burn-in product must be consistent with the assumptions of the failure rate shape. The lifetime distributions can be classified into *uni-modal* and *bimodal* distribution.

Uni-modal Distribution

Researchers who believe that the early failure and useful life come from a single distribution choose *uni-modal* distributions, which are usually of compound forms.

Weibull type distributions are popular choice among literature [see, for example, Kim and Kuo (1998), Chien and Kuo (1995) and Nicolato and Runggaldier (1999)]. Kim and Kuo (1998) use *Competing Weibull distribution*. This distribution is a popular choice to describe the breakdown time of gate oxide in the semiconductor manufacturing industry. Its cumulative distribution function has the form:

$$F(t) = 1 - \exp\left[-\left(\frac{t}{\alpha_e}\right)^{\beta_e}\right]\exp\left[-\left(\frac{t}{\alpha_i}\right)^{\beta_i}\right] \qquad (6.1)$$

Subscripts 'i' and 'e' denote intrinsic and extrinsic breakdown respectively. Thus the failure rate function can be expressed as

$$r(t) = \frac{\beta_e}{\alpha_e}\left(\frac{t}{\alpha_e}\right)^{\beta_e-1} + \frac{\beta_i}{\alpha_i}\left(\frac{t}{\alpha_i}\right)^{\beta_i-1} \tag{6.2}$$

Chien and Kuo (1995) use *Weibull distribution* for integrated circuit (IC) components and *exponential distribution* for non-IC ones. This has the following failure rate form.

$$r(t) = \begin{cases} \beta(1/\alpha)^\beta t^{\beta-1} & 0 \le t \le t_1 \\ \beta(1/\alpha)^\beta t_1^{\beta-1} & t_1 \le t \end{cases}, \quad 0 < \beta < 1 \tag{6.3}$$

In this compound distribution, a two parameter Weibull distribution with shape parameter less than 1 is defined for the infant mortality period and an exponential distribution having constant failure rate used to model the second period. The useful phase is supposed to be long-term based on the fact that the wear-out period for most electronic components is hardly reached in actual situation. Yun, Lee and Ferreira (2002) also use this distribution. Kuo (1984) and Genadis (1996) also apply this distribution, but describe the failure rate for Weibull part as: $r(t) = \alpha_1 t^{-\beta_1}$. If the wear-out period cannot be neglected, a Weilbull distribution with shape parameter larger than 1 is a popular choice to model this phase. Yun, Lee and Ferreira (2002) use the failure rate form for this distribution as:

$$r(t) = \begin{cases} \beta_1(1/\alpha_1)^{\beta_1} t^{\beta_1-1} & 0 \le t \le t_1 \\ \beta_1(1/\alpha_1)^{\beta_1} t_1^{\beta_1-1} & t_1 \le t < t_2 \\ \beta_1(1/\alpha_1)^{\beta_1} t_1^{\beta_1-1} + \beta_2(1/\alpha_2)^{\beta_2}(t-t_2)^{\beta_2-1} & t_2 \le t \end{cases}, \tag{6.4}$$

where $0 < \beta_1 < 1, \beta_2 > 1$.

Other uni-modal models have been suggested in the literature as well. Whitbeck and Leemis (1989) use a pseudo-component to model the effect of assembly defects. Similar to the mixed population approach where normal and weak components are assumed, the authors use a *2-parameter exponential power distribution* with cdf.

$$F(t) = 1 - \exp\left(1 - \exp\left(\alpha t^\beta\right)\right) \tag{6.5}$$

for the normal components and assume that the pseudo-component have a 30% chance of occurring and that, if its failure distribution is defined as

$$F(t) = 1 - \left[(1-w) + w\exp\left(-\alpha_p t^{\beta_p}\right)\right] \quad w = 30\% . \tag{6.6}$$

Based on the uni-modal point of view, Chang (2000) states that only a few single lifetime distributions used in the literature have an exact bathtub failure rate, except those composite ones. Therefore, he uses the special *unimodal (hump) failure rate* which some lifetime distributions follow, such as lognormal distribution, inverse

Gaussian distribution and Birnbaum–Saunders distribution. For the model used in the paper, the author uses the lognormal distribution.

Among the above papers, the Weibull distribution is most frequently used to model the infant mortality stage. According to Kuo and Kuo (1983), other distributions have also been used, such as log-normal distribution, gamma distribution, non-homogeneous Poisson process and empirical distribution. It is noteworthy that different distribution types may be assumed and sometimes fitted to the real data. However, different distribution assumptions could lead to vastly different optimal results for burn-in times. Because of this fact, Washburn (1970) suggests using some more general distributions, such as the *generalized gamma distribution (GGD)*, to accommodate an appropriate fit of data. The GGD has greater representational capability and includes, as special cases, such distributions as normal, Rayleigh, Maxwell, chi, chi^2, Weibull, exponential, ordinary gamma, *etc.*

Bimodal Distribution

In contrast to the uni-modals, some papers state that in general, a small proportion of components will be weak because of some imperfection in the control of the production process or in the design of the components. These components are assumed to deteriorate faster than the strong (normal) components. In such situation, the early failures come from a weak subpopulation of a *bimodal* component lifetime distribution. This suggests using a bimodal distribution which is a mixed distribution of two subpopulations. The most popular models are the mixed-exponential and mixed-Weibull distributions. This idea is used by Jensen and Petersen (1982), Perlstein, Jarvis and Mazzuchi (2001), Yun, Lee and Ferreira (2002), Kuo and Kuo (1983), Yuan and Shih (1991), Sheu and Chien (2005), Kim (1998), and Sultan (1986).

Suppose p is the mixture parameter which represents the proportion of the weak population. Then the mixed exponential distribution has the cdf:

$$F(t) = p(1 - \exp(-\lambda_1 t)) + (1 - p)(1 - \exp(-\lambda_2 t)) \tag{6.7}$$

Perlstein, Jarvis and Mazzuchi (2001) use this distribution and in this paper it has the solely DFR, which is a still a special case for bathtub failure rate.

Yun, Lee and Ferreira (2002) consider both mixed exponential and mixed Weibull distributions. Similar to the above, it has the cdf:

$$F(t) = p\left[1 - \exp\left(-(t/\alpha_1)^{\beta_1}\right)\right] + (1 - p)\left[1 - \exp\left(-(t/\alpha_2)^{\beta_2}\right)\right] \tag{6.8}$$

Kim (1998) noted that "researchers discovered that the two-mixed Weibull distribution is a good model for describing the time-to-failure of many products." Sultan (1986), Yuan and Shih (1991), and Yun, Lee and Ferreira (2002) also use this *mixed two-parameter Weibull distribution.* Yuan and Shih (1991) suggest that to do a better prediction for the wear-out period, a *three-parameter Weibull distribution* for the strong population could be used. Sheu and Chien (2005) also apply a mixed Weibull distribution in the example. They choose the shape parameter of the weak population less than 1 to make the failure rate of this population decreases over time and the shape parameter for the normal population larger than 1 to make the failure rate of the normal population increases over time since the normal unit fails due to wear out. The combined model is assumed to have a bathtub failure rate.

Model Selection

Since various models have been presented and used, *model selection* could be done among different models to compare and choose the best one for the data. Some references just specify a distribution and then make investigation, but some others, like Baskin (2002) and Wu and Su (2002), investigate model appropriateness as well. For example, the Akaike Information Criterion (AIC) has been proposed to be the criterion for the model selection, which is defined as:

$$\text{AIC} = -2\log(\text{maximum likelihood}) + 2(\text{number of parameters fitted}) \qquad (6.9)$$

The lower the AIC value is, the better the model is for the data. Baskin (2002) uses this criterion to select the appropriate model. But he also states that the criterion constructed on the basis of the root-mean-square error (RMSE) is more reliable and advantaged than AIC. He also reminds us that the *goodness-of-fit test* should be done for the selected model because some models that have better AIC or RMSE values may be rejected by this test. Wu and Su (2002) make no assumptions about the distribution at the beginning but by fitting real data to the model and then computing the measures of how well the data forms a straight line for different distribution types, they then choose the distribution type that has the highest value of the measure.

6.2.2 Inference

6.2.2.1 Bayesian Method vs Classical Method (Parameter Estimation)

In Genadis (1996), Mi (1996), Mi (1997), Yun, Lee and Ferreira (2002), Whitbeck and Leemis (1989), Mi (1994a), Sheu and Chien (2005), Lu and Hui (1998), Kim (1998), Yan and English (1997), Hui and Lu (1996), Mi (1995), and Kuo (1984), product lifetime distributions and their parameters are considered to be known before doing the burn-in test. However, in actual situations, uncertainties in the distributions and the values of the parameters always exist. In such cases Bayesian methods may be more appropriate.

Either partial or fully Bayesian methods have been discussed by Perlstein, Jarvis and Mazzuchi (2001), Nicolato and Runggaldier (1999), and Yuan and Shih (1991). Nicolato and Runggaldier (1999) update the parameters of the lifetime distribution sequentially using Bayesian method. The updated information is then used to determine the burn-in time for next *maintenance induced burn-in*. Yuan and Shih (1991) only use Bayesian to determine the mixture parameter p. Perlstein, Jarvis and Mazzuchi (2001) use a full Bayesian procedure. The uncertainty in all parameters of the model is considered and expressed through a multivariate prior distribution, in which the mixture parameter is set to be a beta distribution and the failure rate parameters for the two subpopulations have independent but stochastically ordered gamma distributions. Furthermore, posterior development when additional failure data are available prior to the burn-in process is also considered.

Different classical methods have also been used for parameter estimation. In Kim and Kuo (1998), lifetime parameters are fitted through the maximum likelihood estimation (MLE). In Yuan and Shih (1991), a Bayesian method is used for the mixture parameter, while for the other parameters, the MLE method and the Weibull plotting method are used. Baskin (2002) concludes that estimation of model parameters could be conducted by a method of the moments, a method of maximum likelihood, a method of least squares, a method of quantiles, regression and graphical methods. But he also proved in the paper that if only graphical methods are used to draw the conclusions, erroneous solutions are possible to be made.

6.2.3 Model the Process

6.2.3.1 Static Burn-in, Dynamic Burn-in and Sequential Burn-in

Static burn-in is at a constant input voltage and a constant load for a predetermined time while *dynamic burn-in* includes all the features of the static burn-in plus active control of loads during burn-in, and burn-in chamber temperature control, *etc.* Kuo and Kuo (1983) state that by applying the monitoring powers on the board in dynamic burn-in, test patterns are applied to the inputs and the outputs can be monitored.

Dynamic burn-in can exercise the device under high temperature and other severe conditions, stressing the internal components of a complex system as much as possible. It is very effective at detecting earlier failures especially for large device, but also expensive. The static burn-in is much cheaper and simpler, but it is less effective than the dynamic burn-in for large-scale integration devices.

Spizzichino (1991) gives the definition of *sequential burn-in strategy*, which is actually a special case of the dynamic burn-in. In this strategy, the burn-in duration may change in after each failure has been observed during the burn-in process and these changes follow certain rules defined by the author. He also gives the definition for optimal burn-in strategies and proves the existence of the result.

Lu and Hui (1998) apply a dynamic burn-in which is called "two-stage burn-in" test. The first stage involves a high stress level burn-in to detect the *material defects,* and the second stage involves a low stress level burn-in to screen out *manufacturing defects*. Both two-stage burn-in stresses are higher than the field use stress and the determination of the high and low burn-in stresses is also discussed in the paper.

The authors conclude that the two-stage burn-in saves more money than single-stage burn-in for some stress levels and their associated costs. Yan and English (1997) had a similar "two-stage test" investigation where in the first stage they use an excessive stress accelerated production testing *"environmental stress screening"* (ESS) to force the latent defects to surface as early failures and detect them. In the second stage, burn-in is used to detect the patent defects. Wu and Su (2002) also give some description of ESS. They state that ESS accelerates the aging of latent defects by applying severe stresses without shortening their useful life and yields more economic benefits of a higher quality and lower cost product. They also state that ESS and burn-in are occasionally confused in the literature, but despite the differences, ESS and burn-in processes can be modeled similarly. Finally, they conclude that by combining the characteristics of ESS and burn-in, burn-in test with high stress is effective in eliminating the latent failures. Despite all the descriptions and classifications about ESS, it can just be regarded as a special burn-in test with severe stress levels.

Wu and Su (2002) categorize burn-in in a more specific way. They define and divide burn-in into four types: *during burn-in* which frequently used for DRAM (dynamic random access memory) and SRAM (Static random access memory) processes; *static burn-in* which applies stresses to the samples at either a fixed level or in an elevated pattern; *dynamic burn-in*, which exercised on the samples by stressing them to simulate operating environments; and *stress burn-in*, which is conducted under an extremely stressful environment, and is always more effective than dynamic and static burn-in for defects resulting from corrosion or contamination. They also stress that *stress burn-in* can save both time and energy.

Whitbeck and Leemis (1989), Mi (1994a), Mi (1994b), Mi (1996), Mi (1997), Nicolato and Runggaldier (1999), Yun, Lee and Ferreira (2002), and Sheu and Chien (2004) discuss the case for *static burn-in*. One situation to be mention is that, in Nicolato and Runggaldier (1999), the parameters of the lifetime distribution are updated sequentially based on the failure data in the each time interval, but it is not the sequential burn-in as defined by Spizzichino (1991), because the burn-in times are not updated in-time during each burn-in process. The failure information collected is used to update the burn-in time for the next burn-in process. Perlstein, Jarvis and Mazzuchi (2001) consider the similar situation, the optimal time is selected in advance and remains fixed through burn-in process. When new failure data are available, they are used to choose new burn-in time.

6.2.3.2 Level of Production

An electronic device usually has multi-levels, for example, according to Chien and Kuo (1995), a device can be categorized into three levels: systems, subsystems and components. Among these levels, component burn-in has been investigated widely by Chi and Kuo (1989), Spizzichino (1991), Mi (1994a), Mi (1995), Mi (1996), Mi (1997), Kim (1998), Yun, Lee and Ferreira (2002) and proved to be effective. But when the components are assembled into systems, potential damages will possibly occur, and the incompatibilities among different components or subsystem might exist. Therefore, component burn-in alone will be not enough. In this case, burn-in tests on different levels, such as subsystem levels or system level should be considered.

Kuo (1984) and (1989) use only component burn-in, but from a system viewpoint. He uses system cost optimization and system reliability constraints, while the decision variables are only component burn-in periods.

Whitbeck and Leemis (1989), Chien and Kuo (1995), Genadis (1996), Kim and Kuo (1998) make more effort on this topic. Genadis (1996) considers both module and system burn-ins, the cost function has been built for both cases and the optimal burn-in time is then given under each case. Whitbeck and Leemis (1989) introduce a *pseudo-component* into the system to represent the potential damage occurring during assembly process that will affect the system mean residual life (MRL). Using simulation technique, investigations are then made on how long burn-in should be performed at each subassembly levels to optimize system MRL. According to the levels of product in the semiconductor industries, Kim and Kuo (1998) consider four burn-in policies: no burn-in, Wafer-level burn-in (WLBI), package-level burn-in (PLBI) and wafer-level burn-in (WLBI) prior to package-level burn-in (PLBI). Comparison is then made among these burn-in policies to choose an optimal one. Chien and Kuo (1995) point out that the higher-level burn-in is more effective in terms of getting rid of more potential defects and incompatibilities. They implement component, subsystem and system burn-ins. Furthermore, they also suggest that when the system reliability requirement still cannot be achieved after all the in-process burn-in and the cost for a *redundant* subsystem is acceptable, the subsystem redundancy instead of subsystem burn-in can be considered. Another point should be made here is that, Genadis (1996) considers burn-in tests at different levels separately while Whitbeck and Leemis (1989), Chien and Kuo (1995), and Kim and Kuo (1998) consider the combination.

Yan and English (1997) consider both *part-level* and *unit-level* production testing. But the speciality for this paper is that the authors use the *environmental-stress-screening test* (ESS) for the latent failures. ESS is applied on the *part-level*, which is the lowest level of materials and the ordinary burn-in test is applied on the *unit-level* which is the assembly of parts. ESS could accelerate the aging of latent defects by applying excessive stress without damaging the units. On the system level, there is neither ESS nor burn-in process. A simple flow-chart is:

$$\text{part} \xrightarrow[\text{ESS}]{\text{latent failure}} \text{unit} \xrightarrow[\text{burn-in}]{\text{patent failure}} \text{system}.$$

6.2.3.3 Repair vs. Scrap (During Burn-in)

In-time Failure Treatment (IFT)

Components that fail during burn-in test will be either repaired or scrapped. There are two types of repair: *minimal repair* and *complete repair*. *Minimal repair* means that after repair the component is just the same as it was immediately before the failure and *complete repair* means that after repair the component is as good as new. Non-repairable components are usually scrapped after failure. After *scrap* treatment is always followed by replacement and here we only consider the case of scrap with a replacement by a new product. In Kuo (1984), Sultan (1986), Mi (1994b), Genadis (1996), Mi (1996), Mi (1997), Kim and Kuo (1998), Chang (2000), Yun, Lee and Ferreira (2002), Sheu and Chien (2004), failure can be detected and treatment can be done in time after it happens. During its failure status, the burn-in test is usually stopped. If minimal repair is applied as in Kuo (1984), Sultan (1986), Genadis (1996), Mi (1997), Kim and Kuo (1998), Chang (2000), and Wu and Su (2002), the repair time is set to be negligible and after repair the burn-in timer will start from the exact time point when the component fails. If complete repair (Mi (1994b), Mi (1996), Mi (1997)) or scrap (Mi (1997), Yun, Lee and Ferreira (2002)) with replacement apply, the timer will be reset to zero after each repair/replacement treatment. All these situations have the characteristics that the component failures during burn-in are followed immediately by a failure treatment, and we defined them as "in-time failure treatment" (IFT).

Under IFT *minimal repair* treatment, the total burn-in time until the first component survives the burn-in test is just t_b and if under *complete repair/ replacement* treatment, the total burn-in time until one product survives the burn-in time can be best represented by Mi (1996):

$$\sum_{1}^{\eta-1} X_i + t_b \,,$$

where X_i is iid random variable with lifetime X, and $\eta - 1$ is the number of shop repairs/replacements until the first device survives burn-in. Mi (1996) shows that

$$E\left[\sum_{1}^{\eta-1} X_i + t_b\right] = \frac{\int_0^{t_b} R(t)\,dt}{R(t_b)}\,. \tag{6.10}$$

If the burn-in process is multi-stage or includes more steps, different burn-in failure treatments will be used. Lu and Hui (1998) investigate the two-stage burn-in procedure. For the material defects detected in the first stage (high stress), the author states that "usually material defects and other defects detected at high stress levels will not be repaired" and so in this paper the failed products from this first period are scrapped. The second stage is under low stress level and is designed for screening the manufacturing defects which assumed to be repaired. In Yan and English (1997), replacement is taken place to the failures in ESS process and minimal repair to the failures in burn-in process.

Most papers investigate the repairable and non-repairable products separately, but Sheu and Chien (2004, 2005) state that in practical situations, a general repairable product with a *general failure model* seems more realistic. In these two papers, the failures of the products may be divided into two types. Occurrence probabilities w and $1 - w$ are also assigned to each of the two types of failures. Type I failure is assumed to be minor and can be removed by a minimal repair, whereas type II failure is catastrophic and can only be removed by complete repair/replacement. The burn-in procedure for this kind of products stops when no type II failure occurs during time t_b for the first time. All failures are instantly detected and repaired is also assumed.

After-Burn-in Failure Treatment (AFT)

Most works concern the IFT situation, but there are some practical situations that the burn-in failure treatments will not take place until the end of the preset burn-in period, which is defined as "after burn-in failure treatment" (AFT). This is intuitively not an economic way of dealing with the failures in both cost and facility capacity concerns. It may, however, be appropriate under several situations, such as when a batch of components are burned-in together, or the failure cannot be detected during burn-in. Limited papers (Washburn, 1970; Hui and Lu, 1996; Kim, 1998; Perlstein, Jarvis and Mazzuchi, 2001) concern AFT situations. In these papers, the authors calculate the expected failure number for a single product during burn-in as $F(t_b)$ which is an approximate estimation for failure numbers. Further and more thorough discussion about AFT will be presented in the cost criteria section.

Expected Failure Numbers

For all the situations (IFT and AFT) discussed above, the expected failure numbers for each burn-in board during burn-in process are calculated as follows (Table 6.1):

1. IFT Minimal repair: $\int_0^{t_b} r(s)\,ds$

2. IFT Complete repair or scrap with replacement: $\dfrac{F(t_b)}{R(t_b)}$

3. AFT repair/scrap: $F(t_b)$

6.2.3.4 Warranty Treatment and Strategy

Instead of discussing only the strategy for burn-in process, some papers also suggest that the warranty policy will also drastically influence the profitability of the

Table 6.1 Failure treatments and the expected failure numbers during burn-in for each case

Treatment		Mean failure number for IFT (until one survives burn-in)	Mean failure number for AFT (one component put into burn-in)
Repair	Minimal repair	$\int_0^{t_b} r(s)\,ds$	$F(t_b)$
	Complete repair	$\dfrac{F(t_b)}{R(t_b)}$	
Scrap			

manufacture, and therefore the cost function should be formulated and optimized by considering both burn-in and warranty strategies.

Determination for Warranty Period Length

First of all, the direct problem is how long should the sold products be guaranteed. Sultan (1986) states that few electronic manufacturers guarantee the products for more than two years and some manufacturers usually use one year warranty period. Sultan (1986) uses the so-called *saving ratio* $S(t_w)$ to determine the "optimal warranty time." The saving ratio is defined as

$$S(t_w) = \frac{[C(0, t_w) - C(t_b^*, t_w)]}{C(0, t_w)} \tag{6.11}$$

where $C(0, t_w)$ denotes the total cost with no burn-in for warranty time t_w, and $C(t_b^*, t_w)$ denotes the total cost with optimal burn-in time t_b^* for warranty time t_w. The saving ratios $S(t_w)$ are computed over all t_w values, then the optimal warranty time t_w^* is chosen where $S(t_w)$ reaches its maximum value. Sheu and Chien (2005) also suggest using this index to study the variation in the magnitude of saving in the expected total cost with changing t_w. Instead of investigating a specific warranty period duration, setting a specific warranty period stopping condition can also be adopted.

Basic Warranty Policies

In most other cases, the warranty time t_w is set to be known in order to simplify the burn-in related calculations. However, specific warranty failure treatments or the warranty policies have been discussed widely. The simplest warranty failure treatments are similar to the ones in burn-in stage, such as repair treatment used in Mi (1994a), Genadis (1996), Lu and Hui (1998), Chang (2000), Perlstein, Jarvis and Mazzuchi (2001), and Wu and Su (2002), and replacement with burned-in iid products used in Mi (1994a), Mi (1996), and Lu and Hui (1998), but there are also some special warranty policies:

Failure-free policy and *rebate policy* are two common types of warranty policies. A *Failure-free policy* obligates the manufacturer to maintain the product free of charge during the warranty period, which means that all failed products are repaired or replaced by an iid product during the warranty period for free. The failure-free policy can be further divided into two categories, *renewing* and *non-renewing* policies. *Renewing policy* means that if an item fails within the warranty time, it is replaced by a new item with a new warranty, *i.e.*, warranty begins anew with each replacement. *Non-renewing policy* means that replacements of a failed item do not alter the original warranty. A *rebate policy* obligates the manufacturer to refund a fraction of the purchase price if the product fails within the warranty period. In the rebate policy, the rebate amount $A(t)$, is a function of the failure time for the product under warranty. In Sheu and Chien (2005), the authors chose $A(t)$ to be a linear function of t, *i.e.*,

$$A(t) = \begin{cases} kC_s(1 - \frac{\alpha t}{T}), & \text{for} \quad 0 \le t \le T \\ 0, & \text{for} \quad t > T \end{cases} \tag{6.12}$$

where $0 < k \le 1, 0 \le \alpha \le 1$, when $\alpha = 0$, it is called the *lump sum rebate policy* and when $\alpha = 1, k = 1$, it is called the *pro rata rebate policy*. Mi (1997), and Sheu and Chien (2005) discuss this kind of policies. Mi (1997) considers different combinations of the burn-in type (minimal repair or replacement) and the warranty type (failure-free or rebate), and finally concluded that the optimal burn-in time to minimize the total mean cost function never exceeds the first change point of the bathtub failure rate curve. Sheu and Chien (2005) also derive the expected total cost per unit for various warranty policies (failure-free policies with and without renewing and rebate policy).

Yun, Lee and Ferreira (2002) consider another warranty policy, the *cumulative free replacement warranty* (*CFRW*), which is similar to the non-renewing failure free policy. Instead of covering each product for period t_w, a lot of n items which are used one at a time are warranted for a total time nt_w. Free replacements will be applied if the total lifetime of the whole batch is less than nt_w, but the warranty time will not be renewed after each replacement.

Maintenance Strategy

Maintenance as another mission time strategy is also discussed by Mi (1994b) and Nicolato and Runggaldier (1999). Sheu and Chien (2005) note that Mi (1994b) is the first to build the model to consider burn-in and maintenance simultaneously for nonrepairable and repairable devices. Two combinations of maintenance and repair are considered: *age replacement with complete repair* and *block replacement with minimal repair*. The cost structures related to all the burn-in, maintenance and repair strategies are then investigated. Later on, Nicolato and Runggaldier (1999) also apply maintenance in warranty time and furthermore introduce the maintenance-induced-burn-in based on the fact that after each maintenance implementation there is a "similar infant mortality period" for the component.

6.2.4 Model Optimization

6.2.4.1 Criteria

One major task related to burn-in is to decide how long the burn-in process should be continued. Kuo (1984) states "with insufficient burn-in, high initial failure rates cause high field repair costs. Yet, with excessive burn-in, the reduced failure rate will be at the cost of increased capital and recurring costs, optimal burn-in is necessary for devices in critical applications and for non-repairable systems." This statement is mainly concerned with profits. Reliability level for the field use is another major concern as low reliability level will result in the credibility loss for the manufacturer. The "best" burn-in time (usually called the *optimal burn-in time*) is calculated based on certain specific optimization criteria given by the manufacturers and the decision makers according to their specific situation and interest. There are mainly two types of optimization criteria: cost and reliability.

Cost Function

Kuo and Kuo (1983) state that as profit is so important to the manufacturers, cost reduction is regarded as the central role for burn-in. Washburn (1970) presented a mathematical model for optimal burn-in time to minimize total cost and sub-sequently many researchers have presented burn-in procedures for various cases. A cost criterion is represented by a cost function. In different papers, the specific forms for the cost function may be quite different. Some typical cost models can be found in Kuo and Kuo (1983). In this chapter, the cost function is divided into basic items (which are the main focus in most of the papers) and extra items (which have received some attention). Some typical forms for each of these components will be discussed.

Basic Form

A cost objective function mainly includes three parts: fixed costs including manufacturing and burn-in set-up costs, cost during burn-in period (including burn-in costs and repair/scrap cost for failure components) and cost within warranty time in field use. It usually has the form:

$$C(t_b) = c_0 + c_1(t_b) + c_2(FDBI) + c_3(FDWT) \tag{6.13}$$

where c_0 denote the manufacturing and burn-in set-up costs, $c_1(t_b)$ is the burn-in cost function which depends on the burn-in time, $c_2(FDBI)$ is the burn-in failure treatment cost function which depends on the failure treatment type during burn-in, (FDBI: failure during burn-in) and $c_3(FDWT)$ is the warranty cost function, which depends on the failure and warranty policy during warranty time (FDWT: failure during warranty time). Some papers have the simple cost functions exactly following this basic pattern, such as Genadis (1996), Mi (1997), and Yun, Lee and Ferreira (2002).

Some of the typical forms for the above can be obtained using the following definitions. Some cost factors:

C_0 fixed costs including manufacturing and burn-in set-up costs
C_1 burn-in cost per unit time per unit
C_2 shop repair/replacement cost per failure

C_3 repair/replacement cost per failure during warranty period

$N(t_b)$ number of failures during burn-in

$N_{t_b}(t_w)$ number of failures during warranty time t_w, with burn-in time t_b

Of course these are only the most simplistic descriptions for the cost factors. In actual situations they are more complicated. Washburn (1970) describes these cost factors in much more details. For his model, C_1 is the prorated cost of the facility, which is amortized over its service life and includes capital investment, interest, overhead maintenance, and all related costs for the physical plant and equipment. Added to these are the cost of all direct and indirect labour, and a reasonable rate of return on investment. C_2 is the cost per unit that failed during the burn-in period. It includes all the associated costs of producing a unit and placing it in the burn-in facility and includes material costs but not a profit. Genadis (1996) also describes that the repair costs are more difficult to deal with, as they will depend on the nature of the failure, and in such cases he suggest that the repair costs, like C_2C_3, should be considered as a mean cost over all types of repairs and replacement. Furthermore, since C_0 is constant, it usually has no influence on the result of the optimal burn-in time. Therefore, in some papers like Washburn (1970), Spizzichino (1991), and Mi (1996), it is simply neglected by the authors.

Most works concern the "in-time failure treatment" (IFT). A characteristic of these existing cost models is that they calculate the cost based on "per-item-output" point of view: if minimal repair is applied, the expected burn-in and failure treatment costs until the first unit survives burn-in are:

$$c_1(t_b) = C_1 t_b \tag{6.14}$$

$$c_2(FDBI) = C_2 E[N(t_b)] = C_2 \int_0^{t_b} r(s)\,ds \tag{6.15}$$

Kuo (1984), Sultan (1986), Genadis (1996), Mi (1997), Chang (2000), and Wu and Su (2002) use these forms.

If complete repair or scrap with replacement is applied, the expected burn-in and failure treatment costs until the first unit survives burn-in can be best represented by:

$$c_1(t_b) = C_1 E\left[\sum_{i=1}^{N(t_b)} X_i + t_b\right] = C_1 \frac{\int_0^{t_b} R(s)\,ds}{R(t_b)} \tag{6.16}$$

$$c_2(FDBI) = C_2 E[N(t_b)] = C_2 \frac{F(t_b)}{R(t_b)} \tag{6.17}$$

where X_i is iid random variable with product lifetime X, and $\eta - 1$ is the number of shop repairs/replacements until the first device survives burn-in. Mi (1994b), Mi (1996), Mi (1997), and Yun, Lee and Ferreira (2002) follow these forms.

As previously mentioned, there are some situations defined as "after burn-in failure treatment"(AFT) where the burn-in failure treatments will not take place until the end of the preset burn-in period. AFT is appropriate under several actual situations: (a) due to the limitation of the burn-in facility there is no in-time failure detector or monitor in the burn-in board; (b) for some performance-based failure types, it is hard to detect the failure during burn-in process; (c) when a batch of components is burned-in together, it is inconvenient to find and remove the failed products instantly during burn-in process. When any of these situations occurs, the failures occurring during burn-in process must be treated by the so defined AFT. Unfortunately, very limited cost models (Washburn (1970), Hui and Lu (1996), Kim (1998), and Perlstein, Jarvis and Mazzuchi (2001)) can be used for this situation.

In determining the expected cost for the AFT cases, there are mainly two different situations. In the first situation, the burn-in process is terminated due to product failure, and the effective burn-in time is just the lifetime of that component, so the expected total cost when one component is put into burn-in test at the beginning is (Hui and Lu (1996)):

$$c_1(t_b) = C_1 \left\{ E[X|X < t_b]P(X < t_b) + t_b P(X \geq t_b) \right\} = C_1 \left(\int_0^{t_b} s\,dF(s) + t_b R(t_b) \right)$$

(6.18)

In the second situation, even after product fails during burn-in, the burn-in process will still be continued for the failed component until the end of the preset period t_b. This case is a bit wasteful, but is quite appropriate in a batch burn-in test or that the failure cannot be detected during burn-in. The mean burn-in cost when one component is put into burn-in test at the beginning is (by Washburn (1970) and Perlstein, Jarvis and Mazzuchi (2001)):

$$c_1(t_b) = C_1 t_b$$

(6.19)

For both of the situations, the probability of failure when one component is put into burn-in test at the beginning is the same, $F(t_b)$. So, the expected failure treatment cost when one component is put into burn-in test at the beginning is:

$$c_2(FDBI) = C_2 E[N(t_b)] = C_2 F(t_b)$$

(6.20)

Washburn (1970), Hui and Lu (1996), and Perlstein, Jarvis and Mazzuchi (2001) all apply this form, but Washburn (1970) and Perlstein, Jarvis and Mazzuchi (2001) use $C_2 N F(t_b)$ instead of $C_2 F(t_b)$ for batch burn-in with the batch size N. Instead of using the above forms for $c_1(t_b)$ and $c_2(FDBI)$, Spizzichino (1991) use a general item $cF(t_b)$ representing the whole burn-in cost where c is the general loss when component fails during burn-in.

The above cost models for AFT are constructed on the basis of a "fixed input" product number (one or N) that are tested under the burn-in process. This is quite different to the IFT cost models which are calculated on the "per-item-output" point of view. Some properties that will be resulted from this are discussed in the next section.

The item $c_3(FDWT)$ usually has the form of $c_3(FDWT) = C_3 E[N_{t_b}(t_w)]$. When discussing this item, it is assumed that the warranty policy is failure free, and the treatments are the ones similar to the during-burn-in failure treatment (*i.e.*, minimal/complete repair or replacement), but in the warranty period there will be some differences. Other situations, such as rebate policy, good will gain/loss items will be discussed in the extra item section or the special case section. Similar to the discussion above, the failure treatment cost $c_3(FDWT)$ will be represented in the following forms based on different types of failure treatments and warranty policies during warranty time:

Minimal Repair During Warranty Time

Similar to the minimal repair during burn-in period, the expected failure treatment cost until the first unit survives warranty time, $c_3(FDWT)$ is:

$$c_3(FDWT) = C_3 E[N_{t_b}(t_w)] = C_3 \int_0^{t_w} r_{t_b}(s)\,\mathrm{d}s \tag{6.21}$$

Kuo (1984), Sultan (1986), Genadis (1996), Mi (1997), Chang (2000), and Wu and Su (2002) use this form. But in Wu and Su (2002), the failure rate used in the integral also includes the latent defects failure rate during warranty.

Complete Repair/Replacement During Warranty

Different from the situation during burn-in, when complete repair/replacement is applied during the warranty period, there are two different situations: renewing and non-renewing failure free policy. When *renewing policy* is applied, warranty time begins anew after each failure repair/replacement. This is the same case with the during-burn-in period and the calculation for expected failure treatment cost is also similar:

$$c_3(FDWT) = C_3 \frac{F_{t_b}(t_w)}{R_{t_b}(t_w)} \tag{6.22}$$

where the function $F_{t_b}(t)$ represents the cumulative distribution of $X - t_b | X > t_b$ (denoted as X_{t_b}), and $R_{t_b}(t)$ is the survival function of X_{t_b}. The above formula is actually suitable for the complete repair situation. When replacement treatment is applied, a new component which has been burned-in and has an iid distribution with X_{t_b} is needed, so the above cost should also include the manufacturing and burn-in cost for this new component. In this case, the above expected failure treatment cost will be:

$$c_3(FDWT) = (C_3 + v(t_b)) \frac{F_{t_b}(t_w)}{R_{t_b}(t_w)} \tag{6.23}$$

where $v(t_b)$ represents the mean manufacturing cost per item including burn-in.

When *non-renewing policy* is applied, which is usually the case; the original warranty time will not be changed after each failure treatment. Thus, the expected failure number during warranty period t_w is computed as

$$E(N_{t_b}(t_w)) = \sum_{i=1}^{\infty} P(Z_i \leq t) = \sum_{i=1}^{\infty} F_{t_b}^{(i)}(t_w) \tag{6.24}$$

where $Z_i = X_{t_b,1} + X_{t_b,2} + \ldots + X_{t_b,i}$, $\{X_{t_b,j}\}$ is a random variable sequence iid with X_{t_b}, $i.e.$, $X - t_b | X > t_b$, and $F_{t_b}^{(i)}(t_w)$ is the i-fold convolution of $F_{t_b}(t_w)$ defined as:

$$F_{t_b}(t_w) = \frac{F(t_b + t_w) - F(t_b)}{R(t_b)} \tag{6.25}$$

Thus, the expected failure treatment cost when complete repair is applied during warranty is:

$$c_3(FDWT) = C_3 \sum_{i=1}^{\infty} F_{t_b}^{(i)}(t_w) \tag{6.26}$$

and the expected failure treatment cost when replacement is applied during warranty period is

$$c_3(FDWT) = (C_3 + v(t_b)) \sum_{i=1}^{\infty} F_{t_b}^{(i)}(t_w) \tag{6.27}$$

Yun, Lee and Ferreira (2002) follow this form. This form uses the cumulative free replacement warranty, $i.e.$, instead of covering each product for period t_w, a lot of N items which are used one at a time are warranted for a total time Nt_w. So the expected failure treatment cost in this paper is

$$c_3(FDWT) = (v(b) + C_3)\frac{\sum_{N}^{\infty} F_{t_b}^{(i)}(Nt_w)}{N} \tag{6.28}$$

After Warranty Treatment

Similar to the after burn-in treatment, for this situation the expected failure treatment cost is mainly represented in the following form:

$$c_3(FDWT) = C_3[F(t_b + t_w) - F(t_b)] \tag{6.29}$$

Hui and Lu (1996), Spizzichino (1991), and Perlstein, Jarvis and Mazzuchi (2001) follow this form. But in Hui and Lu (1996), C_3 is not regarded as a constant but a function of time t, so it has the form of

$$c_3(FDWT) = \int_{t_b}^{t_b + t_0} C_3(s - t_b)dF(s) \tag{6.30}$$

Extra Items

Besides the four parts in the basic form, there are some extra items discussed in some of the papers. These extra items may seem not so necessary as the ones mentioned before, but sometime play very important role. The extra items discussed in the following are: *gain/loss of the goodwill, sale price,* and *rebate amount for rebate warranty policy.*

Gain/Loss of the Goodwill

Reputation is always important for manufacturers. Good reputation always brings good profit, especially for long-run business. Thus, the gain or the loss of the good-will should also be involved in the cost function, and should be represented by the quality and the performance of the product being used by the customer. Kuo (1984), Spizzichino (1991), Genadis (1996), Mi (1996), Hui and Lu (1996), Kim and Kuo (1998), Lu and Hui (1998), and Sheu and Chien (2004) discuss this item. Genadis (1996) gives an general idea of what the goodwill depends. It states that the *gain/loss of goodwill* depends on when the failure occurs, and also on how many failures any customer experiences.

Goodwill gain/loss is represented in different forms. Kuo (1984) and Kim and Kuo (1998) use a parameter l to represent the level of loss-of-credibility. The bigger l is, the higher penalty for the credibility loss is. The goodwill loss item in this case is then:

$$(1+l)C_3 \int_0^{t_w} r_{t_b}(s)\, ds\,.$$

Mi (1996) express the goodwill the manufacturer will gain in the cost function to be proportional to the component's mean residual life (MRL) in the field use, *i.e.*, $K \cdot MRL$. Sheu and Chien (2004) also express the gain part as proportional to the mean residual life (but for a specific kind of failure type) in the cost function as $K \cdot MRL$. Hui and Lu (1996) considers the loss of the goodwill in the cost function, which has the form $\int_{t_b}^{t_b+t_w} L(s-t_b)\, dF(s)$, where $L(s-t_b)$ is a function of time, the authors also give the typical functions for $L(t)$:

Constant cost:

$$L(t) = \begin{cases} C, & 0 \le t \le t_w \\ 0, & t > t_w \end{cases} \tag{6.31}$$

Linear decreasing cost

$$L(t) = \begin{cases} C(1 - t/t_w), & 0 \le t \le t_w \\ 0, & t > t_w \end{cases} \tag{6.32}$$

Exponentially decreasing cost

$$L(t) = C e^{-\beta t^{\alpha}}, \quad t \geq 0 \tag{6.33}$$

The authors make more effort in Lu and Hui (1998) where repairable and non-repairable failures in field use are both considered. For the repairable field use, loss of goodwill is proportional to the mean number of failures during warranty which has the form $LN_{t_b}(t_w)$, and for non-repairable field use it is a function of field failure time, $L(X_{t_b})$.

Spizzichino (1991) considers both gain and loss parts according to the operative life time of the component. A general loss C will be incurred if component fails during the warranty time, and a gain K will be incurred if component survives the warranty time. The cost function has the form: $C \cdot P(t_b < t \leq t_b + t_w) - K \cdot P(t > t_b + t_w)$. Though these items are not directly interpreted as related to goodwill, they can be considered as another possible choice for representing the goodwill gain/loss.

Sale Price

Another item is the *sale price* of the units that survive the burn-in process. Washburn (1970) considers the sale price in the utility model:

$$C(t_b) = C_1 t_b + C_2 N F(t_b) + C_s \frac{K_r}{P_E(t_b)} \tag{6.34}$$

where C_s is the sale price of the units that survive the burn-in process and supposed to be essentially the same as C_2 with the additional cost of any handling, testing and shipping associated with delivering the end product to the customer, K_r represents the minimum number of units required for the customer to satisfy the requirements under the assumption that the system effectiveness $P_E(t_b)$ of each unit is 1.0. $\frac{K_r}{P_E(t_b)}$ is then the number of devices required by the customer, and it is easy to derive that the minimum number of units N to be placed on burn-in is $N = \frac{K_r}{P_E(t_b) R(t_b)}$. This model is also reviewed later by Kuo and Kuo (1983). It is worthwhile to mention that this part should especially be considered when the number of the components that survive the burn-in test in the cost function is not fixed and dependent on the burn-in time t_b.

Rebate Amount for Rebate Policy

As stated before, different from the failure free warranty policy, the rebate policy obligates the manufacturer to refund a fraction of the purchase price if the product fails within the warranty period. The rebate amount function $A(t)$ may be defined in varied ways. Mi (1997) involves the rebate amount warranty cost in the cost function as $\int_0^{t_w} A(t) \, dF_{t_b}(t)$.

Special Cases and Cost Models

The above basic and extra items are suitable for most of the usual cases, while practically there are many different real situations so that many different cost functions may be built. Now consider the cost models from another point of view. We do not decompose the cost functions into items and analysis them separately, but discuss them as a whole, to see what different special situations they are suitable to and what the characteristics they have.

Compared with the above usual forms, the cost functions in Yan and English (1997), Kim (1998), Lu and Hui (1998), Sheu and Chien (2004), and Sheu and Chien (2005) seem more complicated.

In Yan and English (1997) and Lu and Hui (1998), the total costs are related to a *two-stage burn-in* procedure. The first stage is with high stress level and the second with low stress. The purpose of the two-stage burn-in is to detect different type of failures in each stage. For example, the high stress level burn-in is to detect the material defects in Lu and Hui (1998) and latent defects in Yan and English (1997), while low stress level burn-in is to detect manufacturing defects in Lu and Hui (1998) and patent failures in Yan and English (1997).

Another special cost function is built by Kim (1998) and concerns the *burn-in error* situation. It is said to be the first article which explicitly modelled the burn-in error. The *burn-in error* may occur for various reasons, such as it is impossible to eliminate all weak components through burn-in due to a nonzero proportion of defectives of the components or that some of the components that fail during burn-in will be from the weak population while others will be from the main population. Associated probability models and cost models are formulated, and the cost function is built based on them.

Sheu and Chien (2004) and Sheu and Chien (2005) apply the *general failure model*, where two types of failures (type I and II) may occur with probabilities w and $1 - w$. Different types of failures can be removed by different treatments (minimal repair for type I and complete repair/replacement for type II). Based on this model, they investigate in Sheu and Chien (2004) two cost functions; one considers the "stop upon type II failure occurs" warranty policy (the warranty period continues until the first occurrence of type II failure) and the gain of goodwill part, the other considers the long-run average cost case. The optimal burn-in times are then proved to be less than the first change point t_1. Later on, in Sheu and Chien (2005) further extends the cost model to the case in which various warranty policies are considered. The conditions when the burn-in test is needed are also derived.

Based on the fact that in many cases, assumptions will be made in order to apply the cost model and many of them are inappropriate due to practical concerns. Su and Wu (2001) study this issue. They apply a new way of building the cost model without any assumptions. Based on the fact that in the actual manufacturing process, a new electronic product is always extended from an old product, the authors apply a neural net work-based approach to determine the optimal burn-in time and cost without any assumptions and the effectiveness of the proposed approach is demonstrated by a case study.

In most papers, the cost model is optimized as a 1-variable problem, the burn-in time t_b is the only interest variable. But sometimes it may be multi-variable optimization problem, like in Sultan (1986) and Mi (1994). Sultan (1986) builds a 2-variable optimization problem where the two variables are burn-in time t_b and the warranty time t_w, and then the optimal values for the two variables are determined sequentially. Firstly t_w is determined by maximizing *saving ratio* $S(t_w)$ and then t_b is determined by minimizing the cost function. Mi (1994) constructs the objective cost function considering both the burn-in time and the maintenance age/interval, and optimizes it as a *2-variable*-problem.

Before searching for the optimal burn-in time for the cost function, it should be determined if there is a necessity of doing a burn-in test. To determine this, check the value of $C(t) - C(0)$; if it is always bigger than zero for any t, *i.e.*, the cost of doing a burn-in test with time t is always more than the cost without burn-in, then a burn-in test should not be applied. Sheu and Chien (2004) also state that it is intuitive that burn-in is useful for a unit with a high infant mortality rate. As an evidence, they prove for their cost model that when the initial failure rate $r(0)$ is large, the burn-in procedure is really beneficial (*i.e.*, $t_b^* > 0$)

Cost Functions Overview

Although varied cost functions have been constructed and used to determine the optimal burn-in time, it can hardly be said that any one is omnipotent and is an exact model. One should anyway build the cost functions according to the practical situations and test them in practical use. Table 6.2 shows different cost functions used in the references.

Papers Sultan (1986), Spizzichino (1991), Mi (1994b), Genadis (1996), Mi (1996), Mi (1997), Yan and English (1997), Lu and Hui (1998), Nicolato and Runggaldier (1999), Perlstein, Jarvis and Mazzuchi (2001), Yun, Lee and Ferreira (2002), Sheu and Chien (2004), Sheu and Chien (2005) take the cost criteria as the *only* criteria to determine the optimal burn-in time.

Reliability Criteria:

Reliability Criteria Overview

Although very rarely is it true that "money is no object" in the business world as stated by Washburn (1970), in some cases, product reliability instead of cost may become the main concern. Such cases occur in the military or aerospace industry where the "high quality design and production" is needed and so the requirement for the reliability is very high. In these cases, burn-in test is mainly being done to serve the purpose of enhancing the reliability and the cost reduction may turn into the secondary concern.

But "reliability criteria" is a very general expression. Different reliability measures can be considered for the burn-in test. Guess, Walker and Gallant (1992)

Table 6.2 Cost models in literature

Reference	C_0	$C_1(t_b)$	$C_2(FDBI)$	$C_3(FDWT)$ Failure free policy	Gain/loss of the goodwill (G/L); rebate policy amount (R); Sale price (S)
Sultan (1986); Mi (1997)	✓	$C_1 t_b$	$C_2 \int_0^{t_b} r(s)\,\mathrm{d}s$	$C_3 \int_0^{t_w} r_{t_b}(s)\,\mathrm{d}s$	✗
Kuo (1984) $C_s = \sum C_c$	✓	$C_1 t_b$	$C_2 \int_0^{t_b} r(s)\,\mathrm{d}s$	$C_3 \int_0^{t_w} r_{t_b}(s)\,\mathrm{d}s$	$(1+l)*$ $C_3 \int_0^{t_w} r_{t_b}(s)\,\mathrm{d}s$ (L)
Genadis (1996) multilevel burn-in	✓	$C_1 t_b$	$C_2 \int_0^{t_b} r(s)\,\mathrm{d}s$	$C_3 \int_0^{t_w} r_{t_b}(s)\,\mathrm{d}s$	✗
Chang (2000)	✓	$C_1 t_b$	$C_2 \int_0^{t_b} r(s)\,\mathrm{d}s$	$C_3 \int_0^{t_w} r_{t_b}(s)\,\mathrm{d}s$	✗
Wu and Su (2002)	✓	$C_1 t_b$	$C_2 \int_0^{t_b} r(s)\,\mathrm{d}s$	$C_3 \int_{t_b}^{t_b+t_w} s(t)\,\mathrm{d}t,$ $s(t)$ is the failure rate including *latent defects* during warranty period.	✗
Mi (1996)	✗	$C_1 \dfrac{\int_0^{t_b} R(t)\,\mathrm{d}t}{R(t_b)}$	$C_2 \dfrac{F(t_b)}{R(t_b)}$	✗	$-K \dfrac{\int_{t_b}^{\infty} R(t)\,\mathrm{d}t}{R(t_b)}$ (G)
Mi (1997)	✓	$C_1 \dfrac{\int_0^{t_b} R(t)\,\mathrm{d}t}{R(t_b)}$	$C_2 \dfrac{F(t_b)}{R(t_b)}$	✗	$\int_0^{t_w} A(t)\,\mathrm{d}F_{t_b}(t)$ (R)
Yun, Lee and Ferreira (2002) CFRW batch N	✓	$C_1 \dfrac{\int_0^{t_b} R(t)\,\mathrm{d}t}{R(t_b)}$	$C_2 \dfrac{F(t_b)}{R(t_b)}$	$(v(b)+C_3)\dfrac{\sum\limits_{N}^{\infty} F_{t_b}^{(i)}(Nt_w)}{N}$	✗
Washburn (1970)	✗	$C_1 t_b$	$C_2 N F(t_b)$	✗	$C_s \dfrac{K_r}{P_{1:}(t_b)}$ (S)
Spizzichino (1991) Sequential burn-in	✗	$cF(t_b)$, c is the general loss when component fails during burn-in		$C[F(t_b+t_w)-F(t_b)]-KR(t_b+t_w)$, C is the general loss when failure occurs during warranty, and K is the gain factor. (G/L)	
Hui and Lu (1996)	✓	$C_1\left(\int_0^t s\,\mathrm{d}F(s)C_2 F(t_b)+t_b R(t_b)\right)$		$\int_{t_b}^{t_b+t_o} C_3(s-t_b)\,\mathrm{d}F(s)$	$\int_{t_b}^{t_b+t_w} L(s-t_b)\,\mathrm{d}F(s)$ (L)
Perlstein, Jarvis and Mazzuchi (2001) batch N burn-in	✓	$C_1 N t_b$	$C_2 N F(t_b)$	$C_3 N[F(t_b+t_w)-F(t_b)]$	✗
Yan and English (1997)		Special case for *two-stage test*: ESS and burn-in. ESS: part-level, latent failures, replacement Burn-in: unit-level, patent failures, minimal repair Based on basic form: $C_1(t_b)+C_2 E[N(t_b)]+C_3 E[N_{t_b}(t_w)]$			✗
Lu and Hui (1998)		Special case for *two-stage burn-in*: high and low stress tests. High stress burn-in: material defects (non repairable). Low stress burn-in: manufacturing defects (repairable). Two cost functions are for non-repairable field failure and repairable field failure separately.			$L(X-t_b)$ for non-repairable field failure; $LN_{t_b}(t_w)$ for repairable field failure. (L)

Table 6.2 (continued)

Reference	C_0 $C_1(t_b)$	$C_2(FDBI)$	$C_3(FDWT)$ Failure free policy	Gain/loss of the goodwill(G/L); rebate policy amount(R); Sale price(S)
Sheu and Chien (2004)	Special case for *general failure model*, burn-in until first component survives burn-in without type II failure; warranty until first type II failure happens		$C_3(\frac{1}{w}-1)$ w is the proportion of type II failure in general failure model	$-K\frac{\int_{t_b}^{\infty}R(t)dt}{R(t_b)}$, here $R(t)$ is type II failure time survival function. (G)
Sheu and Chien (2004)	Special case for general failure model, burn-in until first component survives burn-in without type II failure; warranty until t_w ends. Long-run average cost $C(t_b) = \lim_{t_w \to \infty} \frac{EC(t_b,t_w)}{t_w} = \frac{v(t_b)+C_3(\frac{1}{w}-1)+C_r}{EY}$, C_r is replacement cost		✗	
Sheu and Chien (2005)	✓ $C_1 t_b$	$C_2 \int_0^{t_b} r(s)\,ds$	Special cases for *renewing (and non-) failure free warranty* and *rebate warranty* with linear rebate function $A(t)$, under *general failure model*(two types of failures)	
Kim (1998)	Special case considering the *burn-in error*: weak components survive burn-in, main components fail during burn-in: $C = (1-p)(1-p_G)C_B + pp_B(C_B+C_{BP})$ $+(1-p)p_G(C_B+C_{GF})+p(1-p_B)(C_B+C_{BF})$, (definition for p_i, C_i see ref)		✗	
Mi (1994b)	Multi-objective optimization $C(t_b,T)$ Warranty policies: Age replacement and complete repair Block replacement and minimal repair			

pointed out that each one of the measures is assessing reliability from a different point of view, but when different measures of reliability are used, the results of optimal burn-in times obtained from each measures are not necessarily the same and sometimes quite contradictory. In such a situation, designers should be quite aware when they devise the burn-in plan and choose the reliability criteria for it. The Reliability Criteria have been represented in the following forms:

Delivered Reliability (DR):

$$R^D(t_w \mid t_b) = \frac{R(t_w + t_b)}{R(t_b)} = \exp\left(-\int_{t_b}^{t_w+t_b} r(t)\right) \qquad (6.35)$$

Optimization objective is usually t_b^* to find such that:

$$R^D(t_w \mid t_b^*) = \max_{t_b \geq 0}\left\{\frac{R(t_w + t_b)}{R(t_b)}\right\} \Leftrightarrow \min_{t_b \geq 0}\left\{\int_{t_b}^{t_b+t_w} r(t)\right\} \qquad (6.36)$$

Mean Residual Life (MRL):

$$\mu(t_{\mathrm{b}}) = \frac{\int_{t_{\mathrm{b}}}^{\infty} R(s)\,\mathrm{d}s}{R(t_{\mathrm{b}})} = \frac{1}{R(t_{\mathrm{b}})} \int_{t_{\mathrm{b}}}^{\infty} s \cdot f(s)\,\mathrm{d}s - t_{\mathrm{b}} \tag{6.37}$$

Optimization objective is usually to find t_{b}^{*} such that:

$$\mu(t_{\mathrm{b}}^{*}) = \max_{t_{\mathrm{b}} \geq 0}\{\mu(t_{\mathrm{b}})\} \tag{6.38}$$

Guess, Walker and Gallant (1992) state that the time point that maximizes the MRL is usually equal or strictly earlier than the time point that minimize the failure rate. Mi (1995) then gives more specific properties about the MRL optimization:

1. If the component has a bathtub failure rate, then the associated mean residual life (MRL) has an *upside-down bathtub* shape, but the converse is not necessarily true;
2. If the component has a bathtub failure rate, then a proper burn-in can extend the MRL of the product;
3. The optimal burn-in time t_{b}^{*} never exceeds the first change point t_1, i.e., $t_{\mathrm{b}}^{*} \leq t_1$. Sheu and Chien (2004) also state that when $r(t)$ is a bathtub curve, MRL is decreasing in $t > t_1$.

Chang (2000) mentions that if the device has a unimodal failure rate, then the according MRL function has an *upside-down unimodal* pattern for some single lifetime distributions. In this case, maximizing the MRL cannot be the goal of optimal burn-in decision because the MRL is unbounded in extreme point.

Mean number of failures in field use (MNFU): $E(N_{t_{\mathrm{b}}}(t_{\mathrm{w}}))$

As mentioned before, if minimal repair is applied for the warranty time,

$$E(N_{t_{\mathrm{b}}}(t_{\mathrm{w}})) = E\left(\int_{0}^{t_{\mathrm{w}}} r(t_{\mathrm{b}} + s)\,\mathrm{d}s\right) \tag{6.39}$$

If complete repair is applied for the warranty time,

$$E(N_{t_{\mathrm{b}}}(t_{\mathrm{w}})) = \sum_{i=1}^{\infty} P(Z_i \leq t) = \sum_{i=1}^{\infty} F_{t_{\mathrm{b}}}^{(i)}(t_{\mathrm{w}}) \tag{6.40}$$

where $Z_i = X_{t_{\mathrm{b}},1} + X_{t_{\mathrm{b}},2} + \ldots + X_{t_{\mathrm{b}},i}$, $\{X_{t_{\mathrm{b}},j}\}$ is a random variable sequence iid with $X_{t_{\mathrm{b}}}$, and $F_{t_{\mathrm{b}}}^{(i)}(t_{\mathrm{w}})$ is the i-fold convolution of $F_{t_{\mathrm{b}}}(t_{\mathrm{w}})$.

Optimization objective is usually to find t_{b}^{*} such that:

$$E\left(N_{t_{\mathrm{b}}^{*}}(t_{\mathrm{w}})\right) = \min_{t_{\mathrm{b}} \geq 0}\{E(N_{t_{\mathrm{b}}}(t_{\mathrm{w}}))\} \tag{6.41}$$

Mi (1994a) noted that if the component lifetime has a bathtub failure rate and denote B^{*} as the set of all optimal burn-in times defined by the above objective function, and $t_{\mathrm{b}}^{*} \equiv \inf B^{*}$, then t_{b}^{*} never exceeds the first change point t_1, i.e., $t_{\mathrm{b}}^{*} \leq t_1$ under either replacement or repair treatment.

Besides the above three criteria, the survival function $R(t)$ and the failure rate $r(t)$ are also regarded as reliability measures in some papers as stated by Guess, Walker and Gallant (1992). For example, Guess, Walker and Gallant (1992) noted that the survival function and the delivered reliability (DR) are relevant when the interest is in the probability of surviving mission time, while mean number of failures in field use (MNFU) is relevant when the performance of surviving the mission time becomes the interest; the failure rate (FR) is a measure of "local" reliability that is related to the number of failures in a short time interval after the burn-in time, while the mean residual life (MRL) represents more of 'long term' reliability. Since delivered reliability has the form $\mathrm{DR} = \exp(-\int_{t_b}^{t_w+t_b} r(t))$, there is a close relation between FR and DR as pointed out by Guess, Walker and Gallant (1992): when the products with very short mission length t_w, the DR becomes $\mathrm{DR} \approx \exp(-t_w r(t))$ and the burn-in time t_b for minimum FR will also yield a maximum DR. This is especially true when bathtub FR is used and the interval $(t_b, t_b + t_w)$ is within the constant period.

Literature Using Reliability Criteria

Whitbeck and Leemis (1989), Yuan and Shih (1991), Mi (1994a), Baskin (2002) and Mi (1995) take reliability as the *only* decision consideration regarding to certain requirements by the manufacturers. Whitbeck and Leemis (1989) take the *system mean residual life (system MRL)* as the optimization objective; Mi (1994a) maximizes the *delivered reliability* (DR) for a given mission time and also considers the problem of minimizing the *mean number of failures* in the field use under both replacement and minimal repair cases. The latter problem of minimizing the mean number failures is then proved to be the same optimization problem as the former of maximizing the DR when the minimal repairs will be applied to the field failures. In Yuan and Shih (1991) the optimal burn-in time can only be determined after setting the goal of the *delivered reliability* (DR); Baskin (2002) uses *mean residual life* (MRL) criterion for the examples; and Mi (1995) takes MRL maximization as the objective and draws some important properties about it.

Combination of Cost and Reliability Optimizations

In a lot of papers the cost minimization and the reliability optimization are treated as unconstrained problems which are independent of each other. But solely optimizing the cost may always be reached by choosing a short burn-in time which results in a very low reliability level for the mission time use. Only considering the reliability requirement may make the cost exceed the manufacturer's budget. In such a situation, Kuo and Kuo (1983) point out that a tradeoff between cost and reliability enhancement due to burn-in is of primary concern.

Washburn (1970) combined the economic and reliability criteria into a utility model (cost function) in an implicit way: the value of the cost function is influenced by the sales price item, and the sales profit is determined by the system effectiveness

$P_E(t_b)$. Therefore, the cost and the system reliability have very close relationship and constrain each other.

Kuo (1984), Chi and Kuo (1989), Chien and Kuo (1995), Kim and Kuo (1998) discuss the problem of combining the two optimization criteria in a more direct way. The objectives in Kuo (1984) and Chi and Kuo (1989) are to minimize the system cost, which is the sum of all the components life-cycle costs, under the constraints of both system and component reliabilities. Chi and Kuo (1989) have all the constraints as in Kuo (1984) plus the facility capacity constraint. The constraints, minimal values that the reliabilities and maximum value for capacity are set up by the decision maker. The constraint for cost optimization in Kim and Kuo (1998) is only the mission reliability $R(t_w|t_b)$. Chien and Kuo (1995) combines cost optimization with system reliability.

Kuo (1984) also points out that the constraints in the optimization problem do not always play active roles and therefore solving the problem with no constraints and check the validity of the constraints afterward is preferable. Kim (1998) investigates the multi-objective optimization problem and points out that in many practical situations, decision making is complicated because of mutually conflicting goals and thus multi-objective model should be considered. Two cases are presented in this paper: maximizing delivered reliability (DR) with minimizing total average cost and maximizing mean residual life (MRL) with minimizing total average cost.

If the products have the unimodal failure rate function, some constraints should be considered when the cost optimization problem is formulated. Such a model is built and analyzed by Chang (2000). Firstly, the burn-in time should be at least past the critical time to avoid the highest hazard rate and not exceed a certain fraction of the mean time to failure (MTTF). Secondly, the post-burn-in MRL should exceed its mean time to failure MTTF, *i.e.*, $\text{MRL}(t_b) \geq \text{MTTF} = \text{MRL}(0)$. The necessity for the latter constraint is because the MRL after burn-in has the upside-down unimodal shape and may be less than the mean time to failure MTTF which is the case that the burn-in may be unnecessary.

Hui and Lu (1996) use another way of considering both criteria. They minimize the cost function as the main purpose, but finally compare the optimal burn-in time which minimizes the expected total cost with the optimal burn-in time which maximizes the mean residual lifetime. They conclude that the cost model is more beneficial based on the fact that the change in the expected cost is much bigger than the change in the mean residual lifetime when the burn-in time is large.

6.2.4.2 Optimization Method (How to Get the Optimal Burn-In Time)

There are two extreme cases where the optimal burn-in time is easy to determine as stated by Kim (1998): if the shop replacement and repair costs for components are much lower than field or downstream repair cost, then the minimal burn-in cost is obtained at the optimal burn-in time that begins its useful life region in the bathtub curve, *i.e.*, $t_b^* = t_1$, and if shop replacement and repair costs for components are higher than field or downstream repair costs, which is rarely the case, then the minimal burn-in costs is achieved with no burn-in, *i.e.*, $t_b^* = 0$.

Other than these extreme cases, the optimal burn-in time is investigated by many researchers. Washburn (1970) said that by experiential speaking, the most popular arbitrary figure for burn-in is 168 h, which happens to be exactly 1 week. Obviously, this seems too simple and arbitrary especial for nowadays the varied electronic products. Anyway, the purpose of burn-in is to eliminate infant mortality, hence Hui and Lu (1996) suggest that it is not uncommon to test the product until it reaches the change point where the product failure rate decreases to the constant level stage, even though it is not under first extreme situation. Mi (1996) and Mi (1997) conclude that the optimal burn-in times occur no later than the first change point t_1 for the criteria used in these two papers. Sheu and Chien (2004) also prove the similar result for the two cost models they proposed that $t_b^* \leq t_1$.

But in order to get the more exact result, there are several methods that have been used in the papers to search for the optimal burn-in time, such as *first and second derivatives method, graphic method, simulation technique, dynamic programming, decision tree diagram and surrogate worth trade-off method*:

For some "unconstrained cost optimization" models, the classical calculus method "*first and second derivatives*" is a common choice. Sultan (1986), Genadis (1996), Kim (1998), Sheu and Chien (2005) apply this method on the cost function to get the optimal solution. Differentiating the cost function with respect to t_b and equating it to zero yield a necessary condition for extreme cost value, and the second derivative is greater than zero is the condition for minimal., *i.e.*:

$$\begin{cases} C'(t_b^*) = 0 \\ C''(t_b^*) > 0 \end{cases} \tag{6.42}$$

Washburn (1970) states that this method is "more tedious and less rewarding than graphic method" which graphs the cost function directly to find its minimum. Washburn (1970) states that the advantage of the graphic method is that it reveals all of the relative extrema, inflection points, and other critical points that exist in the interested range, and hence it can be seen not only where t_b^* lies but also how critical it is. Several references apply the graphic method for either finding the optimal burn-in time or other analytic purposes. For example, Yun, Lee and Ferreira (2002) plot the "burn-in time–mean cost" graphs for each distribution case, the optimal times can be chosen according to the plots.

Whitbeck and Leemis (1989) use *simulation technique*, system lifetimes at each burn-in time level are simulated for the decision makers and can be chosen as an optimal result. Spizzichino (1991) and Nicolato and Runggaldier (1999) use the *dynamic programming approach*. Spizzichino (1991) proves the existence of the result relating to the optimal burn-in strategy by applying general ideas of dynamic programming and in Nicolato and Runggaldier (1999) both burn-in and maintenance timing are considered, the optimization problem is defined and solved by a two-level Dynamic Programming approach. Chi and Kuo (1989) follow the *decision tree diagram*, solving the cost optimization problem with no constraints first and then check the reliability and capacity constraints step by step. Kim and Kuo (1998) propose a three-stage approach which calculates the costs of the four burn-in policies (no burn-in, WLBI, PLBI and WLBI prior to PLBI) successively and makes decision for burn-in time.

For the multi-objective optimization model used in Kim (1998), the decision makers must compromise in their preferred solution. In this paper, the author states that the *surrogate worth trade-off* (SWT) method is very powerful for solving the multiple objective function burn-in problems and thus suggests using it to find the preferred optimal solution.

Due to the fact that new products are often developed on the basis of the old (base) products and therefore have close relationship with the base ones, Su and Wu (2001) adopt a back-propagation (BP) neural network-based approach to determine the optimal burn-in time and the cost without any assumptions.

6.3 Development of "After Burn-in Failure Treatment" (AFT) Cost Model

6.3.1 Why the New AFT Model?

From the previous review, especially the cost function part, we know that most works are concerned with the "in-time failure treatment" (IFT). But in the actual manufacturing process, "after burn-in failure treatment" (AFT) is appropriate under several situations.

Unfortunately, very limited cost models can be used for this situation. Moreover, none of these papers has explicitly emphasized speciality for this failure treatment. By reviewing the few existing AFT cost models (Washburn, 1970; Hui and Lu, 1996; Kim, 1998; Perlstein, Jarvis and Mazzuchi, 2001), we found that AFT models are constructed which has a "fixed input" product number (one or N) that are tested under the burn-in process. This is quite different to the IFT cost models which are calculated on the "per-item-output" point of view. Using a fixed input number may have some disadvantages: when the input product number is fixed, number of products that survive burn-in in the cost function is a variable which is a function of burn-in time t_b. Thus the profit from selling these products that survive burn-in is also dependent on time t_b. As this sales profit has influence on the total cost, it may have significant influence on t_b through the cost function. Not including the sales profit item will quite likely result in the overestimation of the optimal burn-in time t_b^*. On the contrary, the cost model constructed on "per-item-output" basis, need not consider this extra item.

Furthermore, the expected total cost in Perlstein, Jarvis and Mazzuchi (2001) has the form:

$$C(t_b) = C_0 + C_1 N t_b + C_2 N F(t_b) + C_3 N [F(t_b + t_w) - F(t_b)] \qquad (6.43)$$

This is a batch burn-in example where N is the batch size. It uses the item $N[F(t_b + t_w) - F(t_b)]$ for the expected number of failures during the warranty period. This is actually only calculating the number of the "first-time failure". If products are repaired/replaced, the repaired/replaced product may fail again during warranty period. This "second" failure is not considered in this cost model and therefore this expected warranty failure number cannot deal with the case of the "multi-failure" situation. It is obviously unrealistic that the manufacturers will not repair/replace a field failure during the warranty period.

From all the discussion above, we can see that the cost model for AFT situation is in need of further investigation. A cost model that could cover all the above disadvantages will now be developed.

In addressing the above issues, first a new cost function is constructed which has a *fixed* output product number (one unit) that survive the burn-in process. By taking this approach, sales profit problem can be addressed. Secondly, the "multi-failure" situation for warranty period can then be addressed by applying the traditional form for the warranty failure treatment, *i.e.*, minimal repair will be performed on the failures.

6.3.2 New Cost Model Construction

The new cost model is constructed on the basis of "per-item-output" point of view which calculates how much one should pay for one product surviving both burn-in process and the warranty time. This study considers a repairable product. Complete repair is assumed to be applied to each burn-in failures and minimal repair is assumed for warranty time failure.

Assumptions for the new cost model:

General

1. Sales price C_s for the product is determined and fixed by the market.
2. All cost functions are twice differentiable.
3. The delivery time is long enough.

The burn-in period

4. The burn-in time is t_b. Products are burned in one (or N) at a time until one (or N) product(s) survives the burn-in time without failure. Thus the entire burn-in process is of random length kt_b where k is the number of products required to achieve the first (or Nth) successful burn-in.
5. Failed products cannot be taken out of the board during burn-in time (AFT situation).
6. The failed products still consume burn-in costs until the end of the fixed burn-in period.
7. Complete repair for the products failing during the burn-in process.
8. The repair cost for failures during burn-in test is always less than the warranty failure cost: $C_2 < C_3$.

The warranty period

9. Failure free, non-renewing warranty policy.
10. Minimal repair for the products failing during warranty period.

Notation:

C_i	are similar to the ones in Sect. 6.2:
C_0	fixed costs including manufacturing and burn-in set-up costs;
C_1	burn-in cost per unit time per unit;
C_2	shop *complete repair* cost per failure;
C_3	field *minimal repair* cost per failure during warranty period, based on the assumption 8, $C_2 < C_3$
C_s	sale price for one unit
t_b	burn-in time, $t_b \geq 0$
t_w	warranty period
X	product lifetime without burn-in
$F(t)$	cumulative distribution function of X
$R(t)$	survival function of X
$r(t)$	failure rate function of X
$N(t_b)$	number of failures during burn-in until one survives the process;
$N_{t_b}(t_w)$	number of failures during warranty time t_w, with burn-in time t_b
X_{t_b}	lifetime during warranty after burn-in time t_b, *i.e.*, $X - t_b \mid X > t_b$
$r_{t_b}(t)$	failure rate function of X_{t_b}, so $r_{t_b}(t) = r(t_b + t)$

Based on the "per-item-output" point of view, let $C(t_b)$ be the expected cost per unit for a product surviving both burn-in and warranty period. Then, it has the form:

$$C(t_b) = C_0 + C_1 E[N(t_b) + 1]t_b + C_2 E[N(t_b)] + C_3 E[N_{t_b}(t_w)] \tag{6.44}$$

where $E[N(t_b) + 1]$ denotes the expected total number of product burn-in times until the first product survives the process (including the final successful one). $C_2 E[N(t_b)]$ is the total repair cost during burn-in and $C_3 E[N_{t_b}(t_w)]$ is the (minimal) repair cost during warranty. Under the complete repair assumption for failure treatment during burn-in, the expected failure number in burn-in is:

$$E[N(t_b)] = \frac{F(t_b)}{R(t_b)} \tag{6.45}$$

which is just the proportion of failed cases to successful cases, and the expected total burn-in times is

$$E[N(t_b) + 1] = \frac{F(t_b)}{R(t_b)} + 1 = \frac{1}{R(t_b)} \tag{6.46}$$

Under the non-renewing failure free policy and the minimal repair failure treatment assumptions for warranty period, the expected failure number until the end of the warranty use is:

$$E[N_{t_b}(t_w)] = \int_0^{t_w} r_{t_b}(s)\,ds = \int_0^{t_w} r(t_b + s)\,ds \tag{6.47}$$

Introducing these three items into (6.44), the expected total cost until the one product survives both burn-in and warranty period has the form:

$$C(t_b) = C_0 + C_1 \frac{t_b}{R(t_b)} + C_2 \frac{F(t_b)}{R(t_b)} + C_3 \int_0^{t_w} r_{t_b}(s)\,ds \qquad (6.48)$$

The above is new AFT cost model, built based on 'per-item output' point of view, and with the assumptions of complete repair for burn-in failures and minimal repair for warranty failures.

The sales price for each cost function becomes constant and thus can be omitted for the cost optimizations. If the total expected cost is also of concern, the cost function will become

$$C(t_b) = (C_0 - C_s) + C_1 \frac{t_b}{R(t_b)} + C_2 \frac{F(t_b)}{R(t_b)} + C_3 \int_0^{t_w} r_{t_b}(s)\,ds \qquad (6.49)$$

Also, the warranty cost part can now deal with multi-failure situation which is more realistic. In this paper, we just use the cost model (3.6). If the sale price C_s must be counted, it can just be regarded as included in the fixed cost C_0.

6.3.3 Properties and Optimization of the New Model

The *optimal burn-in time* in this chapter is the time corresponding to the minimum total cost given in (6.48).

Lemma 6. *Let $r(t)$ be a continuous bathtub-shaped failure rate function with change points t_1 and t_2, $t_w > 0$ be a given mission time, define*

$$B^* = \left\{ t_b \geq 0, \quad \int_{t_b}^{t_b+t_w} r(s)\,ds = \min_{t_b \geq 0} \int_{t_b}^{t_b+t_w} r(s)\,ds \right\} \qquad (6.50)$$

1. If $t_w \leq t_2 - t_1$, then $B^* = [t_1, t_2 - t_w]$;
2. If $t_w > t_2 - t_1$, then $B^* = \{t_b^*\}$ and $t_b^* \in [0, t_1]$.

Clearly, if we are interested in $\inf B^*$, for both of the situations $\inf B_1^*$.

The above lemma can be use to prove the following theorem.

Theorem 5. *Suppose the failure rate function $r(t)$ has a bathtub shape. Then for the cost function $C(t_b)$ given in (3.6), the optimal burn-in time t_b^* never exceeds the first change point t_1, i.e., $t_b^* \leq t_1$.*

Proof: In the above cost function, the first item is constant, and the second and the third items $C_1 \frac{t_b}{R(t_b)}$ and $C_2 \frac{F(t_b)}{R(t_b)}$ are both increasing function of burn-in time t_b.

Thus the sum of the first three items $C_0 + C_1 \frac{t_b}{R(t_b)} + C_2 \frac{F(t_b)}{R(t_b)}$ is strictly increasing in $t_b \geq 0$. Therefore, the infimum of set of the optimal burn-in time for the last item $C_3 \int_0^{t_w} r_{t_b}(s)\,ds$ is of special concern. From the above lemma, if the failure rate curve has a bathtub shape, no matter how long is the warranty period, the infimum of the optimal burn-in time set B^* for the last item never exceeds the first change point t_1 of the bathtub curve, *i.e.*, $\inf B^* \leq t_1$. Therefore, for the cost function given in (3.6), the minimum value cannot be obtained in the interval (t_1, ∞), *i.e.*, the optimal burn-in time $t_b^* \leq t_1$.

Proposition 3. *Suppose the product lifetime X has the exponential distribution, then for the cost function $C(t_b)$ given in (6.48), the optimal burn-in time t_b^* equals zero, i.e., no burn-in for exponential lifetime distribution.*

Proof: This is easy to prove by regarding the failure rate of exponential distribution as a special bathtub shape which has only the second constant period. Actually, this proposition coincides with our intuition: since exponential distribution has a constant failure rate which means that the lifetime distribution starting from any age point is the same as new, so burn-in is not necessary at all.

Remark 1. It is also easy to show that the optimal burn-in time for the products with increasing failure rate is also zero which is also intuitively true.

Optimization of the model

Applying the first and second derivatives method, the optimal burn-in time t_b^* must satisfy:

$$\begin{cases} C'(t_b^*) = 0 \\ C''(t_b^*) < 0 \end{cases}$$

Taking the first derivative of (3.6) with respect to the burn-in time t_b, yields:

$$C'(t_b) = C_1 \frac{R(t_b) - t_b R'(t_b)}{R^2(t_b)} + C_2 \frac{F'(t_b)R(t_b) - F(t_b)R'(t_b)}{R^2(t_b)} + C_3 \frac{d}{dt_b}\left(\int_0^{t_w} r_{t_b}(s)\,ds\right)$$

$$= \frac{C_1[R(t_b) - t_b(-f(t_b))] + C_2[f(t_b)(1 - F(t_b)) - F(t_b)(-f(t_b))]}{R^2(t_b)}$$

$$+ C_3 \frac{d}{dt_b}\left(\int_{t_b}^{t_b+t_w} r(s)\,ds\right)$$

$$= \frac{C_1[R(t_b) + t_b f(t_b)] + C_2 f(t_b)}{R^2(t_b)} + C_3 \frac{d}{dt_b}\left(\int_{t_b}^{t_b+t_w} r(s)\,ds\right)$$

$$= \frac{C_1[R(t_b) + t_b f(t_b)] + C_2 f(t_b)}{R^2(t_b)} + C_3\left[\frac{d(t_b+t_w)}{dt_b}r(t_b+t_w) - \frac{dt_b}{dt_b}r(t_b)\right]$$

$$= \frac{C_1 R(t_b) + (C_1 t_b + C_2)f(t_b)}{R^2(t_b)} + C_3[r(t_b+t_w) - r(t_b)]$$

$$= \frac{C_1 + (C_1 t_b + C_2)r(t_b)}{R(t_b)} + C_3[r(t_b+t_w) - r(t_b)] \tag{6.51}$$

Remark 2. In fact, Theorem 5 can also be proved by (6.51), because for a bathtub failure rate, $r(t_b + t_w) - r(t_b) \geq 0$ for all the $t_b \geq t_1$, so $C'(t_b) > 0$ in the interval (t_1, ∞)

and the optimal burn-in time t_b^* cannot exist in this interval. To be more general, for an arbitrary failure rate $r(t)$, if there exists a time point t_A, s.t. $r(t_b + t_w) - r(t_b) \geq 0$ for all the $t_b \geq t_A$ then the optimal burn-in time $t_b^* \leq t_A$.

Equating this function (3.9) to zero, we get

$$C_1 + (C_1 t_b + C_2)r(t_b) + C_3 R(t_b)[r(t_b + t_w) - r(t_b)] = 0 \qquad (6.52)$$

This is a necessary condition for the optimal solution, *i.e.*, the value of t_b from this equation gives the optimal value of burn-in time t_b^*. The first condition can then be expressed as:

$$t_b^* = \frac{C_3 R(t_b^*)\left[r(t_b^*) - r(t_b^* + t_w)\right] - C_1 - C_2 r(t_b^*)}{C_1 r(t_b^*)} \qquad (6.53)$$

To check the minimal optimality, t_b^* must also satisfy the second derivative condition. Take the derivative of (6.51) we get,

$$\frac{[C_1 r(t_b^*) + (C_1 t_b^* + C_2)r'(t_b^*)]R(t_b^*) + [C_1 + (C_1 t_b^* + C_2)r(t_b^*)]f(t_b^*)}{R^2(t_b^*)}$$
$$+ C_3\left[r'(t_b^* + t_w) - r'(t_b^*)\right] < 0 \qquad (6.54)$$

introducing $r(t_b) = \frac{f(t_b)}{R(t_b)}$ and $r'(t_b) = \frac{f'(t_b)R(t_b) - f(t_b)R'(t_b)}{R^2(t_b)} = \frac{f'(t_b)R(t_b) + f^2(t_b)}{R^2(t_b)}$ into the above inequality yields

$$\frac{\left[C_1 f(t_b^*) + (C_1 t_b^* + C_2)\frac{f'(t_b)R(t_b) + f^2(t_b)}{R(t_b)}\right] + \left[C_1 f(t_b^*) + (C_1 t_b^* + C_2)\frac{f^2(t_b)}{R(t_b)}\right]}{R^2(t_b^*)}$$
$$+ C_3\left[r'(t_b^* + t_w) - r'(t_b^*)\right] < 0 \qquad (6.55)$$

so

$$\frac{(C_1 t_b^* + C_2)f'(t_b) + 2\left[C_1 f(t_b^*) + (C_1 t_b^* + C_2)\frac{f^2(t_b)}{R(t_b)}\right]}{R^2(t_b^*)} + C_3\left[r'(t_b^* + t_w) - r'(t_b^*)\right] < 0$$
$$(6.56)$$

Thus, the optimal burn-in time must satisfy (6.53) and (6.56). The optimal burn-in time t_b^* can be found by solving (6.52), or by directly minimizing function (3.6) with numerical or graphical methods. Besides, when $t_b \to \infty$, $C(t_b) \to \infty$. So the optimal burn-in time t_b^* is either zero or some finite value. For this reason, if (3.10) has no solution, then there is no burn-in, *i.e.*, $t_b^* = 0$.

Furthermore, it is easy to get from (3.9) that

$$C'(0) = C_1 + C_2 r(0) + C_3[r(t_w) - r(0)]$$
$$= C_1 + C_3 r(t_w) + (C_2 - C_3) r(0) \tag{6.57}$$

Since the failure cost $C_2 < C_3$, we know that if

$$r(0) > \frac{C_1 + C_3 r(t_w)}{C_3 - C_2} \tag{6.58}$$

Then, $C'(0) < 0$ which means that the total cost function $C(t_b)$ is decreasing in a very small interval $(0, \varepsilon)$, and so that the optimal burn-in time $t_b^* > 0$. Finally, we get the following theorem:

Theorem 6. *Suppose for the product failure rate function $r(t)$, there exists a time point t_A, s.t. $r(t_b + t_w) - r(t_b) \geq 0$ for all the $t_b \geq t_A$. Then for the cost function $C(t_b)$ given in (3.6), the optimal burn-in time t_b^* never exceeds the time point t_A, i.e., $0 \leq t_b^* \leq t_A$. Furthermore, if $r(0) > \frac{C_1 + C_3 r(t_w)}{C_3 - C_2}$, then $t_b^* > 0$.*

For proof, see Remark 2.

Remark 3. Usually, the product failure rate keeps a constant level during the useful life period, and later goes up when entering the wear-out stage. Thus during these two lifetime periods, the failure rate is always non-decreasing and the condition $r(t_b + t_w) - r(t_b) \geq 0$ is easy to achieve for all t_b during these two stages. For this reason, $0 \leq t_b^* \leq t_A$ just implies that the optimal burn-in time should never exceed the end of the infant mortality period.

Remark 4. From the above theorem, we also notice that when the initial failure rate is higher than a certain value, the burn-in test is especially needed and beneficial. This is to be expected because the burn-in test is just to be designed and conducted for the products with high infant mortality risk.

6.3.4 Numerical Example

6.3.4.1 Given Data and Distribution Parameters

The purpose of this section is to compare the new AFT model with the one in Perlstein, Jarvis and Mazzuchi (2001) whose given data and the lifetime distribution parameters will be adopted for calculation here. Cost functions are the only criteria for the optimization problem. The optimal burn-in times which minimize the two cost functions will then be derived and compared, followed by sensitivity analysis and other discussions.

As mentioned before, there are mainly two types of lifetime distribution for the burn-in products, one is uni-modal distribution, and the other is bimodal (mixture) distribution. A mixed exponential distribution is used in Perlstein, Jarvis and Mazzuchi (2001) to describe the product's time-to-failure:

$$f(t) = pf_1(t) + (1-p)f_2(t)$$
$$= p\lambda_1 e^{-\lambda_1 t} + (1-p)\lambda_2 e^{-\lambda_2 t} \tag{6.59}$$

the new products are composed of two subpopulations: weak ones and strong (normal) ones, the early failures in the infant mortality period are mainly from the weak subpopulation of the products, and the useful operating life is mainly from the strong subpopulation of the products. λ_1 and λ_2 are the failure rates for weak and strong populations, p is the mixture parameter, which represent the proportion for the weak components (a small proportion). Under the mixture distribution assumption, the new AFT cost model has the form:

$$C(t_b) = C_0 + \frac{C_1 t_b + C_2(1 - pR_1(t_b) - (1-p)R_2(t_b))}{pR_1(t_b) + (1-p)R_2(t_b)}$$
$$+ C_3 \int_{t_b}^{t_b + t_w} \frac{pf_1(s) + (1-p)f_2(s)}{pR_1(s) + (1-p)R_2(s)} \, ds \tag{6.60}$$

Knowledge of the model parameters p, λ_1 and λ_2 for the above functions is assumed to be inexact in the reference and therefore must be quantified with Bayesian updating method. In this example, we just simply assume that the true values for the mixture and failure rate parameters p, λ_1, and λ_2 are known, and take the posterior estimations of these parameters in Example 2 of the reference paper as the given values. In addition, it is reasonable to assume that $\lambda_1 > \lambda_2$ as the weak population represents those units which should exhibit shorter life lengths.

The values for p, λ_1 and λ_2 are then:

$$p = 0.2855 , \quad \lambda_1 = 0.0322 , \quad \text{and} \quad \lambda_2 = 0.0001$$

The given cost data for C_i are:

$$C_0 = \$15 , \quad C_1 = \$3 , \quad C_2 = \$300 , \quad \text{and} \quad C_3 = \$1500 ,$$

The warranty period is of 5000 hours, *i.e.*, $t_w = 5000$.

6.3.4.2 Numerical Comparison of Two AFT Optimal Burn-In Models

For comparison, the cost function in Eq. (6.43) is built for computing the batch burn-in cost with a batch size $n = 50$ and the new cost model (Eq (6.48)) is built for "per-unit-output" cost, so that we will make some modification to make them comparable. Our strategy is to make that the numbers of units (denoted as M) that could survive both burn-in and warranty period for the two cost functions, which are also the major concern of the manufacturers, are equal.

The expected costs for both cost models equations (6.48) and (6.43) are plotted in Fig. 6.5. We see in this graph that the expected cost for the old cost model is always higher than the new cost model. This cost difference can be interpreted as mainly from the $N - M$ failures cost amount. The first derivative curves for the two cost models are also shown in the graph: at the beginning, both of the derivative curves have negative values which mean that the expected total costs of the two models decrease when the burn-in time t_b is short; during this period, the first derivative curve of the new cost model has a higher value than that of the old one, and it reaches the zero point first. The two curves have an intersection at around $t_b = 50\,\text{h}$. The first derivative pictures clearly shows that the optimal burn-in time of the old cost model is overestimated compared to the new one. One possible reason for this is that the sales profit is not included in (6.43). It confirms our surmise before that "not including sales profit item will quite likely result in the overestimation of t_b^*, if the cost model is constructed which has a 'fixed input' product number". The optimal (minimal) values of the new cost model equation (6.48) and the old cost model equation (6.43), the optimal burn-in times and some related results are shown in Table 6.3.

Let's now test the criteria in Theorem 6 which says:

When the initial failure rate satisfies $r(0) > \frac{C_1 + C_3 r(t_w)}{C_3 - C_2}$, a proper burn-in test is definitely beneficial.

For this example,

$$r(0) = 0.0093$$

$$r(t_w) = 1.0000 \times 10^{-4}$$

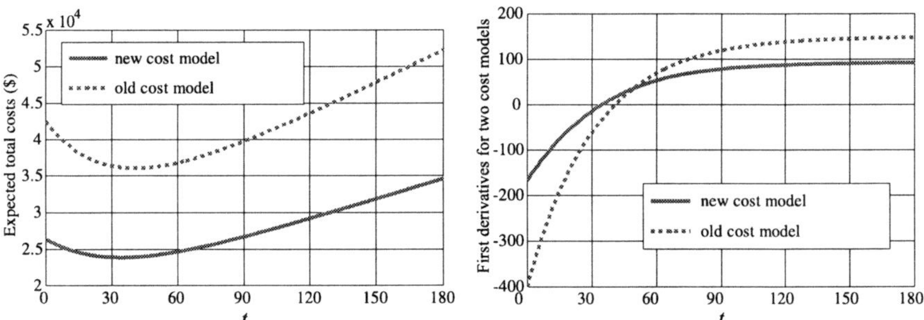

Fig. 6.5 a Expected total cost for two models; **b** first derivative curves for two models

Table 6.3 Comparison results for new and old cost models

	New model equation (6.48)	Old model equation (6.43)
No burn-in	$C(0) = 2.6354 \times 10^{+4}$	$C(0) = 4.2513 \times 10^{+4}$
Optimal burn-in times	$t_b^* = 34.30\,\text{h}$	$t_b^* = 40.65\,\text{h}$
Optimal costs	$C(t_b^*) = 2.388 \times 10^{+4}$	$C1(t_b^*) = 3.6065 \times 10^{+4}$

$$r(0) = 0.0093 > \frac{C_1 + C_3 r(t_w)}{C_3 - C_2} = 0.0026 \,.$$

So the optimal burn-in time should be $t_b^* > 0$, which is consistent with the above result.

The sensitivity analysis results are finally listed in Table 6.4, which shows that C_i is the most influential to the optimal burn-in time, λ_1 (weak population) has more impact on the optimal burn-in time than λ_2 (strong population), and t_w has very little influence on the optimal burn-in time.

In order to check the effects of the parameters, we do the sensitivity tests: C_1 and C_2 are decreased by 20% separately, and C_3 is increased by 20%, λ_1 and λ_2 are decreased by 20% separately, t_w is increased and decreased by 20% separately. The results are shown in the following Figs. 6.6–6.8 and Table 6.4. In Fig. 6.6, among the three cost parameters, C_3 has the greatest influence to both the optimal burn-in time t_b^* and the optimal burn-in cost $C(t_b^*)$ of the two cost models, and C_1 is secondarily important to the optimal results; furthermore, according to the changing percent, the optimal burn-in time of the new cost model compared to the old one is more sensible to the changes of the cost parameters. From Fig. 6.7, we see for

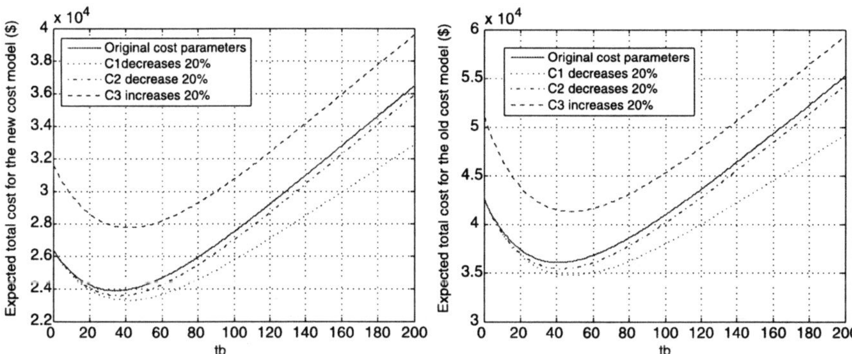

Fig. 6.6 Effect of cost parameters on total expected cost under new (*left*) and old (*right*) models

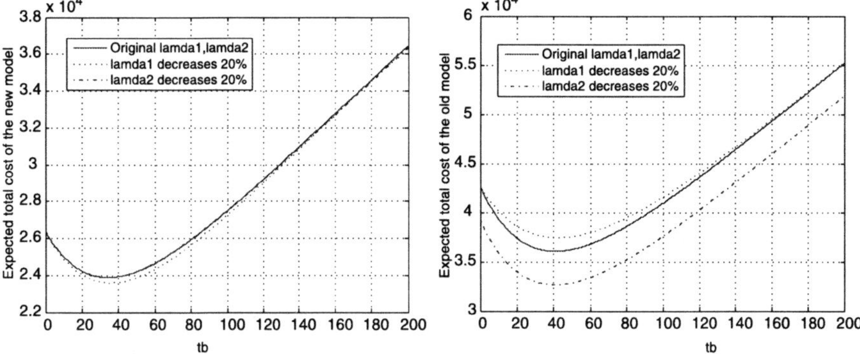

Fig. 6.7 Effect of hazard rate factors on new (*left*) and old (*right*) cost models

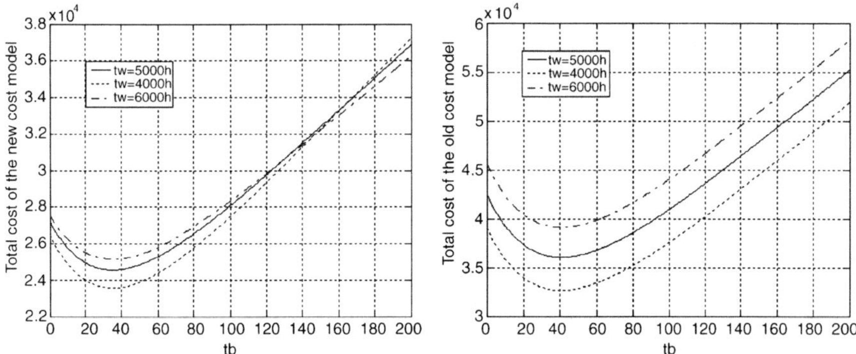

Fig. 6.8 Effect of warranty period on new (*left*) and old (*right*) cost models

both models λ_1 has more impact on the optimal burn-in time t_b^*. That is easy to explain because the burn-in process aims to eliminate the weak populations, and so that the lifetime distribution of the weak subpopulation, which has the hazard rate parameter λ_1, is more influential to the optimal burn-in time. And in Fig. 6.7, the left picture for new cost model, the decreased λ_1 results in decreased cost, but in the right picture which is for old cost model, decreases λ_1 results in increased cost. In this sense, the new model proves to be more realistic: because λ_1 is the failure rate for the weak products, and lower product quality will result in higher burn-in cost.

Table 6.4 Sensitivity analysis for new and old cost models

Increased/decreased parameters (%)	New model equation (6.48)		Old model equation (6.43)	
	t_b^*	$C(t_b^*)$	t_b^*	$C(t_b^*)$
C_1 decreases 20%	19.33	-2.47	17.20	-3.66
C_2 decreases 20%	5.71	-1.27	3.85	-1.77
C_3 increases 20%	20.82	16.16	17.27	14.50
λ_1 decreases 20%	1.02	-1.24	3.57	3.80
λ_2 decreases 20%	0.29	-0.03	-0.22	-9.42
t_w decreases 20%	-0.31	-4.07	-0.16	-9.44
t_w increases 20%	0.31	2.41	0.17	8.54

6.3.5 Application of AFT Model to Updating Strategy Policy Making

In this final short section, we will make a further demonstration on how the new constructed AFT model could help the manufacturers do the practical policy making. We will show that the new AFT model cannot only help decide the optimal burn-in time as shown before, but is also quite helpful in the decision making of other related cost issues.

Suppose that the manufacturers only have static burn-in equipment and after the product fails the equipment cannot detect the failure in time and stop the process. We all know that in this case burning-in the failed components is very costly. If there are some advanced burn-in facility that can detect and stop the burn-in process for the failed component, the manufacturer could save some money, and so they should think about changing the old burn-in board to the new one. For this purpose, it is better to calculate of how much can be saved by adopting the new facility. If it is even more than the money of changing the burn-in board, we should change one.

Compared to the new AFT cost model (3.6), the cost model of applying the new dynamic burn-in board is only different in the burn-in cost item:

$$C_{\text{new}}(t_b) = C_0 + C_1 \left\{ \frac{F(t_b)}{R(t_b)} E[X|X < t_b] + t_b \right\} + C_2 \frac{F(t_b)}{R(t_b)} + C_3 \int_0^{t_w} r_{t_b}(s)\, ds \quad (6.61)$$

$$= C_0 + C_1 \left\{ \frac{F(t_b)}{R(t_b)} \int_0^{t_b} s \frac{f(s)}{F(t_b)}\, ds + t_b \right\} + C_2 \frac{F(t_b)}{R(t_b)} + C_3 \int_0^{t_w} r_{t_b}(s)\, ds$$

$$= C_0 + C_1 \left\{ \frac{1}{R(t_b)} \left[t_b F(t_b) - \int_0^{t_b} F(s)\, ds \right] + t_b \right\} + C_2 \frac{F(t_b)}{R(t_b)}$$
$$+ C_3 \int_0^{t_w} r_{t_b}(s)\, ds$$

$$= C_0 + C_1 \left\{ \frac{t_b F(t_b) - \int_0^{t_b} F(s)\, ds + t_b R(t_b)}{R(t_b)} \right\} + C_2 \frac{F(t_b)}{R(t_b)} + C_3 \int_0^{t_w} r_{t_b}(s)\, ds$$

$$= C_0 + C_1 \left\{ \frac{t_b - \int_0^{t_b} F(s)\, ds}{R(t_b)} \right\} + C_2 \frac{F(t_b)}{R(t_b)} + C_3 \int_0^{t_w} r_{t_b}(s)\, ds \quad (6.62)$$

The money can be saved for "per-item-output" by adopting the new technology is then:

$$\overline{C(t_b)} = C(t_b) - C_{\text{new}}(t_b)$$

$$= C_1 \frac{t_b}{R(t_b)} - C_1 \left(\frac{t_b - \int_0^{t_b} F(s)\, ds}{R(t_b)} \right)$$

$$= C_1 \frac{\int_0^{t_b} F(s)\, ds}{R(t_b)} \quad (6.63)$$

This is usually trivial for one-unit computation, but if multiplied by the production volume in the near future, *e.g.* annual production value A, it will become larger and can be compared to the new facility purchasing cost. Suppose the price for the new dynamic burn-in facility is G, then if

$$A^* C_1 \frac{\int_0^{t_b} F(s)\,ds}{R(t_b)} > G \tag{6.64}$$

i.e., if our production volume in the near future satisfies:

$$A > \frac{GR(t_b)}{C_1 \int_0^{t_b} F(s)\,ds} \tag{6.65}$$

we should think about installing the new burn-in facility.

Back to the above example, suppose we are doing the optimal burn-in test with the burn-in time $t_b^* = 34.3$ h, then the saving cost for producing one unit using the new facility is $C_1 \frac{\int_0^{t_b} F(s)\,ds}{R(t_b)} = 14.5281\$$. If the cost for purchasing and installing the new facility is $G = 10^8$ \$, then we should consider changing the old facility with the advanced one when our near-future production volume $A > 6.8832 \times 10^6$ (unit).

6.4 Conclusions

The first aim of this chapter has been to present a thorough literature review on the different aspects of burn-in process, in particular focus on the different forms of cost functions. The cost functions have been divided into basic item part and extra item part. For each part, typical items are given according to different failure treatment, warranty strategy, etc. Various cost functions are simply the different combination of the typical items. In this way of classification, the cost models are no longer chaotic, similarities and differences can be easily seen by this kind of analysis. This part of work has been done in Sect. 6.2.

The second and main task of this paper is the AFT cost model investigation. Although various cost models have been built for different situations, the practical situation AFT is hardly considered in most of cost models. A new AFT model is therefore constructed in Sect. 6.3 for this situation using the concept of "per-item-output." The maximum value of the optimal burn-in time is derived according to this model with respect to certain failure rate functions. The numerical example as given illustrates that the optimal burn-in time of the old model constructed on basis of "fixed input" is longer than that of the new model. Lack of considering about the sales profit due to "fixed input" may contribute to this overestimation. Sensitivity analysis on different model parameters is also made and the new cost model proves to be more realistic based on some results of the test. Finally a discussion about the trade-off of substituting the old burn-in board with the advanced facility is considered, which is of particular interest to the manufacturers in decision-making.

Further research suggestion is that in case the parameters of the product distribution are not known, the application of Bayesian inference on this AFT cost model should be investigated.

References

1. Baskin EM (2002) Analysis of burn-in time using the general law of reliability. Microelectronics Reliability 42:1967–1974
2. Chang DS (2000) Optimal burn-in decision for products with an unimodal failure rate function. European Journal of Operational Research 126:534–540
3. Chi D, Kuo W (1989) Burn-in optimization under reliability and capacity restrictions. IEEE Transactions on Reliability 38:193–198
4. Chien WK, Kuo W (1995) Modeling and maximizing burn-in effectiveness. IEEE Transactions on Reliability 44:19–25
5. Genadis TC (1996) A cost optimization model for determining optimal burn-in times at the module/system level of an electronic product. International Journal of Quality and Reliability Management 13:61–74
6. Guess F, Walker E, Gallant D (1992) Burn-in to improve which measure of reliability. Microelectronics and Reliability 32:759–762
7. Hui YV, Lu WL (1996) Cost optimization of accelerated burn-in. International Journal of Quality and Reliability Management 13:69–78
8. Jensen F, Petersen NE (1982) Burn-in: an engineering approach to the design and analysis of burn-in procedures. Wiley, New York
9. Kim KN (1998) Optimal burn-in for minimizing cost and multiobjectives. Microelectronics Reliability 38:1577–1583
10. Kim T, Kuo W (1998) Optimal burn-in decision making Qual. Reliab. Engng. Int 14:417–423
11. Kuo W (1984) Reliability enhancement through optimal burn-in. IEEE Transactions in Reliability R-33:145
12. Kuo W, Kuo Y (1983) Facing the headaches of early failures: a state-of-the-art review of burn-in decision. Proc IEEE 71:1257–1266
13. Launer RL (1993) Graphical techniques for analyzing failure data with the percentile residual-life function. IEEE Transactions on Reliability 42:71–80
14. Lu WL, Hui YV (1998) Economic design of high stress first burn-in processes. International Journal of Production Research 36:181–196
15. Mi J (1994a) Maximization of a survival probability and its application. J Appl Prob 31:1026–1033
16. Mi J (1994b) Burn-in and maintenance policy. Advances in Applied Probability 26:207–221
17. Mi J (1995) Bathtub failure rate and upside-down bathtub mean residual life. IEEE Transactions on Reliability 44:388–391
18. Mi J (1996) Minimizing some cost functions related both burn-in and field use. Operations Research 44:497–500
19. Mi J (1997) Warranty policies and burn-in. Naval Research Logistics 44,:199–209
20. Nicolato E, Runggaldier WJ (1999) A Bayesian dynamic programming approach to optimal maintenance combined with burn-in. Ann Oper Res 91:105–122
21. Perlstein D, Jarvis WH, Mazzuchi TA (2001) Bayesian calculation of cost optimal burn-in test duration for mixed exponential populations. Reliab Engng Syst Safety 72:265–273
22. Sheu S-H, Chien Y-H (2004) Minimizing cost-functions related to both burn-in and field-operation under a generalized model. IEEE Transactions on Reliability 53:435–439
23. Sheu S-H, Chien Y-H (2005) Optimal burn-in time to minimize the cost for general repairable products sold under warranty. European Journal of Operational Research 163:445–461
24. Spizzichino F (1991) Sequential burn-in procedures. J Statis Planning Infer 29:187–197
25. Su C-T, Wu C-L (2001) Intelligent approach to determining optimal burn-in time and cost for electronic products. International Journal of Quality and Reliability Management 18:549–559
26. Sultan TI (1986) Optimum burn-in time: model and application. Microelectronics and Reliability 26:909–916
27. Washburn LA (1970) Determination of optimum burn-in time: a composite criterion. IEEE Transactions on Reliability R-19:134–140

28. Whitbeck CW, Leemis LM (1989) Component vs. system burn-in techniques for electronic equipment. IEEE Transactions on Reliability 38:206–209
29. Wu C-L, Su C-T (2002) Determination of the optimal burn-in time and cost using an environmental stress approach: a case study in switch mode rectifier. Reliability Engineering and System Safety 76:53–61
30. Yan L, English JR (1997) Economic cost modelling of environmental-stress-screening and burn-in. IEEE Transactions on Reliability 46:275–282
31. Yuan J, Shih SW (1991) Mixed hazard models and their applications. Reliab Engng Syst Safety 33:115–129
32. Yun WY, Lee YW, Ferreira L (2002) Optimal burn-in time under cumulative free replacement warranty. Reliability Engineering and System Safety 78:93–0100

Part II
Reliability Engineering in Design

Chapter 7
Optimum Threshold Level of Degrading Systems Based on Sensor Observation

Elsayed A. Elsayed, Hao Zhang

Department of Industrial and Systems Engineering, Rutgers University, USA

Notation

CDF	cumulative distribution function
pdf	probability density function
$X(t)$	gamma process describing the degrading system state at time t
D_{F}	failure threshold level of the degradation process
D_{PM}^{*}	optimum predictive maintenance threshold level
D_{PM}	true degrading state when the average measurement of sensors equals to D_{PM}^{*}
n	number of sensors used to monitor the system degradation
$\overline{RS}$	average measurement of sensors
i	index of maintenance action
R_i, R_i^{+}	time instants immediately prior to and after the maintenance action i
M_i	$R_i^{+} - R_i$, the ith maintenance time
T_{R}	time required to perform replacement
$I\{A\}$	indicator function, which equals to 1 if A is satisfied, or 0 otherwise
$F(\cdot), \overline{F(\cdot)}$	CDF and its complement of a random variable
N	total number of maintenance actions in one maintenance cycle
ξ	scale parameter of the gamma process
$\eta(t)$	shape function of the gamma process
$\hat{\sigma}_s^2$	estimated variance of sensor measurement errors
α_i, β_i	parameters of beta distributions
γ_0, γ_1	parameters used to model maintenance time
$\Phi(\cdot)$	standard normal CDF
c	critical value for hypothesis test
CM_i	cost of the ith maintenance action
ω, A_0, A_1	Parameters used to model maintenance cost

7.1 Introduction

Degradation models are normally used to predict the system's failure under condition-based predictive maintenance policies. Repairs or replacements of the system are performed once the degradation level reaches a predetermined threshold level. This results in a significant time and cost savings compared to the situation when the system is repaired upon failure. In the former case, the maintenance is planned and the necessary spare parts and manpower requirements are readily available, while in the latter case it is difficult to predict the failure time and plan the necessary resources to perform the maintenance action immediately upon failure. Recent developments in sensors, chemical and physical nondestructive testing, and sophisticated measurement techniques have facilitated the continuous monitoring of the system condition. A condition parameter could be any critical characteristic such as crack growth, vibration, corrosion, wear and lubricant condition. With the measured data, the predictive maintenance policy determines the optimum threshold level at which maintenance action is performed to bring the system to a "better" condition, if not as good as new, in order to maximize system availability or minimize the average maintenance cost.

Gamma process is usually used to model the degradation of the system due to the fact that system's deterioration occurs gradually over time in a sequence of small increments [1]. Grall *et al.* [2] and Dieulle *et al.* [3] study a maintenance strategy for systems exhibiting continuous degradation modeled by gamma processes. However, most of the literature on system degradation assumes that the measurements of the degradation characteristics are accurate and free from measurement errors. This assumption is impractical since measurement errors of sensors always exist. These errors include systematic errors, which could be corrected by analysis of the physical mechanism of the sensors as discussed by Verinaud [4] and Xu [5], and the random measurement errors, which are usually modeled as Gaussian white noise with zero mean and constant standard deviation [6–8]. In this chapter, we develop an approach for determining the optimum maintenance threshold level taking into consideration the random measurement errors.

The remainder of this chapter is organized as follows. Section 7.2 presents a brief description of the gamma process degradation model. In Sect. 7.3, we present the predictive maintenance model under multiple imperfect maintenance actions. The sensor measurement errors and accuracy are presented and addressed in Sect. 7.4. In Sect. 7.5 we estimate the expected system uptime taking into consideration the imperfect maintenance actions and the sensors' measurement errors. Section 7.6 describes the formulation of the maintenance optimization problem and provides an optimization procedure to determine the optimum system maintenance threshold level that achieves the maximum system availability. A numerical example is given in Sect. 7.6 to demonstrate the proposed maintenance policy. In Sect. 7.7, we develop a maintenance policy that minimizes the maintenance cost. Conclusions are summarized in Sect. 7.8.

7.2 Gamma Process Degradation Model

Failure time data are usually analyzed to estimate product reliability. Failures of highly reliable units are very rare and are mostly censored. Therefore, other information should be used in addition to censored failure time data. One way of obtaining this complementary reliability information is to use higher levels of covariates or stresses (temperature, voltage, etc.) to accelerate the life testing and, hence, to obtain failure time data in short test durations. Another way is to measure some parameters (such as crack length, vibration rate, particle counts in lubricants), which characterize degradation (aging) of the product with time. The measurements of these parameters over time are then utilized to model product degradation. The gamma process is a natural process for degradation modeling in which deterioration takes place gradually over time in a sequence of small independent non-negative increments. It can be thought of as arising from a compound Poisson process of gamma-distributed increments in which the Poisson rate tends to infinity while the size of the increments tend to zero in proportion.

We consider a non-negative process $\{X(t), t \geq 0\}$, where $X(t)$ represents the measured degradation state of the system at time t. A gamma process has the following properties:

1. The increments $\Delta X(t) = X(t + \Delta t) - X(t)$ are independent;
2. The increments $\Delta X(t)$ have a gamma distribution $Ga\{\xi, \eta(t + \Delta t) - \eta(t)\}$, where $\xi > 0$ is a scale parameter, and $\eta(t)$ is a given monotone increasing function, which determines the shape of the gamma process.

A special case of the gamma process is when $\eta(t) = \eta t$. Thus the increments $\Delta X(t)$ follow gamma distribution $Ga\{\xi, \eta \Delta t\}$ with mean $\xi \eta \Delta t$, variance $\xi^2 \eta \Delta t$, and *pdf* $f(x; \xi, \eta \Delta t)$ which is defined by

$$f(x; \xi, \eta \Delta t) = \frac{1}{\Gamma(\eta \Delta t)} \left(\frac{x}{\xi}\right)^{\eta \Delta t - 1} \exp\left(-\frac{x}{\xi}\right). \tag{7.1}$$

Figure 7.1 shows the path of the state of a degradation system, which is modeled by gamma process described in (7.1), without maintenance actions. In Fig. 7.1, D_F is the system failure threshold; when the state of system is equal to or greater than D_F, the system fails and requires maintenance or replacement.

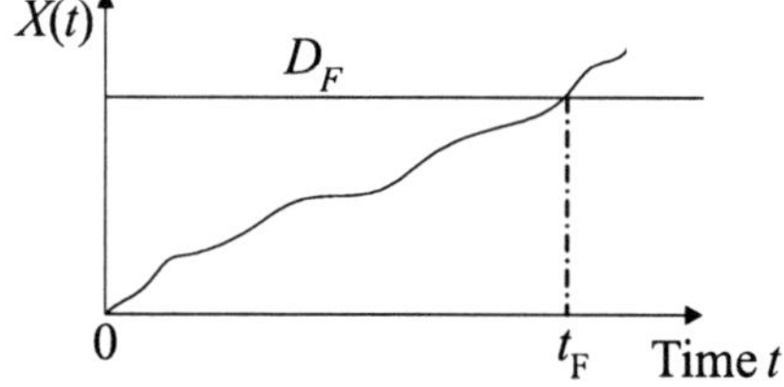

Fig. 7.1 Gamma process model of system degradation without maintenance actions

The relevance of modeling the degradation process as a gamma stochastic process has been justified by several authors [9–11] and has been applied by other authors in maintenance models such as those described in [12–14].

7.3 Imperfect Maintenance Model

We investigate the predictive maintenance model under the following assumptions.

1. States of a degrading system are modeled by a continuous-state gamma process, $X(t)$ as in (7.1), starting from the initial state $X(0) = x_0 = 0$.
2. Monitoring of the system state is continuous and the sensor measurements that define the state have measurement error ε, which follows a normal distribution. Multiple sensors are used to increase the measurement accuracy.
3. Maintenance actions are imperfect. In other words, the system does not return to its original as-good-as-new state after maintenance.
4. The maintenance times are positively correlated to the system state. In other words, the maintenance time increases as the system "ages".
5. Replacement brings the system to as-good-as-new state.

7.3.1 Maintenance Policy

The maintenance policy is depicted in Fig. 7.2 [13]. The system starts from the initial state $X(0) = 0$, which means the system is perfect. The deterioration takes place gradually over time in non-negative increments. When the system state reaches a predictive maintenance threshold, $D_{PM} < D_F$, a preventive maintenance action is carried out (*e.g.*, at time points R_1, R_2, R_3 as shown in Fig. 7.2). The maintenance is imperfect, and it brings the system to an intermediate state (not as good as the previous working state). Due to aging of the system, the uptime before each main-

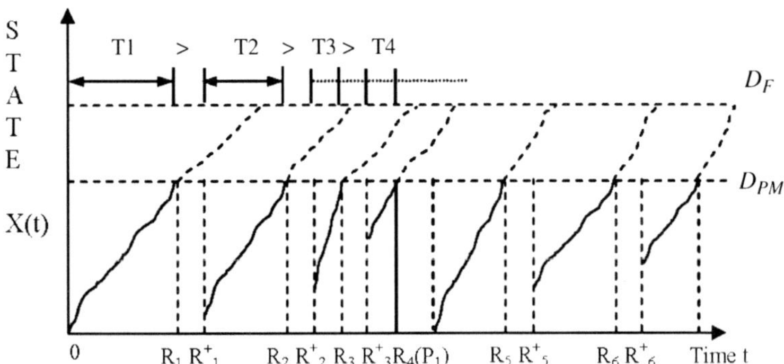

Fig. 7.2 The maintenance policy of the degradation system

tenance action deceases ($T_1 > T_2 > T_3 > \cdots$) as stated in assumption 4. Finally at some point of time, maintenance is not economically feasible and replacement is carried out instead to bring the system to as-good-as-new state.

We define a maintenance cycle as the period of time between two consecutive replacements. The effect of the maintenance policy is assessed by evaluating the system availability. We define two types of availabilities:

$$A1 = \frac{\text{expected total uptime/cycle}}{(\text{expected total uptime} + \text{expected total downtime})/\text{cycle}} \tag{7.2}$$

$$A2(i) = \frac{\text{expected uptime/maintenance action } i}{(\text{expected uptime} + \text{expected downtime})/\text{maintenance action } i} \tag{7.3}$$

where A_1 is the achieved system availability, and downtime may include the time of maintenance and replacement; $A_2(i)$ is the short-term availability of the ith maintenance action within a maintenance cycle.

7.3.2 Imperfect Maintenance Model

Let $f_{X(R_i^+)}(x)$ be the *pdf* of $X(R_i^+)$, the state of the degrading system immediately after the ith imperfect maintenance action. A viable model for the imperfect maintenance is described in Fig. 7.3, where $X(R_i^+)$ falls randomly in the interval $[0, D_{\text{PM}}]$ with an exponentially increasing mean and a constant variance, namely:

$$E\left[\frac{X(R_i^+)}{D_{\text{PM}}}\right] = 1 - \exp(-i\mu) \quad \text{and} \quad Var\left[\frac{X(R_i^+)}{D_{\text{PM}}}\right] = \sigma^2 \tag{7.4}$$

where $\mu \geq 0$ and $\sigma^2 \geq 0$.

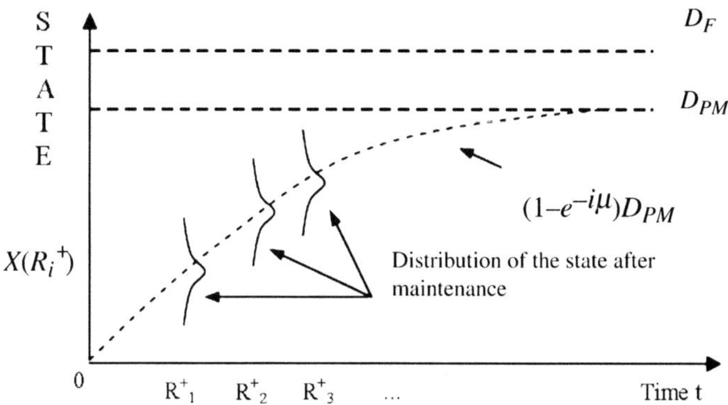

Fig. 7.3 Imperfect maintenance model

Beta distribution could be used to depict $X(R_i^+)$, such that the *pdf* of $X(R_i^+)$ is defined by [13]:

$$f_{X(R_i^+)}(x) = \frac{1}{D_{\mathrm{PM}}} \frac{\Gamma(\alpha_i + \beta_i)}{\Gamma(\alpha_i)\Gamma(\beta_i)} \left(\frac{x}{D_{\mathrm{PM}}}\right)^{\alpha_i - 1} \left(1 - \frac{x}{D_{\mathrm{PM}}}\right)^{\beta_i - 1} I\{0 \le x \le D_{\mathrm{PM}}\},$$

$$(7.5)$$

where the parameters $\alpha_i > 0$ and $\beta_i > 0$ are related to μ and σ^2 by:

$$E\left[\frac{X(R_i^+)}{D_{\mathrm{PM}}}\right] = \frac{\alpha_i}{\alpha_i + \beta_i} = 1 - \exp(-i\mu),$$

$$(7.6)$$

$$Var\left[\frac{X(R_i^+)}{D_{\mathrm{PM}}}\right] = \frac{\alpha_i \beta_i}{(\alpha_i + \beta_i)^2 (\alpha_i + \beta_i)} = \sigma^2.$$

$$(7.7)$$

The parameters μ and σ can be estimated using the Maximum Likelihood Estimation procedure.

7.3.3 Modeling Maintenance Time

The maintenance time usually depends on the initial system state, the predictive maintenance threshold level and the entire evolution of the system states. It is plausible to assume that the *i*th maintenance time M_i is exponentially distributed and independent of M_j ($j \neq i$); and its expectation is an increasing function of D_{PM} and the number of maintenance actions:

$$E[M_i] = \gamma_0 D_{\mathrm{PM}} \exp(i\gamma_1 D_{\mathrm{PM}}),$$

$$(7.8)$$

where $\gamma_0 > 0$ and $\gamma_1 \geq 0$ can be estimated by fitting exponential distribution to maintenance time observations.

7.4 Sensor Errors and Accuracy

The sensors' measurement errors are usually modeled as Gaussian white noise with zero mean and constant standard deviation [6–8], since the systematic measurement errors of sensors usually can be corrected by the analysis of physical mechanisms of the sensors [4,5]. In order to increase the measurement accuracy, we use multiple sensors and assume that the sensor measurements are statistically independent and identical. The average reading $\overline{RS}$ of n sensors is the monitored system state, which could be different from the true system state because of the measurement errors.

The system starts from an initial state $X(R_i^+)(0 \leq X(R_i^+) < D_{PM})$ after the ith maintenance action. We continuously monitor the system state using n sensors. When the average reading $\overline{RS}$ crosses the optimal threshold D_{PM}^*, the $(i+1)$th maintenance action is performed to bring the system state to a smaller degradation value. Since the sensor measurement include errors, when the average reading $\overline{RS}$ reaches the optimal threshold D_{PM}^*, the true system state $X(R_{i+1})$ has an unknown random deviation from the optimum threshold level $|X(R_{i+1}) - D_{PM}^*|$. The objective is to determine the minimum number of sensors in order to meet the requirement that the deviation of the true state from the optimum threshold level $|X(R_{i+1}) - D_{PM}^*|$ is equal to or less than a required value D. In other words, we are interested in finding the relationship between the minimum number of sensors and the required measurement accuracy, which is defined by D. Therefore the minimum number of sensors is defined by $n^* = \inf_{n \in Z^+} \{n : |X(R_{i+1}) - D_{PM}^*| \leq D\}$.

Hypothesis

Consider the hypothesis test $H_0 : X(R_{i+1}) = D_{PM}^*$ vs. $H_1 : X(R_{i+1}) < D_{PM}^*$. The test rule is: reject H_0 in favor of H_1 if $\overline{RS} \leq c$.

There are two types of errors associated with this hypothesis test. Type I error is the error in which H_0 is rejected when H_0 is true. Let α denote the probability of this event: $\alpha = P(\text{reject} H_0 | H_0 \text{ is true})$. It is desired α to be very small since when type I error event happens the system may have already failed. So in practice, we usually choose $\alpha = 0.01, \text{or} 0.05$. Given α value, we then determine the critical value c as follows:

When H_0 is true, we have $\overline{RS} \propto N(D_{PM}^*, \hat{\sigma}_s^2/n)$, where $\hat{\sigma}_s^2$ is the estimated variance of the sensor errors. We have the following relation after standardization.

$$\frac{\overline{RS} - D_{PM}^*}{\hat{\sigma}_s^2/\sqrt{n}} \propto N(0, 1) . \tag{7.9}$$

Thus, we obtain

$$\alpha = P\left(\overline{RS} \leq c | H_0 \text{ is true}\right) = P\left(\frac{\overline{RS} - D_{PM}^*}{\hat{\sigma}_s^2/\sqrt{n}} \leq \frac{c - D_{PM}^*}{\hat{\sigma}_s^2/\sqrt{n}}\right) = \Phi\left(\frac{c - D_{PM}^*}{\hat{\sigma}_s^2/\sqrt{n}}\right) , \tag{7.10}$$

where $\Phi(\cdot)$ is the *CDF* of standard normal distribution. Rewriting the above expression, we obtain

$$\frac{c - D_{PM}^*}{\hat{\sigma}_s^2/\sqrt{n}} = \Phi^{-1}(\alpha) = -\Phi^{-1}(1 - \alpha) , \tag{7.11}$$

and

$$c = D_{PM}^* - \Phi^{-1}(1 - \alpha)\hat{\sigma}_s^2/\sqrt{n} . \tag{7.12}$$

The second type of error is type II error, in which H_0 is accepted when H_1 is true. Let β denote the probability of this event: $\beta = P(\text{accept } H_0 | H_1 \text{ is true})$.

When this event happens, the maintenance action will be performed too early resulting in a decrease in system availability. In practice, type II error probability should be chosen between 0.1 and 0.2.

When H_1 is true, we have

$$\overline{RS} \propto N\left(X(R_{i+1}), \, \hat{\sigma}_s^2/n\right), \tag{7.13}$$

or

$$\frac{\overline{RS} - X(R_{i+1})}{\hat{\sigma}_s^2/\sqrt{n}} \propto N(0, 1). \tag{7.14}$$

Therefore, we obtain

$$\beta = P\left(\text{accept } H_0 | H_1 \text{ is true}\right) = P\left(\overline{RS} > c | H_1 \text{ is true}\right) = 1 - P\left(\overline{RS} \leq c | H_1 \text{ is true}\right), \tag{7.15}$$

or

$$1 - \beta = P\left(\overline{RS} \leq c | H_1 \text{ is true}\right) = P\left(\frac{\overline{RS} - X(R_{i+1})}{\hat{\sigma}_s^2/\sqrt{n}} \leq \frac{c - X(R_{i+1})}{\hat{\sigma}_s^2/\sqrt{n}}\right)$$

$$= \Phi\left(\frac{c - X(R_{i+1})}{\hat{\sigma}_s^2/\sqrt{n}}\right), \tag{7.16}$$

Since $c = D_{PM}^* - \Phi^{-1}(1 - \alpha)\hat{\sigma}_s^2/\sqrt{n}$, we have

$$1 - \beta = \Phi\left[\frac{D_{PM}^* - \Phi^{-1}(1 - \alpha)\hat{\sigma}_s^2/\sqrt{n} - X(R_{i+1})}{\hat{\sigma}_s^2/\sqrt{n}}\right]$$

$$= \Phi\left[\frac{D_{PM}^* - X(R_{i+1})}{\hat{\sigma}_s^2/\sqrt{n}} - \Phi^{-1}(1 - \alpha)\right], \tag{7.17}$$

Rewriting the above equation yields

$$\frac{D_{PM}^* - X(R_{i+1})}{\hat{\sigma}_s^2/\sqrt{n}} - \Phi^{-1}(1 - \alpha) = \Phi^{-1}(1 - \beta), \tag{7.18}$$

or

$$|D_{PM}^* - X(R_{i+1})| = [\Phi^{-1}(1 - \alpha) + \Phi^{-1}(1 - \beta)]\hat{\sigma}_s^2/\sqrt{n}. \tag{7.19}$$

Since it is desired that $|X(R_{i+1}) - D_{PM}^*| \leq D$, then when we have

$$\left[\Phi^{-1}(1 - \alpha) + \Phi^{-1}(1 - \beta)\right]\hat{\sigma}_s^2/\sqrt{n} \leq D, \tag{7.20}$$

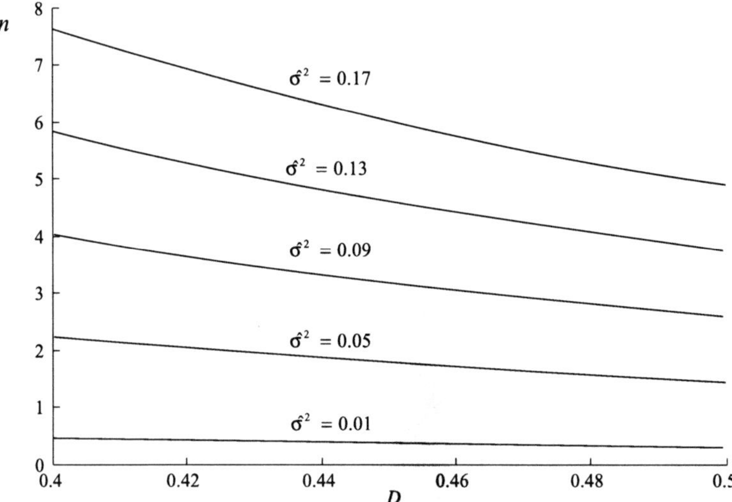

Fig. 7.4 Relationship between n and D for different $\hat{\sigma}^2$ ($\alpha = 0.05$, $\beta = 0.15$)

or

$$n \geq \frac{\left[\Phi^{-1}(1-\alpha) + \Phi^{-1}(1-\beta)\right]^2 \hat{\sigma}_s^2}{D^2} . \tag{7.21}$$

The above equation determines the minimum number of sensors that meets the required accuracy. As shown, the minimum number is a function of the variance of sensors' errors and the acceptable deviation. Figure 7.4 shows the relationship between the minimum number of sensors, the variance of sensors' errors and the acceptable deviation.

7.5 Uptime Modeling

The system state is modeled by gamma process as given in (7.1). The expected $(i+1)$th uptime is the expected time period when system state reaches its true value $X(R_{i+1})$ (at that time the $\overline{RS} = D_{\text{PM}}^*$) starting from $X(R_i^+)$ after ith maintenance action. Because of the white noise of sensors' errors, $X(R_{i+1})$ is normally distributed, i.e., $X(R_{i+1}) \propto N(D_{\text{PM}}^*, \hat{\sigma}_s^2/n)$.

The expected $(i+1)$th uptime is calculated by conditioning on $X(R_{i+1})$:

$$E[T_{i+1}] = E\left[E[T_{i+1}|X(R_{i+1})]\right] = \int_{-\infty}^{+\infty} E\left[T_{i+1}|X(R_{i+1}^{)}\right] f_{X(R_{i+1})}(X(R_{i+1}))\,\mathrm{d}X(R_{i+1}) .$$
$$\tag{7.22}$$

Without loss of generality, we let $X(R_{i+1}) = y$ to simplify the calculation. Then the above equation is expressed as

$$E[T_{i+1}] = \int_{-\infty}^{+\infty} E[T_{i+1}|y] f_{X(R_{i+1})}(y) \, dy \,, \tag{7.23}$$

where

$$f_{X(R_{i+1})}(y) = \frac{1}{\sqrt{2\pi(\hat{\sigma}_s^2/\sqrt{n})}} \exp\left[-\frac{1}{2\hat{\sigma}_s^2/n}(y - D_{PM}^*)^2\right] \,, \tag{7.24}$$

and $E[T_{i+1}|y]$ is calculated by conditioning on $X(R_i^+)$, which follows *beta* distribution given by (7.5),

$$E[T_{i+1}|y] = E\left\{E\left[(T_{i+1}|y)|X(R_i^+)\right]\right\} = \int_0^y \frac{y-x}{\eta\xi} f_{X(R_i^+)}(x) \, dx \,, \tag{7.25}$$

where

$$E\left[(T_{i+1}|y)|X(R_i^+)\right] = \frac{y - X(R_i^+)}{\eta\xi} = \frac{y-x}{\eta\xi} \quad (\text{where} \quad X(R_i^+) = x) \,. \tag{7.26}$$

Finally, the expected $(i+1)$th uptime is obtained as

$$E[T_{i+1}] = \int_{-\infty}^{+\infty} \left\{\int_0^y \frac{y-x}{\eta\xi} f_{X(R_i^+)}(x) \, dx\right\} f_{D_{PM}}(y) dy \quad i = 0, 1, 2, \ldots \tag{7.27}$$

7.6 Threshold Level: System Availability Maximization

7.6.1 Formulation of the Availability Maximization Problem

The objective of the proposed predictive maintenance policy is to achieve maximum system availability by controlling the predictive maintenance threshold degradation level. The optimum threshold D_{PM}^* is obtained by solving the following optimization problem:

$$\text{Max } A_1(D_{PM}^*)$$

Subject to

$$0 < D_{PM}^* \le D_F$$
$$A_2(i) > A_{2\min} \,; \quad i = 1, 2, \ldots, N-1$$
$$A_2(N) \le A_{2\min}$$

where

$$A_1 = \frac{\sum\limits_{i=1}^{N} E[T_i]}{\sum\limits_{i=1}^{N} E[T_i] + \sum\limits_{i=1}^{N-1} E[M_i] + E[T_R]},$$

$$A_2(i) = \frac{E[T_i]}{E[T_i] + E[M_i]},$$

$$N = \inf_{i \in Z^+} \{i : A_2(i) \le A_{2\min}\}.$$

and $E[M_i]$, $E[T_i]$ are given by (7.8) and (7.27) respectively.

The last constraint ensures that the system availability will not be less than $A_{2\min}$, which results in the determination of the maintenance cycle length before the replacement of system.

The above optimization problem can be easily solved by numerical search algorithm.

7.6.2 Numerical Example

In this example, the gamma degradation process parameters are $\eta = 4$ and $\xi = 0.25$. A replacement action is carried out when the degradation criterion (system state) reaches the failure threshold $D_F = 10$, or when the short-term availability $A_2(i) \le A_{2\min} = 0.6$; and a maintenance action is performed when system state crosses the optimum maintenance threshold D_{PM}^*. The expected replacement time and maintenance time are characterized by $T_R = 2$, $\gamma_0 = 0.02$, and $\gamma_1 = 0.05$. In addition, the system state after each maintenance action is characterized by $\mu = 0.5$ and $\sigma^2 = 0.005$. Furthermore, there are three sensors used, and the variance of sensor error is $\sigma_s^2 = 0.01$. The optimum predictive maintenance threshold D_{PM}^* is determined by solving the following non-linear optimization problem:

$$\text{Max } A_1(D_{PM}^*)$$

Subject to

$$0 < D_{PM}^* \le 10$$
$$A_2(i) > 0.6; \quad i = 1, 2, \ldots, N - 1$$
$$A_2(N) \le 0.6$$

The optimum solution is obtained by numerical search method:

$$D_{PM}^* = 7.2, \quad N^* = 5, \quad A_1^* = 0.7934.$$

7.7 Threshold Level: Maintenance Cost Minimization

The model discussed in Sect. 7.6 determines the optimum predictive maintenance
threshold level of the degradation system that maximizes the system availability.
There are many situations where the resources for the maintenance actions are limited. Therefore the minimization of the maintenance cost might be more important
than the maximization of the system availability. In this section, we present the optimum predictive maintenance model with the objective of minimizing the total cost
per unit time.

The degrading system is subject to failure when the system state reaches the
failure threshold. When failure occurs, the degrading system must be replaced.
Since failure is unexpected, it is reasonable to assume that a failure replacement is
more costly than a predictive maintenance/replacement. A predictive maintenance
is planned and necessary spare parts and manpower requirements are readily available such that arrangements are made to perform the maintenance without unnecessary delays. Similarly, a failure may cause significant losses and higher repair cost.
Therefore, we can readily modify the objective function of the maintenance policy
discussed in Sect. 7.3 to determine the optimum predictive maintenance threshold
level such that the total maintenance cost per unit time is minimized.

7.7.1 Formulation of the Cost Minimization Problem

The objective of the predictive maintenance model in this section is to achieve the
minimized total cost per unit time by determining the optimum predictive maintenance threshold degradation level. The expected total cost per unit time for a predictive maintenance threshold level D_{PM} is $C(D_{\mathrm{PM}})$

$$C(D_{\mathrm{PM}}) = \frac{\text{Total expected cost in the maintenance cycle}}{\text{Expected length of the maintenance cycle}}, \qquad (7.28)$$

where

$$\begin{array}{l}\text{Total expected cost during} \\ \text{the maintenance cycle}\end{array} = \text{Expected cost of maintenance actions}$$

$$+ \text{Expected cost of replacement}$$

$$\text{Expected cost of maintenance actions} = \sum_{i=1}^{N-1} E[CM_i]$$

$$\text{Expected cost of replacement} = E[C_{\mathrm{R}}]$$

$$\text{Expected length of the maintenance cycle} = \sum_{i=1}^{N} E[T_i] + \sum_{i=1}^{N-1} E[M_i] + E[T_{\mathrm{R}}]$$

Therefore the total expected cost per unit time $C(D_{\text{PM}})$ is given by

$$C(D_{\text{PM}}) = \frac{\sum_{i=1}^{N-1} E[CM_i] + E[C_R]}{\sum_{i=1}^{N} E[T_i] + \sum_{i=1}^{N-1} E[M_i] + E[T_R]} \,. \tag{7.29}$$

The denominator has already been determined in Sects. 7.3 and 7.5. The expected cost of replacement has a fixed value. The expected maintenance cost usually depends on the initial system state, the predictive maintenance threshold level, and the entire evolution of the system states. It is plausible to assume that the ith maintenance cost CM_i is exponentially distributed and independent of $CM_j (j \neq i)$; and its expectation is an increasing function of D_{PM} and the number of maintenance actions as described follows

$$E[CM_i] = A_0 D_{\text{PM}} \left[1 - \omega \exp(-iA_1 D_{\text{PM}})\right] , \tag{7.30}$$

where $A_0 > 0$, $0 \leq \omega \leq 1$, and $A_1 \geq 0$ can be estimated by fitting exponential distribution to maintenance cost observations.

The optimum threshold D_{PM}^* is obtained by solving the following optimization problem:

$$\text{Max } C(D_{\text{PM}})$$

Subject to

$$0 < D_{\text{PM}} \leq D_F$$
$$E[CM_i] < E[C_R] ; \quad i = 1, 2, \ldots, N - 1$$
$$E[CM_N] \geq E[C_R]$$

where

$$N = \inf_{i \in Z^+} \{i : E[CM_i] \geq E[C_R]\} .$$

7.7.2 Numerical Example

In this example, we have the same data as the numerical example in Sect. 7.6.2. Furthermore, the parameters related to the replacement and maintenance cost are: $E[C_R] = \$100$, $A_0 = 18.5$, $\omega = 0.9$, $A_1 = 0.02$. Therefore the optimum predictive maintenance threshold D_{PM}^* is determined by solving the following nonlinear optimization problem:

$$\text{Max } C(D_{\text{PM}})$$

Subject to

$$0 < D_{\text{PM}} \leq 10$$
$$E[CM_i] < 100 ; \quad i = 1, 2, \ldots, N - 1$$
$$E[CM_N] \geq 100$$

where

$$N = \inf_{i \in Z^+} \{i: E[CM_i] \geq 100\} .$$

The optimum solution is obtained as:

$$D_{\text{PM}}^* = 7.5 , \quad N^* = 7 , \quad C(D_{\text{PM}}^*) = 8.56 .$$

The corresponding system availability $A_1 = 0.7254$.

7.8 Conclusions

In this chapter, we present a continuous condition-based predictive maintenance policy for degrading systems which is continuously monitored by sensors. This predictive maintenance policy considers the imperfect maintenance actions and the random measurement errors of sensors. The number of sensors and its effect on the accuracy of reliability estimate are investigated. The optimum predictive maintenance thresholds are obtained for two situations: maximization of system availability and minimization of maintenance cost per unit time. Numerical examples are given to demonstrate that this predictive maintenance policy is effective and realistic, especially with consideration of the random measurement errors of the sensors.

This work serves as a nucleus for further research on sensors and their role in reliability engineering. One of the important issues to be investigated is the determination of the optimum threshold of sensors in order to minimize the effect of type I error, also known as α error, when a sensor observes a difference when in truth there is none (or more specifically – no statistically significant difference) and type II error, also known as a β error, when a sensors fails to detect a difference when in truth there is one. Another important issue to be investigated is the optimal location of sensors so that obtained observations do actually present the status of the system.

References

1. Lawless J, Crowder M (2004) Covariates and random effects in a gamma process model with application to degradation and failure. Lifetime Data Analysis 10:213–228
2. Grall A, Dieulle L, Bèrenguer C, Roussignol M (2002) Continuous-time predictive-maintenance scheduling for a deteriorating system, IEEE Transactions on Reliability 51:141–150

3. Dieulle L, Bèrenguer C, Grall A, Roussignol M (2003) Sequential condition-based maintenance scheduling for a deteriorating system, European Journal of Operational Research 150:451–461

4. Verinaud C (2004) On the nature of the measurements provided by a pyramid wave-front sensor, Optics Communications 223:27–38

5. Xu Y, Jones NB, Fothergill JC, Hanning CD (2001) Error analysis of two-wavelength absorption-based fiber-optic sensors, Optics and Lasers in Engineering 36:607–615

6. Wang G-H, Mao S-Y, He Y (2003) Analytical performance evaluation of association of active and passive tracks for airborne sensors, Signal Processing 83:973–981

7. Wilson DR, Apreleva MV, Eichler MJ, Harrold FR (2003) Accuracy and repeatability of a pressure measurement system in the patellofemoral joint, Journal of Biomechanics 36:1909–1915.

8. Ding Y, Elsayed EA, Kumara S *et al.* (2006) Distributed sensing for quality and productivity improvements, IEEE Transactions on Automation Science and Engineering 3:344–359

9. Singpurwalla N (1997) Gamma processes and their generalizations: an overview. In: Cooke M, Medel M, Vrijling H (eds) Engineering probabilistic design and maintenance for flood protection. Kluwer Academic Publishers, Dordrecht, pp. 67–73

10. Van der Weide H (1997) Gamma processes. In: Cooke M, Medel M, Vrijling H (eds) Engineering probabilistic design and maintenance for flood protection. Kluwer Academic Publishers, Dordrecht, pp. 77–82

11. Van Noortwijk J, Klatter HE (1999) Optimal inspection decisions for the block mats of the Eastern-Scheldt barrier, Reliability Engineering and System Safety 65:203–211

12. Crowder M, Lawless J (2007) On a scheme for predictive maintenance, European Journal of Operational Research 176:1713–1722

13. Liao H, Chan L-Y, Elsayed EA (2006) Maintenance of continuously monitored degrading systems, European Journal of Operational Research 175:821–835

14. Grall A, Dieulle L, Bèrenguer C, Roussignol M (2006) Asymptotic failure rate of a continuously monitored system, Reliability Engineering and System Safety 91:126–130

Chapter 8
Weibull Data Analysis with Few or no Failures

Ming-Wei Lu, Cheng Julius Wang

Daimler Chrysler Corporation, 800 Chrysler Drive, Auburn Hills, MI 48326, USA

8.1 Introduction

Laboratory testing is a critical step in the development of vehicle components or systems. It allows the design engineer to evaluate the design early in the reliability development phase. A good lab test will shorten the product development cycles and minimizes cost and part failures at the PG or field testing before the vehicle volume production. Appropriate testing is available to correlate test time in the lab (or lab test bogey) to the real world survival time (or field design life). The testing must be in some accelerated fashion or typically called accelerated testing. The failure mechanism(s) that the accelerated test will bring out is of great importance. No one test can surface all potential failure mechanisms of the part. Certain failure mechanisms dominate throughout the useful lifetime of the part, and some may never occur.

To verify a new product design meeting a reliability target requirement, one can perform data analysis on the life testing data by using Weibull life distribution. However, in fitting a Weibull distribution to reliability data, one may have only few or no failures. This paper presents method to estimate the reliability and confidence limits that apply to few or no failures with an assumed Weibull slope value of β.

The Weibull slope is known before the testing. This assumption can be (and will be) misused by people to achieve one goal – fewer samples for testing. The known Weibull slope should be estimated from similar component testing data, experience and engineering knowledge.

To judge the effect of this assumed slope value, we can repeat the calculations to see how much the reliability estimates changes with other reasonable slope values. We may also use the Excel Solver program to find the minimum reliability value subject to certain constraints of slope values.

Section 8.2 describes the Weibull analysis method with few or no failures based on the paper by Nelson [1]. In Sect. 8.4, the Monte Carlo simulation technique with only three failure samples was performed to investigate the properties of the estimates from a two-parameter Weibull distribution.

8.2 Theory

The reliability function at time x from a two-parameter Weibull distribution is given by

$$R(x) = \exp\left[-\left(\frac{x}{\theta}\right)^{\beta}\right] \quad x > 0 \tag{8.1}$$

8.2.1 Nelson's Method [1]

For a given sample of size n, suppose that the $r \geq 0$ failure times and the $(n-r)$ times of non-failures are $t_1, t_2, \ldots, t_n$. Assuming that the β is given, the corresponding lower C confidence limit for the true θ is given by

$$\theta_C = \left\{2t' / \chi^2_{(C;2r+2)}\right\}^{1/\beta} \tag{8.2}$$

where $t' = t_1^{\beta} + t_2^{\beta} + \ldots + t_n^{\beta}$, and $\chi^2_{(C;2r+2)}$ is the $(100C)$th percentile of the Chi-square distribution with $(2r+2)$ degrees of freedom.

When the part is tested to time t, then the reliability at t_0 with C confidence limit is given by

$$\begin{aligned} R(t_0) &= \exp\left[-(t_0/\theta_C)^{\beta}\right] \\ &= \exp\left[-t_0^{\beta}\chi^2_{(C;2r+2)}/(2t')\right] \end{aligned} \tag{8.3}$$

Equation (8.3) can be rewritten as

$$R(t_0) = \exp\left[-\frac{\chi^2_{(C,2r+2)}}{2\sum\limits_{i=1}^{n}\left(\frac{t_i}{t_0}\right)^{\beta}}\right] \tag{8.4}$$

Under the condition

$$(t_1, t_2, \ldots, t_n)^{1/n} < t_0 < \max(t_1, t_2, \ldots, t_n) \tag{8.5}$$

there exists a unique β_0 such that [2]

$$R(t_0) \text{ is minimum at } \beta = \beta_0 \text{ for } \beta > 0$$

Equation (8.4) suggests that each part can be stopped at different times. The usefulness of this equation is in trading off on the number of parts and test times when certain parts are no longer available during the testing. This may be due to the fact that certain testing fixtures are broken; certain parts may be required for other purposes and being pulled out of the testing.

When small number of failures (say less than or equal than 3 parts) occurred during the testing, (8.4) can be used. If large numbers of failures occurred, then we can apply the Weibull plot to find the underlying Weibull parameters.

8.2.2 Extended Test Method [3]

One of the situations encountered frequently in testing involves a trade-off between sample size and testing time. If the test item is expensive, the number of test items can be reduced by extending the time of testing on fewer items. Extended testing is a method of reducing sample size by testing the samples to a time that is higher than the test bogey requirement. Some limitations of using the extended test method are: (1) it requires knowledge of Weibull slope, and (2) there is an extended test time and no failures are allowed. The approach is based on the Weibull distribution with parameters β and θ (8.1). It is assumed that all parts on test must complete the planned extended test time without failure to successfully demonstrate the reliability target requirement. If failures occur, the demonstration is no longer valid. However, if failures do occur, one can use Weibull analysis to assess the status.

The relationship between sample size, extended testing time, confidence level, and reliability is given by [3]

$$R(t_0) = \exp \left[\frac{\ln(1-C)}{n\left(\frac{T}{t_0}\right)^{\beta}} \right] \tag{8.6}$$

where

$R(t_0) = $ reliability at time t_0
$t_0 = $ test bogey
$T = $ extended testing time
$n = $ number of items running without failure to T
$\beta = $ Weibull shape parameter or Weibull slope
$C = $ confidence level specified in percentage

Since $dR(t_0)/d\beta > 0$, then $R(t_0)$ is an increasing function of β ($\beta > 0$). Hence, $R(t_0)$ is minimum at $\beta = a$ when β ranges from a to b (where $0 < a < b$).

8.3 Examples

Example 1. For component life testing, a total of five tests were conducted with no failures (5 suspended cycles: 100, 700, 1000, 2000, 2500 cycles). The test bogey is

2000 cycles. From past testing history, the slope ranges from 2.5 to 5 for similar parts. What is the reliability value at 2000 cycles with $C = 50\%$?

Solution 1. Assuming that the $\beta = 2.5$ (conservative), from (8.4), the reliability at 2000 cycles with $C = 50\%$ is given by

$$R(2000) = 79.35\%$$

where $n = 5$, $r = 0$, $C = 50\%$, $\beta = 2.5$, $t_0 = 2000$, $t_1 = 100$, $t_2 = 700$, $t_3 = 1000$, $t_4 = 2000$, $t_5 = 2500$, and

$$\chi^2_{(C;2r+2)} = \chi^2_{(0.5;2)} = 1.386.$$

Note that since $(t_1, t_2, \ldots, t_5)^{1/n} = 811 < t_0 = 2000 < \max(t_1, t_2, \ldots, t_5) = 2500$, by (8.4) and the Excel Solver program, there exists a unique $\beta_0 = 1.894$ such that $R(t_0)$ is minimum. The minimum reliability value is

$$R(2000) = 78.97\%$$

Example 2. A bearing engineer has developed a new bearing. The new bearing must have a B_5 life of 1000 h. The company has set a requirement of a 50% confidence level for all decisions. Only three bearings are available for testing and three bearings run without failure by the end of 3000 h. We also know that the Weibull slope is about 1.5. Does the bearing meet $R = 95\%$ with $C = 50\%$ reliability requirement?

Solution 2. By using (8.6) with $n = 3$, $t = 1000$, $T = 3000$, $\beta = 1.5$, and $C = 50\%$, then

$$R(1000) = 95.65\%$$

Hence, three bearings must run 3000 h without failure to meet the $R = 95\%$ with $C = 50\%$ reliability requirement.

Example 3. Supposed we would like to demonstrate reliability to be 95% at design life "$t_0 = 1000\,h$" at 90% statistical confidence level. We can test each part to 1000 h. We also know that Weibull slope is about 1.0. How many parts do we need without failure to demonstrate this target?

Solution 3. Use (8.6), we obtain

$$R(1000) = 0.95 = \exp\left[\frac{\ln(1 - 0.9)}{n\left(\frac{1000}{1000}\right)^1}\right]$$

Solving for n, we obtain $n = 45$. That means we need to test 45 parts without failure to 1000 h each to demonstrate 95% reliability at the design life of 1000 h of laboratory testing at 90% statistical confidence level.

Example 4. Continuing from Example 3, what if we can test each part to 1500 hours, then how many samples do we need?

Solution 4. Using (8.6, we obtain

$$R(1000) = 0.95 = \exp\left[\frac{\ln(1 - 0.9)}{n\left(\frac{1500}{1000}\right)^{1}}\right]$$

Solving for n, we obtain $n = 30$.

From Example 3 and 4, we know if we test longer, we can cut down the sample size.

Example 5. Continuing on Example 3, suppose we have a different Weibull slope, say the β is 3.0, not 1.0. What is the required sample size?

Solution 5. Applying (2.6) with $\beta = 3.0$, we obtain

$$R(1000) = 0.95 = \exp\left[\frac{\ln(1 - 0.9)}{n\left(\frac{1500}{1000}\right)^{3}}\right]$$

Solving for n, we have $n = 14$.

From Example 4 and 5, we learned that if the Weibull slope is larger, we can cut down the sample size.

Example 6. Continuing from Example 5, the plan is to test 14 parts without failure to 1500 h to demonstrate 95% reliability at 1000 h (design life in laboratory testing) at 90% statistical confidence level. The Weibull slope is assumed to be 3.0. Right in the middle of the testing, two parts were stopped at 450 h due to a broken connector. At this time, another department requires these two parts for other use so you lose these two parts. How long should the other 12 parts be tested to maintain the reliability demonstration requirement?

Solution 6. Use (8.4), we obtain

$$R(1000) = 0.95 = \exp\left[-\frac{4.605}{2\left(2\left(\frac{450}{1000}\right)^{3} + 12\left(\frac{T}{1000}\right)^{3}\right)}\right]$$

Solving for T, we obtain $T = 1550$ h. Therefore, the remaining 12 parts will be tested to a total of 1550 h, rather than 1500 h as originally planned.

Example 7. Continuing from Example 6, one failure occurs around 1200 h. Engineering examined the failure, and concluded that it might be caused by a component batch problem (prototype build) and they do not want to repair the part and restart the test. The question is how many more testing hours are needed for the rest of the 11 parts to pass without failure so we can still demonstrate the reliability in time.

Solution 7.

$$R(1000) = 0.95 = \exp\left[-\frac{7.779}{2\left(2\left(\frac{450}{1000}\right)^3 + \left(\frac{1200}{1000}\right)^3 + 11\left(\frac{T}{1000}\right)^3\right)}\right]$$

Solving for T, we obtain $T = 1887$ h. This means that the remaining 11 parts will be tested to 1887 h without failure, rather than the 1550 h as planned before.

8.4 Simulation Study with Only Three Failures

In this section, the Monte Carlo simulation technique was used to investigate the properties of the estimates of the two parameter Weibull distribution. The simulation study was based on 1000 Monte Carlo trials for a sample size of $n = 3$ failures. The histograms on the generated 1000 $B5$ life, $B10$ life, $B50$ life, and theta (θ) values from a Weibull distribution with $\beta = 2$ and $\theta = 100$ by using rank regression (RR) method are shown in Figs. 8.1–8.4. The $B5$ life, $B10$ life, and $B50$ life are defined as the 5% failure life, 10% failure life and 50% failure life respectively.

The mean and standard deviation by $\beta = 1, 2, 3, 4$ and $\theta = 100$ are shown in Table 8.1.

The error values by $\beta = 1, 2, 3, 4$ and $\theta = 100$ are shown in Table 8.2.

For $x = 5, 10, 50$, Error(Bx) and Error(θ) are defined as

$$\text{Error}(Bx) = \sqrt{\frac{1}{1000}\sum\left(\frac{\text{Estimated } Bx - \text{True } Bx}{\text{True } Bx}\right)^2}$$

$$\text{Error}(\theta) = \sqrt{\frac{1}{1000}\sum\left(\frac{\text{Estimated } \theta - \text{True } \theta}{\text{True } \theta}\right)^2}$$

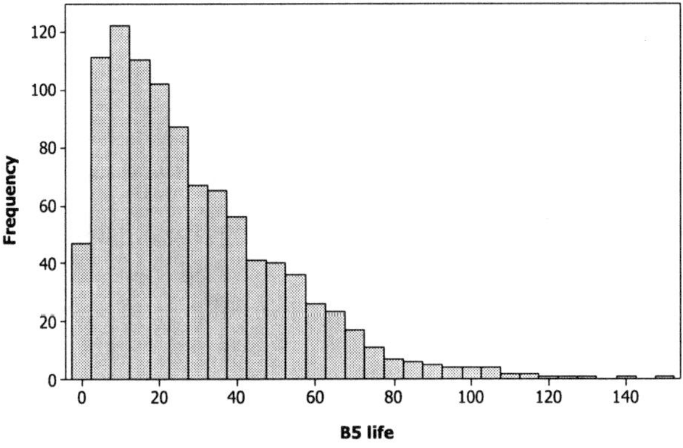

Fig. 8.1 Histogram of $B5$ life

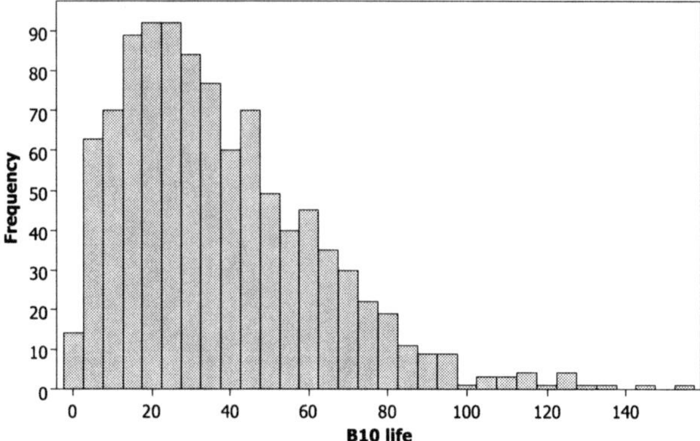

Fig. 8.2 Histogram of *B*10 life

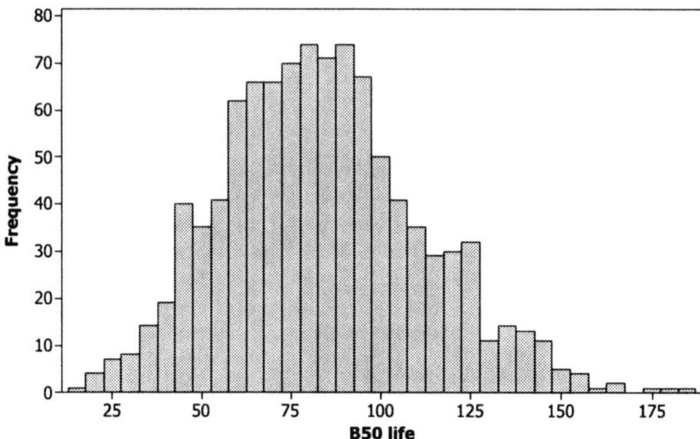

Fig. 8.3 Histogram of *B*50 life

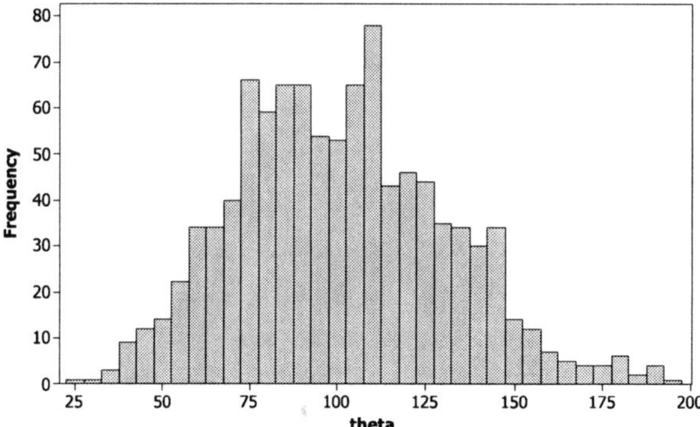

Fig. 8.4 Histogram of theta

As can be seen from the above table, the estimation of *B*5 life and *B*10 life were unstable at different beta values. The *B*5 life had the largest error and the θ had the smallest error (*i.e.*, the most stable).

Table 8.1 Mean and standard deviation (SD) by $\beta = 1, 2, 3, 4$ and $\theta = 100$

| | β | 1 | 2 | 3 | 4 |
	θ	100	100	100	100
$B5$ life	Average	13.9	28.9	41.3	50.3
	SD	21.2	23.3	23.6	22.0
$B10$ life	Average	20.6	37.2	49.8	58.5
	SD	27.2	24.8	23.5	20.9
$B50$ life	Average	76.8	83.7	87.5	90.3
	SD	49.7	27.6	20.1	15.5
θ	Average	109.6	100.8	99.6	99.1
	SD	60.5	29.9	20.3	15.6

Table 8.2 Error values by $\beta = 1, 2, 3, 4$ and $\theta = 100$

| β | 1 | 2 | 3 | 4 |
θ	100	100	100	100
True $B5$	5.1	22.6	37.2	47.6
Error ($B5$)	446.4%	106.3%	64.4%	46.5%
True $B10$	10.5	32.5	47.2	57.0
Error ($B10$)	275.5%	77.9%	50.0%	36.8%
True $B50$	69.3	83.3	88.5	91.2
Error ($B50$)	72.5%	33.2%	22.8%	17.0%
True θ	100	100	100	100
Error (θ)	61.2%	29.9%	20.3%	15.6%

We may have the following suggestion to calculate $B5$ or $B10$ life based on the three failure data if the β value is known.

1. Estimate the θ value by rank regression method.
2. Estimate $B5$ or $B10$ life by the following formula

$$B5 \text{ life} = \theta \left[\ln(0.95)/\ln(0.632)\right]^{1/\beta} \tag{8.7}$$

$$B10 \text{ life} = \theta \left[\ln(0.90)/\ln(0.632)\right]^{1/\beta} \tag{8.8}$$

8.4.1 Weibull Parameter Estimation Method

Most commercially available Weibull analysis software packages utilize two basic parameter estimation methods: (1) rank regression (RR) (conventional method of least squares curve fit to obtain the least error between time-to-failure data and the fit line) and (2) maximum likelihood estimate (MLE) (iterative solution procedure to obtain the fit line that will "most likely" produce the actual data value). Based on 1000 Monte Carlo trials for a sample size of $n = 5, 10, 20, 30, 50, 100$, Error($\beta$), Error($\theta$), and Error($B10$) are shown below by different $\beta = 0.5, 1, 2, 3, 4$ and $\theta = 100$.

Both methods have associated advantages and disadvantages. Rank regression tends to provide more accurate estimates for $B10$ life for small sample sizes. MLE is considered by statisticians to have excellent mathematical qualities but tends to overestimate β with small sample sets. It is recommended that the rank regression method be used for all sample sizes of 15 or less. For sample sizes greater than 15, the MLE method tends to be more accurate (Table 8.3).

Table 8.3 Maximum likelihood estimate (MLE) method

Sample size	β	Error (β)		Error (θ)		Error (B10)	
		RR	MLE	RR	MLE	RR	MLE
5	0.5	0.7063	0.9379	1.3356	1.1841	10.9015	14.9165
	1	0.6811	0.8970	0.4921	0.4909	1.7411	2.1885
	2	0.6706	0.8904	0.2342	0.2397	0.5726	0.6648
	3	0.6479	0.9998	0.1544	0.1557	0.3728	0.3934
	4	0.6741	0.8329	0.1237	0.1156	0.2758	0.2750
10	0.5	0.3595	0.3981	0.8307	0.8016	4.1884	5.5120
	1	0.3623	0.3741	0.3437	0.3402	0.9809	1.1032
	2	0.3970	0.4022	0.1655	0.1746	0.4143	0.4373
	3	0.3568	0.4171	0.1110	0.1103	0.2706	0.2699
	4	0.3771	0.4121	0.0854	0.0831	0.2056	0.1939
20	0.5	0.2296	0.2119	0.5324	0.5480	1.9158	2.2941
	1	0.2324	0.2182	0.2341	0.2336	0.5994	0.6179
	2	0.2251	0.2103	0.1202	0.1146	0.2808	0.2642
	3	0.2407	0.2199	0.0814	0.0747	0.1992	0.1742
	4	0.2218	0.2233	0.0608	0.0591	0.1416	0.1356
30	0.5	0.1818	0.1671	0.4104	0.4049	1.2414	1.4012
	1	0.1882	0.1654	0.1962	0.1945	0.4988	0.4966
	2	0.1820	0.1700	0.1028	0.0945	0.2376	0.2258
	3	0.1877	0.1655	0.0663	0.0631	0.1578	0.1450
	4	0.1859	0.1661	0.0480	0.0480	0.1205	0.1079
50	0.5	0.1380	0.1170	0.3171	0.3146	0.8829	0.9021
	1	0.1469	0.1188	0.1564	0.1509	0.3886	0.3690
	2	0.1437	0.1281	0.0744	0.0737	0.1809	0.1744
	3	0.1379	0.1244	0.0499	0.0513	0.1217	0.1181
	4	0.1424	0.1256	0.0377	0.0368	0.0925	0.0819
100	0.5	0.1011	0.0848	0.2222	0.2088	0.5726	0.5822
	1	0.1005	0.0809	0.1091	0.1023	0.2628	0.2385
	2	0.1016	0.0789	0.0527	0.0504	0.1305	0.1125
	3	0.1047	0.0845	0.0360	0.0360	0.0909	0.0804
	4	0.1046	0.0814	0.0260	0.0266	0.0673	0.0581

8.5 Conclusions

When the reliability data have few or no failures, the following concerns should be considered for calculating reliability with a given confidence level.
1. Do we have the assumed Weibull slope value?

The Weibull slope can often be approximated from similar component testing data or engineering experience. If we have the assumed Weibull slope value, then we can compute the reliability values by using either (8.4) or (8.6).

2. To judge the effect of this assumed slope value, we can repeat the calculations to see how much the reliability estimates changes with other reasonable slope values. We may also use Excel Solver program to find the minimum reliability value subject to certain constraints of slope values.

From the simulation study with only three failures in Sect. 8.4, the estimate of the $B5$ life and $B10$ life should be viewed with care, since they have significant errors. If the β value is known, one can estimate $B5$ or $B10$ life by (8.7) or (8.8) along with the estimated θ value from the three failure data points.

From the simulation study with different sample sizes in Sect. 8.4, it is recommended that the Rank Regression method be used for all sample sizes of 15 or less.

References

1. Nelson W (1985) Weibull analysis of reliability data with few or no failures. Journal of Quality Technology 17:140–146
2. Huang Z, Porter AA (1991) Lower bound on reliability for Weibull distribution when shape parameter is not estimated accurately. Annual Reliability and Maintainability Symposium Proceedings, pp. 183–189
3. Lee Y-L, Pan J, Hathaway R, Barkey M (2004) Fatigue testing and analysis (theory and practice). Elsevier/Butterworth-Heinemann, Burlington, Mass., USA

Chapter 9
A Load-weighted Statistical Average Model of Fatigue Reliability

Liyang Xie, Zheng Wang

Department of Mechanical Engineering,
Northeastern University,
Shenyang, 110004, China

9.1 Introduction

Stress-strength interference (SSI) analysis method has been applied to reliability estimation of a broad range of structural components under a variety of loading conditions [1–6]. This method is successful for static strength failure. For fatigue failure, some of the current methods use SSI technique directly, assuming that the distributions of applied stress and fatigue strength corresponding to specific number of cycles to failure are known. However, the exact distribution of fatigue strength at specific number of cycles to failure cannot be obtained from a test [3]. Alternatively, methods to determine fatigue strength distribution from fatigue life distribution were proposed [4].

To derive the distribution of fatigue strength which cannot be exactly obtained through tests, Murty *et al.* [4] and others took fatigue strength as a function of fatigue life to make use of the advantage that fatigue life distribution can be obtained directly from tests. In Murty's work, the fatigue life distribution was derived from S–N curve, which was developed from the fatigue life test data at two stress levels, the number of cycles to failure was assumed to follow a log-normal distribution.

Ni and Zhang [7] studied the distribution of residual fatigue life under two-stage loading. Based on the two-dimensional probabilistic Miner's rule, a method of fatigue reliability analysis under two-stage loading was established, and verification was made in terms of eight large samples of low-high and high-low two-stage fatigue experimental data given by Tanaka *et al.* [8].

Xie [9] analyzed the equivalent fatigue life distributions under two-level cyclic stress experimentally and presented a model of describing equivalent life distribution. Based on the equivalent life distribution model, an "equivalent damage-equivalent life distribution" method was developed to predict fatigue failure probability under variable amplitude loading. The underlying principles were the fatigue damage accumulation rule and the "load cycles-fatigue life" interference theorem. In the equivalent damage-equivalent life distribution method, two parameters, *i.e.*, mean and standard deviation of the (equivalent) life distribution were used to describe cumulative fatigue damage effect. A basic feature of the equivalent damage-equivalent life distribution method was its capability of reflecting the loading history dependent change of the equivalent life distribution.

Focusing on time-variant reliability assessment of deteriorating structures under fatigue conditions, Petryna *et al.* [10] proposed a phenomenological fatigue damage model of reinforced concrete. It was emphasized that reliability assessment of structures under fatigue conditions is a highly complicated problem, which implies interaction of different scientific fields such as damage and continuum mechanics, non-linear structural analysis and probabilistic reliability theory, while the majority of publications still concerns simple cases of local damage in a single cross-section and associate them with a global limit state, such as in typical fatigue crack growth models.

Karadeniz [11] presented a procedure of modeling uncertainties in the spectral fatigue analysis of offshore structures with reference to the reliability assessment. The inherent uncertainties of the damage were categorized in two main groups as being those of introduced by the stress statistical characteristics and those that may be encountered in the damage model of the fatigue phenomenon self.

Returning to the situation where only stress and strength are concerned, the two parameters have the same units (MPa) and the relative magnitudes of the two variables determine whether or not failure will occur, the simplest method to calculate reliability (*i.e.*, the probability that strength exceeds stress) is using the well-known SSI model

$$R = \int_0^\infty h(y) \left[\int_y^\infty f(x)\,\mathrm{d}x \right] \mathrm{d}y \qquad (9.1)$$

Where, $h(y)$ – pdf of load or stress random variable y, $f(x)$ – pdf of component strength random variable x.

Similarly, fatigue reliability can be easily calculated through stress cycles-fatigue live interference model in the situation of deterministic cyclic loading of constant amplitude. Obviously, such a model is not suitable for other situations of stochastic cyclic load [9].

Conventional method to calculate component fatigue reliability, which applies directly the well known stress-strength interference (SSI) model, often encounters the difficult to determine the fatigue strength distribution under an assigned number of stress cycles to failure. In the present article, traditional SSI model is interpreted as a statistical average of the probability that strength exceeds stress over the whole range of the random stress. Thus, the same model configuration, which traditionally can only be used to the case of the same system-of-units parameters (*e.g.*, both the stress and strength are measured in MPa), can be applied to the more general situation of different system-of-units parameters by putting into different contents. In other words, the traditional model is extended to the situation of any two variables, as long as one of the variables can be expressed as a function of the other. This extended interference analysis method can be applied to calculate fatigue reliability under cyclic load with uncertainty in stress amplitude, needing not to know the strength distribution of the object at assigned number of cycles to failure. With a specific load amplitude distribution, fatigue reliability is calculated by the statistical average of the probabilities that fatigue life is greater than an assigned value with respective the stochastically distributed constant amplitude cyclic loads.

The application of the load-weighted statistical average model is illustrated by an example, and the effect of load uncertainty on reliability is shown.

9.2 Statistical Average Interpretation of SSI Model

As we know, the SSI model (Eq. (9.1)) is normally applied in the situation that both of the stress and strength are random variables. In the condition of deterministic stress Y (denoted by italic capital letter), the model degenerates into

$$R(Y) = \int_Y^\infty f(x)\,dx \tag{9.2}$$

Obviously, the above equation is equal to the probability that the strength x (a random variable) exceeds the specific stress Y (a deterministic value). Moreover, either component reliability or failure probability can be taken as a function of stress y. Here, we define component failure probability condition to stress y as

$$\pi(y) = \int_0^y f_{(x)}\,dx \tag{9.3}$$

and define component reliability condition to stress y as

$$\psi(y) = \int_y^\infty f_{(x)}\,dx \tag{9.4}$$

For any y

$$\pi(y) + \psi(y) = 1 \tag{9.5}$$

In the general condition that both stress and strength are random variables, the conditional component failure probability is also a random variable, since it is a function of random stress y. The individual probability density functions can be derived from the probability density function of the stress y.

Contrasting to the expression of the conditional failure probability is the cumulative distribution function (cdf) of a random variable in probability theory. If $f(x)$ denotes the probability density function of a random variable x, then the corresponding cdf is

$$F(x) = \int_0^x f(x)\,dx \tag{9.6}$$

Different from the cdf (Eq. (9.6)), there are two different variables in the conditional failure probability or conditional reliability ((9.3) or (9.4)), *i.e.*, the variable x is strength, while the variable y is stress in these two equations.

Thus, the SSI model (*i.e.*, Eq. (9.1)) can be rewritten as

$$R = \int_0^\infty h(y)\psi(y)\,dy \tag{9.7}$$

The above model can be interpreted as the statistical average of the conditional component reliability $\psi(y)$ weighted by the random stress y over its whole range of definition $(0 - \infty)$.

Based on the statistical average interpretation to the SSI model, we can develop a general success/failure probability model to which the traditional SSI model is a special case. As we know, the traditional SSI model is only suitable for two comparable variables of the same system-of-units, such as stress and strength, number of stress cycles and number of fatigue life cycles, and so on. In terms of the statistical-average interpretation, the formula in the typical form can be used to calculate the statistical average of a more general function of an arbitrary argument y (a random variable) with respect to the random variable y over its possible region. For instance, let

$$\pi(y) = \int_0^{H(y)} f(x, y)\,dx$$

where $H(y)$, the upper limit of the integral, is also a function of y.

Obviously, it is not necessary for the variable x to have the same units as the variable y in this equation.

For the reliability problem related to static strength failure,

$$H(y) = y$$

$$\pi(y) = \int_0^y f(x)\,dx$$

Apparently, here $\pi(y)$ is the probability that component strength x is less than a specific stress value y.

For the reliability problem related to fatigue failure under constant amplitude cyclic stress,

$$\pi(Y) = \int_0^N f(n, Y)\,dn$$

where, $f(n, Y)$ stands for the pdf of the fatigue life n under the constant amplitude cyclic stress Y, N is an arbitrarily assigned fatigue life.

This $\pi(Y)$ is the probability that the fatigue life n is less than a specific value N under a specific cyclic stress amplitude Y.

Accordingly,

$$\int_0^\infty h(y)\pi(y)\,dy = \int_0^\infty h(y)\left[\int_0^N f(n, y)\,dn\right] dy$$

is equal to the fatigue failure probability under the random stress y, and

$$\int_0^\infty h(y)\psi(y)\,\mathrm{d}y = \int_0^\infty h(y)\left[\int_N^\infty f(n,y)\,\mathrm{d}n\right]\mathrm{d}y$$

is equal to the fatigue reliability under the random stress y. Details are as follows.

9.3 A Statistical Load-weighted Average Model of Fatigue Reliability

To extend the traditional SSI analysis methodology and its application as well, illustrated here is the fatigue reliability calculation in the condition of constant amplitude cyclic stress with uncertainty in stress amplitude. In other words, the amplitudes of the cyclic stress population follow a certain distribution and the amplitude of any cyclic stress sample is a cyclic stress with deterministic amplitude. The approach is to calculate component fatigue reliability according to stress distribution and fatigue life distribution directly. With a extended SSI model, *i.e.*, a stress-weighted conditional reliability statistical average model, no fatigue strength distribution at specific fatigue life is required when calculating fatigue reliability, so that the difficulty in determining fatigue strength distribution is avoided.

Let $f(n,Y)$ denotes the pdf of fatigue life under a deterministic cyclic stress with amplitude Y, then the probability that the fatigue life n is greater than a specific number of stress cycles N is (see Fig. 9.1).

$$R(N,Y) = \int_N^\infty f(n,Y)\,\mathrm{d}n \qquad (9.8)$$

Evidently, the only uncertainty involved in Eq. (9.8) is in the aspect of fatigue life under deterministic load arising from material performance. On the other hand, the uncertainty in cyclic stress is also one of the most important factors to affect fatigue reliability. The uncertainty in stress amplitude is not negligible and stress amplitude has to be taken as a random variable. As to the approach to introduce the uncer-

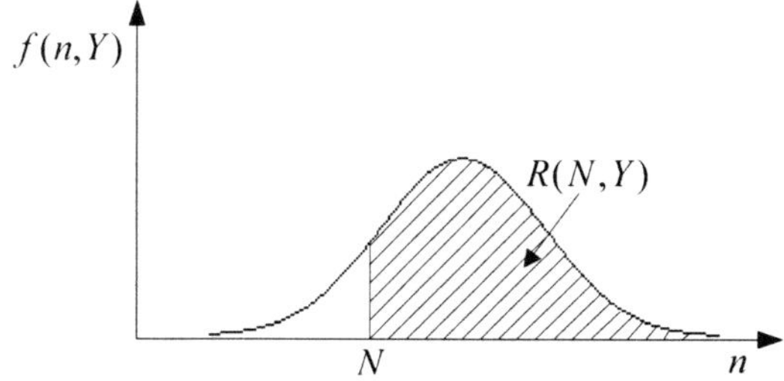

Fig. 9.1 Conditional reliability to a deterministic amplitude cyclic stress Y at specified cycle number N

tainty of operating environments for reliability prediction, Pham [12] presents a new mathematical function called systemability, which highlights the effect of the status of the system such as vibration level, efficiency, or number of random shocks on the system on the failure rate of a system.

Under random cyclic stress of constant amplitude, the fatigue reliability model can be developed as below.

First, let us consider a simple situation that the stress amplitude takes only on two possible values of Y_1 and Y_2 with the probability of p_1 and p_2 respectively ($p_1 + p_2 = 1$). The fatigue reliability to a specific life N at the stress amplitude Y_1 is

$$R_1(N, Y_1) = \int_N^\infty f(n, Y_1)\,dn\,,$$

and that at the stress amplitude Y_2 is

$$R_2(N, Y_2) = \int_N^\infty f(n, Y_2)\,dn\,.$$

By the total probability principle, under the loading condition that the stress amplitude takes on the two values of Y_1 and Y_2 with the probability of p_1 and p_2 respectively, the fatigue reliability is

$$R(N) = p_1 R_1(N, Y_1) + p_2 R_2(N, Y_2)$$

i.e.,

$$R(N) = p_1 \int_N^\infty f(n, Y_1)\,dn + p_2 \int_N^\infty f(n, Y_2)\,dn\,. \tag{9.9}$$

Extending Eq. (9.9) to the situation of stochastic constant amplitude cyclic stress that the stress amplitude takes on m values of Y_i ($i = 1, 2, \ldots, m$) with the respective probabilities of p_i ($i = 1, 2, \ldots, m, \sum p_i = 1$), the fatigue reliability is then (refer to Fig. 9.2)

$$R(N) = \sum_{i=1}^m p_i \int_N^\infty f(n, Y_i)\,dn\,. \tag{9.10}$$

When the stress amplitude is a random variable with the pdf of $h(y)$, then the p_i in Eq. (9.10) should be replaced by $h(y_i)\Delta y_i$, i.e.

$$p_i = h(y_i)\Delta y_i \quad (i = 1, 2, \ldots, n)$$

where, y_i are the arbitrarily determined discrete values of the stress y; Δy_i are the associated stress intervals.

Accordingly, Eq. (9.10) should be rewritten as

$$R(N) = \sum_{i=1}^m h(y_i)\Delta y_i \int_N^\infty f(n, y_i)\,dn \tag{9.11}$$

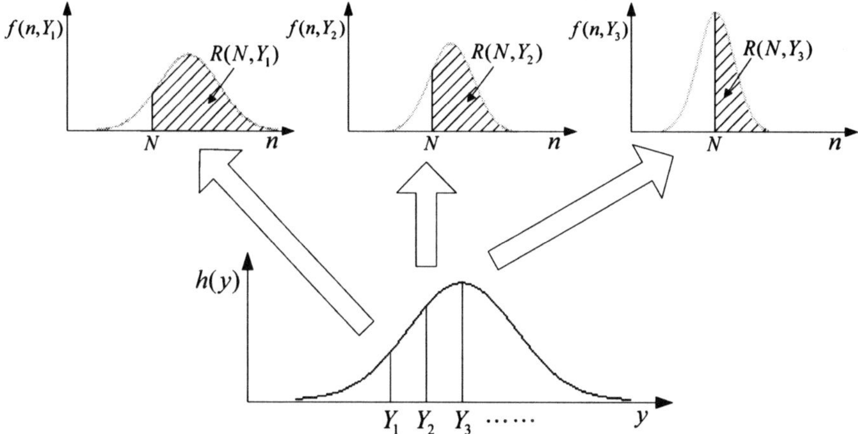

Fig. 9.2 Random stress and the individual reliability related to different stress samples

Let $\Delta y \rightarrow 0$, $m \rightarrow \infty$, obtained is the equation of fatigue life reliability in the situation of randomly distributed constant amplitude cyclic stress

$$R(N) = \int_0^\infty h(y) \int_N^\infty f(n,y)\,\mathrm{d}n\,\mathrm{d}y \qquad (9.12)$$

Thus, fatigue reliability can be calculated by such statistical stress-weighted average model.

It is also well known that the stress cycles-fatigue life interference model for a predetermined stress level takes on the form

$$R(t) = \int_0^{+\infty} l(c,t) \left[\int_c^{+\infty} f(n,Y)\,\mathrm{d}n \right] \mathrm{d}c \qquad (9.13)$$

where, $f(N,Y)$ is again the pdf of fatigue life under cyclic stress Y, $l(c,t)$ is the pdf of stress cycle c during the time range $(0 \sim t)$.

Contrasting to the traditional SSI model (Eq. (9.1)) or the stress cycles-fatigue life interference model (Eq. (9.13)), Eq. (9.12) can also be called as an interference model of different units-of-system, since the two variables involved in the model have different units-of-system (stress/MPa and fatigue life/times of stress cycles), which is different from the conventional models where only the same units-of-system variables are involved (*e.g.*, stress/MPa and strength/MPa or applied stress cycles/number of stress cycles and fatigue life/number of stress cycles).

As to fatigue reliability under complex loading history, equivalent amplitude concept can be applied to characterize the damage effect of the complex amplitude stress history. Accordingly, the associated fatigue life distribution should be that under the complex loading history. That is to say, if an equivalent constant amplitude cyclic load can be found to produce the same fatigue life distribution as a complex loading history does, it can be used for the extended interference model to predict the fatigue reliability to take the place of the complex loading history. To determine the range and distribution of the amplitude of the equivalent cyclic load, both the intensity of the complex load and the loading sequence effect should be considered, since different combination of load intensity and sequence will produce different life distribution.

9.4 Fatigue Life Distribution Under Constant Amplitude Cyclic Stress and Fatigue Reliability Calculation

In order to determine the relationship between fatigue life distribution and cyclic stress level, fatigue tests were conducted at rotated-bending fatigue test machine, using smooth specimens made of normalized 0.45% carbon steel and hot rolled alloy steel 16 Mn respectively [9]. For the normalized 0.45% carbon steel, the stress amplitudes used in the constant amplitude loading test are 366 MPa, 331 MPa, and 309 MPa respectively. All of the three stress amplitudes are less than the yield limit (380 MPa) of the material. Test results (expectations and standard deviations of fatigue lives) of the normalized 0.45% carbon steel are shown in Table 9.1 and Fig. 9.3.

Table 9.1 Life distribution parameters of the carbon steel

Stress	Sample size	Expectation	Std
y_1	15	45027	9917
y_2	18	151422	52286
y_3	16	658944	209045

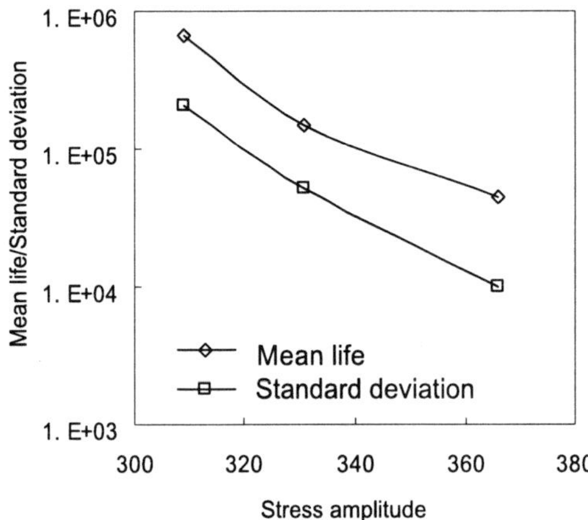

Fig. 9.3 Life distribution parameters of the carbon steel

For the test of the rolled alloy steel 16 Mn, three chosen stress amplitudes are 394 MPa, 373 MPa and 344 MPa respectively. All of them are less than the yield limit of the material, 410 MPa, too. Test results of the alloy steel 16 Mn are shown in Table 9.2 and Fig. 9.4.

The test results illustrate that both the relationship between stress level y and the logarithm mean life μ and the relationship between stress y and the logarithm standard deviation σ can be approximately expressed by linear equations, $i.e.$

$$\ln \mu = ay + b \tag{9.14}$$

$$\ln \sigma = cy + d \tag{9.15}$$

where a, b, c, d are the coefficients.

According to the cyclic stress-fatigue life equation normally applied to describe the relationship between fatigue life and cyclic stress amplitude, $i.e.$

$$\sigma_a = \sigma_f'(2N)^b$$

or

$$N = \frac{1}{2}\left(\frac{\sigma_a}{\sigma_f'}\right)^{\frac{1}{b}}$$

where, σ_a = cyclic stress amplitude, σ_f' = fatigue strength coefficient, b = fatigue strength exponent, N = fatigue life (number of stress cycles).

Table 9.2 Life distribution parameters of the alloy steel

Stress	Sample size	Expectation	Std
y_1	15	113893	15130
y_2	15	196720	27322
y_3	15	722200	125800

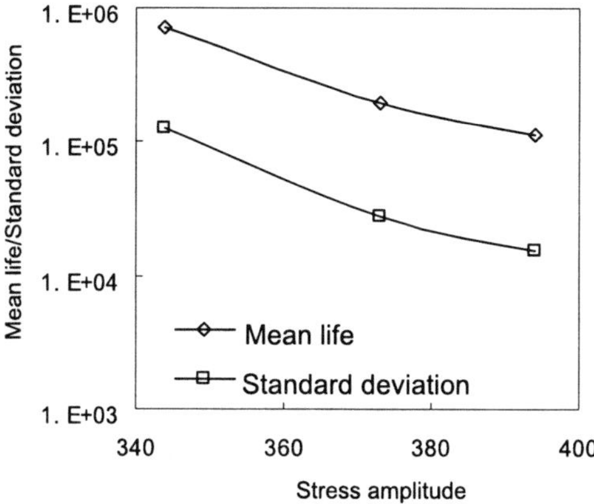

Fig. 9.4 Life distribution parameters of the alloy steel

There should be the relationship between logarithm life index and logarithm stress. The semi-logarithm relationships described by Eq. (9.14) and Eq. (9.15) are for the sake of simplicity.

By least squares fitting, it is obtained that for the material St.45, $a_{45} = -0.046$, $b_{45} = 27.4$, the correlation factor $r_{45\mu} = -0.983$; and $c_{45} = -0.053$, $d_{45} = 28.7$, $r_{45\sigma} = -0.999$. For the material 16 Mn alloy, $a_{16} = -0.047$, $b_{16} = 29.9$, and the correlation factor $r_{16\mu} = -0.98$; $c_{16} = -0.043$, $d_{16} = 26.4$, $r_{16\sigma} - 0.988$.

With such equations, the relationship between fatigue life distribution and stress level can be regressed by means of the mean lives μ_i and standard deviations σ_l at different stress amplitudes. For instance, if fatigue life follows the normal distribution, the relationship between fatigue life distribution and stress amplitude can be described as

$$f(n, y) = \frac{1}{\sqrt{2\pi}\sigma(y)} \, e^{-\frac{1}{2}\left(\frac{n-\mu(y)}{\sigma(y)}\right)^2} \tag{9.16}$$

Thus, the fatigue reliability under randomly distributed constant amplitude cyclic stress y (the pdf of the stress amplitude is denoted by $h(y)$) is

$$R(N) = \int_0^\infty h(y) \int_N^\infty \frac{1}{\sqrt{2\pi}\sigma(y)} \, e^{-\frac{1}{2}\left(\frac{n-\mu(y)}{\sigma(y)}\right)^2} \, dn \, dy \tag{9.17}$$

Following this arithmetic, an equivalent constant amplitude cyclic load can be found for a variable amplitude load history, which will lead to the same fatigue life as the variable amplitude load history does, and the equivalent constant amplitude load can be used in the extended interference model to predict the fatigue reliability corresponding to the variable amplitude load history.

9.5 Examples of Application

The following is an example to illustrate the procedure of applying the proposed model to calculate fatigue reliability according to the fatigue performance of the normalized 45% carbon steel.

Let the stress amplitude y follows the normal distribution

$$h(y) = \frac{1}{\sqrt{2\pi}\sigma_y} \, e^{-\frac{(y-\mu_y)^2}{2\sigma_y^2}}$$

and the fatigue life at specific stress amplitude y follow the normal distribution

$$f(N) = \frac{1}{\sqrt{2\pi}\sigma_N(y)} \, e^{-\frac{(N-\mu_N(y))^2}{2\sigma_N(y)^2}}$$

where $\mu_N(y) = e^{-0.046y+27.4}$, $\sigma_N(y) = e^{-0.053y+28.7}$.

According to Eq. (9.12), fatigue reliabilities under specifically distributed cyclic stress can be calculated. Illustrated in Fig. 9.5 is the stress distribution, and in Fig. 9.6 the relationship between fatigue reliability and life index in the situation of the stochastic cyclic stress with expectation of 300 MPa and standard deviation of 60 MPa.

In the following, we consider the situation that fatigue life at specific stress amplitude follows the Weibull distribution

$$f(N) = \frac{\beta}{\eta} \left(\frac{N - \gamma}{\eta} \right)^{\beta - 1} e^{-\left(\frac{N-\gamma}{\eta} \right)^{\beta}}$$

where the shape parameter β, scale parameter η and position parameter γ are the arbitrarily constructed functions of the stress amplitude as shown below respectively:

$$\beta = 2.0 - 0.0008y,$$
$$\eta = 10^7 - 9y^2,$$
$$\gamma = 10^5 - 100y.$$

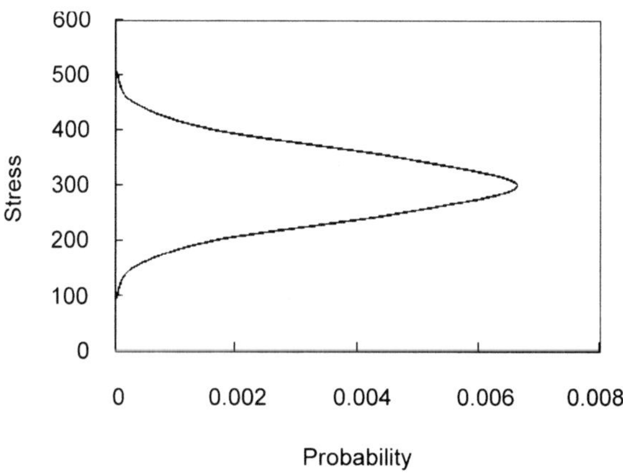

Fig. 9.5 Cyclic stress amplitude distribution

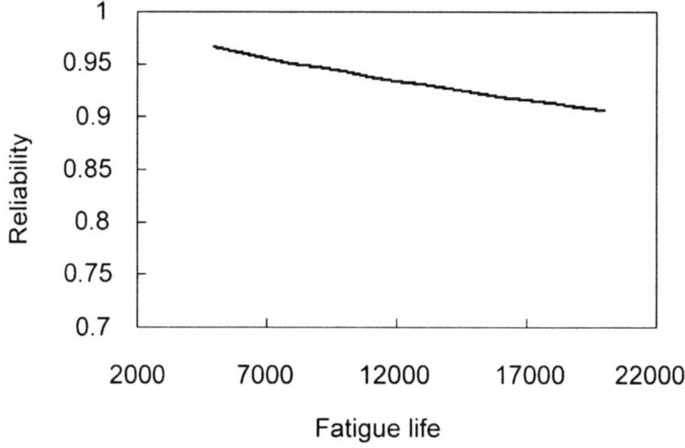

Fig. 9.6 Relationship between fatigue reliability and life

The constants in the above functions are also arbitrarily selected to obtain a reasonable fatigue life distribution. Figure 9.7 illustrates some of the distributions with different stress amplitude, where the three curves are correspond to the middle level stress (300 MPa), the lowest stress (260 MPa), and the highest stress (340 MPa) respectively.

According to Eq. (9.12), fatigue reliabilities at different stress levels and those at different life indexes can be calculated.

Illustrated in Fig. 9.8 are the relationship between fatigue reliability (to the life index 10^6 cycles) and stress amplitude in the situation of deterministic cyclic stress, and the relationship between fatigue reliability (to the life index 10^6 cycles) and the expectation of the stress amplitude in the situation of stochastic cyclic stress respectively. Six stress levels are considered that the expectations of stress amplitude are 300, 305, 310, 315, 320, 325, 330 MPa respectively, the standard deviation is 200 MPa for all these six stress levels.

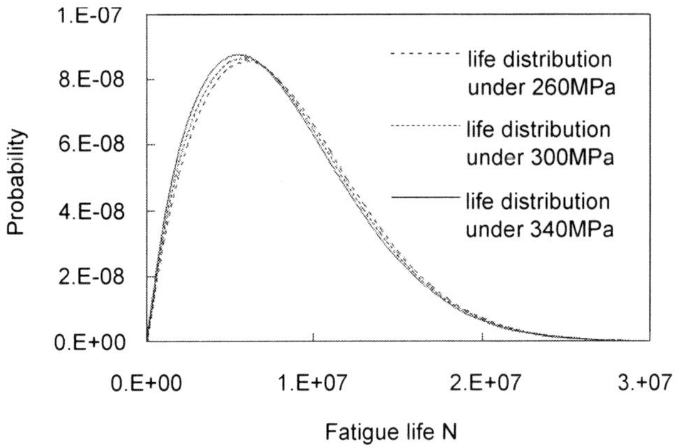

Fig. 9.7 Life distributions at different stress levels

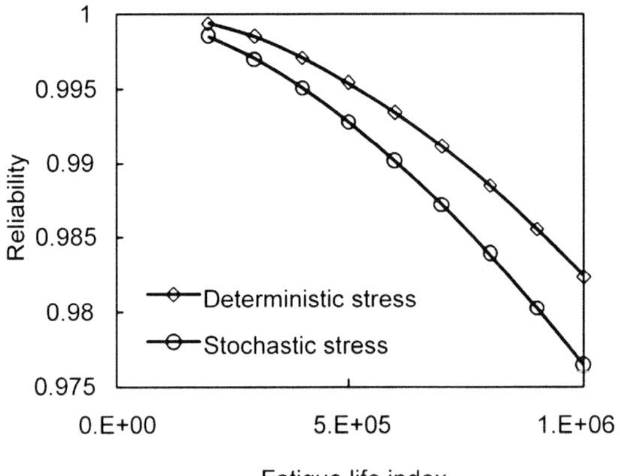

Fig. 9.8 Relationship between reliability and stress level

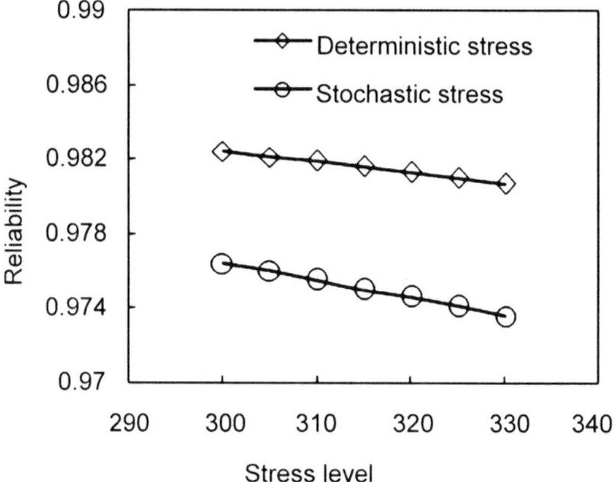

Fig. 9.9 Relationship between reliability and life index

Illustrated in Fig. 9.9 are the relationships between fatigue reliability and life index in the situation of deterministic cyclic stress with amplitude of 300 MPa and the situation of stochastic cyclic stress with expectation/standard deviation of 300 MPa/200 MPa.

In these two graphs, the relationship between reliability and stress level and the relationship between reliability and life index are compared, in the situation of deterministic cyclic stress and stochastic cyclic stress. The considerable differences between the reliability-stress level curves and the differences between the reliability-life index curves in the situation of deterministic stress and that of stochastic stress indicate the significant effect of the uncertainty in stress amplitude to fatigue reliability.

9.6 Conclusions

The traditional SSI model, which expresses the probability that a random strength is greater or less than a random stress, is explained as a stress-weighted statistical average arithmetic to calculate the average of the probabilities that component strength is greater or less then the individual stress samples over the whole range of the random stress.

Furthermore, an extended interference analysis model is presented to calculate fatigue reliability under constant amplitude cyclic load with uncertainty in stress amplitude. With a specific load amplitude distribution, the method is capable to calculate the statistical average of the probability that fatigue life random variable is greater than an assigned value.

By such a model, fatigue failure probability or fatigue reliability can be calculated with known load/stress distribution and fatigue life distribution under constant amplitude cyclic loading, thus avoid the difficulty to obtain fatigue strength distribution at assigned life that will be encountered when the conventional SSI model is applied.

Acknowledgements This research is subsidized by the Special Funds for the Major State Basic Research Projects 2006CB605000 and the Hi-Tech Research and Development Program (863) of China with the grant No. 2006AA04Z408.

References

1. Kececioglu D (1972) Reliability analysis of mechanical components and systems. Nuclear Engineering and Design 19:259–290
2. Witt FJ (1985) Stress-strength interference methods. Pressure Vessel and Piping Technology – A Decade of Progress. American Society of Mechanical Engineers, New York, pp. 761–769
3. Chen D (1991) A new approach to the estimation of fatigue reliability at a single stress level. Reliability Engineering and System Safety 33:101–113
4. Murty ASR, Gupta UC, Krishna AR (1995) A new approach to fatigue strength distribution for fatigue reliability evaluation. Int J Fatigue 17:85–89
5. Kam JPC, Birkinshaw M (1994) Reliability-based fatigue and fracture mechanics assessment methodology for offshore structural components. Int J Fatigue 16:183–199
6. Siddiqui NA, Ahmad S (2001) Fatigue and fracture reliability of TLP tethers under random loading. Marine Structures 14:331–352
7. Ni K (2000) Shengkun Zhang. Fatigue reliability analysis under two-stage loading. Reliability Engineering and System Safety 68:153–158
8. Tanake S, Ichikawa M, Akita SA (1984) Probabilistic investigation of fatigue life and cumulative cycle ratio. Engineering of Fracture Mechanics 20:501–513
9. Xie LY (1999) Equivalent life distribution and fatigue failure probability prediction. International Journal of Pressure Vessels and Piping 76:267–273
10. Petryna YS, Pfanner D, Stangenberg F *et al.* (2002) Reliability of reinforced concrete structures under fatigue. Reliability Engineering and System Safety 77:253–261
11. Karadeniz H (2001) Uncertainty modeling in the fatigue reliability calculation of offshore structures. Reliability Engineering and System Safety 74:323–335
12. Pham H (2005) A new generalized systemability model. International Journal of Performability Engineering 1:145–155

Chapter 10
Markovian Performance Evaluation for Software System Availability with Processing Time Limit

Masamitsu Fukuda, Koichi Tokuno, Shigeru Yamada

Department of Social System Engineering,
Tottori University, 4-101 Minami, Koyama-cho,
Tottori-shi, Tottori 680-8552, Japan

10.1 Introduction

Today the engineering system of the service reliability engineering is receiving growing attention [1, 2]; this aims at the establishment of the evaluation methods for the quality of service created by the use of artificial industrial products as well as the inherent quality of the products. Considering that software systems are just the industrial products to provide the services for the users, especially in computer network systems, it is meaningful to discuss the performance evaluation methods for software systems oriented to service reliability engineering. Recently, the consortium of the Service Availability Forum [3] has been created to develop the computing framework and the interface between hardware and software systems with high service availability.

Studies on performance evaluation considering reliability for computing systems have much been discussed from the viewpoint of the hardware configuration [4, 5]. On the other hand, from the viewpoint of software system, the discussions on inherent quality/reliability evaluation such as the estimation of the residual fault content and the prediction of software failure time have much been conducted [6, 7], while there exist few studies on the reliability-related performance evaluation. Kimura *et al.* [8, 9] have discussed the evaluation methods of the real-time property for the N-version programming and the recovery block software systems; these are well known as the methodologies of fault-tolerant software systems. However, Kimura's studies have simply applied the framework for the analyzing from the aspect of the hardware configuration to the fault-tolerant software systems and have not included the characteristics particular to software systems such as the reliability growth process and the upward tendency of difficulty in debugging.

In this paper, we discuss the performance evaluation method of the software systems considering the real-time property; this is a different approach from Kimura's studies. The real-time property is defined as the attribute that the system can complete the task within the stipulated response time limit [10, 11]. We assume that the software system can process the plural tasks simultaneously. Then the time-dependent behavior of the software system alternating between up and down states in the dynamic environment are described by the Markovian software availability model [12]. The stochastic behavior of the number of tasks whose processes can be complete within the prespecified processing time limit is modeled with the infinite sever queueing model [13].

The organization of the rest of the paper is as follows. Section 10.2 outlines the software availability model used in the paper. Section 10.3 describes the stochastic processes of the numbers of tasks whose processes are complete within the prespecified processing time limit and canceled out of the tasks arriving up to a given time point. Section 10.4 derives several software performance measures considering the real-time property. The measures are given as the functions of time and the number of debuggings. Section 10.5 presents several numerical examples of software performance analysis. Finally, Sect. 10.6 summarizes the results obtained in this paper.

10.2 Markovian Software Availability Model

10.2.1 Model Description

The following assumptions are made for software availability modeling [12].

A1. The software system is unavailable and starts to be restored as soon as a software failure occurs, and the system cannot operate until the restoration action is complete.

A2. The restoration action includes the debugging activity; this is performed perfectly with the perfect debugging rate a $(0 < a \leq 1)$ and imperfectly with probability $b(= 1 - a)$. When the debugging activity succeeds, one fault is corrected and the software reliability growth occurs.

A3. The next software failure time, X_n, and the restoration time, U_n, when n faults have already been corrected from the system, follow the exponential distributions with means $1/\lambda_n$ and $1/\mu_n$, respectively. λ_n and μ_n are non-increasing functions of n.

The state space of the stochastic process $\{Z(t), t \geq 0\}$ representing the state of the software system at the time point t is defined as follows:

W_n: the system is operating,
R_n: the system is inoperable and restored,

where n denotes the cumulative number of corrected faults. Figure 10.1 illustrates the sample state transition diagram of $Z(t)$.

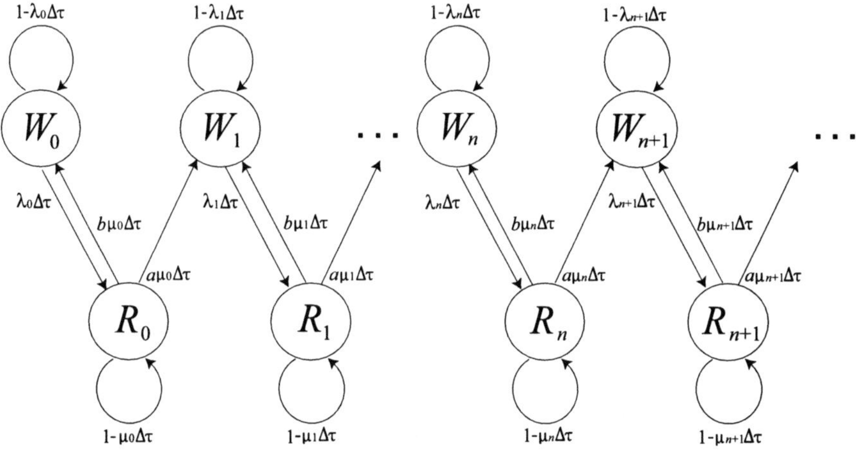

Fig. 10.1 A sample state transition diagram of $Z(t)$

10.2.2 Software Availability Measures

Let $S_{i,n}$ $(i \leq n)$ be the random variable representing the transition time of $Z(t)$ from state W_i to state W_n, and $G_{i,n}(t)$ be the distribution function of $S_{i,n}$, respectively. Then, we obtain the following renewal equation with respect to $G_{i,n}(t)$:

$$G_{i,n}(t) = Q^*_{W_i,R_i} Q^*_{R_i,W_{i+1}} G_{i+1,n}(t) + Q^*_{W_i,R_i} Q^*_{R_i,W_i} G_{i,n}(t) , \qquad (10.1)$$

where $Q_{A,B}(\tau)$ $(A, B \in \{W_n, R_n; n = 0, 1, 2, \ldots\})$ denotes the one-step transition probability that after making a transition into state A, the process $\{Z(t), t \geq 0\}$ makes a transition into state B by time τ, and * denotes a Stieltjes convolution.

We use Laplace–Stieltjes (L–S) transforms to solve Eq. (10.1), where the L–S transform of $G_{i,n}(t)$ is defined as

$$\tilde{G}_{i,n}(s) \equiv \int_0^\infty e^{-st} \, dG_{i,n}(t) . \qquad (10.2)$$

Accordingly, we obtain $\tilde{G}_{i,n}(s)$ as

$$\tilde{G}_{i,n}(s) = \prod_{m=i}^{n-1} \frac{a\lambda_m \mu_m}{(s+x_m)(s+y_m)}$$

$$= \sum_{m=i}^{n-1} \left(\frac{A^1_{i,n}(m)x_m}{s+x_m} + \frac{A^2_{i,n}(m)y_m}{s+y_m} \right) , \qquad (10.3)$$

228 M. Fukuda, K. Tokuno, S. Yamada

where

$$\left.\begin{array}{c} x_m \\ y_m \end{array}\right\} = \frac{1}{2}\left[(\lambda_m + \mu_m) \pm \sqrt{(\lambda_m + \mu_m)^2 - 4a\lambda_m\mu_m}\right]$$

(double signs in same order), $\qquad$ (10.4)

$$A^1_{i,n}(m) = \frac{\prod_{j=i}^{n-1} x_j y_j}{x_m \prod_{j=i}^{n-1}(y_j - x_m)\prod_{\substack{j=i \\ j \neq m}}^{n-1}(x_j - x_m)} \qquad (m = i,\ i+1,\ \ldots,\ n-1), \qquad (10.5)$$

$$A^2_{i,n}(m) = \frac{\prod_{j=i}^{n-1} x_j y_j}{y_m \prod_{j=i}^{n-1}(x_j - y_m)\prod_{\substack{j=i \\ j \neq m}}^{n-1}(y_j - y_m)} \qquad (m = i,\ i+1,\ \ldots,\ n-1), \qquad (10.6)$$

respectively. By inverting Eq. (10.3), we obtain the distribution function of $S_{i,n}$ as

$$G_{i,n}(t) \equiv \Pr\{S_{i,n} \leq t\}$$
$$= 1 - \sum_{m=i}^{n-1}\left[A^1_{i,n}(m)\,\mathrm{e}^{-x_m t} + A^2_{i,n}(m)\,\mathrm{e}^{-y_m t}\right]. \qquad (10.7)$$

Let $P_{A,B}(t) \equiv \Pr\{X(t) = B | X(0) = A\}$ $(A, B \in \{W_n, R_n;\ n = 0, 1, 2, \ldots\})$ be the state occupancy probability that the system is in state B at the time point t on the condition that the system was in state A at time point $t = 0$.

First, we derive $P_{W_i,W_n}(t)$. We obtain the following renewal equations with respect to $P_{W_i,W_n}(t)$:

$$P_{W_i,W_n}(t) = G^*_{i,n} P_{W_n,W_n}(t), \qquad (10.8)$$
$$P_{W_n,W_n}(t) = \mathrm{e}^{-\lambda_n t} + Q^*_{W_n,R_n} Q^*_{R_n,W_n} P_{W_n,W_n}(t). \qquad (10.9)$$

Substituting the L–S transforms of Eq. (10.9) into that of Eq. (10.8) yields

$$\tilde{P}_{W_i,W_n}(s) = \frac{s\tilde{G}_{i,n+1}(s)}{a\lambda_n} + \frac{s^2\tilde{G}_{i,n+1}(s)}{a\lambda_n\mu_n}. \qquad (10.10)$$

By inverting Eq. (10.10), $P_{W_i,W_n}(t)$ is obtained as

$$P_{W_i,W_n}(t) \equiv \Pr\{Z(t) = W_n | Z(0) = W_i\}$$
$$= \frac{g_{i,n+1}(t)}{a\lambda_n} + \frac{g'_{i,n+1}(t)}{a\lambda_n\mu_n}, \qquad (10.11)$$

where $g_{i,n}(t)$ is the probability density function associated with $G_{i,n}(t)$ and $g'_{i,n} \equiv \mathrm{d}g_{i,n}(t)/\mathrm{d}t$.

Using a similar procedure for the derivation of $P_{W_i,W_n}(t)$, we obtain the following renewal equations with respect to $P_{W_i,R_n}(t)$:

$$P_{W_i,R_n}(t) = G^*_{i,n} Q^*_{W_n,R_n} P_{R_n,R_n}(t) , \tag{10.12}$$

$$P_{R_n,R_n}(t) = e^{-\mu_n t} + Q^*_{R_n,W_n} Q^*_{W_n,R_n} P_{R_n,R_n}(t) . \tag{10.13}$$

Substituting the L–S transform of Eq. (10.13) into that of Eq (10.12) yields

$$\tilde{P}_{W_i,R_n}(s) = \frac{s\tilde{G}_{i,n+1}(s)}{a\mu_n} . \tag{10.14}$$

By inverting Eq. (10.14), $P_{W_i,R_n}(t)$ is obtained as

$$P_{W_i,R_n}(t) \equiv \Pr\{Z(t) = R_n | Z(0) = W_i\}$$

$$= \frac{g_{i,n+1}(t)}{a\mu_n} . \tag{10.15}$$

The instantaneous and the average software availabilities are given by

$$A(t;l) \equiv \sum_{i=0}^{l} \binom{l}{i} a^i b^{l-i} \sum_{n=i}^{\infty} P_{W_i,W_n}(t)$$

$$= 1 - \sum_{i=0}^{l} \binom{l}{i} a^i b^{l-i} \sum_{n=i}^{\infty} \frac{g_{i,n+1}(t)}{a\mu_n} , \tag{10.16}$$

$$A_{\mathrm{av}}(t;l) \equiv \frac{1}{t} \int_0^t A(x;l)\,dx$$

$$= 1 - \frac{1}{t} \sum_{i=0}^{l} \binom{l}{i} a^i b^{l-i} \sum_{n=i}^{\infty} \frac{G_{i,n+1}(t)}{a\mu_n} , \tag{10.17}$$

respectively, where $\binom{l}{i} \equiv l!/[(l-i)!i!]$ denotes the binomial coefficient and we use the identical equation $\sum_{n=i}^{\infty} [P_{W_i,W_n}(t) + P_{W_i,R_n}(t)] \equiv 1$ for arbitrary time t. Equations (10.16) and (10.17) represent the probability that the system is operating at the time point t and the expected proportion of the system's operating time to the time interval $(0, t]$, given that the l-th debugging activity was complete at time point $t = 0$ respectively.

10.3 Model Analysis

We make the following assumptions for the system's task processing.

B1. The process $\{N(t), t \geq 0\}$ representing the number of tasks arriving at the system up to the time t follows the homogeneous Poisson process with the arrival rate θ.

B2. The processing time of a task, Y, follows a general distribution whose distribution function is denoted as $H(t)$. Each of the processing times is independent.

B3. When the system causes a software failure in task processing or the processing times of tasks exceed the prespecified processing time limit, T_r, the corresponding tasks are canceled.

B4. The number of tasks the system can process simultaneously is sufficiently large.

Figure 10.2 illustrates the configuration of the system task processing. In the discussion below, consider i faults have already been corrected at time point $t = 0$.

Let $\{X_i(t|T_r), \ t \geq 0\}$ be the random variable representing the cumulative number of tasks whose processes can be complete within the processing time limit T_r out of the tasks arriving up to the time t.

By conditioning with $\{N(t) = k\}$, we can obtain the probability mass function of $X(t|T_r)$ as

$$\Pr\{X_i(t|T_r) = j\} = \sum_{k=0}^{\infty} \Pr\{X_i(t|T_r) = j|N(t) = k\}\, e^{-\theta t}\frac{(\theta t)^k}{k!} \quad (j = 0, 1, 2, \ldots).$$

$$(10.18)$$

From Fig. 10.2, given that $\{Z(t) = W_n\}$ and that an arbitary task arrives at the system at the time point t, the probability that the process of an arbitrary task is complete within the processing time limit T_r is given by

$$\beta_n(T_r) \equiv \Pr\{T_r > Y, \ X_n > Y | Z(t) = W_n\} = \int_0^{T_r} e^{-\lambda_n y}\, \mathrm{d}H(y)\,; \qquad (10.19)$$

this equation is independent of time t, since X_n has a memoryless property. Then unconditioning Eq. (10.19) with respect to the cumulative number of corrected faults, n, yields

$$\pi_i(x|T_r) = \sum_{n=i}^{\infty} \beta_n(T_r) \cdot \Pr\{Z(x) = W_n | Z(0) = W_i\}$$

$$= \sum_{n=i}^{\infty} \beta_n(T_r)\left\{\frac{g_{i,n+1}(x)}{a\lambda_n} + \frac{g'_{i,n+1}(x)}{a\lambda_n\mu_n}\right\}. \qquad (10.20)$$

Equation (10.20) means that the probability that the process of a task is complete within T_r on the condition that the task has arrived at the system at the time point x $(0 \leq x \leq t)$. Furthermore, the arrival time of an arbitrary task out of ones arriving up to the time t is distributed uniformly over the time interval $(0, t]$ [13], i.e., the density function of this time is given by $f(x) = 1/t$ $(0 \leq x \leq t)$. Therefore, the probability that the process of the task having arrived up to the time t is complete within the processing time limit T_r is obtained as

$$p_i(t|T_r) = \int_0^t \pi_i(x|T_r) f(x)\,\mathrm{d}x = \frac{1}{t}\sum_{n=i}^{\infty}\left[\beta_n(T_r)\left\{\frac{G_{i,n+1}(t)}{a\lambda_n} + \frac{g_{i,n+1}(t)}{a\lambda_n\mu_n}\right\}\right]. \quad (10.21)$$

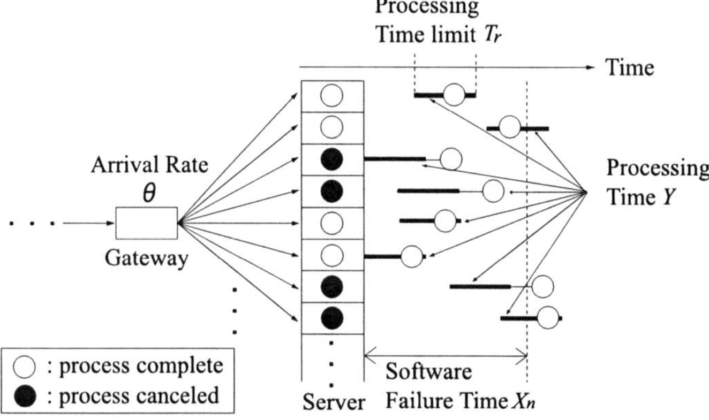

Fig. 10.2 Configuration of system's task processing

Then from assumption B2,

$$\Pr\{X_i(t|T_r) = j | N(t) = k\} =$$

$$\begin{cases} \dbinom{k}{j} [p_i(t|T_r)]^j [1 - p_i(t|T_r)]^{k-j} & (j = 0, 1, 2, \ldots, k) \\ 0 & (j > k) \end{cases} . \qquad (10.22)$$

That is, given that $\{N(t) = k\}$, the number of tasks whose processes can be complete within the processing time limit T_r follows the binominal process with mean $k p_i(t|T_r)$. Accordingly, from Eq. (10.18) the distribution of $X_i(t|T_r)$ is given by

$$\Pr\{X_i(t|T_r) = j\} = \sum_{k=j}^{\infty} \binom{k}{j} [p_i(t|T_r)]^j [1 - p_i(t|T_r)]^{k-j} e^{-\theta t} \frac{(\theta t)^k}{k!}$$

$$= e^{-\theta t p_i(t|T_r)} \frac{[\theta t p_i(t|T_r)]^j}{j!} \qquad (j = 0, 1, 2, \ldots). \qquad (10.23)$$

Equation (10.23) means that $\{X_i(t|T_r), \ t \geq 0\}$ follows the non-homogeneous Poisson process (NHPP) with the mean value function $\theta t p_i(t|T_r)$.

Let $\{W_i(t|T_r), \ t \geq 0\}$ be the random variable representing the cumulative number of tasks whose processes are interrupted out of the tasks arriving up to the time t. Then we can apply the same discussion as above to the derivation of the distribution of $W_i(t|T_r)$, i.e., we can obtain $\Pr\{W_i(t|T_r) = j\}$ as

$$\left. \begin{array}{l} \Pr\{W_i(t|T_r) = j\} = e^{-\theta t q_i(t|T_r)} \dfrac{[\theta t q_i(t|T_r)]^j}{j!} \qquad (j = 0, 1, 2, \ldots) \\[2em] q_i(t|T_r) = 1 - \dfrac{1}{t} \sum_{n=i}^{\infty} \left[\beta_n(T_r) \left\{ \dfrac{G_{i,n+1}(t)}{a\lambda_n} + \dfrac{g_{i,n+1}(t)}{a\lambda_n \mu_n} \right\} \right] \end{array} \right\} . \qquad (10.24)$$

Equation (10.24) means that $\{W_i(t|T_r), \ t \geq 0\}$ follows the NHPP with mean value function $\theta t q_i(t|T_r)$.

10.4 Derivation of Software Performance Measures

Based on the above analysis, we can obtain several measures for software performance evaluation.

The expected numbers of tasks completable and incompletable within the processing time limit T_r out of the tasks arriving up to the time t are given by

$$\Lambda_i(t|T_r) \equiv \mathrm{E}[X_i(t|T_r)]$$
$$= \theta \sum_{n=i}^{\infty} \left[\beta_n(T_r) \left\{ \frac{G_{i,n+1}(t)}{a\lambda_n} + \frac{g_{i,n+1}(t)}{a\lambda_n\mu_n} \right\} \right], \tag{10.25}$$

$$M_i(t|T_r) \equiv \mathrm{E}[W_i(t|T_r)]$$
$$= \theta t - \theta \sum_{n=i}^{\infty} \left[\beta_n(T_r) \left\{ \frac{G_{i,n+1}(t)}{a\lambda_n} + \frac{g_{i,n+1}(t)}{a\lambda_n\mu_n} \right\} \right], \tag{10.26}$$

respectively.

The instantaneous numbers of tasks completable and incompletable at the time point t are given by

$$\lambda_i(t|T_r) \equiv \frac{\mathrm{d}\Lambda_i(t|T_r)}{\mathrm{d}t}$$
$$= \theta \sum_{n=i}^{\infty} \left[\beta_n(T_r) \left\{ \frac{g_{i,n+1}(t)}{a\lambda_n} + \frac{g'_{i,n+1}(t)}{a\lambda_n\mu_n} \right\} \right], \tag{10.27}$$

$$m_i(t|T_r) \equiv \frac{\mathrm{d}M_i(t|T_r)}{\mathrm{d}t}$$
$$= \theta - \theta \sum_{n=i}^{\infty} \left[\beta_n(T_r) \left\{ \frac{g_{i,n+1}(t)}{a\lambda_n} + \frac{g'_{i,n+1}(t)}{a\lambda_n\mu_n} \right\} \right], \tag{10.28}$$

respectively.

Furthermore, the instantaneous task completion and incompletion ratios are given by

$$h_i(t|T_r) \equiv \frac{\mathrm{d}\Lambda_i(t|T_r)}{\mathrm{d}t} \Big/ \theta$$
$$= \sum_{n=i}^{\infty} \left[\beta_n(T_r) \left\{ \frac{g_{i,n+1}(t)}{a\lambda_n} + \frac{g'_{i,n+1}(t)}{a\lambda_n\mu_n} \right\} \right], \tag{10.29}$$

$$\chi_i(t|T_r) \equiv \frac{\mathrm{d}M_i(t|T_r)}{\mathrm{d}t} \Big/ \theta$$
$$= 1 - \sum_{n=i}^{\infty} \left[\beta_n(T_r) \left\{ \frac{g_{i,n+1}(t)}{a\lambda_n} + \frac{g'_{i,n+1}(t)}{a\lambda_n\mu_n} \right\} \right], \tag{10.30}$$

respectively. These represent the ratios of the numbers of tasks completed and canceled to that of tasks arriving at the system per unit time at the time point t, respectively. We note that Eqs. (10.29) and (10.30) have no bearing on the arrival rate of the tasks, θ.

As to $p_i(t|T_r)$ in Eq. (10.21) and $q_i(t|T_r)$ in Eq. (10.24), we can give the following interpretations:

$$p_i(t|T_r) = \frac{E[X_i(t|T_r)]}{E[N(t)]} , \tag{10.31}$$

$$q_i(t|T_r) = \frac{E[W_i(t|T_r)]}{E[N(t)]} . \tag{10.32}$$

That is, $p_i(t|T_r)$ and $q_i(t|T_r)$ are the cumulative task completion and incompletion ratios in the time interval $(0, t]$ respectively. Equations (10.31) and (10.32) also have no bearing on θ.

We should note that it is too difficult to use Eqs. (10.25)–(10.30) practically, since this model assumes the imperfect debugging environment and the initial condition i appearing in the above equations, which represents the cumulative number of corrected faults, cannot be observed immediately. However, applying the similar idea to $A(t;l)$ and $A_{\mathrm{av}}(t;l)$ in Sect. 10.2, we can convert Eqs. (10.25)–(10.30) into the functions of the number of debuggings, l, i.e., we obtain

$$\Lambda(t,l|T_r) = \theta \sum_{i=0}^{l} \binom{l}{i} a^i b^{l-i} \sum_{n=i}^{\infty} \left[\beta_n(T_r) \left\{ \frac{G_{i,n+1}(t)}{a\lambda_n} + \frac{g_{i,n+1}(t)}{a\lambda_n\mu_n} \right\} \right], \tag{10.33}$$

$$M(t,l|T_r) = \theta \sum_{i=0}^{l} \binom{l}{i} a^i b^{l-i} \left[t - \sum_{n=i}^{\infty} \left[\beta_n(T_r) \left\{ \frac{G_{i,n+1}(t)}{a\lambda_n} + \frac{g_{i,n+1}(t)}{a\lambda_n\mu_n} \right\} \right] \right], \tag{10.34}$$

$$\lambda(t,l|T_r) = \theta \sum_{i=0}^{l} \binom{l}{i} a^i b^l {}^{i} \sum_{n=i}^{\infty} \left[\beta_n(T_r) \left\{ \frac{g_{i,n+1}(t)}{a\lambda_n} + \frac{g'_{i,n+1}(t)}{a\lambda_n\mu_n} \right\} \right], \tag{10.35}$$

$$m(t,l|T_r) = \theta \sum_{i=0}^{l} \binom{l}{i} a^i b^{l-i} \left[1 - \sum_{n=i}^{\infty} \left[\beta_n(T_r) \left\{ \frac{g_{i,n+1}(t)}{a\lambda_n} + \frac{g'_{i,n+1}(t)}{a\lambda_n\mu_n} \right\} \right] \right], \tag{10.36}$$

$$h(t,l|T_r) = \sum_{i=0}^{l} \binom{l}{i} a^i b^{l-i} \sum_{n=i}^{\infty} \left[\beta_n(T_r) \left\{ \frac{g_{i,n+1}(t)}{a\lambda_n} + \frac{g'_{i,n+1}(t)}{a\lambda_n\mu_n} \right\} \right], \tag{10.37}$$

$$\chi(t,l|T_r) = \sum_{i=0}^{l} \binom{l}{i} a^i b^{l-i} \left[1 - \sum_{n=i}^{\infty} \left[\beta_n(T_r) \left\{ \frac{g_{i,n+1}(t)}{a\lambda_n} + \frac{g'_{i,n+1}(t)}{a\lambda_n\mu_n} \right\} \right] \right], \tag{10.38}$$

$$p(t,l|T_r) = \frac{1}{t} \sum_{i=0}^{l} \binom{l}{i} a^i b^{l-i} \sum_{n=i}^{\infty} \left[\beta_n(T_r) \left\{ \frac{G_{i,n+1}(t)}{a\lambda_n} + \frac{g_{i,n+1}(t)}{a\lambda_n\mu_n} \right\} \right], \tag{10.39}$$

$$q(t,l|T_r) = \sum_{i=0}^{l} \binom{l}{i} a^i b^{l-i} \left[1 - \frac{1}{t} \sum_{n=i}^{\infty} \left[\beta_n(T_r) \left\{ \frac{G_{i,n+1}(t)}{a\lambda_n} + \frac{g_{i,n+1}(t)}{a\lambda_n\mu_n} \right\} \right] \right], \tag{10.40}$$

respectively. Eqs. (10.33)–(10.40) represent the expected numbers of tasks completable and incompletable, the instantaneous numbers of tasks completable and incompletable, the instantaneous task completion and incompletion ratios, the cumulative task completion and incompletion ratios at the time point t, given that the l-th debugging was complete at the time point $t = 0$, respectively.

10.5 Numerical Examples

We show several numerical examples of software performance analysis based on the above measures. Here, we use Moranda's model [14] to the hazard rate $\lambda_n \equiv Dc^n$ $(D > 0,\ 0 < c < 1)$ and the restoration rate $\mu_n \equiv Er^n$ $(E > 0,\ 0 < r \leq 1)$ respectively. We cite the estimates of the parameters associated with λ_n and μ_n from Tokuno [15], i.e., we use the following values:

$$\hat{D} = 0.246,\ \hat{c} = 0.940,\ \hat{E} = 1.114,\ \hat{r} = 0.960,$$

where we set $a = 0.8$. These values have been estimated based on the simulated data set generated from data cited by Goel and Okumoto [16]; this consists of 26 software failure-occurrence time interval data and the unit of time is day.

For the distribution of the processing time of a task, Y, we apply the gamma distribution, the density of which is given by

$$\frac{dH(t)}{dt} \equiv \frac{\alpha^v t^{v-1} e^{-\alpha t}}{\displaystyle\int_0^\infty e^{-x} x^{v-1}\,dx} \qquad (t \geq 0; v > 0, \alpha > 0), \tag{10.41}$$

where v and α are the shape and the scale parameters, respectively. Then the mean and the variance of the processing time are given by $E[Y] = v/\alpha$ and $\mathrm{Var}[Y] = v/\alpha^2$ respectively.

Figures 10.3 and 10.4 show the instantaneous task completion ratio, $h(t, l|T_r)$, in Eq. (10.37) and the cumulative task completion ratio, $p(t, l|T_r)$, in Eq. (10.39) for the various numbers of debuggings, l respectively, where $T_r = 0.005$, $v = 2.0$, $\alpha = 1000.0$. We can see that software performance also improves as the debugging processes.

Figure 10.5 shows the time-dependent behaviors of $h(t, l|T_r)$ and the instantaneous task incompletion ratio, $\chi(t, l|T_r)$ in Eq. (10.38) along with the instantaneous software availability, $A(t; l)$, in Eq. (10.16). This figure tells us that the new measure considering the real-time property $(h(t, l|T_r))$ gives more pessimistic evaluation than the traditional one $(A(t; l))$.

Figure 10.6 shows the dependence of the instantaneous number of task completable, $\lambda(t, l|T_r)$, in Eq. (10.35) on the distribution of the processing time of a task, $H(t)$, where we set the parameters v and α as equalize the means of the processing time for $H_1(t)$ and $H_2(t)$ in Fig. 10.6, i.e.,

$$H(t) \equiv H_1(t) = 1 - e^{-\alpha_1 t}$$

$$\text{(exponential distribution: } v \equiv v_1 = 1.0,\ \alpha \equiv \alpha_1 = 500.0\text{)}, \tag{10.42}$$

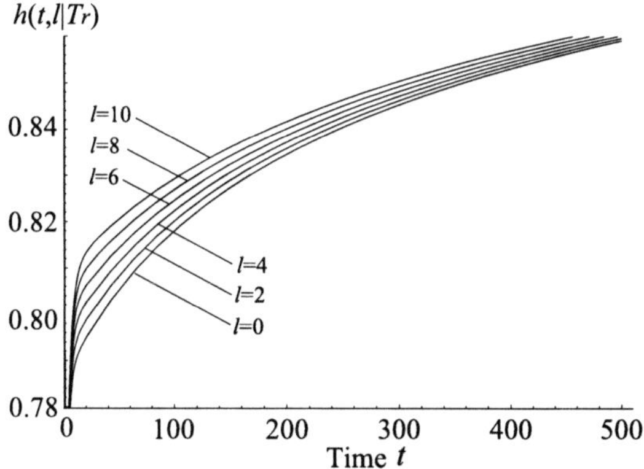

Fig. 10.3 $h(t, l|T_r)$ for various numbers of debuggings, l ($T_r = 0.005$; $v = 2.0$, $\alpha = 1000.0$)

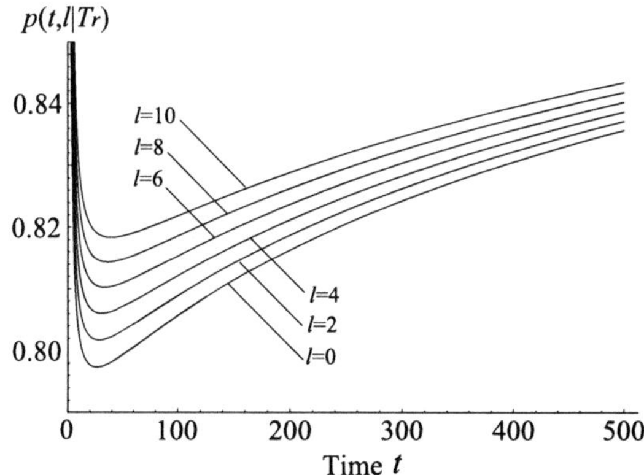

Fig. 10.4 $p(t, l|T_r)$ for various numbers of debuggings, l ($T_r = 0.005$; $v = 2.0$, $\alpha = 1000.0$)

$$H(t) \equiv H_2(t) = 1 - (1 + \alpha_2 t)\,e^{-\alpha_2 t}$$

$$\text{(gamma distribution of order two: } v \equiv v_2 = 2v_1, \ \alpha \equiv \alpha_2 = 2\alpha_1) \,, \quad (10.43)$$

respectively. This figure indicates that the performance evaluation in the case of the gamma distribution is higher than that of the exponential distribution. As to the variances of the processing time, the cases of $H_1(t)$ and $H_2(t)$ are $1/\alpha_1^2$ and $2/\alpha_2^2 = 1/(2\alpha_1^2) < 1/\alpha_1^2$ respectively. We can see that the smaller dispersion-degree of the processing time rises the software performance evaluation.

Figure 10.7 shows the dependence of $h(t, l|T_r)$ on r representing the decreasing ratio of the restoration rate, μ_n. According to Tokuno and Yamada [12], the behavior of the maintenance factor, $\rho_n \equiv \lambda_n/\mu_n$, decides whether the instantaneous and the average software availabilities improve or degrade with time, *i.e.*, the traditional software availability improves (degrades) if ρ_n is a decreasing (increasing) function of n. From Fig. 10.7, we can see that ρ_n has a similar impact on software performance evaluation considering real-time property.

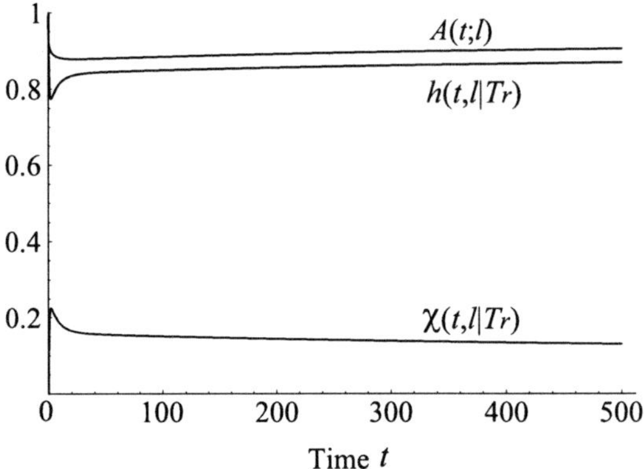

Fig. 10.5 $h(t, l | T_r)$, $\chi(t, l | T_r)$, and $A(t; l)$ ($T_r = 0.005$; $l = 26$, $v = 2.0$, $\alpha = 1000.0$)

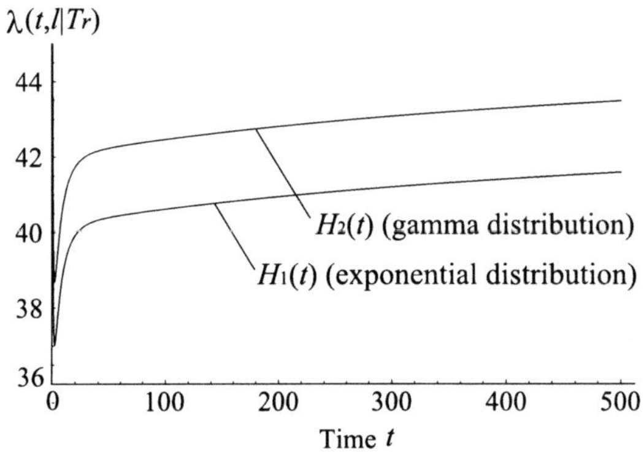

Fig. 10.6 Dependence of $\lambda(t, l | T_r)$ on the distribution of Y, $H(t)$ ($T_r = 0.005$; $l = 26$, $\theta = 50.0$)

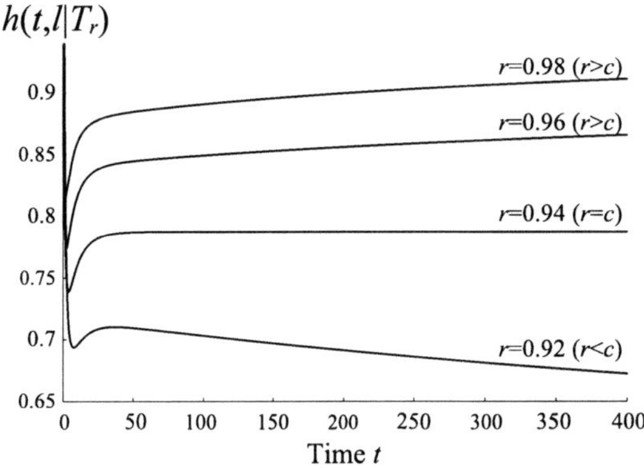

Fig. 10.7 Dependence of $h(t, l | T_r)$ on r ($T_r = 0.005$; $l = 26$, $v = 2.0$, $\alpha = 1000.0$)

Figures 10.8 and 10.9 show the dependence of $p(t, l | T_r)$ on the perfect debugging rate, a, in the cases of $r > c$ and $r < c$, respectively. These figures tell us that

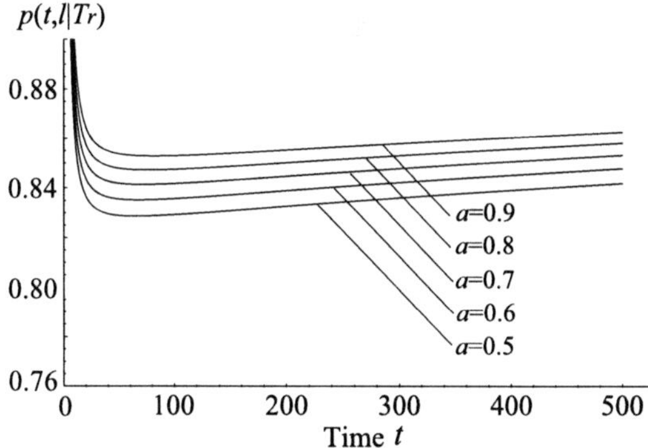

Fig. 10.8 Dependence of $p(t, l|T_r)$ on perfect debugging rate, a, in case of $r > c$ ($T_r = 0.005$; $r = 0.960$, $l = 26$, $v = 2.0$, $\alpha = 1000.0$, $\theta = 50.0$)

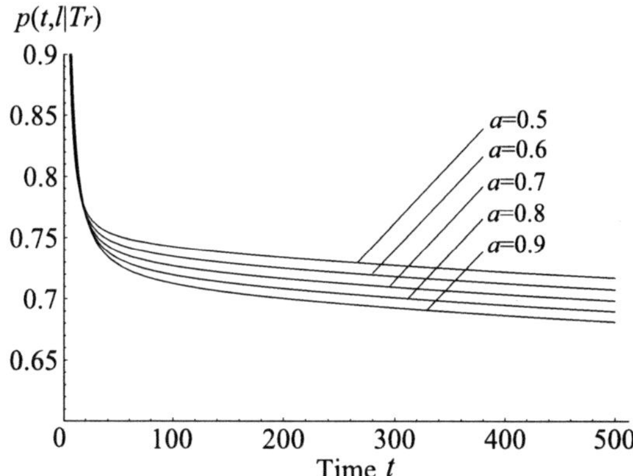

Fig. 10.9 Dependence of $p(t, l|T_r)$ on perfect debugging rate, a, in case of $r < c$ ($T_r = 0.005$; $r = 0.920$, $l = 26$, $v = 2.0$, $\alpha = 1000.0$, $\theta = 50.0$)

the software performance becomes higher (lower) as the perfect debugging rate becomes larger when $r > c$ ($r < c$). The case of $r < c$ may be a paradoxical result that the software performance decreases more slowly with decreasing a. This reasoning is that the proposed measure is related to the ratio of the software failure time and the restoration time, *i.e.*, ρ_n increases more slowly with decreasing a since smaller a means that it is more difficult to increase the cumulative number of corrected faults.

10.6 Concluding Remarks

In this paper, we have discussed the performance evaluation method for software systems considering real-time property. The stochastic behavior to the software system alternating between up and down states have been described by the Markovian availability model. Assuming that the cumulative number of the tasks arriving at the system up to a given time point follows the homogeneous Poisson process, we have analyzed the distribution of the number of tasks whose processes can be complete within the processing time limit with the infinite server queueing model. From the model, we have derived the quantitative measures for software performance assessment, which have been given as the functions of time and the number of debugging activities. We have also illustrated the several numerical examples of these measures to show that these measures are useful for grasping the relationship between software performance evaluation and the number of debuggings. In particular, it has been meaningful to correlate the real-time property evaluation with the software reliability and restoration characteristics.

We have assumed that the number of tasks arriving at the system has stationary increments. However, there are cases where this process has non-stationary increments in the actual situation. The investigation into the case where the number of tasks arriving at the system follows an NHPP remains as one of the interesting issues.

References

1. Tortorella M (2005) Service reliability theory and engineering, I: Foundations. Quality Technology and Quantitative Management 2:1–16
2. Tortorella M (2005) Service reliability theory and engineering, II: Models and examples. Quality Technology and Quantitative Management 2:17–37
3. http://www.saforum.org
4. Beaudry MD (1978) Performance-related reliability measures for computing systems. IEEE Transactions on Computing C-27:540–547
5. Meyer JF (1980) On evaluating the performability of degradable computing systems. IEEE Transactions on Computing C-29:720–731
6. Pham H (2000) Software Reliability. Springer-Verlag, Singapore
7. Yamada S (2002) Software reliability models. Stochastic models in reliability and maintenance, Chapter 10. Springer-Verlag, Berlin, pp 253–280
8. Kimura M, Yamada S (1995) Performance evaluation modeling for redundant real-time software systems (in Japanese). Trans IEICE J78-D-I:708–715
9. Kimura M, Yamamoto M, Yamada S (1998) Performance evaluation modeling for fault-tolerant software systems with underline underline processing time limit (in Japanese). J Reliab Eng Assoc Japan 20:422–432
10. Muppala JK, Woolet SP, Trivedi KS (1991) Real-time-systems performance in the presence of failures. Computer 24:37–47
11. Ihara H (1994) A review of real time systems (in Japanese). Journal of Information Processing Society of Japan 35:12–17
12. Tokuno K, Yamada S (2000) Markovian software availability measurement based on the number of restoration actions. IEICE Transactions on Fundamentals E83-A:835–841
13. Ross SM (1992) Applied Probability Models with Optimization Applications. Dover Publication, New York
14. Moranda PB (1979) Event-altered rate models for general reliability analysis. IEEE Transactions on Reliability R-28:376–381
15. Tokuno K (1999) A study on Markovian Software Reliability Modeling for Availability and Safety Assessment. Doctoral Dissertation, Tottori University
16. Goel AL, Okumoto K (1979) Time-dependent error-detection rate model for software reliability and other performance measures. IEEE Transactions on Reliability R-28: 206–321

Chapter 11
Failure Probability Estimation of Long Pipeline

Liyang Xie, Zheng Wang, Guangbo Hao, Mingchuan Zhang

Department of Mechanical Engineering,
Northeastern University,
Shenyang, 110004, China

11.1 Introduction

It is well known that, for the majority of pressurized pipelines, both the load and the resistance parameters show evident uncertainty, and a probabilistic approach should be applied to assess their behaviors. Concerning reliability estimation of passive components such as pressure vessel and pipeline, there are two kinds of approaches – direct estimation using statistics of historical failure event data, and indirect estimation using probabilistic analysis of the failure phenomena of consideration. The direct estimation method can be validated relatively easily. However, it suffers statistical uncertainty due to scarce data. Indirect estimation method relies on the statistics of material property and those of environment load which are more readily available. As to systems composed of passive components, statistical dependence among component failures is a complex issue that cannot be ignored in reliability estimation.

Estimating pipeline failure probability by traditional reliability engineering and statistical analysis principles is complicated also because no generally applicable passive component partition approach exists [1]. Consequently, it is conventional to take the whole object as a component. To evaluate the failure frequency of a long pipeline in this way, a coefficient was used to reflect size effect [2].

The validity to use a size factor to characterize size effect is questionable, since a long pipeline normally contains many susceptible sites, and the failures of the different susceptible sites are not absolutely correlated to each other. For the same reason, it does not seem feasible to take the whole pipeline as one component when estimating its failure probability.

Another important technique for the reliability evaluation of a large-scale system is the asymptotic approach [3–6]. There, the initial complex formula of system reliability is approximated by assuming that the numbers of system components tends to infinity and deriving the respective limit reliability function of the system. Mathematically, such an approach is based on the limit theorems of order statistics distribution. However, the statistical dependences among component failures (*i.e.*, the so-called common cause failure phenomenon), which play an important role in system safety and reliability, may violate the statistically independent failure assumption underlying the limit theorems of order statistics distribution.

One of the acceptable methods for modeling a run of a pipe in probabilistic risk assessment is to divide the pipe run into segments. Piping segment is defined as portions of piping for which the potential degradation mechanism is the same, and a failure at any point in the segment results in the same consequence [7].

Based on parametric statistical model for the strength of individual wires, MH Faber investigated cable failure probability [8]. It was thought that the fluctuations along wire are responsible for the pronounced weakest-link effect, *i.e.*, reduction in strength as the wire length increases. Individual wires were considered as a weakest link structural system (a series system). The numbers of components in the system depends not only on the length of the wire, but also on the statistical characteristics of the material parameters together with the defects (cracks, corrosion pits, etc.) in the wires [8]. Under this framework, one important parameter is "correlation length", defined as the length over which the material parameters and/or the defects in the wires are correlated. If the correlation length is shorter compared with the length of the wire, the numbers of components in the series system will be very large, as in the situation of old or damaged wires where the correlation length may be in the order of the diameter of the wire [8].

With probabilistic risk assessment as the scenario, a pipeline was treated as a large-scale series system composed of a great number of segments, and the issues of pipeline discretization, segment failure dependence and system failure probability estimation were concerned in the present paper. First of all, pipeline strength distribution was described with respect to different segment partitions. Then, the statistical dependence among segment failures was expounded with respect to the virtual segments. Finally, a series system failure probability model was presented and pipeline failure probability was demonstrated as a function of the number of segments. It was shown that there exists strong dependence among segment failures, and in particular, there exists an upper bound for the failure probability of a large-scale series system such as a long pipeline.

In addition, a time-dependent (or load action number-dependent) series system reliability model was introduced to reflect the effect of the multiple application of randomly repeated load. It presented the decrease of reliability over time, or more exactly, with the increase of loading history owing to the fact that higher load is more likely to appear among larger samples.

11.2 Segment Partition and System Strength Distribution

For the purpose of pipeline failure probability evaluation, a pipeline can be taken as a series system composed of a great number of virtual elements. All the following analysis and calculation are under this framework.

First of all, we divide the pipeline into n segments (virtual components), meaning that the pipeline is an n-segment series system.

For system failure probability evaluation, an important issue is the dependence among element failures. System failure probability is determined by environment load and the strengths of the individual, given the system configuration. Generally, there are significant uncertainties in both the load and component strength. Therefore, both load and strength are normally expressed by random variables. In such a situation, component failures are not independent of each other [13, 15]. Firstly, we discuss two special cases, *i.e.*, deterministic load and deterministic component strength.

Taking a series system as an example, if environment load is a deterministic constant, *i.e.*, the loads subjected to the system population are exactly the same, while the strengths of the individual components are independently identical distributed random variables, the events of component failures under such a deterministic load environment are statistically independent of each other, since individual component failures are entirely determined by its strength. In this case, the load-component strength interference relationship is as shown in Fig. 11.1, and series system failure probability is equal to

$$p_S = 1 - (1 - p)^n,$$

where, p_S-failure probability of a system composed of n identical component in series

p component failure probability
n number of components in system.

In the other case, load is a random variable, while component strength is deterministic, *i.e.*, all the components in a system are exactly the same. Obviously, all the components will fail to or survive to the common load meanwhile, since no differences between the resistances of the individual components to load. The load-strength interference relationship for this case is as shown in Fig. 11.2. This is the case that component failures are perfectly dependent on each other, and system failure probability is equal to component failure probability, *i.e.*, $p_S = p$, no matter how many components there are in the system and what the system configuration is like.

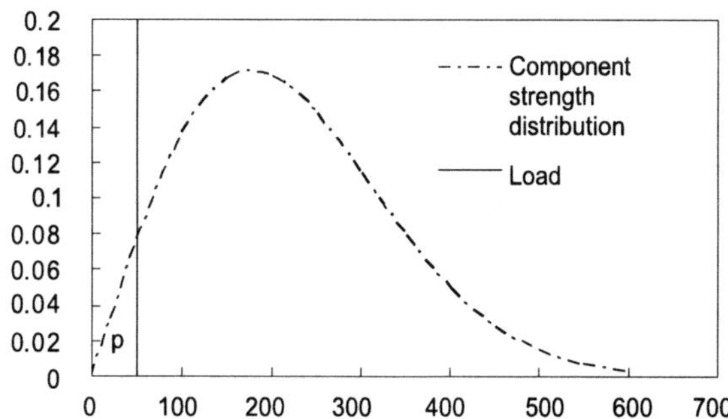

Fig. 11.1 Load-strength interference relationship in the case of deterministic load

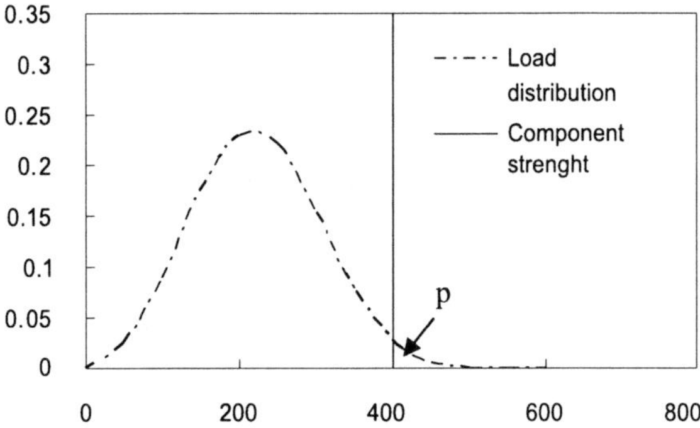

Fig. 11.2 Load-strength interference relationship in the case of deterministic component strength

The normal situation for system failure probability is that both the load and the component are random variables, and the load-strength relationship is as shown in Fig. 11.3.

Usually, the strengths of the individual components, denoted by X_i ($i = 1 \sim n$), are independent and identically distributed random variables. Thus, the strength of a weakest link system (*i.e.*, a series system composed of n elements), denoted by X, is equal to the smallest component strength in statistical sense, *i.e.*

$$X = \min(X_1, X_2, \ldots, X_n) . \tag{11.1}$$

It indicates that the system strength equals to the smallest sample value among the n element strengths X_i ($i = 1, 2, \ldots, n$). Obviously, it is more likely to achieve low system strength X if the component numbers n is large.

Taking a pipeline as a series system, the relationship between pipeline (a series system) failure probability P_n and its segment failure probability p can be expressed, in the condition that all the segment failures are independent of each other, as:

$$P_n = 1 - (1 - p)^n . \tag{11.2}$$

Consider a different partition to the pipeline by using larger segment, *e.g.*, let k segments mentioned above make up a larger segment, then the pipeline can be consid-

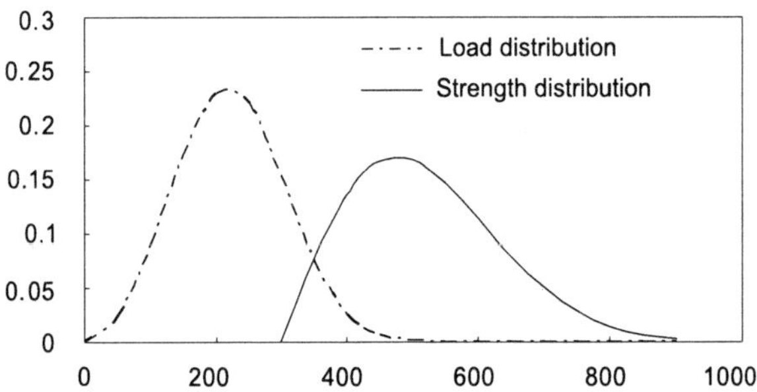

Fig. 11.3 Load-strength interference relationship

ered as composed of m ($m = n/k$) larger segments. Obviously, the failure probability of the larger segment equals:

$$p_k = 1 - (1 - p)^k .$$ (11.3)

Then the relationship between pipeline failure probability and segment failure probability is:

$$P_m = 1 - (1 - p_k)^m = 1 - (1 - p)^{k \times m} = 1 - (1 - p)^n .$$ (11.4)

It turns out that the partition to the pipeline does not influence the estimation of its failure probability in the condition of statistically independent failures.

As a system made up of statistically identical segments, the segments can be taken as the samples coming from the same population, and the strengths $X_1, X_2, \ldots, X_n$ of the segments in the system are independent and identically distributed random variables. The order statistic $X_{(k)}$ stands for the strength of the kth weakest segment. Obviously, the strength of the series system can be described by the minimum order statistic $X_{(1)}$. According to probability theory [9], the cumulative probability function of the minimum order statistic $X_{(1)}$ is:

$$G_1(x) = 1 - (1 - F(x))^n ,$$ (11.5)

where, $G_1(x)$ and $F(x)$ are the cumulative distribution function (cdf) of the minimum order statistic $X_{(1)}$ and that of the population X respectively.

The probability density function of the minimum order statistic $X_{(1)}$, which is just the probability density function (pdf) of the series system, is then

$$g_1(x) = n[1 - F(x)]^{n-1} f(x) ,$$ (11.6)

where, $g_1(x)$ and $f(x)$ are the pdf of the minimum order statistic $X_{(1)}$ and that of the population X respectively.

Considering a different partition of m ($m = n/k$) larger segments, the cdf and pdf of the strength of the larger segment (made up k small elements) are respectively:

$$J_1(x) = 1 - (1 - F(x))^k$$ (11.7)

$$j_1(x) = k[1 - F(x)]^{k-1} f(x) .$$ (11.8)

Therefore, the pdf of system strength can be expressed as:

$$g_1(x) = \frac{n}{k}[1 - J_1(x)]^{\frac{n}{k}-1} j_1(x)$$
$$= n[1 - F(x)]^{n-1} f(x) .$$ (11.9)

This proved that the segment partition scheme does not influence the system strength distribution. Nevertheless, one should remember in mind that such a conclusion is only hold true with the underlying precondition that the pipe material is continuous and its strength is uniform in the length direction.

For a pipeline with defects randomly distributed along its length, the segment size should be reasonably chosen to let the strengths of the segments are continuously distributed, so as to make the pipeline failure probability estimation easier.

11.3 Pipeline Failure Probability Estimation and Failure Dependence Analysis

The conventional assumption "component failures are statistically independent of each other in a system" is not usually valid since common cause failure exists in the majority of systems [10–14]. Subsequently, failure dependence has to be taken into account when estimate system failure probability.

In order to estimate system failure probability, the conventional method is to determine component failure probability first, then calculate system failure probability via a system failure probability model developed under the assumption of "component failures are statistically independent of each other".

By means of order statistics, we can develop a series system failure probability model without any assumption on component failure dependence. As mentioned above, pipeline can be taken as an n-segment series system, and its failure probability is equal to the probability that system strength (*i.e.*, the minimum order statistic of segment strengths) is less than the related load. That is, the failure probability of a series system made up of n segments equals to

$$P_{\text{series}} = \int_{-\infty}^{\infty} h(y) \left[\int_0^y g_1(x)\,\mathrm{d}x \right] \mathrm{d}y, \tag{11.10}$$

where $h(y)$ is the pdf of the load subject to the pipeline or all its segments.

In contrast, the system failure probability model can also be developed through the load-strength interference analysis at system-level. In the condition that component strengths (denoted by x) are independent and identically distributed random variables, and all the components are subjected to the same random load y, the series system failure probability model can be derived by means of system-level load-strength interference analysis [15, 16]:

$$P_{\text{series}} = 1 - \int_{-\infty}^{\infty} h(y) \left[\int_y^{\infty} f(x)\,\mathrm{d}x \right]^n \mathrm{d}y. \tag{11.11}$$

It can be proved that the order statistics model (11.10) and the system-level load-strength interference model (11.11) are the same.

To use the proposed model, the size of the pipe segment should be carefully chosen to assure the statistical uniformity of the segment strength and that of the associate load environment, so that the failure probability of a pipeline can be reasonably evaluated.

The differences between the statistically dependent system failure probability model (11.10) and the traditional statistically independent system failure probability model (11.2) will be shown through an example in the later part. The degree of the difference depends on both the load dispersion and the segment strength dispersion.

11.4 Pipeline Failure Probability Estimation

In the following, three types of models, *i.e.*, the order statistics model, the system-level load-strength interference model, and the conversional statistically independent system failure probability model are applied to estimate the failure probability of large-scale series system respectively.

The segments (components in a series system) is assumed to have normal-distributed strength and subjected to normal-distributed load. Illustrated in Fig. 11.4a is the situation of load expectation $\mu_l = 400\,\mathrm{MPa}$ load standard deviation (std) $\sigma_l = 30\,\mathrm{MPa}$ segment strength expectation $\mu_s = 600\,\mathrm{MPa}$ and segment strength std $\sigma_s =$

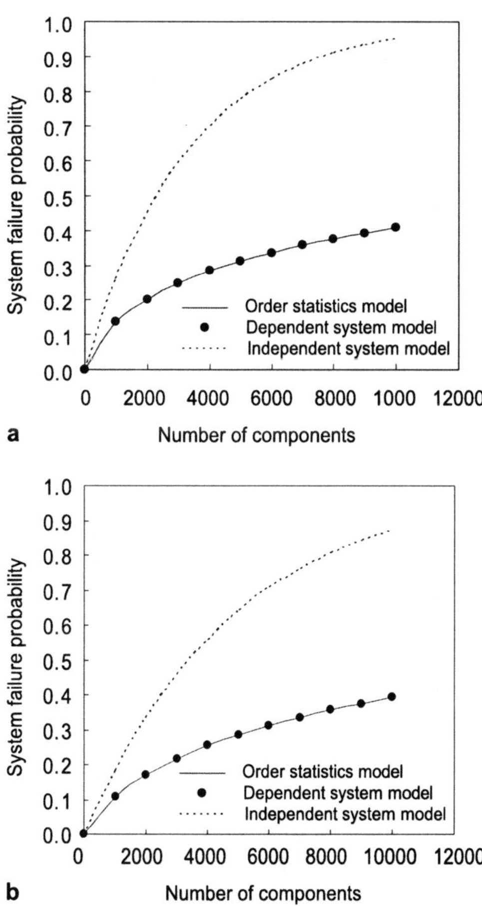

Fig. 11.4 Series system failure probabilities estimated by different models. **a** $\mu_l = 400\,\mathrm{MPa}$, $\sigma_l = 30\,\mathrm{MPa}$, $\mu_s = 600\,\mathrm{MPa}$, $\sigma_s = 50\,\mathrm{MPa}$. **b** $\mu_l = 200\,\mathrm{MPa}$, $\sigma_l = 50\,\mathrm{MPa}$, $\mu_s = 600\,\mathrm{MPa}$, $\sigma_s = 100\,\mathrm{MPa}$

50 MPa. In this situation, the segment failure probability is equal to 2.5×10^{-3}. The system failure probability is estimated by the three models respectively. Illustrated in Fig. 11.4b is the situation of load expectation $\mu_l = 200$ MPa load standard deviation (std) $\sigma_l = 50$ MPa segment strength expectation $\mu_s = 600$ MPa and segment strength (std) $\sigma_s = 100$ MPa. In this situation, the segment failure probability is equal to 1.7×10^{-4}. Likewise, the system failure probability is estimated by the three models respectively.

The results indicate that the order statistics model and the system-level load-strength interference model are the same, while the conventional statistically independent system model is quite conservative.

11.5 Upper Limit of Large-scale Series System Failure Probability

In the following, discussed is the effect of component numbers on system failure probability. For a system such as pipeline, if the size of the segment is small, then the numbers of segment in the system will be very large. According to conventional statistically independent series system failure probability model, system failure probability will approach 1 quickly with the increase of component number. It means that for a series system composed of a huge number of components, *e.g.*, 10,000 or more, its failure probability will approach 1, even though the individual component failure probability is very low. Obviously, the estimated failure probability (closed to one) is not reasonable even for a very long pipeline. Fortunately, different conclusions can be drawn with the statistically dependent system failure probability model.

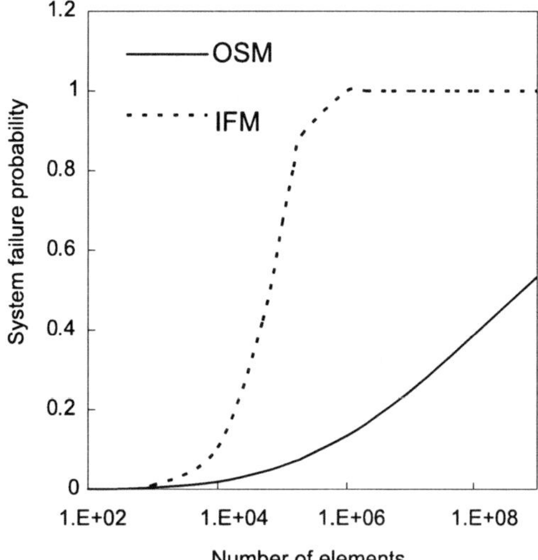

Fig. 11.5 Relationships between series system failure probability and segment number in condition of normal-distributed load and segment strength

For instance, let load pdf $h(y) \sim N(300, 50)$ (*i.e.*, the load y follows the normal distribution with the expectation of 300 MPa and the std. of 50 MPa), segment strength pdf $f(x) \sim N(600, 50)$, which yields segment failure probability of 1.105×10^{-5}, the relationship between the failure probability of series system and segment numbers is shown in Fig. 11.5. It shows that system failure probability increases with the increase of segment number, the system failure probability estimated by the conventional statistically independent system model (IFM in short) approaches 1 quickly, but that estimated by the statistically dependent system model (the order statistic model, OSM in short) does not show such a strong tendency.

When load pdf $h(y) \sim W(200, 3, 100)$ (*i.e.*, load follows the Weibull distribution of which the minimum load equals 100 MPa), and segment strength pdf $f(x) \sim W(300, 3.5, 420)$ (*i.e.*, segment strength follows the Weibull distribution

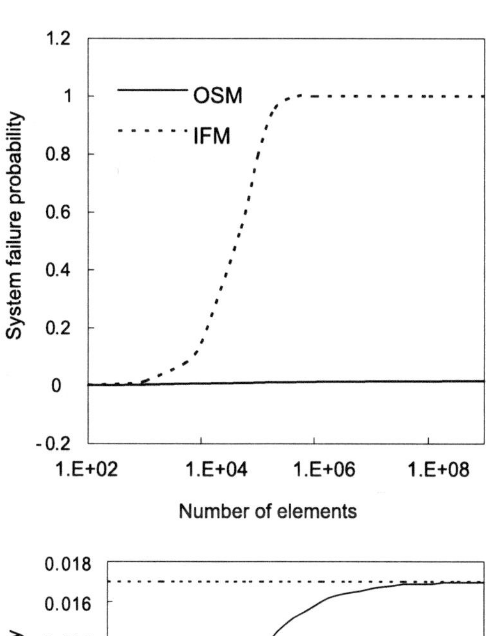

Fig. 11.6 Relationships between series system failure probability and segment number in condition of Weibull-distributed load and segment strength

Fig. 11.7 The upper limit of large-scale series system failure probability in the condition of Weibull-distributed load and segment strength (the upper limit of system failure probability equals to 0.017)

of which the minimum strength equals 420 MPa), which yields a segment failure probability of 1.575×10^{-5}, the relationship between series system failure probability and segment number is shown in Fig. 11.6. In the situation in which the minimum of the segment strength random variable is greater with minimum load, system failure probability will not exceed a limit of less than one with the number of segment approaching infinity (Fig. 11.7). The limit equals the probability of load random variable exceeding the minimum order statistic of segment strength. The reason is simply that, in the situation of infinite segments, the strength minimum order statistic becomes into a deterministic constant that is equal to the minimum strength parameter of the Weibull-distributed random variable.

11.6 Pipeline Reliability Under Randomly Repeated Load

Most of pipelines are subjected to dynamic load in service. In other words, a pipeline in use will subject to many times the load actions that appear stochastically both in time and intensity. In terms of the likelihood of failure, the pipeline failure probability will be affected by the number of load actions, and the degree of the effect depends on the stochastic characteristics of the load.

In the condition that strength does not degrade during operation, the reliability that a pipeline survives w times of randomly repeated load is equal to the reliability that it survives the maximum of the w load samples. A reliability model of a pipeline subjected to dynamic load can be developed according to this principle.

Based on the reliability model under single load action, the reliability of an n component series system subjected to multiple load actions can be modeled by means of the order statistic of load samples [17]:

$$
\begin{aligned}
R_{\text{series}}^{(w)} &= \int_{-\infty}^{+\infty} \left(\int_{y}^{+\infty} f(x)\,\mathrm{d}x \right)^{n} w\,[H(y)]^{w-1}\,\mathrm{d}H(y) \\
&= \int_{-\infty}^{+\infty} [1 - F(y)]^{n}\,w\,[H(y)]^{w-1}\,\mathrm{d}H(y) ,
\end{aligned}
\tag{11.12}
$$

where $f(x)$ stands for the probability density function of strength x, $H(y)$ stands for the cumulative distribution function of load y.

For most mechanical equipment and systems, operation load can be described by a stochastic process [18, 19]. Let $N(t)$ denote the times of stochastic load subjected to a component in the time interval $(0, t)$ which is assumed to present on the following properties:

(1) $N(0) = 0$;
(2)   For any $0 < t_1 < t_2 < \cdots < t_w$, $N(t_1)$, $N(t_2) - N(t_1), \ldots, N(t_w) - N(t_{w-1})$ are independent of each other;

(3)   The number of load action depends only on the time interval and has nothing to do with the beginning point of the time, *i.e.*

$$\forall s, t \geq 0, w \geq 0,$$
$$P[N(s+t) - N(s) = w] = P[N(t) = w]$$

(4)   For $t > 0$ and sufficient small $\Delta t > 0$,

$$\begin{cases} P[N(t+\Delta t) - N(t) = 1] = \lambda \Delta t + o(\Delta t) \\ P[N(t+\Delta t) - N(t) \geq 2] = o(\Delta t) \end{cases}.$$

A loading process that satisfies the above conditions can be described by the homogeneous Poisson process with parameter λ. Subsequently, the probability that load acts for w times ($N(t) = w$) during the time interval $(0 - t)$ is equal to

$$P[N(t) - N(0) = w] = \frac{(\lambda t)^w}{w!} e^{-\lambda t}.$$

The probability that a component is subjected to exactly w times the load action in the time interval $(0 - t)$ and successively survives these load actions is then

$$\begin{aligned} R_{\text{series}}(w, t) &= P[N(s+t) - N(s) = w] R_{\text{series}}^{(w)} \\ &= \frac{(\lambda t)^w}{w!} e^{-\lambda t} \int_{-\infty}^{+\infty} [1 - F(y)]^n w [H(y)]^{w-1} \, dH(y). \end{aligned} \tag{11.13}$$

According to the total probability formula, series system reliability at time t is

$$\begin{aligned} R_{\text{series}}(t) &= \sum_{w=0}^{+\infty} R_{\text{series}}(w, t) \\ &= \sum_{w=0}^{+\infty} \frac{(\lambda t)^w}{w!} e^{-\lambda t} \int_{-\infty}^{+\infty} [1 - F(y)]^n w [H(y)]^{w-1} \, dH(y) \\ &= e^{-\lambda t} \left\{ 1 + \int_{-\infty}^{+\infty} [1 - F(y)]^n \sum_{w=1}^{+\infty} \frac{(\lambda t)^w}{w!} w [H(y)]^{w-1} \, dH(y) \right\} \\ &= e^{-\lambda t} \left\{ 1 + \lambda t \int_{-\infty}^{+\infty} [1 - F(y)]^n \sum_{w=1}^{+\infty} \frac{(\lambda t)^{w-1}}{(w-1)!} [H(y)]^{w-1} \, dH(y) \right\}. \end{aligned}$$
$$\tag{11.14}$$

Using the Taylor expansion of an exponential function, the above equation can be simplified as

$$\begin{aligned} R_{\text{series}}(t) &= e^{-\lambda t} \left(1 + \lambda t \int_{-\infty}^{+\infty} [1 - F(y)]^n e^{\lambda t H(y)} \, dH(y) \right) \\ &= \int_{-\infty}^{+\infty} e^{[H(y)-1]\lambda t} n [1 - F(y)]^{n-1} \, dH f(y) \end{aligned} \tag{11.15}$$

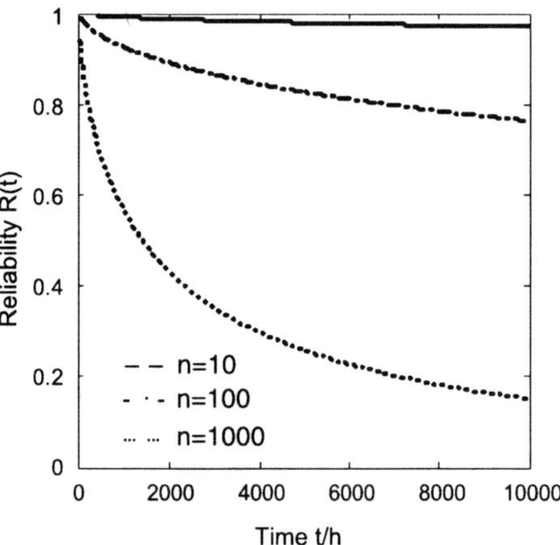

Fig. 11.8 Relationships between pipeline reliability and service time

Taking the series system with different number of segments (*i.e.*, 10, 100, 1000 respectively) as an example, when the Poisson process parameter λ is taken as $0.5\,\mathrm{h}^{-1}$, the strength of component follows the normal distribution with mean $\mu_x = 600\,\mathrm{MPa}$ and standard deviation $\sigma_x = 50\,\mathrm{MPa}$, the stress follows the normal distribution with mean $\mu_y = 330\,\mathrm{MPa}$ and standard deviation $\sigma_y = 35\,\mathrm{MPa}$, the relationships between the reliability of the pipeline and service time are shown in Fig. 11.8.

11.7 Conclusion

Aiming at failure probability estimation of pipeline or similar type of object, the present paper takes a long pipeline as a large-scale series system, presents failure probability models for statistically dependent series system, and illustrates the limit situation of large-scale series system failure probability.

The investigation showed that segment size/numbers selection does not affect system failure probability estimation. In the situation that segment strength distribution has a finite lower limit, series system failure probability has an upper limit of less than 1 when the segment number approaches to infinity. More precisely, with the increase of segment number, series system failure probability will not approach 1, but to a limit value much less than 1, which is equal to the probability that the lower limit of the segment strength distribution is less than the related load random variable.

It was also shown that there are considerable differences between the independent system failure probability model and the statistically dependent system failure probability model. It means that failure dependence plays an important role in system of which the components are subjected to the same random load. Obviously, here the load has a general sense, it can be mechanical stress, temperature, corrosion intensity, and so on. Correspondently, the component strength will be the property against mechanical stress, temperature, corrosion, and so on.

In addition, a reliability model for a pipeline to subject to multiple load actions is presented, where the Poisson process is used to describe the load history. The relationship between the reliability of a pipeline and its service time is demonstrated.

Acknowledgements This research was subsidized by the Special Funds for the Major State Basic Research Projects 2006CB605000 and the Hi-Tech Research and Development Program (863) of China with the grant No. 2006AA04Z408.

References

1. Nyman R, Erixon S, Tomic B *et al.* (1995) Reliability of piping system components. SKI Report 95:58, Vol. 1: Piping reliability – a resource document for PSA applications, December, ISSN 1104-1374, ISRN SKI-R-95/58-SE, Swedish Nuclear Power Inspectorate
2. Taylor JR (1994) Risk analysis for process plant, pipelines and transport. E&F N Spon, London, pp. 3–5
3. Kolowrocki K (2003) Asymptotic approach to reliability evaluation of large multi-state systems with application to piping transportation. Pressure Vessels and Piping 80:59–73
4. Sutherland LS (1997) Review of probabilistic models of the strength of composite materials. Reliability Engineering and System Safety 56:183–196
5. Cichocki A (2001) Limit reliability functions of some homogeneous regular series-parallel and parallel-series systems of higher order. Applied Mathematics and Computation 117:55–72
6. Bozena K-S (2001) A remark on limit reliability function of large series-parallel systems with assisting components. Applied Mathematics and Computation 122:155–177
7. Draft standard review plan for the review of risk informed in-service inspection of piping, Draft SRP Chapter 3.9.8, October 1997
8. Faber MH, Engelund S, Rackwitz R (2003) Aspects of parallel wire cable reliability. Structural Safety 25:201–225
9. Mao SS, Wang JL, Pu XL (1998) Advanced statistics. High Education Press, Beijing
10. Hughes RP (1987) A new approach to common cause failure. Reliability Engineering 17:211–236
11. Gupta RC (2002) Reliability of a k out of n system of components sharing a common environment, Applied Mathematics Letters 15:837–844
12. Kvam PH, Miller JG (2002) Common cause failure prediction using data mapping. Reliability Engineering and System Safety 76:273–278
13. Xie LY (1998) A knowledge based multi-dimension discrete CCF model J. Nuclear Engineering Design 183:107–116
14. Goble WM, Brombacher AC, Bukowski JV *et al.* (1998) Using stress-strain simulations to characterize common cause. In Mosley A, Bari RA (eds) Probabilistic safety assessment and management (PSAM4). Springer, New York, pp. 519–403
15. Xie LY, Zhou JY (2004) System-level load-strength interference based reliability modeling of k-out-of-n dependent system. Reliability Engineering And System Safety 84:311–317
16. Xie LY, Zhou JY (2005) Load-strength order statistics interference models for system reliability evaluation. International Journal of Performability Engineering 1:23–36
17. Xie LY, Zhou JY (2005) Data mapping and the prediction of common cause failure probability. IEEE Transactions on Reliability 54:291–296
18. Ditlevsen O (2002) Stochastic model for joint wave and wind loads on offshore structures. Structural Safety 24:139–163
19. Li JP, Thompson G (2005) A method to take account of in-homogeneity in mechanical component reliability calculations. IEEE Transactions on Reliability 54:159–168

Part III
Software Reliability and Testing

Chapter 12
Software Fault Imputation in Noisy and Incomplete Measurement Data

Andres Folleco, Taghi M. Khoshgoftaar, Jason Van Hulse

Department of Computer Science and Engineering,
Florida Atlantic University, Boca Raton, Florida

12.1 Introduction

Reliability is an important characteristic of all software products, and is especially significant for high-assurance and/or mission-critical applications and systems. Obtaining highly reliable software is an extremely difficult challenge. Early software fault prediction is a proven strategy to achieve satisfactory software reliability, with quality enhancement efforts directed towards modules most likely to contain faults. Fault prediction efforts rely on obtaining obtaining and quantifying measures and methods of software quality, such as software measurement data and historical fault information.

Recent work [1] has demonstrated that software quality estimates based on software measurement data and metrics [2, 3] can generate accurate predictions. These techniques can be used to predict the response attribute, which can be the class of a module, *e.g.*, fault-prone or not fault-prone, or a quality factor such as the number of faults for a particular module. When the response variable is the class of a module, the software estimation strategy is known as a classification model [4]. If the response variable is a quality factor then prediction models and methods are used for software quality estimation [5].

This work demonstrates the importance of ensuring the quality of software measurement data before software quality estimation efforts can be conducted[1]. It is our belief that for any given software measurement data, improving the quality of the data will improve the performance and accuracy of most software quality imputation techniques. Therefore, software managers should pay close attention to collecting reliable, high-quality software measurement data as well as gathering software metrics that can represent the underlying software quality. To our knowledge, this is one of the first studies to focus specifically on the impact of noise on the calcu-

---

[1] "Quality" has two distinct meanings in this work. The *quality of software mesurement data* is related to data noise or errors in the data characterizing the module, *i.e.*, the software metrics and measurements. *Software quality estimation*, on the other hand, refers to the quality of the software itself, *i.e.*, if the software module itself contain faults.

lation of the imputation performance metric, the average absolute error (*aae*). The *aae* is a commonly used measure of the accuracy of software quality imputation techniques. The objective of this work is to demonstrate the potential difficulties that may be encountered when evaluating imputation techniques using data of suspicious quality (*i.e.*, noisy data). In other words, using noisy data in the evaluation of imputation techniques can mistakenly lead to erroneous conclusions.

In addition, we evaluate the impact of noisy data on the effectiveness of various imputation techniques. By examining different levels of noise, we can quantitatively measure the relationship between noise and imputation. Our results demonstrate that poor quality data have a detrimental impact on imputation, further confirming the importance of collecting high-quality measurement data when performing empirical analysis. Further, this study is one of the first in the domain of software quality to analyze Bayesian multiple imputation, a robust and theoretically sound methodology for handling missing values.

12.2 Empirical Datasets

12.2.1 CCCS Dataset Description

The CCCS dataset comprises measurement data taken from a total of 282 Ada software program modules [6]. CCCS has eight attributes representing software metrics (presented in Table 12.1), along with a continuous dependent attribute, *nfaults*, indicating the number of faults recorded for each module during the system integration phase, the test phase, and the first year of deployment. In all, 136 of the 282 instances in the CCCS dataset contained at least one known fault. The original CCCS dataset is denoted O in this work.

The dataset O along with three other derived CCCS datasets were used for the experiments presented in this work. As explained in Sect. 12.2.2.1, O contains 20 instances with *inherent* noise, as determined by a software engineering domain expert.

Table 12.1 CCCS Software Metrics Description

Independent attributes
Unique operators
Total operators
Unique operands
Total operands
Cyclomatic complexity
Logical operators
Total lines of code
Executable LOC

Dependent attribute
Number of faults (*nfaults*)

In addition to the inherent noise already contained in CCCS, two derived datasets (denoted $05p$ and $10p$) contain 5% and 10% injected noise in the dependent variable *nfaults* as explained in Sect. 12.2.2.2. The third derived dataset was denoted C and is introduced in Sect. 12.2.3.

12.2.2 Inherent and Simulated Noise

12.2.2.1 Inherent Noise Detection

Dataset O has been used previously for software quality estimation studies by our research group [6]. Because it is a real-world dataset, O already contained noise. We refer to noise that naturally exists in a dataset as *inherent* noise, in contrast to noise injected artificially into a dataset. The only way to accurately determine whether a real-world dataset contains inherent noise is for a domain expert to perform detailed inspections of the data, which is a tedious and time-consuming task. In our study, an expert with over 15 years of experience working with software engineering measurement data inspected O and labeled 20 instances as noisy with respect to the dependent variable *nfaults*. The domain expert also made use of numerous statistical and data analysis techniques to help locate the noisy data. Note that the expert maintained a very conservative noise-labeling strategy. Instances were tagged as noisy only if the evidence of noise in *nfaults* was absolutely conclusive. In addition to locating the noisy instances in the dataset, the expert also assigned a cleansed value for *nfaults* for the 20 noisy instances. The combination of expert input and several data analysis methods identified, with a high degree of confidence, 20 instances in O with severe inherent noise in the dependent variable *nfaults*.

12.2.2.2 Simulated Noise Injection

Two additional datasets used in our experiments, called $05p$ and $10p$, were derived from O by injecting simulated noise into the dependent variable *nfaults* of a selected set of instances. Noise was injected into these datasets using expert input in a manner that was similar to the types of noise found in software engineering metrics data. We consider this type of noise injection *domain-realistic*, as opposed to the random selection of instances and the injection of random noise (which may or may not occur in real-world data). When studying the effects of noise with artificial noise injection, it is critically important to make the corruption process reasonable for the given application domain. If this is not the case, then it is unclear if the results translate to real-world datasets, and hence the conclusions may be dubious. In this study, the domain expert located instances that were relatively clean from O and corrupted the dependent variable *nfaults* to a noisy value reasonable for this application domain.

The expert identified 14 relatively clean instances and corrupted *nfaults* for these 14 instances, creating the $05p$ dataset. Next, the expert found 14 additional relatively clean instances and corrupted the dependent variable *nfaults*, generating the $10p$ dataset. The datasets $05p$ and $10p$ therefore contain 5% and 10% simulated noise, respectively. In addition, $05p$ and $10p$ include the 20 inherently noisy instances described previously that exist in O. Therefore in total, there are 20 noisy instances in O, 34 noisy instances (14 with simulated noise) in $05p$, and 48 noisy instances (28 with simulated noise) in $10p$.

12.2.3 Relatively Clean CCCS Dataset

The third derived dataset considered in this study is a clean (relative to the dependent variable *nfaults*) version of CCCS. Expert input and several data analysis procedures were used to further inspect O in order to locate additional instances with noise in *nfaults* besides the 20 previously identified. Sixty-one additional instances (and 81 in total) were identified and clean values were determined for all 81 noisy observations. These 61 additional noisy instances were not as noisy, in a relative sense, as the first 20 identified. The objective in creating this dataset, however, was for it to be as clean as possible with respect to the dependent variable for the purposes of experimental analysis. The cleaned CCCS dataset was denoted C, and by construction was very clean with respect to the dependent variable *nfaults*. Note that the independent variables were not cleansed in C.

In summary, this work utilizes four datasets, C, O, $05p$, and $10p$, with different noise characteristics. C is the cleanest dataset relative to *nfaults*, with no noisy instances. O contains 20 instances with inherent noise. Next, dataset $05p$ has 20 instances with inherent noise and 14 instances with injected noise. The noisiest dataset is $10p$, which has 20 inherently noisy instances and 28 instances with injected noise.

12.3 Imputation Techniques

12.3.1 Regression Imputation

Multi-variable linear regression [7] was used in this work as implemented within the Weka data mining tool [8]. The regression model was built using the observed data from all the independent variables. Missing values in the dependent variable are replaced with predicted ones built on complete data observations. Given independent variables $X_1, \ldots, X_n$ and dependent variable Y, the linear regression model $\hat{Y} = \beta_0 + \beta_1 X_1 + \ldots + \beta_n X_n$ is constructed and the parameters $(\beta_0, \ldots, \beta_n)$ are estimated such that the squared error $\sum (\hat{Y}(x_i) - Y(x_i))^2$ is minimized.

12.3.2 REPTree Decision Tree Imputation

This algorithm as implemented in Weka [8] builds a decision tree by evaluating the predictor attributes against a quantitative target attribute. This building process is accomplished by using variance reduction to derive balanced tree splits and minimize error corrections. The process is recursively repeated to simplify and prune the tree by comparing it against already established trees. This process is repeated until a stopping criterion is satisfied.

12.3.3 Nearest Neighbor Imputation

As implemented in Weka, nearest neighbor imputation, also denoted as IBk imputation, searches for the k most similar complete cases to the instance with the missing value and replaces the missing value with the mean of the corresponding attributes of these similar cases. IBk is known to preserve the population sample distribution by substituting the average of several of the most similar observed values [8]. The value for the k parameter [9] has been subjected to intense study because it can critically affect the method's performance. In our experiments, we utilized the Weka data mining software tool instance-based classifier to 'automatically' determine an optimal value for k from a range of possible values. Jonsonn [9] recommended that the maximum value for k be the square root of the number of observations with complete data. The instance-based classifier in Weka utilizes leave-one-out cross-validation to determine a value for k before the IBk imputation takes place.

12.3.4 Mean Imputation

This technique imputes each missing value with the mean of the observed values for the corresponding attribute. This technique is fast and simple but under-estimates the sample variance.

12.3.5 Bayesian Multiple Imputation

The version of Bayesian multiple imputation (also referred to as multiple imputation or BMI) used in our study was designed by Schafer [10] and implemented in SAS [11]. BMI was first proposed by Rubin [12] and can produce estimates that are consistent, asymptotically efficient, and normal when the data is MAR or MCAR (explained further in Sect. 12.4). BMI can be used with virtually any kind of model and the analysis can be done with available software. Input preparation and pre-processing requirements are minimal when compared to other techniques.

A normal model implies that all variables have normal distributions and each variable can be represented as a linear function of all the other variables, together with a normal, homoscedastic error term [13]. The error term is an adjustment factor needed because of the uncertainty about predicting missing values. Even though these are strict requirements, in practice thiss model works very well even if some of the variables are far from normality [10]. For variables that do not have normal distributions, normalizing transformations can significantly improve the imputation quality.

BMI is used to derive m complete datasets from one with missing values. These complete datasets, with missing values filled in, can be combined for the purpose of performing inference. More details of BMI can be found in numerous references [10, 13–15]. Using Schafer's [10] notation, a dataset D with missing values can be split into two datasets, D_{obs} and D_{mis}, where D_{obs} is the completely observed portion of D and D_{mis} is the portion of D with missing values. The m datasets are created by sampling the posterior predictive distribution of the missing data $P(D_{\text{mis}}|D_{\text{obs}})$. Directly sampling from this distribution may not be practical, so Markov chain Monte Carlo (MCMC) methods [16] such as data augmentation (DA) can be applied.

Data augmentation is used to obtain imputed datasets D_{mis}^k by repeatedly sampling from the distribution $P(D_{\text{mis}}|D_{\text{obs}})$. Suppose θ represents the unknown population parameters, which in the case of multivariate normal data is the mean and covariance matrix. DA proceeds iteratively through two steps known respectively as the I-(or Imputation) Step and the P-(or Posterior) Step [17] as follows:

- (I) Suppose that the current estimate for parameters θ are given by θ^t. Draw an estimate of the missing data $D_{\text{mis}}^{(t+1)}$ from the distribution $P(D_{\text{mis}}|D_{\text{obs}}, \theta^t)$.
- (P) Applying the estimate for the missing data, draw $\theta^{(t+1)}$ from $P\left(\theta|D_{\text{obs}}, D_{\text{mis}}^{(t+1)}\right)$.

The P-Step augments the observed data with assumed value of D_{mis}, resulting in a complete data posterior $P(\theta|D_{\text{obs}}, D_{\text{mis}})$ which is much easier to use. A sequence of datasets $D_{\text{mis}}^1, \ldots, D_{\text{mis}}^m$ is generated that has $P(D_{\text{mis}}|D_{\text{obs}})$ as its target distribution. Hence, having i large enough, D_{mis}^i is a random draw from the distribution $P(D_{\text{mis}}|D_{\text{obs}})$, as needed. The sequence $\{\theta_i\}$ can be ignored as we only need the imputed datasets.

Applying the DA algorithm and retaining the first m imputed datasets is problematic for two reasons. First, it must run long enough for convergence to occur $\left(i.e.,\text{ for } D_{\text{mis}} \text{ to really be a random draw from } P(D_{\text{mis}}|D_{\text{obs}})\right)$. Second, as DA is an iterative procedure where the current draw from $P(D_{\text{mis}}|D_{\text{obs}})$ relies on the previous draw, D_{mis}^i and D_{mis}^j are generally dependent when i and j are close. Schafer [10] defines multiple imputations that are independent draws from the posterior predictive distribution of the missing data $P(D_{\text{mis}}|D_{\text{obs}})$ as *Bayesian proper*. Sequential or parallel chains can be used to generate Bayesian proper multiple imputations with data augmentation.

A *sequential* (or single) chain consists of one chain of imputation cycles. A start-up period must be included where the first b iterations are ignored, allowing the chain to more surely converge to the target distribution. An imputed dataset is then stored after every k iterations, where k is known as the *lag*. If m is the desired number of complete datasets, a total of $b + (m-1)k$ iterations must be completed. Using a sequential chain, the sequence of stored imputed datasets is $\left\{ D_{\text{mis}}^{b}, D_{\text{mis}}^{(b+k)}, \ldots, D_{\text{mis}}^{(b+(m-1)k)} \right\}$, where D^{α} is the imputed dataset that results from the αth iteration of BMI, $1 \leq \alpha \leq b + (m-1)k$. BMI will generate imputed datasets $D_{\text{mis}}^{1}, D_{\text{mis}}^{2}, \ldots, D_{\text{mis}}^{(b-1)}$, all of which will be discarded after the next dataset in the sequence is generated. At the next iteration, imputed dataset D_{mis}^{b} is created and *stored*. Datasets $D_{\text{mis}}^{(b+1)}, D_{\text{mis}}^{(b+2)}, \ldots, D_{\text{mis}}^{(b+k-1)}$ are then created and discarded, while $D_{\text{mis}}^{(b+k)}$ is created and stored. This process is repeated until all m imputed datasets are generated.

Using parallel (or multiple) chains, BMI is executed m distinct times with D_{mis}^{i} selected in the final iteration of the ith run, $1 \leq i \leq m$. The first execution of BMI creates imputed datasets $D_{\text{mis}}^{(1,1)}, D_{\text{mis}}^{(1,2)}, \ldots, D_{\text{mis}}^{(1,b-1)}, D_{\text{mis}}^{(1,b)}$, where b again represents the start-up period. The first superscript indicates the distinct executions of BMI starting with the initialization of parameters, while the second is an index on the imputed datasets within each execution. In the first run of BMI, all datasets except $D_{\text{mis}}^{(1,b)}$ are discarded. BMI is then re-executed, probably with different initial parameters, and another chain of imputed datasets $D_{\text{mis}}^{(2,1)}, D_{\text{mis}}^{(2,2)}, \ldots, D_{\text{mis}}^{(2,b-1)}, D_{\text{mis}}^{(2,b)}$ is derived. Similar to the first chain, only dataset $D_{\text{mis}}^{(2,b)}$ is retained. After repeating this process m times, the imputed datasets used for analysis are $\left\{ D_{\text{mis}}^{(1,b)}, D_{\text{mis}}^{(2,b)}, \ldots, D_{\text{mis}}^{(m,b)} \right\}$. The datasets derived at the final step of each BMI run, of which there are a total of m runs, are used for imputation.

12.4 Missing Data Mechanisms

Related work has established three common types of missing data mechanisms [12]. The accuracy of an imputation technique can be directly impacted by the missing data mechanism. The question of whether or not the missing data can be imputed by the attribute values of the observed dataset may directly affect how missing values are handled and whether or not they can be ignored. Little and Rubin [18] defined three such mechanisms: missing completely at random (MCAR), missing at random (MAR), and non-ignorable (NI) missing data. In this work, the missing data were artificially induced in a completely random fashion, MCAR.

If the input dataset D contains n instances and m attributes, we can let R be an $n \times m$ matrix such that the entry $r_{ij} = 1$ if the jth attribute of instance i is missing, and zero otherwise. R is used to identify the location of the missing values in D. MCAR missingness occurs if $P(R|D_{\text{obs}}, D_{\text{mis}}) = P(R)$. Missingness is MAR if $P(R|D_{\text{obs}}, D_{\text{mis}}) = P(R|D_{\text{obs}})$. In other words, missing data are MAR if the occurrence of missing data depends only on the observed values and not on the missing values. Lastly, if the missingness pattern is related to the missing values themselves, then the missingness is known as non-ignorable or NI.

12.5 Experimental Design

12.5.1 Injection of Missing Data

Six levels of simulated missing values were established in the dependent variable *nfaults* for O, C, $05p$, and $10p$. The number of instances at each missingness level was 14 for 5%, 28 for 10%, 42 for 15%, 56 for 20%, 85 for 30%, and 113 for 40% missingness levels. Missingness was introduced only in the dependent variable *nfaults* and all instances (including noisy ones) were selected randomly (*i.e.*, in MCAR fashion). In other words, this study did not implement MAR or NI missingness.

In order to minimize any potential bias due to the random selection process, five versions of the O, C, $05p$, and $10p$ datasets were generated at each missingness level. At missingness level $\gamma\%$, β versions of the datasets were created for O, C, $05p$, and $10p$ respectively, where $\gamma\% = 5\%$, 10%, 15%, 20%, 30%, or 40% and $\beta = 1, \ldots, 5$. Given a missingness level and a dataset version, the same selected set of instances were used for the O, C, $05p$, and $10p$ datasets. Since the same instances were selected for each of the four datasets, it was possible directly to compare the imputation accuracy of each technique on the same set of instances, while the noise level of the baseline dataset changed. This fact was very important in our empirical work.

Each dataset O, C, $05p$, and $10p$ was partitioned into the instances with fully observed values for *nfaults* and the instances with missing values for *nfaults*. These datasets are denoted with a superscript of 'o' in the first case and 'm' in the second case. For example, O^o and O^m denote the subset of instances with observed and missing values respectively, from the dataset O. Similar notation is used for datasets C, $05p$ and $10p$.

Models were built using regression or REPTree, with the observations in either O^o, C^o, $05p^o$, or $10p^o$. After construction, these models were applied to the observations in either O^m, C^m, $05p^m$, or $10p^m$, imputing the missing values for *nfaults*. The imputation of missing values for the instances in O^m, C^m, $05p^m$, or $10p^m$ using mean imputation used the average value for *nfaults* in O^o, C^o, $05p^o$, or $10p^o$. The imputation of missing values in O^m, C^m, $05p^m$, or $10p^m$ using IBk was based on their nearest neighbors in O^o, C^o, $05p^o$, or $10p^o$. BMI does not require an explicit partitioning of the datasets into subsets and hence data preprocessing was not needed.

12.5.2 BMI Experimental Settings

Experiments were conducted using BMI implemented in the SAS procedure PROC MI [11, 19], which assumes that the underlying distribution of the data is multivariate normal. Hence, the population parameters for BMI are $\theta = (\mu, \Sigma)$ where μ is the mean vector and Σ is the covariance matrix. Empirical research has demonstrated that BMI under the assumption of normality will perform well even with non-normal or categorical data [10, 14]. Schafer et al. [15] discuss a study that reported excellent results using a dataset with highly non-normal variables, where imputation was performed using BMI with the assumption of normality without any variable transformations.

The parameters θ^0 must be initialized for the first run. Listwise deletion, where instances with missing data are simply removed and the initial estimates of μ^0 and Σ^0 are calculated using only the complete cases, is one possibility. Another more robust option is to use the maximum likelihood estimates calculated from the Expectation Maximization or EM algorithm [20]. Indeed, EM was the technique used to obtain start-up BMI parameter values for all our experiments.

In our experiments, BMI used a sequential chain with 15 distinct imputed datasets which were stored every time BMI was run. The sequential chain had a start-up cycle of 100 runs (called the *burn*) and a delay (or *lag*) of 100 iterations. Each of the sequential chains thus consisted of a total of 1500 iterations or cycles. Empirical evidence [10, 14] confirms that a chain of this length is sufficient to assure convergence with minimal error rates.

In our experiments, BMI was invoked 15 times for a given dataset with missing values, so for each dataset there will be 15 sequential chains. Therefore, each missing value had $15 \times 15 = 225$ imputed values generated by BMI, with the final imputed value being the average over all 225 imputations. The parameters applied for the experiments were deemed reasonable for these datasets. Additional experiments have used other parameter values which have produced similar results to those presented here.

12.5.3 Imputation Performance Metric

The average absolute errors were obtained from the combination of all five subsets of instances with missing values for each missingness level and were calculated as follows:

$$aae = \frac{1}{n} \sum_{i=1}^{n} |Y_i - \hat{Y}_i|$$

where Y_i was the original software metric value, $\hat{Y}_i$ was the imputed value rounded to the nearest integer, and n was the number of the imputations for all missing data instances. Any negative imputed value in $\hat{Y}_i$ was set to zero. The average absolute error measures how close the imputed and actual values are to one another. In general, an imputation technique that, on average, imputes the missing values closer to the actual values is considered better[2]. Note that other metrics can also be used to measure imputation performance, but we believe that *aae* is appropriate for this study.

12.6 Statistical Analysis

12.6.1 Imputation Average Absolute Errors (aae)

In this section, we examine the imputation results on the set of noisy instances in the datasets $O, 05p$, and $10p$. After using each imputation method to impute the missing values, we measured the imputed values of *nfaults* relatively to *two different values*, the noisy value denoted *nfaults^n* and the cleansed value *nfaults^c*. Recall from the description of construction of the datasets, in the case of inherently noisy instances, *nfaults^c* was determined using expert inspection and with the help of several data analysis techniques. For the simulated noise, *nfaults^c* was simply the original value in the dataset before the instance was corrupted. The imputation results as measured by the absolute error using the cleansed value *nfaults^c* were denoted 'CLEAN', while the results with the noisy value *nfaults^n* were denoted 'NOISY'. We refer to the analysis related to these two possible values for *nfaults* as test scenarios. *Note that the imputed values in both scenarios were the same, the only difference between these two scenarios was the value that the imputation was measured against when calculating the aae.* These results demonstrate the importance of understanding the underlying data quality when evaluating the effectiveness of the imputation techniques. As will be shown, measuring the imputed value against a noisy attribute value will lead to misleading conclusions.

Table 12.2 contains the average absolute errors (*aae*) of the imputations for each of the five techniques for the four different datasets. The "Tech/Data" column is the name of the imputation techniques and has nearest neighbor imputation as "IBk", regression as "Reg", BMI, mean imputation as "Mea", and REPTree as 'Rep'. This column also contains the name of the CCCS datasets, *i.e.*, C, O, $05p$, or $10p$. The rest of the columns, labeled 'CLEAN' and 'NOISY', contain the respective *aae* over all five subsets (five different random selections of missing values) of imputations for the 5%, 10%, and 30% missingness levels. Only the *aae* at 5%, 10%, and 30% missingness levels are reported in Table 12.2 due to the similarity of empirical conclusions.

---

[2] As we show in this work, the actual value (*i.e.*, the value given in the dataset) can be misleading. If an instance is noisy, then a very good imputation technique will not impute the value close to the actual (*i.e.*, noisy) value. However, if the actual value is clean, then strong imputation techniques should have the imputed value close to the actual (clean) value.

Table 12.2 Average absolute imputation errors

Tech/Data		CLEAN (5%)	NOISY (5%)	CLEAN (10%)	NOISY (10%)	CLEAN (30%)	NOISY (30%)
IBk	C	1.186	–	0.557	–	0.706	–
	O	5.900	10.800	3.714	7.714	2.513	6.179
	$05p$	3.750	6.625	3.333	6.400	2.260	8.240
	$10p$	4.412	6.294	2.609	6.217	2.810	6.899
Reg	C	0.729	–	0.529	–	0.619	–
	O	2.100	11.200	0.143	8.714	2.744	7.538
	$05p$	0.375	8.750	0.733	7.800	1.020	9.200
	$10p$	1.588	8.765	1.130	7.609	1.165	8.013
BMI	C	0.714	–	0.536	–	0.553	–
	O	1.300	11.200	0.143	8.714	2.590	8.872
	$05p$	0.375	9.000	0.933	7.733	1.020	9.080
	$10p$	1.471	8.882	1.087	7.652	1.241	7.962
Mea	C	4.100	–	2.179	–	2.722	–
	O	11.700	11.800	9.714	2.000	7.667	4.256
	$05p$	4.875	5.250	4.400	6.000	5.280	7.100
	$10p$	7.412	4.353	4.913	5.130	4.949	5.468
Rep	C	1.543	–	0.671	–	0.706	–
	O	8.000	11.100	4.714	8.429	4.000	7.564
	$05p$	3.250	7.625	2.600	7.667	2.320	8.820
	$10p$	3.471	6.765	2.565	6.261	2.975	8.354

The differences between the *aae* for the 'CLEAN' and 'NOISY' test scenarios can be clearly seen in Table 12.2. Across the different missingness levels, the *aae* generally increases (and hence the imputation accuracy decreases) when going from the 'CLEAN' to 'NOISY' test scenario. For example, with 5% missingness for dataset O, the *aae* for 'CLEAN' for BMI was 1.3 while for 'NOISY' it was 11.2. In other words, for the subset of noisy instances in dataset O, when measured against the clean value for *nfaults*, the imputation error was 1.3. However, when measured against the noisy value for *nfaults*, the imputation error for BMI was 11.2. BMI is therefore imputing missing values closer to that of the clean value ($nfaults^c$) when compared to the noisy value $nfaults^n$. Mean imputation is the only imputation technique where the *aae* decreases when measuring the imputed value against $nfaults^n$ instead of $nfaults^c$. Very often, the *aae* values for BMI are the smallest amongst the techniques for the 'CLEAN' scenario, but are some of the largest in the 'NOISY' scenario. Under the 'NOISY' scenario (*i.e.*, when the noisy value of *nfaults* is used to measure the *aae*), mean imputation often appears to be the most accurate imputation technique. This result is completely misleading, because if the *aae* is measured against the correct (*i.e.*, clean) value, mean imputation consistently exhibits the worst performance. Note that C does not contain noisy instances relative to *nfaults*, and hence the test scenario 'NOISY' has no *aae* values (denoted "–"). We include the results for C for the 'CLEAN' scenario in Table 12.2 only for reference. More detailed analyses for the C dataset are presented in Sect. 12.6.4.

12.6.2 Three-way ANOVA: Randomized Complete Block Design

Analysis of variance (ANOVA) is a well-known statistical technique which can be used to understand which experimental factors significantly impact a measurement of interest, in this case the imputation errors. Additional information on ANOVA models can be found in Berenson et al. [21]. A three-way randomized complete block design experimental ANOVA analysis was performed to compare the results of the five imputation methods with different CCCS datasets under multiple levels of missingness in *nfaults*. Experimental design models were built for this comparative study involving five imputation methods ("Method"), three CCCS datasets from each test scenario ("Dataset") (note that *C* was excluded from the ANOVA because it had no subset of noisy instances), and six levels of missingness ("Missing") in the dependent variable for each of those datasets. Two different ANOVA models were constructed, first using the *aae* calculated with the noisy values for *nfaults* (*i.e.*, the 'NOISY' test scenario), and second using the *aae* calculated with the clean values for *nfaults* (*i.e.*, the 'CLEAN' test scenario). Note that the logarithmic transformation of the *aae* was done to better approximate normality in the distribution of these values because ANOVA models assume the data is normal. There were no other significant violations of ANOVA assumptions.

The statistical results of the three-way ANOVA model for the 'CLEAN' and 'NOISY' scenarios are presented in Table 12.3. The factors that were significant at a 5% significance level for each of the two ANOVA models are highlighted in bold. The results for the 'CLEAN' scenario from Table 12.3 indicate that the Method variable is the most significant experimental factor, with a *P*-value much less than 1%. The Dataset variable obtained *P*-values of 2.44% and 15.43% for the 'CLEAN' and 'NOISY' test scenarios respectively, meaning that it was a significant factor for 'CLEAN' but not for the 'NOISY' test scenario. The Missing attribute had *P*-values of 5.5% and 1.64% for the 'CLEAN' and 'NOISY' test scenarios, respectively. For the 'NOISY' test scenario, the missingness level is significant at $\alpha = 5\%$ and is further analyzed in the next section. We omit a detailed analysis of the Dataset factor, which is significant in the 'CLEAN' scenario, for space considerations.

Table 12.3 ANOVA analysis for *aae*

	Blocks	DF	SumSq	MeanSq	F	*P*-value
CLEAN	**Method**	4	37.5976	9.3993	42.34	0.0000
	Dataset	2	1.7295	0.8647	3.90	0.0244
	Missing	5	2.5262	0.5052	2.28	0.0550
NOISY	**Method**	4	1.8366	0.4592	12.97	0.0000
	Dataset	2	0.1356	0.0678	1.91	0.1543
	Missing	5	0.5271	0.1054	2.98	0.0164

12.6.3 Multiple Pairwise Comparisons

Once a factor in the ANOVA analysis has been determined to be significant, the next step is to understand which levels of these factors are significantly different from one another. Two very common techniques, Fisher's least-significant-difference (LSD) test, which controls the comparison-wise error rate, and Tukey's studentized range test (HSD), which controls the type I experiment-wise error rate [22, 23], were selected for this experimental analysis [21, 24, 25].

The test results for the factors 'Method' and 'Missing' are presented in Tables 12.4 and 12.5. Those levels that were *not* different at the $\alpha = 5\%$ significance level were placed within a common group, indicated by an identical block letter assigned to each such variable. Different block letters indicate different groups of variables. It is important to recognize that a variable can belong to more than one group. Tests were also conducted at 1% and 10% significance levels with similar or identical results to the 5% level.

12.6.3.1 Analysis of Method Factor for *aae*

Table 12.4 contains the LSD and HSD groupings with respect to the imputation techniques. In the 'CLEAN' scenario, the LSD test had mean imputation in group A with the highest average *aae* value (*i.e.*, the worst imputation performance). REP-Tree and IBk were in group B and regression and BMI were grouped in C. BMI had

Table 12.4 Noise effect on imputation techniques ($\alpha = 5\%$, $N = 18$)

	LSD	HSD	Mean	Methods
CLEAN	A	A	1.8256	Mea
	B	B	1.2165	Rep
	B	B	1.1647	IBk
	C	C	0.1856	Reg
	C	C	0.1571	BMI
NOISY	A	A	2.1673	BMI
	A	A	2.1523	Reg
	B A	A	2.1032	Rep
	B	A	2.0150	IBk
	C	B	1.7780	Mea

Table 12.5 Noise effect on *nfaults* missingness ($\alpha = 5\%$, $N = 15$)

	LSD	HSD	Mean	Missing
NOISY	A	A	2.1074	5%
	A	A	2.1014	15%
	A	B A	2.0876	20%
	A	B A	2.0679	40%
	B A	B A	2.0051	30%
	B	B	1.8896	10%

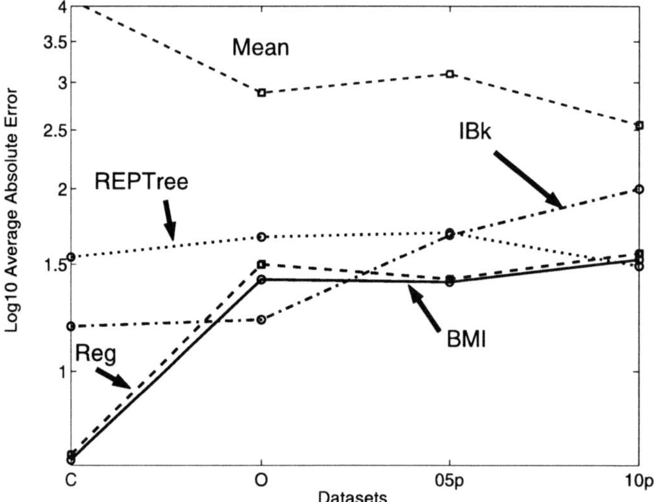

Fig. 12.1 Impact of noise on relatively clean observations at 5% missingness

the lowest average *aae*, while mean imputation had the highest average *aae*. The results for the HSD test in the 'CLEAN' scenario were identical to those of the LSD test.

The 'NOISY' test scenario in Table 12.4 has the HSD test with BMI, regression, REPTree, and IBk in a single group A and mean imputation in its own B group. Based on the LSD grouping, BMI, regression and REPTree are in group A. REPTree and IBk are in group B, while mean imputation is alone in group C. Notice that the best technique in the 'CLEAN' scenario, BMI, is the worst in the 'NOISY' scenario. Likewise, mean imputation is the best technique in the 'NOISY' scenario, while it was the worst in the 'CLEAN' scenario. Clearly, measuring the imputation error using a noisy value will result in a dramatically different assessment of the relative performance of the imputation techniques.

12.6.3.2 Analysis of 'Missing' Factor for *aae*

The ANOVA analysis (Sect. 12.6.2) demonstrated that the missingness level was a significant factor with a *P*-value less than 5% in the 'NOISY' test scenario. Table 12.5 presents a detailed analysis of the Missing factor. The LSD test placed the 5%, 15%, 20%, 30%, and 40% missingness levels in group A, while the 10% and 30% levels were grouped in B. This test shows a lack of significant differences amongst missingness levels of 5%, 15%, 20%, 30%, and 40%. The HSD test in Table 12.5 grouped 5 missingness levels 5%, 15%, 20%, 30%, and 40% in group A, and 10%, 20%, 30%, and 40% in group B.

12.6.4 *Noise Impact on Remaining (Non-noisy) Instances*

The previous sections focused on the subset of instances with noise in the CCCS dataset, as well as the injected noise in datasets $05p$ and $10p$. In this section, we concentrate on the other subset of instances that are non-noisy. In particular, once the subset of noisy instances (relative to the dependent variable) was identified, we measured the *aae* for those instances with missing data that were deemed to be relatively clean. We refer to this set of instances as the *remainder* or *non-noisy* portion of the dataset. In this "controlled" experiment, the objective was to understand the impact of either inherent noise and/or simulated noise, depending on the dataset, on imputing missing values in the remaining observations excluding those identified noisy instances (*i.e.*, the relatively clean subset of the dataset). Recall that C, O, $05p$, and $10p$ have noise distributions in *nfaults* as follows:

- C has no noisy instances relative to *nfaults*, and hence we examine the imputation on any of the 282 instances with missing data.
- O has 20 known instances with *inherent* noise, and thus missing values in 262 instances from the 282 are examined.
- $05p$ has 20 known instances with *inherent* noise and 14 additional ones with simulated noise. Thus, missing values in 248 of the 282 instances are examined.
- $10p$ has 20 known instances with *inherent* noise and 28 (14 added to the 14 from $05p$) with simulated noise. Thus, missing values in 234 instances are analyzed.

The logarithmic scale of the *aae* from the imputation of the instances with missing values in the remaining portion of the dataset for four of the six missingness levels (5%, 10%, 30%, and 40%) in *nfaults* were plotted in Figs. 12.1–12.4. The logarithmic transformation was applied to clarify the figures.

The x-axis identifies each dataset in the following order from left to right: C, O, $05p$, and $10p$. The y-axis contains the logarithm base 10 value of the *aae* from each imputation method. Each of the five imputation methods is plotted as a separate line. It is clear that the upward slopes present in most of the plotted lines indicates a trend by the majority of the imputation methods to have an increasing *aae* as the noise level increases. Only mean imputation exhibited different characteristics, sometimes quite the opposite to the general trend observed in regression, REPTree, IBk, and BMI.

In Fig. 12.1, the *aae* of mean imputation decreases as the noise level increases, indicating an improvement in the results. Mean imputation also has the highest *aae* values amongst all the techniques. The *aae* for REPTree imputation levels off between the O and $05p$ and actually decreases slightly for the noisiest dataset $10p$. BMI, IBk, and regression imputation performances generally worsened as the noise level increased. For C, BMI and regression imputations both performed well with very low *aae* values.

Figure 12.2 shows that with the exception of REPTree, the performance of all techniques deteriorates as the noise level increases. The *aae* for REPTree increased up to dataset $05p$ and then leveled off for dataset $10p$. Mean imputation had the highest *aae* while BMI and regression imputation had the lowest. For C and O, BMI,

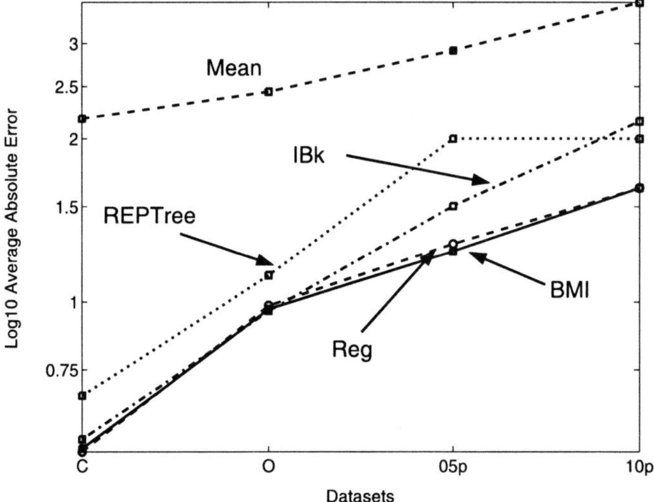

Fig. 12.2 Impact of noise on relatively clean observations at 10% missingness

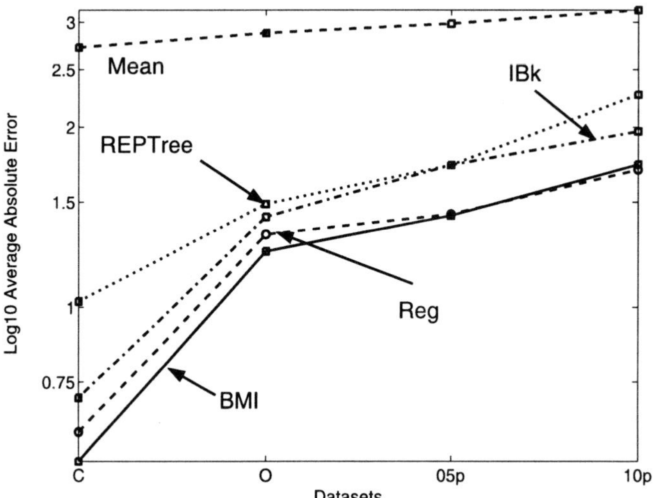

Fig. 12.3 Impact of noise on relatively clean observations at 30% missingness

regression, and IBk imputation obtain nearly identical performance, while BMI and regression are similar for $05p$ and $10p$.

The imputation results for 30% missingness are presented in Fig. 12.3. The performance trend of BMI and regression imputation are very similar to one another. All five imputation techniques demonstrate deteriorating performance with more noise in the dataset, although the decline for mean imputation is minimal.

Figure 12.4 shows the performance of mean imputation slightly improving (*i.e.*, the *aae* decreases) as the noise level increases. Overall, mean imputation obtained

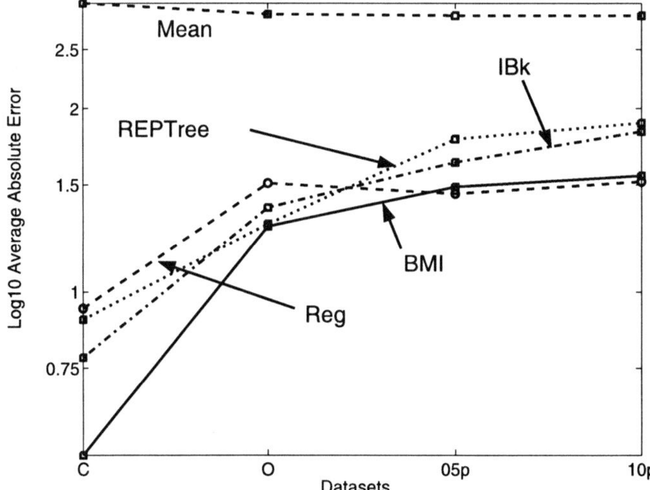

Fig. 12.4 Impact of noise on relatively clean observations at 40% missingness

the highest *aae* values by a wide margin. On the other hand, the rest of the techniques
have their performance deteriorate as the noise level increases. At 40% missingness,
the results from BMI and regression leveled off between $05p$ and $10p$. BMI exhibited lower imputation errors than the other techniques, but had the largest difference
in *aae* amongst all the techniques between the cleansed (C) and original (O) datasets.
Regression performed slightly better than BMI for the $05p$ and $10p$ datasets.

In summary, the objective of this analysis was to demonstrate the impact of noise
on imputation by examining the imputation accuracy of the five techniques on a set
of relatively clean instances. Further, the level of noise in the overall dataset increases when moving from left to right in each figure, due to the construction of
the four datasets C, O, $05p$, and $10p$. It is anticipated that as the level of noise in
the dataset increases, the imputation accuracy will decrease. This is not the case,
however, with mean imputation, which often performs equally poorly regardless of
the noise level in the dataset (see Fig. 12.2).

12.7 Conclusion

The objective of this study was to demonstrate the impact of noise on the accuracy of
software quality imputation techniques. More specifically, we evaluated the effect
of measuring the accuracy of imputation techniques using noisy data. We believe
that ignoring the underlying quality of data when evaluating imputation procedures
can result in misleading conclusions, such as concluding that mean imputation is an
effective technique for handling missing values.

Our case study used a real-world software measurement dataset called CCCS and three derived datasets, two of which contained injected noise. Missing values were injected into each dataset in the dependent variable at six levels. Five imputation techniques were used to impute the missing values. The effectiveness of the techniques was measured using the average absolute (imputation) error *aae*, which is calculated by subtracting the imputed value from the actual value.

First, we specifically examined missing data in the set of noisy instances in each dataset. When measuring the *aae* for these instances, the important question is: *What is the actual value?* Since the instances are noisy, the value for *nfaults* in the dataset is incorrect. Measuring the imputation against this value is clearly undesirable. Instead, the imputation should be measured against the *clean* value for *nfaults*. This leads to the two test scenario's discussed in this work, 'CLEAN' and 'NOISY'.

A three-way randomized complete block ANOVA model constructed for the 'CLEAN' scenario showed that there were significant differences amongst the imputation techniques. Likewise, the results for the 'NOISY' scenario demonstrated significant differences amongst the imputation techniques and on the level of missingness in the dependent variable *nfaults*. It is evident that these results, together with the differences in the *aae* performance metric between the two test scenarios, can be attributed to the presence of noise in the 'NOISY' scenario, since all other factors were the same.

We conducted additional statistical tests to further examine the effects of noise on the accuracy of the imputation techniques and on the missingness level blocking variable. In the 'CLEAN' scenario, these tests revealed that mean imputation had the highest *aae* values whereas BMI had the lowest. The results of the 'NOISY' scenario showed BMI with the highest *aae* values while mean imputation had the lowest. Clearly, the results of these two test scenarios are quite different. The differences in the results of the test scenarios, all else being equal, can be attributed to the presence of noise in the measurement of the *aae*. If data quality was not considered, then the conclusion might be reached that mean imputation is the best imputation technique, when clearly it is the worst.

Figures 12.1–12.4 graphically demonstrated the impact of noise on the imputation of non-noisy or remaining observations. Even though only four of six missingness levels (5%, 10%, 30%, 40%) are presented due to space constraints, 15% and 20% exhibited similar characteristics. These figures demonstrate conclusively the negative impact of noise on the imputation methods (except for mean imputation, which we believe should never be used).

The results presented in this work demonstrate the significant impact of noise on the accuracy of software quality imputation techniques. Accurate imputation techniques (*e.g.* BMI) impute approximately correct values, which are relatively far from the incorrect values in a noisy instance. Future work will examine more expert-supervised and carefully controlled experiments with univariate missingness. Future work can also consider MAR and NI missingness mechanisms (we only consider MCAR in this work). Another important area of future research is multivariate missingness, and detailed experimentation on handling missing values of this type should be performed. Finally, the impact of missing values on the construction of a classifier (for binary target prediction) or an estimation model (such as linear regression) should be explored.

References

1. T. Khoshgoftaar, E. Allen, W. Jones, and J. Hudepohl. Accuracy of software quality models over multiple releases. *Annals of Software Engineering*, 9(1-4):103–116, 2000.
2. T. Khoshgoftaar and N. Seliya. Fault Prediction Modeling for Software Quality Estimation: Comparing Commonly Used Techniques. *Empirical Software Engineering Journal*, 8:255–283, September 2003.
3. N. Schneidewind. Software Metrics Validation: Space Shuttle Flight Software Example. *Annals of Software Engineering*, 1:287–309, 1995.
4. M. Ohlsson and P. Runeson. Experience from Replicating Empirical Studies on Prediction Models. In *Proceedings of the 8th International Symposium on Software Metrics*, pages 217–226, 2002.
5. S. Gokhale and M. Lyu. Regression Tree Modeling for the Prediction of Software Quality. In H. Pham, editor, *Proceedings: 3rd International Conference on Reliability and Quality in Design*, pages 31–36, 1997.
6. T. Khoshgoftaar, A. Folleco, J. Van Hulse, and L. Bullard. Multiple Imputation of Missing Values in Software Measurement Data. Technical report, Florida Atlantic University, February 2006.
7. Y. Haitovsky. Missing data in regression analysis. *Journal Royal Statistical Society*, 30:67–81, 1968.
8. I. Witten and E. Frank. *Data Mining: Practical machine learning tools and techniques*. Morgan Kaufmann, San Francisco, CA, 2nd edition, 2005.
9. P. Jonsson and C. Wohlin. An evaluation of k-nearest neighbour imputation using likert data. *10th IEEE Intl. Symposium on Software Metrics (METRICS'04)*, pages 108–118, 2004.
10. J. Schafer. *Analysis of Incomplete Multivariate Data*. Chapman and Hall/CRC, Boca Raton, FL, 2000.
11. SAS Institute. SAS/STAT User's Guide. 2004.
12. D. Rubin. *Multiple Imputation*. John Wiley and Sons, New York, NY, 1987.
13. J. Schafer and M. Olsen. Multiple imputation for multivariate missing data problems: A data analyst's perspective. *Multivariate Behavioral Research*, 33(4):545–571, 1998.
14. P. Allison. *Missing Data*. Sage University Press, Thousand Oaks, CA, 2002.
15. J. Schafer and J. W. Graham. Missing data: Our view of the state of the art. *Psychological Methods*, 7(2):147–177, 2002.
16. P. Bremaud. *Markov Chains: Gibbs Fields, Monte Carlo Simulation, and Queues*. Springer, 1999.
17. M. A. Tanner and W. H. Wong. The calculation of posterior distributions by data augmentation. *Journal of the American Statistical Society*, 82:528–550, 1987.
18. R. Little and D. Rubin. *Statistical Analysis with Missing Data*. John Wiley and Sons, New York, NY, 2nd edition, 2002.
19. Y. C. Yuan. Multiple imputation for missing data: Concepts and new development. In *Proceedings of the 25th Annual SAS Users Group International Conference*, 2000. Paper No 267.
20. A. Dempster, N. Laird, and D. Rubin. Maximum likelihood estimates from incomplete data via the EM algorithm. *Journal of the Royal Statistical Society*, 39:1–38, 1977. Series B.
21. M. Berenson, D. Levine, and M. Goldstein. *Intermediate Statistical Methods and Applications: A Computer Package Approach*. Prentice Hall, Englewood Cliffs, NJ, USA, 1983.
22. A. Hayter. A Proof of the Conjecture that Tukey-Kramer Methods is Conservative. *The Annals of Statistics*, 12:61–75, 1984.
23. C. Kramer. Extension of Multiple Range Tests to Group Means with Unequal Number of Replications. *Biometrics*, 29(1):4–11, 1956.

24. H. Scheffe. *The Analysis of Variance*. John Wiley and Sons, New York, NY, 1959.
25. R. Waller and D. Duncan. A Bayes Rule for the Symmetric Multiple Comparison Problem. *Journal of the American Statistical Association*, 64:1484–1499, 1969.

Chapter 13
A Linearized Growth Curve Model
for Software Reliability Data Analysis

Mitsuhiro Kimura

Department of Industrial and Systems Engineering,
Hosei University, Japan

13.1 Introduction

There have been many assessment models have been proposed in the literature of
software reliability modeling. They are the achievement of studies performed by
a lot of researchers. Various approaches have been tried to describe software reliability quantitatively through the observation of software development processes
(for example, [1–7]). Namely, there are the models which use stochastic processes,
non-parametric models [8], neural networks [9, 10], and so on. This fact shows that
each model has some advantages for several data sets that are analyzed in the paper
itself; however, the model is not always applicable to all kinds of data. Therefore,
we have a number of software reliability models. In order to overcome this com-
plication of model selection, we discuss a method of generalizing several proposed
models in this study. In particular, we deal with some growth curve models for
software reliability data analysis. These models describe the time behavior of the
cumulative number of detected software faults. We show that an exponential, delayed S-shaped [11], Gompertz, logistic curve models and the logarithmic Poisson
execution time model (*e.g.*, [2, 3]) can be at least included in a linearized growth
curve model, which is proposed in this chapter.

After discussing the modeling, we also propose a method of parameter estimation
by using a two-parameter numerical differentiation method. This method allows us
to use linear regression analysis for the data set.

The following section expresses how we generalize the traditional growth curve
models, and proposes a linearized growth curve model. We show the method of parameter estimation for the proposed model and its small sample code in Mathematica language [12] in Sect. 13.3. Section 13.4 illustrates several numerical examples
of the data analysis by using the proposed model, and we discuss its applicability
and limitations.

13.2 Generalization of Growth Curve Models

In this study, we first assume that a data set forms (t_i, y_i) $(i = 1, 2, \ldots, n)$, where t_i means the i-th testing time recorded and y_i the cumulative number of detected (or removed) software faults up to t_i. We also assume that $0 < y_1 < y_2 < \ldots < y_n$ is satisfied. One of our main concern is to predict the future behavior of (t_j, y_j) (t_j is given, and $j > n$). Figure 13.1 shows the behavior of a sample data set [7].

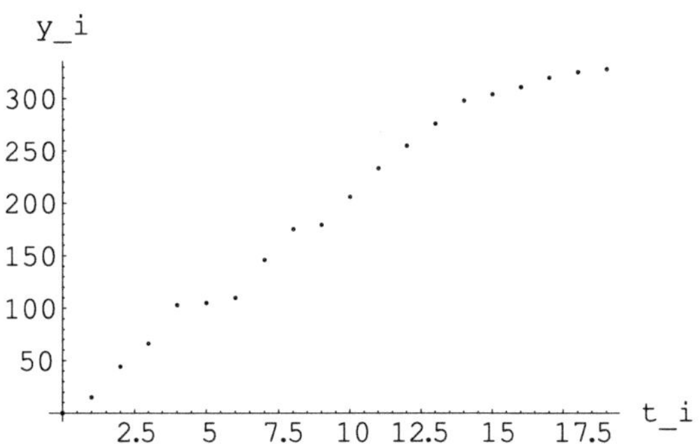

Fig. 13.1 Sample data set of (t_i, y_i) $(i = 1, 2, \ldots, 19)$

In order to investigate an appropriate function to describe the behavior of such data sets, we focus on the following five growth curves.

$$M(t) = m_1(1 - e^{-m_2 t}) \quad (m_1 > 0, \ m_2 > 0), \tag{13.1}$$

$$D(t) = d_1 \left(1 - (1 + d_2 t)e^{-d_2 t}\right) \quad (d_1 > 0, \ d_2 > 0), \tag{13.2}$$

$$G(t) = g_1 g_2^{\exp[-g_3 t]} \quad (g_1 > 0, \ 0 < g_2 < 1, \ g_3 > 0). \tag{13.3}$$

$$L(t) = \frac{l_1}{(1 + l_2 e^{-l_3 t})} \quad (l_1 > 0, \ l_2 > 0, \ l_3 > 0), \tag{13.4}$$

$$P(t) = \frac{1}{p_1} \log[1 + p_1 p_2 t] \quad (p_1 > 0, p_2 > 0). \tag{13.5}$$

Equations (13.1), (13.2) and (13.5) are often used as mean value functions of non-homogeneous Poisson process (NHPP, for short) models, which are widely known as software reliability assessment models [1, 2, 4–6]. These three formulae are referred to as exponential SRGM (software reliability growth model), delayed S-shaped SRGM, and logarithmic Poisson execution time model, respectively. In addition, Equations (13.3) and (13.4) are the so-called Gompertz and logistic curves, which have been applied to describe various growth phenomena in many research areas, not only used for software reliability growth analyses.

Among them, the logarithmic Poisson execution time model, $P(t)$, shows a different aspect in terms of asymptotic behavior with respect to time t. That is, letting $t \to \infty$ for all these models yields

$$\lim_{t \to \infty} M(t) = a_1, \tag{13.6}$$

$$\lim_{t \to \infty} D(t) = d_1, \tag{13.7}$$

$$\lim_{t \to \infty} G(t) = g_1, \tag{13.8}$$

$$\lim_{t \to \infty} L(t) = l_1, \tag{13.9}$$

$$\lim_{t \to \infty} P(t) = \infty . \tag{13.10}$$

Thus we call the first four models *convergence type*, and the last one *divergence type* models.

Based on these functions, we can derive the following relations:

$$\log \left\{ \frac{dM(t)}{dt} \right\} = \log \{m_1 m_2\} - m_2 t, \tag{13.11}$$

$$\log \left\{ \frac{dD(t)}{dt} \Big/ t \right\} = \log \{d_1 d_2^2\} - d_2 t, \tag{13.12}$$

$$\log \left\{ \frac{dG(t)}{dt} \Big/ G(t) \right\} = \log \left\{ g_3 \log \frac{1}{g_2} \right\} - g_3 t, \tag{13.13}$$

$$\log \left\{ \frac{dL(t)}{dt} \Big/ L(t)^2 \right\} = \log \left\{ \frac{l_2 l_3}{l_1} \right\} - l_3 t, \tag{13.14}$$

$$\log \left\{ \frac{dP(t)}{dt} \right\} = \log p_2 - p_1 P(t). \tag{13.15}$$

In these equations, we can see that the left-hand side of (13.11), (13.12), (13.13), and (13.14) is a function of t, and that of (13.15) is a function of $P(t)$. Therefore if we directly obtain the left-hand side values of each model when $t = t_i$ $(i = 1, 2, \ldots, n)$ from the data set to be analyzed, we can use a linear regression scheme to estimate the unknown parameters appeared in the right-hand side of the above equations. We describe it in the following section.

13.2.1 Two-parameter Numerical Differentiation Method

In general, the method of numerical differentiation by using n data pairs of (t_i, y_i), is often applied in such a situation. For instance, the i-th value of $\frac{dM(t)}{dt}$ in (13.11) can be approximately given by

$$\frac{dM(t)}{dt}\Big|_{t=t_i} = \begin{cases} \dfrac{1}{2} \left\{ \dfrac{y_{i+1} - y_i}{t_{i+1} - t_i} + \dfrac{y_i - y_{i-1}}{t_i - t_{i-1}} \right\} & (1 \le i \le n-1) \\[4mm] \dfrac{y_n - y_{n-1}}{t_n - t_{n-1}} & (i = n) \end{cases}, \tag{13.16}$$

where $t_0 \equiv 0$ and $y_0 \equiv 0$. That is, we extract a central difference from the data set. The original idea of using the central difference for software reliability data analysis can be seen in [14].

Based on this method, in order to deal with (13.11), (13.12), (13.13), (13.14), and (13.15) by unit formula, we introduce here a non-negative, increasing, and differentiable function $H(t)$ and $z(\alpha, \beta, t_i)$. By using them, we can describe the following formula.

$$\log\left\{ \frac{dH(t)}{dt} \middle/ t^{\alpha} \middle/ H(t)^{\beta} \right\}\Bigg|_{t=t_i} \equiv z(\alpha, \beta, t_i)$$

$$= \begin{cases} \log\left[\frac{1}{2}\left\{ \frac{y_{i+1} - y_i}{t_{i+1} - t_i} + \frac{y_i - y_{i-1}}{t_i - t_{i-1}} \right\} \middle/ t_i^{\alpha} / y_i^{\beta} \right] & (1 \le i \le n-1) \\ \log\left[\frac{y_n - y_{n-1}}{t_n - t_{n-1}} \middle/ t_i^{\alpha} / y_i^{\beta} \right] & (i = n) \end{cases} \tag{13.17}$$

In the above equations, we introduced two parameters α and β to obtain more applicability of the model. We call this transform the two-parameter numerical differentiation.

Consequently, we model a linear relation among t_i, y_i, and $z(\alpha, \beta, t_i)$ as

Convergence type model:

$$z(\alpha, \beta, t_i) = A - Bt_i + \varepsilon_i \quad (i = 1, 2, \ldots, n; B > 0), \tag{13.18}$$

Divergence type model:

$$z(\alpha, \beta, t_i) = A - Cy_i + \varepsilon_i \quad (i = 1, 2, \ldots, n; C > 0). \tag{13.19}$$

In these equations (13.18) and (13.19), A, B, and C are constant parameters, which can be estimated under the least squares rule, and we assume that ε_i is a standard normal error term with homoscedasticity (equality of variance).

13.2.2 Linearized Growth Curve Model

From (13.17) and (13.18), we have a differential equation as follows.

$$\log\left\{ \frac{dH(t)}{dt} \middle/ t^{\alpha} / H(t)^{\beta} \right\} = A - Bt \quad (B > 0). \tag{13.20}$$

This equation describes the mean behavior of $z(\alpha, \beta, t)$ when t is given. Equation (13.20) can be solved with an initial condition $H(0) = h_0 > 0$ as follows.

$$H(t) = \left[h_0^{1-\beta} + \frac{(1-\beta)e^A}{B^{\alpha+1}} \left\{ \Gamma[\alpha+1, 0] - \Gamma[\alpha+1, Bt] \right\} \right]^{\frac{1}{1-\beta}} \quad (\beta > 1) \tag{13.21}$$

where $\Gamma[m, x]$ is the incomplete gamma function defined as

$$\Gamma[m, x] = \int_x^\infty s^{m-1} e^{-s} ds \,. \tag{13.22}$$

In (13.21), when $\beta \to 1$, we have

$$H(t) = h_0 \exp\left[\frac{e^A}{B^{\alpha+1}}\{\Gamma[\alpha+1, 0] - \Gamma[\alpha+1, Bt]\}\right]. \tag{13.23}$$

If $\beta < 1$, we can choose the initial condition $H(0) = h_0 = 0$, then $H(t)$ can be rewritten as

$$H(t) = \left[\frac{(1-\beta)e^A}{B^{\alpha+1}}\{\Gamma[\alpha+1, 0] - \Gamma[\alpha+1, Bt]\}\right]^{\frac{1}{1-\beta}} \quad (\beta < 1)\,. \tag{13.24}$$

As a special case, setting $(\alpha = 0, \ \beta = 0)$ and $(\alpha = 1, \ \beta = 0)$ respectively yields

$$H(t) = \frac{e^A}{B}(1 - e^{-Bt})\,, \tag{13.25}$$

and

$$H(t) = \frac{e^A}{B^2}(1 - (1 + Bt)e^{-Bt})\,. \tag{13.26}$$

These equations correspond to $M(t)$ and $D(t)$ in (13.1) and (13.2) respectively.

Similarly, we obtain the Gompertz and logistic functions if we set $(\alpha = 0, \beta = 1)$ and $(\alpha = 0, \beta = 2)$ in (13.23) as

$$H(t) = h_0 \exp\left[\frac{e^A}{B}\right] \times \left(\exp\left[-\frac{e^A}{B}\right]\right)^{e^{-Bt}}, \tag{13.27}$$

and

$$H(t) = \left(\frac{h_0}{1 - \frac{e^A}{B}h_0}\right) \Big/ \left(1 + \frac{h_0}{1 - \frac{e^A}{B}h_0} \times \frac{e^A}{B}e^{-Bt}\right), \tag{13.28}$$

respectively.

On the other hand, for the divergence type model denoted by (13.19), we have the following differential equation from (13.17) and (13.19), when the parameter $\beta = 0$.

$$\log\left\{\frac{\mathrm{d}H(t)}{\mathrm{d}t}\Big/t^\alpha\right\} = A - CH(t) \ \ (C > 0). \tag{13.29}$$

In this case, we can obtain the solution $H(t)$ with the initial condition $H(0) = 0$ as

$$H(t) = \frac{1}{C}\log\left[1 + \frac{Ce^A}{\alpha+1}t^{\alpha+1}\right]. \tag{13.30}$$

It is confirmed that $H(t)$ coincides with $P(t)$ in (13.5) if $\alpha = 0$. Thus we call $H(t)$ derived by (13.21) and (13.30) the linearized growth curve model (LGC model, for short), and propose it for the software reliability data analysis.

13.3 Parameter Estimation

This section discusses a method of parameter estimation for the proposed models. Since the structure of the convergence and divergence type models are the same from the viewpoint of the parameter estimation, we only show the method for the former type model as follows.

We here recall Equation (13.18) as

$$z(\alpha, \beta, t_i) = A - Bt_i + \varepsilon_i \quad (i = 1, 2, \ldots, n). \tag{13.31}$$

By using the least squares estimation, the parameters A and B can be analytically obtained by

$$\hat{A} = \frac{1}{n} \sum_{i=1}^{n} z(\alpha, \beta, t_i) + \hat{B} \times \frac{1}{n} \sum_{i=1}^{n} t_i, \tag{13.32}$$

$$\hat{B} = \frac{\sum_{i=1}^{n} (t_i - \frac{1}{n} \sum_{i=1}^{n} t_i) z(\alpha, \beta, t_i)}{\sum_{i=1}^{n} (t_i - \frac{1}{n} \sum_{i=1}^{n} t_i)^2}. \tag{13.33}$$

Note that these $\hat{A}$ and $\hat{B}$ are the functions of the unknown parameters α and β. Therefore, the sum of squared errors, $S(\alpha, \beta)$, is derived by

$$S(\alpha, \beta) = \sum_{i=1}^{n} \varepsilon_i^2$$

$$= \sum_{i=1}^{n} \left[z(\alpha, \beta, t_i) - \bar{z}(\alpha, \beta) + (t_i - \bar{t}) \frac{\sum_{i=1}^{n} (t_i - \bar{t}) z(\alpha, \beta, t_i)}{\sum_{i=1}^{n} (t_i - \bar{t})^2} \right]^2, \tag{13.34}$$

where

$$\bar{z}(\alpha, \beta) = \frac{1}{n} \sum_{i=1}^{n} z(\alpha, \beta, t_i),$$

$$\bar{t} = \frac{1}{n} \sum_{i=1}^{n} t_i. \tag{13.35}$$

Minimizing $S(\alpha, \beta)$ with respect to α and β, we obtain $\hat{\alpha}$ and $\hat{\beta}$ numerically. $\hat{A}$ and $\hat{B}$ can be also estimated via (13.32) and (13.33) with $\hat{\alpha}$ and $\hat{\beta}$.

Mathematica Code

The above estimation scheme can be called an adaptive regression analysis, since the value of the objective (dependent) variable $z[\alpha, \beta, t_i]$ changes adaptively with the perturbation of the parameters α and β. Therefore, one needs to use a kind of mathematical tool providing formula manipulation system in order to find the estimates minimizing in (13.34). Therefore, we present a sample code in Mathematica language [12] as follows.

```
(* data set (fictive), its form is (t[i],y[i]) (i=1,2,...,n). *)
(* n denotes the size of the data set *)

t[0]=0;
t[1]=10;
... (omitted)
t[5]=52;
y[0]=0;
y[1]=23;
... (omitted)
y[5]=32;
n=5;

z[alpha_, beta_, k_] =
    If[k < n,
       Log[((y[k + 1] - y[k])/(t[k + 1] - t[k]) +
        (y[k] - y[k - 1])/(t[k] - t[k - 1]))/2/t[k]^alpha/y[k]^beta],
       Log[(y[n] - y[n - 1])/(t[n] - t[n - 1])/t[n]^alpha/y[n]^beta]];

fittedFunction[x_, alpha_, beta_] =
    Fit[Table[{t[i], z[alpha, beta, i]}, {i, 1, n}], {1, x}, x];
solution =
  FindMinimum[
    Sum[(z[alpha, beta, i] - fittedFunction[t[i], alpha, beta])^2, {i, 1, n}],
      {alpha, 0.5}, {beta, 2}]
(* 0.5 and 2 are the initial values appropriately given for FindMinimum[] *)

regline = fittedFunction[-x, solution[[2, 1, 2]], solution[[2, 2, 2]]]

Print["A=", regline[[1]]];
Print["B=", regline[[2, 1]]];
Print["alpha=", solution[[2, 1, 2]]];
Print["beta=", solution[[2, 2, 2]]];
```

13.4 Examples of Data Analysis and Discussion

13.4.1 Regression Analysis

In this section, we analyze a data set cited from [13] in order to show several results by the regression analysis with the proposed model. We call this data set DS-1 and it forms (t_i, y_i, w_i) $(i = 1, 2, \ldots, 19)$, where t_i is the i-th testing time recorded (measured in days), y_i is the cumulative number of software faults detected by t_i, and w_i is the cumulative number of test cases processed up to t_i. Table 13.1 shows the original data set extracted from [13]. Note that 518^* in the table was treated as 517.9 when we perform the data analysis in order to satisfy the condition $y_i < y_{i+1}$ (for all i). Figures 13.2 and 13.3 are the behavior of (t_i, y_i) and (t_i, w_i) respectively.

Since our model has general versatility, we can analyze both behaviors of (t_i, y_i) and (t_i, w_i) by our model.

Table 13.1 Data set DS-1

i	1	2	3	4	5	6	7	8	9	10
t_i	7	14	21	28	35	42	49	56	63	69
y_i	6	31	62	64	121	138	152	202	283	318
w_i	205	403	951	970	1684	2347	2737	2869	2950	3026

i	11	12	13	14	15	16	17	18	19
t_i	76	84	91	98	105	112	119	126	133
y_i	359	417	443	460	484	497	511	518*	518
w_i	3151	3335	3487	3927	4629	5077	5349	5464	5521

13.4.1.1 Single Regression Analysis

Let us first apply the regression model $z(\alpha, \beta, t_i) = A - Bt_i + \varepsilon_i$ for (t_i, y_i). Table 13.2 summarizes the estimation results of the unknown parameters and ANOVA (analysis of variance) table. Figure 13.4 shows the regression line with the 95% confidence intervals of $z[\hat{\alpha}, \hat{\beta}, t_i]$. In the convergence type model, when the value of

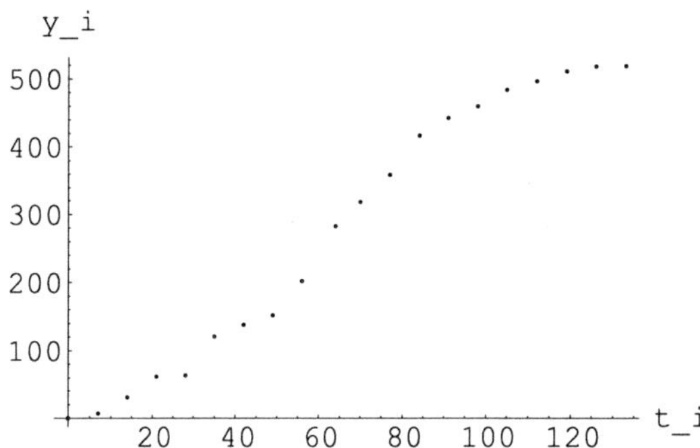

Fig. 13.2 Behavior of (t_i, y_i) of DS-1 $(i = 1, 2, \ldots, 19)$

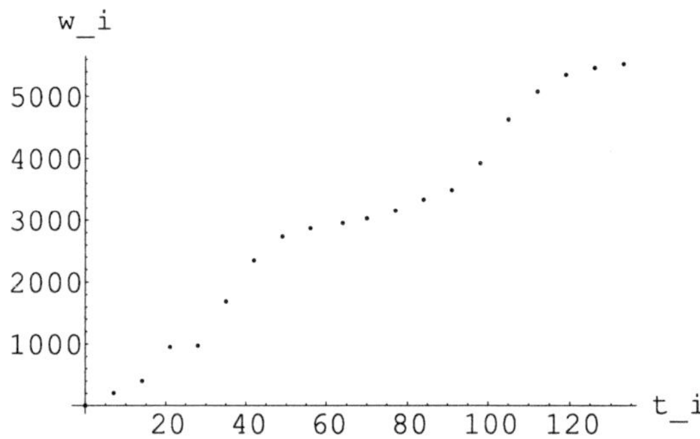

Fig. 13.3 Behavior of (t_i, w_i) of DS-1 $(i = 1, 2, \ldots, 19)$

$z[\hat{\alpha}, \hat{\beta}, t_i]$ becomes smaller, it indicates $\hat{H}(t)$ is near its convergence value. That is, it represents that the software fault detection process is retarded. It can be considered that almost all inherent software faults have been detected for the prepared test cases, if the size of software is sufficiently large and the test cases are well designed. Therefore, we can use this regression analysis and the estimated values of $z[\hat{\alpha}, \hat{\beta}, t_i]$ $(i = 1, 2, \ldots, n)$ as assessment measures for software testing progress monitoring.

The proposed model has many unknown parameters. In particular, the parameter β often makes the estimation results unstable. Hence we now analyze DS-1 again by using the model $z[\alpha, 0, t_i] = A - Bt_i + \varepsilon_i$, i.e., $\beta \to 0$. The results are shown in Table 13.3.

Although this model does not include the factor of Gompertz and logistic curve models because $\beta = 0$, as a result, this model fits better than the former one for this data set. It is therefore shown that the all independent parameters included in the model do not always contribute to gain the goodness-of-fit of the model.

Table 13.2 Estimation results and ANOVA table (I). Model: $z[\alpha, \beta, t_i] = A - Bt_i + \varepsilon_i$ for (t_i, y_i)

$\hat{\alpha}$	$\hat{\beta}$	$\hat{A}$	$\hat{B}$
-0.771217	1.48252	1.06927	0.0373791

	DF	SSE	V	F_0	R^{*2}
Model	1	39.0054	39.0054	105.416**	0.85296
Error	17	6.29025	0.370015		
Total	18	45.2956			

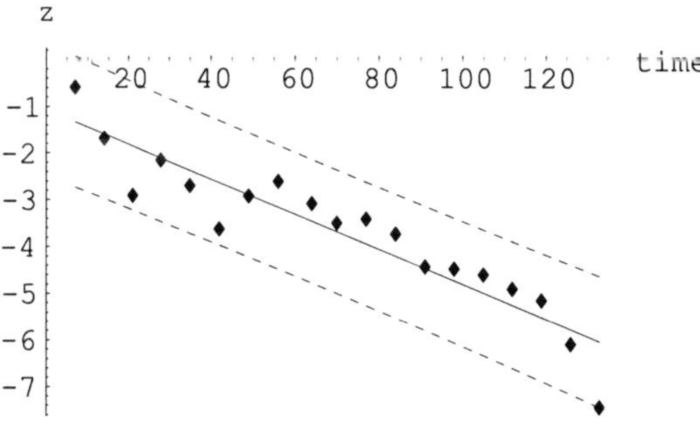

Fig. 13.4 Regression results for (t_i, y_i) of DS-1 $(i = 1, 2, \ldots, 19)$

Table 13.3 Estimation results and ANOVA table (II). Model: $z[\alpha, 0, t_i] = A - Bt_i$ for (t_i, y_i)

$\hat{\alpha}$	$\hat{\beta}$	$\hat{A}$	$\hat{B}$
1.93489	0	−3.21541	0.05010

	DF	SSE	V	F_0	R^{*2}
Model	1	70.0719	70.0719	174.139**	0.905828
Error	17	6.84063	0.40239		
Total	18	76.9126			

13.4.1.2 Multiple Regression Analysis

Three-parameter Model

We can further perform a multiple regression analysis for DS-1 as follows. We consider the following formula.

$$z[\alpha, \beta, t_i] = A - Bt_i - Cy_i + \varepsilon_i \quad (i = 1, 2, \ldots, n). \tag{13.36}$$

In (13.36), although this equation has two independent variables, t_i and y_i for this multiple regression, these variables are not mutually independent in many cases. However, by naturally assuming the relation between t_i and y_i as:

$$y_i = H(t_i) + \varepsilon_{1i} \quad (i = 1, 2, \ldots, n), \tag{13.37}$$

we can understand that the error term ε_i in (13.36) consists of ε_{1i} and the error between $z[\alpha, \beta, t_i]$ and $A - Bt_i$. Consequently, unless $H(t_i)$ is a linear function of t_i, the multicollinearity problem does not appear.

Now (13.36) can be also analyzed to estimate the model parameters in the same manner discussed in Sect. 13.3. The estimation results are given in Table 13.4. In

Table 13.4 Estimation results and ANOVA table (III)
Model: $z[\alpha, \beta, t_i] = A - Bt_i - Cy_i + \varepsilon_i$ for (t_i, y_i)

$\hat{\alpha}$	$\hat{\beta}$	$\hat{A}$	$\hat{B}$	$\hat{C}$
4.94242	−1.81192	−4.41421	0.14348	−0.0177983
			$\hat{B}_s$	$\hat{C}_s$
			2.29253	−1.35754

	DF	SSE	V	F_0	R^{*2}
Model	2	106.766	53.383	330.612**	0.973421
Error	16	2.58349	0.161468		
Total	18	109.350			

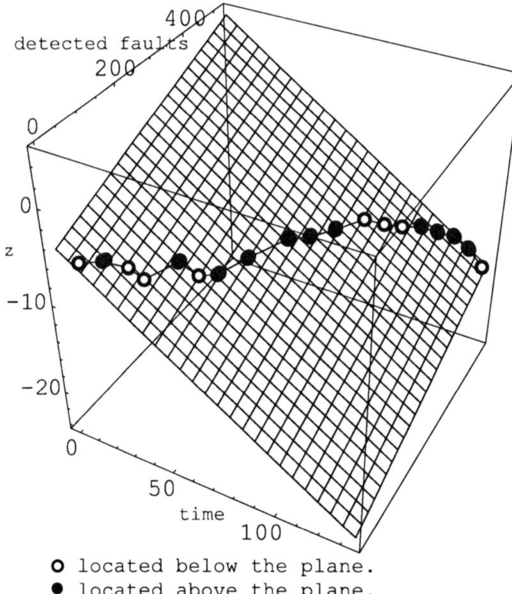

Fig. 13.5 Estimated regression plane and $(t_i, y_i, z[\hat{\alpha}, \hat{\beta}, t_i])$ $(i = 1, 2, \ldots, 19)$

Table 13.4, B_s and C_s are the estimated standardized partial regression coefficient of B and C, respectively.

Among the estimated models in this section, this multiple regression model fits best for DS-1. Figure 13.5 depicts the estimated regression plane and the data points.

Derivation of $H(t)$

Based on the regression formula shown in (13.36), we can derive another LGC model. The mean behavior of (13.36) is shown by

$$z[\alpha, \beta, t] = \log\left\{\frac{dH(t)}{dt}/t^{\alpha}/H(t)^{\beta}\right\} = A - Bt - CH(t).\qquad(13.38)$$

If the conditions $\beta = 0$, $B > 0$ and $C > 0$ hold, we can obtain the solution of this differential equation with the initial condition $H(0) = h_0 > 0$ as

$$H(t) = \frac{1}{C}\log\left[e^{Ch_0} + \frac{Ce^A}{B^{\alpha+1}}\{\Gamma[\alpha + 1, 0] - \Gamma[\alpha + 1, Bt]\}\right].\qquad(13.39)$$

Since the estimated value of C in (13.36) is negative ($\hat{C} = -0.0177983$), we cannot plot $\hat{H}(t)$ by (13.39) for this data set DS-1.

13.4.1.3 Four-parameter Model

In order to include another factor into the model, we apply the following model to DS-1.

$$z[\alpha, \beta, t_i] = A - Bt_i - Cy_i - Dw_i + \varepsilon_i \quad (i = 1, 2, \ldots, 19). \tag{13.40}$$

This model considers the factor of the cumulative number of processed test cases, w_i.

The estimation results are shown in Table 13.5. This result of regression analysis is slightly worse than that of Table 13.4.

Table 13.5 Estimation results and ANOVA table (IV)
Model: $z[\alpha, \beta, t_i] = A - Bt_i - Cy_i - Dw_i + \varepsilon_i$ for (t_i, y_i, w_i)

$\hat{\alpha}$	$\hat{\beta}$	$\hat{A}$	$\hat{B}$	$\hat{C}$	$\hat{D}$
4.35677	-1.62452	-3.47307	0.17382	-0.0201079	-0.000610987
			$\hat{B}_{\mathrm{s}}$	$\hat{C}_{\mathrm{s}}$	$\hat{D}_{\mathrm{s}}$
			3.07001	-1.695338	-0.462164

	DF	SSE	V	F_0	R^{*2}
Model	3	87.2726	29.0909	195.5295**	0.970228
Error	15	2.22034	0.148023		
Total	18	89.4929			

13.4.1.4 Discussion on the Regression Models

In the previous section, we have applied several regression models to DS-1. The LGC model with the two-parameter numerical differentiation method is highly flexible in the sense that we can add several factors into the multiple regression formula in order to investigate how the factor effects the software reliability growth. Also, this model can be used as a testing progress monitoring tool during the software testing phase. When a testing phase progresses on schedule, $z[\hat{\alpha}, \hat{\beta}, t]$ will gradually decrease on the estimated regression line even if the data set (t_i, y_i) shows some strange-shaped growth curve. This fact suggests that the regression results of $z[\hat{\alpha}, \hat{\beta}, t]$ can be used as a measure of software testing progress; however, we should also consider that the prediction based on the regression analysis by extrapolation sometimes results in giving us an inaccurate estimation. It should be noted that this difficulty cannot be avoided in the whole regression analysis.

13.4.2 Curve-fitting Analysis

In this section, we use the LGC model to fit the obtained data set. This curve fitting scheme can provide the prediction of the future behavior of the data set in the sense of time series analysis.

In general, a function which has many independent parameters fits better for a lot of data sets. However, we often have a difficulty of setting the initial values for searching the optimal solution for fitting. In such a situation, the LGC model and its estimation method discussed in Sect. 13.3 can help us, because the results of regression analysis discussed in the section can be used as the initial values for curve fitting calculation.

In order to show examples, we analyze two data sets DS-2 and DS-3. DS-2 consists of 19 pairs of the testing time (measured in month) and cumulative number of detected software faults [11], and DS-3 is 25 pairs.

We show the estimation result for DS-2. We have applied the LGC model by (13.24) denoted by $H_c(t)$, as

$$H_c(t) = \left[\frac{e^A}{B^{\alpha+1}} \left\{ \Gamma[\alpha+1, 0] - \Gamma[\alpha+1, Bt] \right\} \right], \tag{13.41}$$

where we assume $\beta = 0$ in (13.30) to simplify the model.

First we have estimated the regression line for the model. The estimates are $\hat{\alpha} = 0.616812$, $\hat{A} = 2.98269$, and $\hat{B} = 0.164745$. By using these estimates as initial values, we have found the optimal estimates under the least squares rule such that

$$\min_{\alpha, A, B} \sum_{i=1}^{n} (H_c(t_i) - y_i)^2. \tag{13.42}$$

Therefore, the estimates obtained are $\hat{\alpha}^\star - 0.34266$, $\hat{A}^\star - 3.06696$, and $\hat{B}^\star = 0.0924122$. Since $\hat{\alpha}$ is about 0.34, the fitted curve is intermediate shape between exponential and delayed S-shaped curves. Figure 13.6 illustrates the curve and the data points of DS-2.

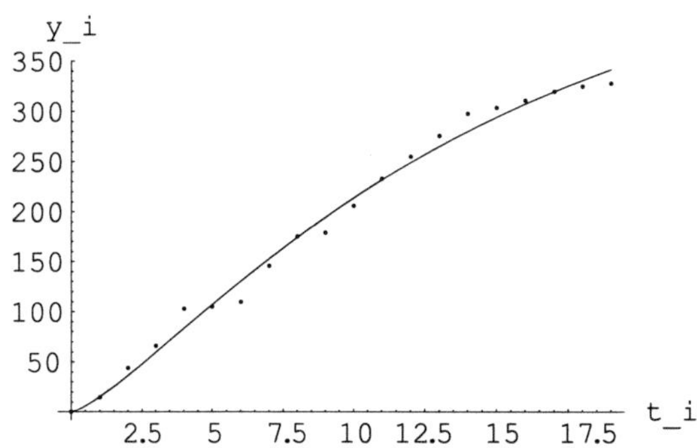

Fig. 13.6 Estimated $\hat{H}_c(t)$ and DS-2

13.4.2.1 Discussion on the Model Selection Problem

Here we compare two kinds of the LGC models which are shown by (13.41) and
(13.30) with DS-3. The latter model by (13.30) is denoted by $H_d(t)$ as follows.

$$H_d(t) = \frac{1}{C} \log \left[1 + \frac{Ce^A}{1+\alpha} t^{\alpha+1} \right] . \tag{13.43}$$

The estimates are $\hat{\alpha} = -0.295477$, $\hat{A} = 1.28942$, and $\hat{B} = 0.000713079$ for
$H_c(t)$, and $\hat{\alpha} = -0.193487$, $\hat{A} = 1.06632$, and $\hat{B} = 0.00220316$ for $H_d(t)$. Figure 13.7 shows the data set and estimated curves, $\hat{H}_c(t)$ and $\hat{H}_d(t)$. This figure illustrates these two models shows almost the same behavior. From the viewpoint of
the AIC (Akaike information criterion [15]), $\hat{H}_d(t)$ is slightly better than $\hat{H}_c(t)$ by
$\text{AIC}(\hat{H}_c(t)) = 97.2777$ and $\text{AIC}(\hat{H}_d(t)) = 94.9732$.

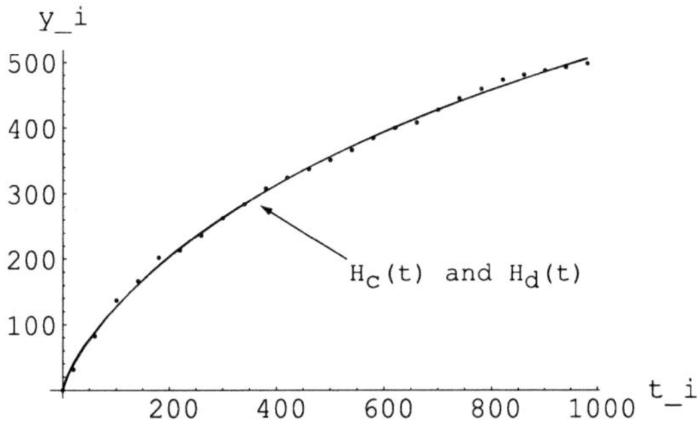

Fig. 13.7 Fitted curves and DS-2

In this situation we should focus on the fact that the characteristics of both two
models are particularly different. In fact, as mentioned in Sect. 13.2, $H_c(t)$ has
a convergence value when $t \to \infty$, but $H_d(t)$ does not, *i.e.*, it means that the model
described by $H_d(t)$ assumes that the number of fault detection could be infinite.
Actually, by using the estimates obtained by DS-3, we have

$$\lim_{t \to \infty} \hat{H}_c(t) = 772.561 , \tag{13.44}$$

$$\lim_{t \to \infty} \hat{H}_d(t) = \infty . \tag{13.45}$$

This result confronts us a model selection problem. That is, if a software testing
manager first uses $H_c(t)$ for software reliability estimation, he/she can estimate
the number of initial faults in the software by $\hat{H}_c(\infty) \approx 773$. And he/she might
additionally estimate several useful software reliability measures based on NHPP
modeling, *e.g.*, cumulative mean time between failures, $MTBF_c(t) = t/H(t)$, and
software reliability $R(x|t) = \exp[-H(t+x) + H(t)]$ [1,2,4–6]. These measures will
give him/her realistic assessment quantities when the $H(t)$ is the convergence type

model, because it cannot be usually considered that the number of inherent software faults takes an infinite value even if there exists a possibility of imperfect debugging environment (see *e.g.*, [6]).

However, if $H_d(t)$ is later found and estimated by him/her, the first estimation result of 773 faults by using $\hat{H}_c(t)$ might not be trustworthy for the software testing control any more, since $\mathrm{AIC}(\hat{H}_c(t)) > \mathrm{AIC}(\hat{H}_d(t))$. That is, in this case, should he/she discard those estimated useful software assessment measures $\widehat{MTBF}_c(t)$ and $\hat{R}(x|t)$ based on $\hat{H}_c(t)$, for the reason that AIC of $\hat{H}_d(t)$ is marginally better than that of $\hat{H}_c(t)$, and is the number of initial fault content of this software, $\hat{H}_d(\infty) = \infty$, really appropriate for this software testing process?

This situation suggests that we have an essential risk for software reliability assessment if we only use limited information such as a data set consists of (t_i, y_i) $(i = 1, 2, \ldots, n)$. Therefore we should gather another related information to software reliability growth phenomena from the actual testing processes, and we need to invent new assessment methodology which can include all the information obtained. In this sense, the multiple regression scheme with the LGC model proposed in this chapter might provide one answer for this issue.

13.5 Concluding Remarks

This chapter has proposed the LGC (linearized growth curve) model and discussed the methodology of data analysis by linear regression schemes. This model can present wide applicability for software fault-count data sets which show various shapes, and the model can also provide an easy method for unknown parameters estimation. In order to illustrate the model performance, we have analyzed several data sets in the section of numerical examples.

In the proposed methodology of data analyses with our LGC model, linear regression analyses give us assessment measures of software testing progress. If we use one of the several types of the derived $H(t)$ in Sect. 13.2 for a non-linear regression analysis based on a data set of (t_i, y_i) $(i = 1, 2, 3, \ldots, n)$, the direct estimation of unknown parameters (*i.e.*, the estimation does not go through a linear regression scheme for finding the initial values for the curve fitting) would be difficult to set the good initial values for searching the numerical optimal solutions. From this point of view, our proposition of using the estimated values by linear regression as the initial values for curve fitting analyses discussed in Sect. 13.4.2 provides one of the realistic approaches for the software reliability data analyses.

In addition, our regression model can consider several factors which influence the testing progress and reliability during the software testing phase by simply introducing such factors into the model as independent variables for multi regression analyses. Therefore as mentioned in Sect. 13.4.2.1, we need to construct some practical systems and/or methods for the collection of software reliability related data throughout software development processes in the future study.

Acknowledgements This research was partially supported by the Japan Society for the Promotion of Science, Grant-in-Aid for Scientific Research (C), 18500066, 2006.

References

1. Pressman R (2001) Software Engineering: A Practitioner's Approach. McGraw-Hill Higher Education, Singapore
2. Musa JD, Iannino A, Okumoto K (1987) Software Reliability: Measurement, Prediction, Application, McGraw-Hill, New York
3. Musa JD (1998) Software Reliability Engineering. McGraw-Hill, New York
4. Bittanti S (1988) Software Reliability Modeling and Identification. Springer-Verlag, Berlin
5. Lyu M (2000) Handbook of Software Reliability Engineering, IEEE Computer Society Press, Los Alamitos
6. Pham H (2000) Software Reliability. Springer-Verlag, Singapore
7. Ohba M (1984) Software reliability analysis models. IBM Journal of Research Development 28:428–443
8. Sofer A, Miller DR (1991) A nonparametric software-relibility-growth model. IEEE Transactions on Reliability 40:329–337
9. Karunanithi N, Whitely D, Malaya K (1992) Prediction of software reliability using connectionist models. IEEE Transactions on Software Engineering 18:563–574
10. Sitte R (1999) Comparison of software-reliability-growth predictions: neural networks vs parametric-recalibration. IEEE Transactions on Reliability 48:285–291
11. Yamada S, Ohba M, Osaki S (1983) S-shaped reliability growth modeling for software error detection. IEEE Transactions on Reliability R-32:475–478, 484
12. http://documents.wolfram.com/mathematica/ (for language reference)
13. Shibata K (1980) Project planning and phase management of software product. Information Process Society Japan 21:1035–1042
14. Yamada S, Somaki H (1996) Statistical methods for software testing-progress control based on software reliability growth models (in Japanese). Transactions of the Japan Society for Industrial and Applied Mathematics 6:33–43
15. Akaike H (1974) A new look at the statistical model identification. IEEE Transactions on Automatic Control AC-19:716–723

Chapter 14
Software Reliability Model Considering Time-delay Fault Removal

Seheon Hwang, Hoang Pham

Department of Industrial and Systems Engineering,
Rutgers University,
96 Frelinghuysen Road,
Piscataway, NJ, 08854, USA

14.1 Introduction

Software reliability has proven to be one of the most useful indices in evaluating software applications quantitatively [1–3]. Among many different methodologies for constructing software reliability models, the software reliability growth models (SRGMs) based on the non-homogeneous Poisson process (NHPP) has been widely used in practical software reliability engineering and, has attracted many engineers and researchers who assess software systems [2–17].

The NHPP model is characterized by its mean value function. Therefore, enormous NHPP models have been developed to obtain unique mean value function formulated with various assumptions reflecting different aspects of testing activities [2, 18–21]. Since the introduction of the simplest initial model with constant fault detection rate and perfect debugging assumption [4], many NHPP models have been developed towards more relaxed assumptions closed to practical testing and debugging process. For instance, the assumption of constant fault detection rate has the substantive relaxation substituted with time dependent fault detection rate due to tester learning effects. The imperfect debugging assumption may allow the new faults inadvertently introduced into the software during debugging [8, 10, 17] and further, the detected faults removed incompletely from the software [3, 22]. Although the models have been progressed successfully by modifying the assumptions regarding the testing/debugging process, most of the existing NHPP models still assume the instantaneous fault removal. This means a fault is removed as soon as it is detected, and to our knowledge, only a few recent articles have considered the time required to remove the detected faults, which motivates this study [22–25].

There are several steps until the debugging activity begins. Since the faults found by testers are fixed by programmers, identifying the detected faults is the first step in conducting the effective debugging. In practice, when a tester detects a fault manifested by a failure he or she has to report it using a form to request a modification [22, 26]. The content of the problem report varies by company and organization, but it generally includes fields such as type of problem, severity, summary of problem, how to reproduce etc. After identifying the detected faults, the review board members including project manager prioritize the removal schedule for the detected faults and assign particular programmers code changes. The removal schedule is usually determined by priority, which is the impact of a detected fault on the business such as stop-shipping or defer-fixing. Based on the level of priority, the removing of a detected fault can be deferred until there is a later version or until the next release [26].

Zhang *et al.* [22] introduce fault removal efficiency and integrate it into software reliability growth models to describe not only imperfect debugging phenomena, but the detected faults that may not be removed immediately. They formulate the failure rate function by the product of the expected number of faults detected by the present time and the fault removal efficiency, the percentage of detected faults completely removed, and as the result, provide some of in-process metrics which offer useful information about the development project management as well as the traditional reliability measures. Jeske and Zhang [24] point out some of the practical problems when using NHPP SRGMs including the mismatch between the test profile and the operational profile, the use of instantaneous repair time assumption, and the poor quality in the failure data that is needed to estimate the mean value function. Gokhale *et al.* [23] propose a number of different fault removal policies considering the delays for fault removal and analyze their effect on the residual number of faults at the end of the testing process using a non-homogeneous continuous time Markov chain model. They also extend this framework to include the imperfections in the fault removal process.

In this paper, we develop a generalized mean value function by considering the time-delay due to identifying, prioritizing the detected faults, and assigning developers for analysis and code changes. In particular, we more focus on describing how to formulate and expand the initial differential equation to derive the mean value function under the time-delay assumption. The time for debugging is not taken into account in the delayed time because it varies with the severity and the priority of fault. Section 14.2 presents the formulation of the NHPP model incorporating the time-delay assumption into generalized software reliability model. The general solution of the mean value function (MVF) for the proposed model is derived and then, it is applied to some of the existing models, which have been formulated based on instantaneous fault removal assumption. Section 14.3 provides numerical examples to illustrate the application of the proposed model to the existing NHPP models using a software failure data set. The conclusions are summarized in Sect. 14.4.

Notation

$m(t)$	expected number of failures occurred by time t
$a(t)$	total fault content rate function, *i.e.*, the sum of expected number of initial software faults and introduced faults by time t
$b(t)$	failure detection rate function, *i.e.*, the rate to detect a fault at time t

$\lambda(t)$ failure intensity function at time t
s a fixed time-delay for fault removal
$m_s(t)$ expected number of faults detected and removed by time t when time-delay is s

14.2 Model Formulation

Pham *et al.* [13] have unified the existing NHPP software reliability models into a generalized NHPP model, which is given by

$$\frac{\mathrm{d}m(t)}{\mathrm{d}t} = b(t)[a(t) - m(t)] . \tag{14.1}$$

The NHPP models have mostly been constructed based on the common assumption that the failure rate at time t is proportional to the number of remaining faults in the software at time t and the fault detection rate per a fault at that moment [1,2,4,16,18]. Since a detected fault is immediately removed in instantaneous fault removal assumption, the number of failures detected up to any time is equivalent to the number of faults removed by that time. Therefore, the term $m(t)$ in (14.1) can be regarded as the number of faults detected and completely removed from the software by time t so that $a(t) - m(t)$ represents the remaining number of faults in the software at time t.

Under the time-delay fault removal assumption, however, $m(t)$ implies the expected number of detected faults by time t, some of which have been removed and some of which have not. Consequently, $m(t)$ in (14.1) should be substituted for its appropriate term, meaning the faults detected and completely removed from the software by time t, which is given as follows:

$$\frac{\mathrm{d}m(t)}{\mathrm{d}t} = b(t)[a(t) - m_s(t)] \quad \text{for} \quad t > 0 , \qquad m(0) = 0 , \tag{14.2}$$

where $m_s(t)$ is the expected number of faults detected and removed by time t when time-delay is s.

The faults that have been detected but not removed may cause other failures attributed to the faults already detected, unless they are eventually removed from the system. This may result in duplicate counting in number of failures [25]. The failure intensity function in (14.2) takes into consideration the duplicated failures possibly incurred by substituting the number of faults just detected by time t with the one removed as well as detected by time t, which is shown as follows. Therefore, the proposed model provides the benefit to compensate the number of failures possibly duplicated in counting the failure data.

14.2.1 Time-delay Fault Removal Model

The formulation of the proposed model is based on the following assumptions:

1. The occurrence of software failures follows an NHPP.
2. A detected fault requires a certain amount of fixed time until it is finally removed.
3. All the faults detected during a certain period are removed at once at the end of that period.
4. No debugging time is considered.

As Fig. 14.1 describes, the time-delay model assumes that the debugging activity is conducted periodically and all the detected faults during the same period are eventually removed at once instead of removing individually. For instance, the faults detected during the weekday would be removed at the end of the week and no debugging activities are conducted during the time period, thus, the software program is updated on the first day of the week. This implies $m_s(t)$ would be a constant during a period; therefore, the expected number of failures might be more than those of the existing models, and consequently, the proposed model can reflect such a phenomenon that the fault reported multiple times before it is finally removed.

Let us define $m_{s,n}(t)$ as expect number of failure occurred during the interval $(n-1)s < t < ns$ with an arbitrary delay s. Based on the assumption 2 and 3, the testing time can be divided into several intervals regarding the time delay s as it is shown in Fig. 14.1.

Now, the general solution of the failure intensity function in (14.2) can be derived from the following procedure by employing the method of steps [27]. The method is applicable to every interval determined by the value of s.

i)   If $t \in (0, 1s]$, then, $m_s(t) = m(0)$.

$$\frac{dm(t)}{dt} = b(t)[a(t) - m_s(t)]$$
$$= b(t)[a(t) - m(0)]$$
$$= a(t)b(t) .$$

Since the faults detected during the interval $t \in (0, 1s]$ is to be removed at $t = 1s$, the number of faults that have removed remains unchanged during this

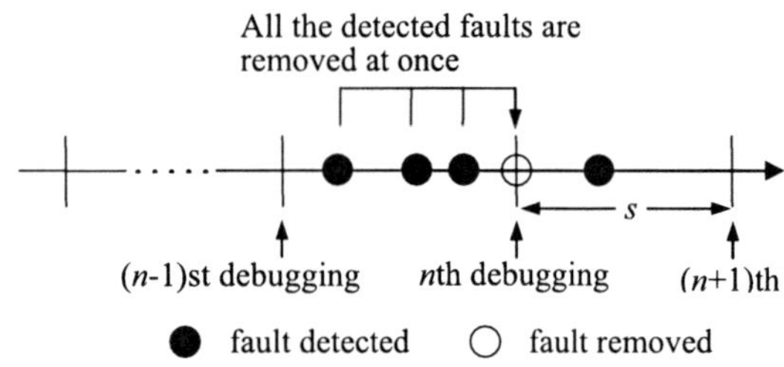

Fig. 14.1 Description of time-delay fault removal assumption

interval, which is zero. Thus, the solution for the interval $(0, 1s]$ is

$$m_{s,1}(t) = \int_0^t a(\tau)b(\tau)\,\mathrm{d}\tau \ . \qquad (14.3)$$

Here, let us define $D_x(t) = \int_x^t a(\tau)b(\tau)\,\mathrm{d}\tau$, then,

$$m_{s,1}(t) = D_0(t) \qquad (14.4)$$

ii)   If $t \in (s, 2s]$, $m_s(t)$ can be replaced with $m(1s)$.

$$\frac{\mathrm{d}m(t)}{\mathrm{d}t} = b(t)[a(t) - m(1s)]$$
$$= a(t)b(t) - b(t)m(1s) \ .$$

Thus,

$$m_{s,2}(t) = \int_s^t [a(\tau)b(\tau) - b(\tau)m(s)]\,\mathrm{d}\tau$$
$$= \int_s^t b(\tau)[a(\tau) - m(s)]\,\mathrm{d}\tau$$
$$= D_s(t) - m(s)B_s(t) \ , \qquad (14.5)$$

where $B_x(t) = \int_x^t b(\tau)\,\mathrm{d}\tau$.

iii)   If $t \in (2s, 3s]$, $m_s(t)$ can be replaced with $m(2s)$.

$$\frac{\mathrm{d}m(t)}{\mathrm{d}t} = b(t)[a(t) - m_s(t)]$$
$$= b(t)[a(t) - m(2s)] \ .$$

Thus,

$$m_{s,3}(t) = \int_{2s}^t [a(\tau)b(\tau) - b(\tau)m(2s)]\,\mathrm{d}\tau$$
$$= \int_{2s}^t b(\tau)[a(\tau) - m(2s)]\,\mathrm{d}\tau \qquad (14.6)$$
$$= D_{2s}(t) - m(2s)B_{2s}(t) \ .$$

iv)   If $t \in (3s, 4s]$,

$$\frac{\mathrm{d}m(t)}{\mathrm{d}t} = b(t)[a(t) - m_s(t)]$$
$$= b(t)[a(t) - m(3s)] \ .$$

Thus,

$$m_{s,4}(t) = \int_{3s}^t [a(\tau)b(\tau) - b(\tau)m(3s)]\,\mathrm{d}\tau$$
$$= \int_{3s}^t b(\tau)[a(\tau) - m(3s)]\,\mathrm{d}\tau$$
$$= D_{3s}(t) - m(3s)B_{3s}(t) \ . \qquad (14.7)$$

Therefore, the expected number of failures found during the time $(n-1)s < t \le ns$ can be obtained from the formula below:

$$m_{s,n}(t) = \int_{(n-1)s}^{t} [a(\tau)b(\tau) - b(\tau)m((n-1)s)]\,\mathrm{d}\tau$$

$$= \int_{(n-1)s}^{t} b(\tau)[a(\tau) - m((n-1)s)]\,\mathrm{d}\tau \qquad (14.8)$$

$$= D_{(n-1)s}(t) - m[(n-1)s]B_{(n-1)s}(t) \, .$$

Finally, we obtain a general solution of $m(t)$ with arbitrary s when the time-delay assumption has been applied to the generalized NHPP software reliability model.

$$m(t) = \sum_{k=1}^{w} m_{s,k}(ks) + m_{s,w+1}(t)$$

$$= \sum_{k=1}^{w} \{D_{(k-1)s}(ks) - m[(k-1)s] \cdot B_{k-1}(ks)\} + D_{ws}(t) - m(ws) \cdot B_{ws}(t) \, ,$$

$$(14.9)$$

where w is denoted by $\{w: w$ is an integer which satisfies $\lfloor \frac{t}{s} \rfloor\}$.

We, now, formulate the mean value function of several existing models such as G–O, delayed s-shaped and inflection s-shaped considering the time-delay assumption using the result of (14.9).

Time-delay Model Based on G–O Model

Since $a(t) = a$ and $b(t) = b$ [4], $D(t)$ and $B(t)$ are as follows:

$$D_x(t) = \int_{x}^{t} ab\,\mathrm{d}\tau$$

$$= ab(t - x)$$

$$B_x(t) = \int_{x}^{t} b\,\mathrm{d}\tau$$

$$= b(t - x) \, .$$

From (14.9), the expected number of failures detected during the time interval $(n-1)s < t \le ns$ can be expressed as follows:

$$m_{s,n}(t) = D_{(n-1)s}(t) - m[(n-1)s]B_{(n-1)s}(t)$$

$$= ab[t - (n-1)s] - m[(n-1)s]b[t - (n-1)s] \qquad (14.10)$$

$$= [ab - bm((n-1)s)]\,[t - (n-1)s] \, .$$

Hence, the expected number of failures found by time t can be expressed by adding $m_n(\tau)$ to n, which is given by

$$m(t) = \sum_{k=1}^{\lfloor \frac{t}{s} \rfloor} [ab - bm((k-1)s)]s + \left[ab - bm\left(\left\lfloor \frac{t}{s} \right\rfloor s\right)\right]\left(t - \left\lfloor \frac{t}{s} \right\rfloor s\right) \qquad (14.11)$$

Time-delay Model Based on Delayed s-Shaped

Since $a(t) = a$ and $b(t) = \frac{b^2 t}{bt+1}$ [14], $D_x(t)$ and $B_x(t)$ are as follows:

$$D_x(t) = \int_x^t a(\tau) \cdot b(\tau)\, d\tau$$

$$= \int_x^t \frac{ab^2 \tau}{b\tau + 1}\, d\tau$$

$$= a[bt - \log(1 + bt)]$$

$$B_x(t) = \int_x^t b(\tau)\, d\tau$$

$$= [bt - \log(1 + bt)]\,.$$

Here, since $D_x(t)$ equals $aB_x(t)$, (14.9) can be simplified to

$$m_{s,n}(t) = a \cdot B_{(n-1)s}(t) - m[(n-1)s] \cdot B_{(n-1)s}(t) = (a - m[(n-1)s]) \cdot B_{(n-1)s}(t)\,.$$

Thus, the expected number of failures found during the time interval $(n-1)s < t \leq ns$ can be expressed as follows:

$$m_{s,n}(t) = (a - m[(n-1)s]) \cdot [b(t - (n-1)s) - \log(1 + b(t - (n-1)s))]\,. \quad (14.12)$$

Therefore, the expected number of failures found by time t for the time-delay model based on delayed s-shaped is given by

$$m(t) = \sum_{k=1}^{\lfloor \frac{t}{s} \rfloor} \{(a - m[(k-1)s]) \cdot [b(k - (k-1)s) - \log(1 + b(k - (k-1)s)]\}$$

$$+ \left(a - m\left[\left\lfloor \frac{t}{s} \right\rfloor s\right]\right) \cdot \left[b\left(t - \left\lfloor \frac{t}{s} \right\rfloor s\right) - \log\left(1 + b\left(t - \left\lfloor \frac{t}{s} \right\rfloor s\right)\right)\right]$$

$$(14.13)$$

Time-delay Model Based on Inflection s-Shaped

This model still restricts to perfect debugging but, has s-shaped fault detection rate, which is $b(t) = \frac{b}{1+\beta e^{-bt}}$ [7]. Therefore, we can construct $D_x(t)$ and $B_x(t)$ as follows:

$$D_x(t) = \int_x^t \frac{ab}{1 + \beta e^{-b\tau}}\, d\tau$$

$$= a[\log(\beta + e^{bt})]$$

$$B_x(t) = \int_x^t \frac{b}{1 + \beta e^{-b\tau}}\, d\tau$$

$$= \log(\beta + e^{bt})\,.$$

Thus, the expected number of failures found during the time interval $(n-1)s < t \leq ns$ is

$$m_{s,n}(t) = (a - m[(n-1)s]) \cdot \log\left(\beta + e^{b[t-(n-1)s]}\right) . \tag{14.14}$$

The expected number of failures found by time t for the time-delay model based on inflection s-shaped model is given by

$$m(t) = \sum_{k=1}^{\lfloor \frac{t}{s} \rfloor} (a - m[(k-1)s]) \cdot \log\left(\beta + e^{b[k-(k-1)s]}\right)$$
$$+ \left(a - m\left[\left\lfloor \frac{t}{s} \right\rfloor s\right]\right) \cdot \log\left(\beta + e^{b[t - \lfloor \frac{t}{s} \rfloor s]}\right) . \tag{14.15}$$

14.3 Numerical Examples

In this section, we have applied the time-delay assumption into the existing software reliability models using the failure data shown in Table 14.1. The failure data was obtained from a real-time control system consisting of about 200 modules with an average 1000 lines of high-level language [3].

14.3.1 General Approach

In this section, we consider three existing software reliability models as basic such as G–O model, delayed s-shaped model, and inflection s-shaped model, which have been being widely used. The mean value function for each existing model and the corresponding time-delay model are summarized in Table 14.2. The time-delay s can be interpreted as an interval that a review board meeting is held, therefore, it would be a given number that can be determined by the developers according to the development schedule. Since we have no clue regarding how frequently the development teams had review board meetings when the failure data was gathered, we assume the delayed time s as an integer in these numerical examples. In order to find the optimal solution for the estimators, we vary the value s from 1 day up to 30 days in most of the cases and conduct parameter estimation for each value of s.

The estimation of the unknown parameters for each model is carried out by using maximum likelihood estimate (MLE) method, the most widely used estimation technique. First, we utilize all 111 data points for estimating the parameters to fit the actual failures with varying the value s, and then we compare the result for each of the existing models with the one for the corresponding model under time-delay assumption. In addition, the first 86 failure data have been employed for the parameter estimation and the remaining data points have been used to verify the goodness-of-fit for the predicting performance. From Table 14.2, we can find no failure occurs during the 6 days between the 81st and 86th. It is considered that the software system

Table 14.1 Real-time software failure data [21]

Day	Faults	Cumulative faults	Day	Faults	Cumulative faults	Day	Faults	Cumulative faults	Day	Faults	Cumulative faults	Day	Faults	Cumulative faults
1	5	5	24	4	234	47	3	417	70	0	467	93	0	475
2	5	10	25	2	236	48	8	425	71	0	467	94	0	475
3	5	15	26	4	240	49	5	430	72	1	468	95	0	475
4	5	20	27	3	243	50	1	431	73	1	469	96	1	476
5	6	26	28	9	252	51	2	433	74	0	469	97	0	476
6	8	34	29	2	254	52	2	435	75	0	469	98	0	476
7	2	36	30	5	259	53	2	437	76	0	469	99	0	476
8	7	43	31	4	263	54	7	444	77	1	470	100	1	477
9	4	47	32	1	264	55	2	446	78	2	472	101	0	477
10	2	49	33	4	268	56	0	446	79	0	472	102	0	477
11	31	80	34	3	271	57	2	448	80	1	473	103	1	478
12	4	84	35	6	277	58	3	451	81	0	473	104	0	478
13	24	108	36	13	290	59	2	453	82	0	473	105	0	478
14	49	157	37	19	309	60	7	460	83	0	473	106	1	479
15	14	171	38	15	324	61	3	463	84	0	473	107	0	479
16	12	183	39	7	331	62	0	463	85	0	473	108	0	479
17	8	191	40	15	346	63	1	464	86	0	473	109	1	480
18	9	200	41	21	367	64	0	464	87	2	475	110	0	480
19	4	204	42	8	375	65	1	465	88	0	475	111	1	481
20	7	211	43	6	381	66	0	465	89	0	475			
21	6	217	44	20	401	67	0	465	90	0	475			
22	9	226	45	10	411	68	1	466	91	0	475			
23	4	230	46	3	414	69	1	467	92	0	475			

Table 14.2 The mean value functions for several existing software reliability models with and without (original model) time-delay assumption

	Original model	Time-delay assumption
Model 1: Goel–Okumoto	$m(t) = a(1 - \mathrm{e}^{-bt})$	$m(t) = \sum_{k=1}^{\lfloor \frac{t}{s} \rfloor} \left([ab - bm[(k-1)s]]\,[k - (k-1)s] \right) + \left[ab - bm\left(\lfloor \frac{t}{s} \rfloor s \right) \right] \left[t - \lfloor \frac{t}{s} \rfloor s \right]$
Model 2: Delayed *s*-shaped	$m(t) = a(1 - (1 + bt)\,\mathrm{e}^{-bt})$	$m(t) = \sum_{k=1}^{\lfloor \frac{t}{s} \rfloor} \{(a - m[(k-1)s]) \cdot [b(k - (k-1)s) - \log(1 + b(k - (k-1)s))]\}$ $+ \left(a - m\left[\lfloor \frac{t}{s} \rfloor s \right] \right) \cdot \left[b\left(t - \lfloor \frac{t}{s} \rfloor s \right) - \log\left(1 + b\left(t - \lfloor \frac{t}{s} \rfloor s \right) \right) \right]$
Model 3: Inflection *s*-shaped	$m(t) = \frac{a(1 - \mathrm{e}^{-bt})}{1 + \beta \mathrm{e}^{-bt}}$	$m(t) = \sum_{k=1}^{\lfloor \frac{t}{s} \rfloor} (a - m[(k-1)s]) \cdot \log\left(\beta + e^{b[k - (k-1)s]} \right) + \left(a - m\left[\lfloor \frac{t}{s} \rfloor s \right] \right) \cdot \log\left(\beta + e^{b\left[t - \lfloor \frac{t}{s} \rfloor s \right]} \right)$

becomes to be stabilized at this point since there is a relatively long time between failures, in fact, more than a week. Therefore, we take the first 86 data points for fitting the model.

After MLE, the estimators and the curves of the mean value functions for each model are obtained based on the actual failure data. As the criterion for evaluating the effectiveness of fitting and predicting performance, we calculate the mean squared error (MSE) and Akaike Information Criterion (AIC) for each of the existing models and each of the corresponding time-delay models. The mean squared error is determined by dividing the sum of squared error (SSE) by the total number of data points minus the number of parameters, expressed as follow:

$$\mathrm{MSE} = \frac{\sum\limits_{i=1}^{n} [y_i - \hat{m}(t_i)]^2}{n - l},$$

where

y_i is total number of actual failures observed up to time t_i according to the failure data;

$\hat{m}(t_i)$ is estimated cumulative number of failures up to time t_i obtained from the fitted mean value function, $i = 1, 2, \ldots n$;

n is number of data points;

l is number of parameters.

Akaike Information Criterion measures the ability of a model to maximize the likelihood function that is directly related to the degrees of freedom during fitting, which means the less the AIC value is, the better the model is.

$$\mathrm{AIC} = -2\log[\text{maximum of likelihood function}] + 2N,$$

where N is the number of parameters in the model.

We focus on comparing the results of fitting performance not among the entire models, including three existing models and the corresponding time-delay models, but between the existing model and the corresponding time-delay model, in this study. Therefore, which model among the entire models compared shows the least MSE is not our interest. Instead we are to compare the result of the time-delay model only with the corresponding existing model.

14.3.2 Analysis of Performance of Models for Fitting Failure Data

14.3.2.1 Time-delay Model Based on G–O Model

In this section, we perform MLE estimation to find the optimal estimator to fit the failure data of the proposed model by varying the integer value s from 1 to 30. The MSE and AIC values for goodness-of-fit as well as the MLE estimators for the time-delay model based on G–O model are listed in Table 14.3. As seen from Table 14.3, the proposed model provides the best fit for this data set when s is equal to 22 days in this scenario, which shows the least AIC value. The result also indicates that the time-delay model shows a good performance to fit the failure data from the viewpoint of SSE value, where the proposed model has 876.78 while G–O model has 1008.32 when s is 22.

We also do MLE estimation considering s as one of the parameters to be estimated, and as the result, we obtain the best fit for this data when s equals 45 days, which shows both MSE and AIC values are less than those of G–O model. As we observe from Table 14.4, the proposed model decreases its MSE value from 1008.32 to 327.63 in this scenario, which provides significant improvement in fitting performance. Note that the number of parameters increases by 1 for calculating MSE and AIC in this case because s is also considered as an estimated parameter.

Table 14.3 The maximum likelihood estimators, MSE, and AIC for time-delay model based on 111 failure data when s is given

	s	$\hat{a}$	$\hat{b}$	MSE_{fit}	AIC
G–O model		497.29	0.0308	1008.32	723.76
Delay model	1	494.29	0.0304	1008.32	723.76
based on G–O	2	497.20	0.0299	1008.03	722.86
	3	497.26	0.0294	1005.15	723.76
	4	497.40	0.0289	1000.70	725.02
	5	494.30	0.0285	997.53	724.38
	6	497.37	0.0281	992.43	724.72
	7	496.54	0.0281	995.65	717.22
	8	496.50	0.0277	988.99	717.27
	9	497.30	0.0269	974.40	724.82
	10	498.04	0.0262	964.99	733.60
	11	497.01	0.0263	962.24	724.40
	12	498.74	0.0253	947.83	739.65
	13	498.27	0.0252	931.75	734.36
	14	493.97	0.0263	964.19	693.34
	15	493.10	0.0264	977.16	678.68
	16	493.59	0.0257	932.97	690.89
	17	496.14	0.0245	875.57	714.64
	18	495.59	0.0243	862.82	714.06
	19	494.95	0.0242	846.34	706.87
	20	493.90	0.0243	839.72	690.45
	21	492.60	0.0244	842.72	677.95
	22	*490.64*	*0.0246*	876.78	*659.86*
	23	491.56	0.0241	832.89	666.50
	24	492.54	0.0235	795.58	675.43
	25	493.91	0.0227	762.12	694.34
	26	*494.62*	*0.0221*	*758.01*	708.48
	27	494.61	0.0217	767.78	717.66
	28	494.93	0.0213	780.90	726.16
	29	497.19	0.0206	798.33	738.26
	30	497.99	0.0203	805.49	740.17

Table 14.4 The maximum likelihood estimators, MSE, and AIC for time-delay model based on 111 failure data when s is estimated

	$\hat{s}$	$\hat{a}$	$\hat{b}$	MSE_{fit}	AIC
G–O	–	497.29	0.0308	1008.32	723.76
Time-delay	45	488.72	0.0186	372.63	652.24

14.3.2.2 Time-delay Model Based on Delay s-Shaped Model

Similar efforts are conducted for existing delay s-shaped models and the corresponding time-delay model. Table 14.5 summarizes the MLE and goodness-of-fit results according to the different given values of s based on 111 date points. In this case, the least AIC value is found when s is 15, but the MSE of the proposed model does not provide a good result. We rather observe a better MSE when s is 4 even though it has a larger AIC value than the existing delay s-shaped model, implying not a good estimator.

Table 14.5 The maximum likelihood estimators, MSE, and AIC for delay s-shaped model and time-delay model based on 111 failure data when s is estimated

	$\hat{s}$	$\hat{a}$	$\hat{b}$	MSE_{fit}	AIC
Delay s-shape	–	483.04	0.0687	346.87	644.03
Time-delay	45	481.51	0.0572	390.58	631.86

14.3.2.3 Time-delay Model Based on Inflection s-Shaped Model

In this case, the lowest AIC is obtained when s has 22 days of delay time and the MSE also shows smaller value comparing to the existing inflection s-shaped model, which are 609.14 and 257.66 respectively. The MSE and AIC values for goodness-of-fit regarding each value of s are listed in Table 14.6. Figure 14.2 depicts the plots of MSE and AIC values for the proposed model and inflection s-shaped model versus the time delay s. As seen from the plots, both MSE and AIC have the similar trend with the delay time and increase rapidly from the lowest points as s increases.

Table 14.6 The maximum likelihood estimators, MSE, and AIC for inflection s-shaped model and time-delay model based on 111 failure data when s is estimated

	$\hat{s}$	$\hat{a}$	$\hat{b}$	$\hat{\beta}$	MSE_{fit}	AIC
Inflection s-shape	–	483.04	0.0687	0.0687	346.87	644.03
Time-delay	45	481.51	0.0572	0.0572	390.58	631.86

Fig. 14.2 Plots of MSE and AIC values regarding the different time delay s for the time-delay model based on inflection s-shaped model

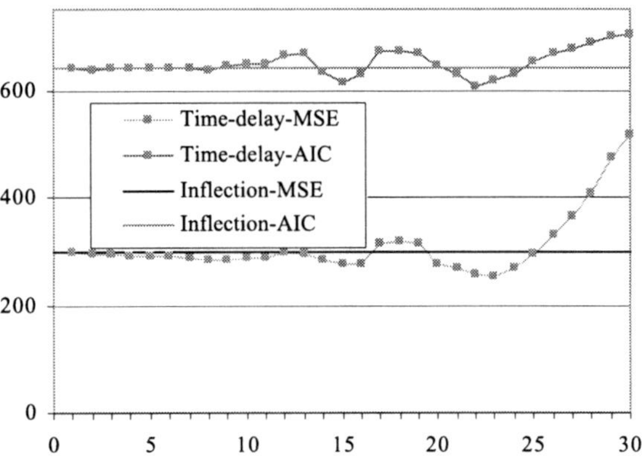

14.3.3 Analysis of Performance for Predicting Future Failure

In this section, we try to test the predictive power of the proposed model and other existing models. We use 86 data points to fit the models, and the rest 25 data points to evaluate the prediction performance. In this case, the mean squared error for both fitting and predicting can be determined using the following formula:

$$\text{MSE}_{\text{fit}} = \frac{\sum\limits_{k=1}^{86} [y_k - \hat{m}(t_k)]^2}{86 - l}$$

$$\text{MSE}_{\text{predict}} = \frac{\sum\limits_{k=87}^{111} [y_k - \hat{m}(t_k)]^2}{25 - l}.$$

Looking into the MSE values in Table 14.7, we observe that the time-delay model based on G–O model shows significant improvements both in fitting the model and in predicting the future failures comparing to G–O model, indicating 1036.49 and 459.66 for fitting and 146.13 and 21.22 for predicting respectively. In the case of the delayed s-shaped model, only tiny differences exist between the existing model

Table 14.7 The comparison of maximum likelihood estimators, MSE, and AIC for several models based on 86 failure data for fitting

	$\hat{s}$	$\hat{a}$	$\hat{b}$	MSE_{fit}	MSE_{pred}	AIC
G–O	–	522.74	0.0274	1036.49	146.13	676.4
Time-delay	45	492.55	0.0185	459.66	21.22	613.36
Delay s-shape	–	481.66	0.0693	461.23	1.05	606.00
Time-delay	15	480.18	0.0672	487.07	0.89	583.73
Inflection s-shape	–	477.42	0.0748	412.25	3.48	598.59
Time-delay	22	473.90	0.0471	285.36	13.92	558.62

and the time-delay model in terms of MSEs for fitting and predicting the values of estimators. The time-delay model based on the inflection s-shaped model has relatively good results in fitting the model, but not a good performance predicting the future failures. The numerical results show that consideration of time-delay fault removal into a generalized NHPP model improves the descriptive properties of the model, and in some cases the predictive properties.

14.4 Concluding Remarks

The time-delay model is developed by integrating the time required to not only identify and prioritize the detected faults but also assign programmers before conducting the actual code change into software reliability assessment, which reflects more realistic conditions on testing and debugging activities. Based on the time-delay assumption, the failure intensity function is newly formulated and consequently, a generalized mean value function is derived by employing method of steps. The proposed model extends its application by allowing the assumptions that the existing models originally include except instantaneous fault removal assumption. Thus, one of the most practical benefits of the time-delay model is that it can be applied to any existing model by simple modification in order to formulate the mean value function under time delay assumption.

We provide the numerical examples by utilizing the real failure data to illustrate the use of the proposed model when it is applied to some of the existing software reliability models. In this paper, we focus our attention on developing a general formula of mean value function by integrating time-delay assumption and estimate parameters considering the time delay as a pre-determined value. For further consideration, we are planning to extend our work to treat the time-delay as a time-varying parameter. The numerical results show, in this paper, that consideration of time-delay removal into a generalized NHPP model in Sect. 2.2 improves the descriptive properties of the model, and in some cases the predictive properties as well. This is interesting and encouraging for further study. It is worth to note that more research in broader validation of this conclusion in a general setting is needed, and encouraged for further study, by using other data sets and other existing NHPP models for both descriptive and predictive modeling.

References

1. Dohi T, Yasui K, Osaki S (2003) Software reliability assessment models based on cumulative Bernoulli trial processes. Mathematical and Computer Modelling 38:1177–1184
2. Pham H, Zhang X (2003) NHPP software reliability and cost models with testing coverage, European Journal of Operational Research 145:443–454
3. Tokuno K, Yamada S (2000) An imperfect debugging model with two types of hazard rates for software reliability measurement and assessment. Mathematical and Computer Modeling 31:343–352

4. Goel AL, Okumoto K (1979) Time-dependent error-detection rate model for software and other performance measures. IEEE Transactions on Reliability 28:206–211

5. Hossain SA, Ram CD (1993) Estimating the parameters of a non-homogeneous Poisson process model for software reliability, IEEE Transactions on Reliability 42:604–612

6. Lyu M (ed) (1996) Handbook of Software Reliability Engineering. McGraw-Hill, New York

7. Musa JD, Iannino A, Okumoto K (1987) Software Reliability: Measurement Prediction Application. McGraw-Hill, New York

8. Ohba M (1984) Software reliability analysis models. IBM Journal of Research Development 28:428–443

9. Ohba M, Yamada S (1984) S-shaped software reliability growth models. In: Proceedings of the 4th International Conference on Reliability and Maintainability, pp. 430–436

10. Pham H (1993) Software reliability assessment: imperfect debugging and multiple failure types in software development, EG&G-RAAM-10737. Idaho National Engineering Laboratory

11. Pham H (1996) A software cost model with imperfect debugging, random life cycle and penalty cost, International Journal of Systems Science 27:455–463

12. Pham H (1999) Software reliability. In: Webster JG (ed) Wiley encyclopedia of electrical and electronics engineering, vol. 19. Wiley, New York, pp. 565–578

13. Pham H, Nordmann L, Zhang X (1999) A general imperfect-software-debugging model with S-shaped fault-detection rate. IEEE Transactions on Reliability 48:169–175

14. Yamada S, Ohba M, Osaki S (1983) S-shaped reliability growth modeling for software error detection. IEEE Transactions on Reliability 12:475–484

15. Yamada S, Ohba M, Osaki S (1984) S-shaped software reliability growth models and their applications. IEEE Transactions on Reliability R-33:289–292

16. Yamada S, Osaki S (1985) Software reliability growth modeling: models and applications, IEEE Transactions on Software Engineering 11:1431–1437

17. Yamada S, Tokuno K, Osaki S (1992) Imperfect debugging models with fault introduction rate for software reliability assessment. International Journal of Systems Science 23:2253–2264

18. Lo JH, Huang CY, Chen I et al. (2005) Reliability assessment and sensitivity analysis of software reliability growth modeling based on software module structure. Journal of Systems and Software 76:3–13

19. Shyur H (2003) A stochastic software reliability model with imperfect-debugging and change-point. Journal of Systems and Software 66:135–141

20. Xie M, Hong GY, Wohlin C (1999) Software reliability prediction incorporating information from a similar project. Journal of Systems and Software 49:43-48

21. Zhang X, Pham H (2000) An analysis of factors affecting software reliability. Journal of Systems and Software 50:43–56

22. Zhang X, Teng X, Pham H (2003) Considering fault removal efficiency in software reliability assessment. IEEE Transactions on Systems, Man, and Cybernetics – Part A Systems and Humans 33:114–120

23. Gokhale SS, Lyu MR, Trivedi KS (2004) Analysis of software fault removal policies using a non-homogeneous continuous time Markov chain. Software Quality Journal 12:211–230

24. Jeske DR, Zhang X (2005) Some successful approaches to software reliability modeling in industry, Journal of Systems and Software 74:85–99

25. Matsuodani T, Tsuda K (2004) Evaluation of debug-testing efficiency by duplication of the detected fault and delay time of repair. Information Sciences 166:83–103

26. Kaner C, Falk J, Nguyen HQ (1999) Testing Computer Software, 2nd edn. Wiley, New York

27. Gorecki H, Fuksa S, Grabowski P, Korytowski A (1989) Analysis and Synthesis of Time Delay Systems. Wiley, New York

Further Reading

Ohtera H, Yamada S (1990) Optimal allocation and control problems for software-testing resources, IEEE Transactions on Reliability 39:171–176

Tohma Y, Yamano H, Ohba M, Jacoby R (1991) The estimation of parameters of the hypergeometric distribution and its application to the software reliability growth model. IEEE Transactions on Software Engineering 17:483–489

Chapter 15
Heuristic Component Placement
for Maximizing Software Reliability

Michael W. Lipton[1], **Swapna S. Gokhale**[2]

[1] IBM Corporation, 150 Kettletown Rd., Southbury, CT 06488, USA
[2] Dept. of CSE, University of Connecticut, Storrs, CT 06269, USA

15.1 Introduction

With the growing size and complexity of software systems[1] and shrinking resources to design, test and maintain them, it is becoming increasingly necessary to determine ways to improve the application reliability in a cost-effective manner. Relating the application reliability to the reliabilities of the components[2] comprising an application can rank components from a reliability perspective, and such ranking can be used to guide an efficient allocation of resources. Further, if such a relationship can be established early in the software life cycle, then it can be used through the software development process to direct effort and resources towards those components that offer the highest return on expended resources. Reliability analysis of a software application based on its architecture can achieve the above two objectives, since it can establish a relationship between the application and component reliabilities from early phases of the software development life cycle [1].

Architecture-based software reliability analysis has received a great degree of attention in the last few years [2–7]. Most of the existing architecture-based analysis techniques, however, ignore interface failures. In modern software applications, interfaces are complex and error-prone. Further, the application components may be deployed across several nodes of a distributed system as opposed to a single node, and the failure of the underlying network connecting these nodes can also lead to application failure. Ignoring interface failures may thus lead to an optimistic estimate of application reliability.

While obtaining an estimate of the application reliability considering its architecture and the component and interface reliabilities for a particular deployment configuration is essential, it is of limited value since it does not provide any guidance

---

[1] The terms system, software system, application and software application are used interchangeably in this chapter.

[2] The terms component and module are used interchangeably in this chapter.

about how to deploy the application components to mitigate the impact of interface failures in order to achieve the highest possible reliability. A systematic method to determine how the components of an application must be deployed, based on the application architecture and component and interface reliabilities to maximize the overall application reliability, is thus essential.

In this chapter, we present a methodology for architecture-based software reliability analysis considering interface failures. The methodology generates an analytical reliability function that expresses application reliability in terms of the reliabilities and visit statistics of the components and interfaces comprising the application. Based on the analytical reliability function, we then present an optimization approach that produces a desirable deployment configuration of the application components given the application architecture and the component and interface reliabilities, subject to two types of constraints. The first type of constraint is the node size constraint and is concerned with the physical limit of the nodes, where a single node cannot accommodate more than a certain maximum number of components. The second type of constraint is the component location constraint, and is concerned with component deployment, where there are restrictions on which components can be deployed on which nodes due to reasons such as architectural mismatch. The optimization framework uses simulated annealing as the underlying optimization technique. We illustrate the value of the analysis and optimization methodologies using several examples.

The layout of this chapter is as follows: Section 15.2 provides an overview of the techniques used in the analysis and optimization methodologies. Section 15.3 presents the analysis and optimization methodologies. Section 15.4 illustrates the methodologies with case studies. Section 15.5 summarizes the related research. Section 15.6 offers concluding remarks and directions for future research.

15.2 Overview

This section provides a brief overview of the techniques used in the analysis and optimization methodologies.

15.2.1 Discrete Time Markov Chains (DTMCs)

A brief overview of DTMCs is provided in this section; a detailed treatment can be obtained from elsewhere [8,9].

A DTMC is characterized by its one-step transition probability matrix, $\mathbf{P} = [p_{i,j}]$. $\mathbf{P}$ is a stochastic matrix since all the elements in a row of $\mathbf{P}$ sum to one, and each element lies in the range $[0, 1]$.

We consider a terminating application, that is, an application which operates on demand. For such an application, it is possible to distinguish one run of the application from the other. An absorbing DTMC may be used to represent the architecture of a terminating application; as a result, we elaborate on absorbing DTMCs in this section.

The transition probability matrix of an absorbing DTMC can be partitioned as:

$$\mathbf{P} = \begin{bmatrix} \mathbf{Q} & \mathbf{C} \\ \mathbf{0} & \mathbf{1} \end{bmatrix}$$

where $\mathbf{Q}$ is a $(n-m) \times (n-m)$ substochastic matrix (with at least one row sum less than 1), $\mathbf{1}$ is a $(m \times m)$ identity matrix, $\mathbf{0}$ is a $m \times (n-m)$ matrix of zeros, and $\mathbf{C}$ is a $(n-m) \times m$ matrix, when there are m absorbing states in the chain with n states.

The k-step transition probability matrix $\mathbf{P}^k$ has the form:

$$\mathbf{P}^k = \begin{bmatrix} \mathbf{Q}^k & \mathbf{C}' \\ \mathbf{0} & \mathbf{1} \end{bmatrix}$$

where the entries of matrix $\mathbf{C}'$ are not relevant. The (i, j)-th entry of matrix $\mathbf{Q}^k$ denotes the probability of arriving in transient state j, starting from transient state i in exactly k steps. It can be shown that $\sum_{k=0}^{t} \mathbf{Q}^k$ converges as t approaches infinity. This implies that the inverse matrix $(\mathbf{I} - \mathbf{Q})^{-1}$, called the fundamental matrix $\mathbf{M}$, exists, and is given by:

$$\mathbf{M} = (\mathbf{I} - \mathbf{Q}^{-1}) = \mathbf{I} + \mathbf{Q} + \mathbf{Q}^2 + \ldots = \sum_{l=0}^{\infty} \mathbf{Q}^l \tag{15.1}$$

Without loss of generality, we assume that the process begins in state 1. We let X_j denote the number of visits to state j starting from state 1 before the process is absorbed. The $(1, j)$-th entry of the fundamental matrix $\mathbf{M}$ represents the expected number of visits to state j starting from state 1 before it is absorbed [9]. The metric denoted v_j is given by:

$$E[X_j] = m_{1,j} = v_j \tag{15.2}$$

The fundamental matrix can also be used to compute the variance of the expected number of visits [8]. Let σ_j^2 denote the variance of the number of visits to state j starting from state 1. Then, we have:

$$\sigma^2 = \mathbf{M}(2\mathbf{M}_{dg} - \mathbf{I}) - \mathbf{M}_{sq} \tag{15.3}$$

where $\mathbf{M}_{dg}$ represents the diagonal matrix of the fundamental matrix $\mathbf{M}$ and $\mathbf{M}_{sq}$ denotes the square of the fundamental matrix $\mathbf{M}$. Thus:

$$Var[X_j] = \sigma_j^2 \tag{15.4}$$

15.2.2 Simulated Annealing

This section provides an overview of simulated annealing and offers motivation for its use in determining a desirable deployment of application components.

Optimization techniques can generally be classified into two categories: exact methods and heuristic methods. Each category has a distinct set of advantages and disadvantages. Exact optimization techniques [10, 11] explore the entire scope of the objective function to identify an optimal solution. In our case, the objective function is the application reliability. Since the application reliability is a function of the number of components in the application and the number of nodes on which these components are to be deployed, the function's scope can grow superlinearly. In the worse case, m^n different configurations must be examined for an application consisting of m components and n nodes. As an illustration, for an application consisting of 15 components and 15 nodes, a total of 4.3789×10^{17} configurations must be examined, which will take a prohibitive amount of time. We further note that this problem does not exhibit the optimal substructure property [12]. That is, we cannot partition the problem into two subsets, each comprising some nodes and components, which could be solved optimally. In this scenario an arbitrary choice of partitioning may lead to inferior solutions, hence all combinations must be explored exhaustively. Thus exact optimization techniques are unsuitable for determining an optimal deployment configuration even for an application of moderate size, which led us to consider heuristic optimization approaches.

We considered four heuristic optimization techniques, namely, Hill climbing, Evolutionary Algorithms, Tabu search, and Simulated Annealing. Hill climbing [13] is one of the more primitive approaches to heuristic optimization. While hill climbing is simple to implement, it often gets "stuck" at local optima that are far from the optimal solution. Evolutionary algorithms [14, 15] maintain a population of structures that evolve according to the rules of selection and other operators, such as recombination and mutation. These algorithms are generally easy to program, but are sensitive to the EA parameters. If the parameters are not chosen correctly, an EA-based technique may not converge to a good solution. Tabu search [16] is an iterative optimization technique in which previous moves are stored in a list, and are referenced to assist in determining future moves. Tabu search is a technique that can be most easily compared to simulated annealing, the technique we selected for our optimization framework. In general, both have been deemed superior to other techniques [17].

Simulated annealing [18] draws its inspiration from the annealing process that many substances undergo while changing state. For our purposes, it is sufficient to say that the simulated annealing algorithm moves iteratively at random across the spectrum of application configurations, unconditionally accepting moves that increase reliability, and accepting moves which decrease reliability with a probability proportional to the simulated temperature value. Simulated annealing is a particularly attractive option since it can be used to optimize objective functions possessing an arbitrary degree of non-linearity, stochasticity, boundary conditions, and constraints. In our case, the degree of non-linearity is a function of the number of nodes and components. Finally, it has been determined through a series of experiments [19, 20] that simulated annealing generally requires less overhead than Tabu search.

15.3 Analysis and Optimization Methodologies

The architecture-based reliability analysis and optimization methodologies incorporating the impact of interface failures are discussed in this section.

15.3.1 Reliability Analysis

The ultimate objective of the reliability analysis methodology is to obtain an analytical function that expresses application reliability in terms of the reliabilities and visit statistics of the components and interfaces. The visit statistics of the components (interfaces) include the mean and the variance of the number of visits to components (interfaces), and these depend on the architecture of the application.

We assume that the application has n components and its architecture is given by its probabilistic control flow graph. We let $w_{i,j}$ denote the probability that control is transferred to component j from component i. The reliability of component i is denoted R_i. If component i can directly transfer control to component j (that is, component j can be executed immediately after the execution of component i, or $w_{i,j} > 0.0$), then interface (i, j) exists in the application, and its reliability is denoted $R_{i,j}$. Conversely, if component i cannot directly transfer control to component j, then interface (i, j) does not exist. In this case $w_{i,j} = 0.0$ and we set $R_{i,j} = 1.0$. We also assume that the components and interfaces fail independently of each other as well as in successive executions.

During a single execution, the application reliability, denoted R, is given by:

$$R = \prod_{i=1}^{n} R_i^{X_i} \prod_{i=1,j=1}^{i=n,j=n} R_{i,j}^{X_{i,j}} \tag{15.5}$$

where X_i is the number of visits to component i, and $X_{i,j}$ is the number of visits to interface (i, j). In (15.5), since the number of visits to components and interfaces, namely X_i's and $X_{i,j}$'s are random variables, R itself is a random variable.

The expected application reliability, denoted $E[R]$, is given by:

$$E[R] = E\left[\prod_{i=1}^{n} R_i^{X_i} \prod_{i=1,j=1}^{i=n,j=n} R_{i,j}^{X_{i,j}}\right] = \prod_{i=1}^{n} E\left[R_i^{X_i}\right] \prod_{i=1,j=1}^{i=n,j=n} E\left[R_{i,j}^{X_{i,j}}\right] \tag{15.6}$$

Based on the Taylor series expression for the mean of a function of a random variable [21], we have:

$$E\left[R_i^{X_i}\right] = R_i^{v_i} + \frac{1}{2}\left(R_i^{v_i}\right)(\log R_i)^2 \sigma_i^2 \tag{15.7}$$

$$E\left[R_{i,j}^{X_{i,j}}\right] = R_{i,j}^{v_{i,j}} + \frac{1}{2}\left(R_{i,j}^{v_{i,j}}\right)(\log R_{i,j})^2 \sigma_{i,j}^2 \tag{15.8}$$

where v_i and σ_i^2 are the mean and the variance of the number of visits to component i, and $v_{i,j}$ and $\sigma_{i,j}^2$ are the mean and the variance of the number of visits to interface (i, j) respectively. A second-order Taylor series is used in (15.7) and (15.8) because the higher orders yield a negligibly small value.

Equation (15.6) can thus be written as:

$$E[R] = \prod_{i=1}^{n} R_i^{v_i} + \frac{1}{2}(R_i^{v_i})(\log R_i)^2 \sigma_i^2 \prod_{i=1,j=1}^{i=n,j=n} R_{i,j}^{v_{i,j}} + \frac{1}{2}\left(R_{i,j}^{v_{i,j}}\right)(\log R_{i,j})^2 \sigma_{i,j}^2 \tag{15.9}$$

In (15.9) the reliability of each interface, similar to the reliability of each component, is represented by a single value, $R_{i,j}$. However, it may be desirable to partition the interface reliability into two portions; $R_{H_{i,j}}$ represents the reliability of the hardware portion of the interface (*i.e.*, the physical connection between the two nodes on which the two components are deployed) and $R_{S_{i,j}}$ represents the reliability of the software portion of the interface (*i.e.*, parameter passing, marshalling and demarshalling of transferred data, *etc.*). Substituting $R_{i,j}$ from (15.10) in (15.9) provides a more fine grained expression expression for the expected reliability of a software system (15.11).

$$R_{i,j} = R_{H_{i,j}} \cdot R_{S_{i,j}} \tag{15.10}$$

$$E[R] = \prod_{i=1}^{n} R_i^{v_i} + \frac{1}{2}\left(R_i^{v_i}\right)(\log R_i)^2 \sigma_i^2 \prod_{i=1,j=1}^{i=n,j=n} (R_{H_{i,j}} \cdot R_{S_{i,j}})^{v_{i,j}}$$
$$+ \frac{1}{2}(R_{H_{i,j}} \cdot R_{S_{i,j}})^{v_{i,j}}(\log R_{H_{i,j}} \cdot R_{S_{i,j}})^2 \sigma_{i,j}^2 \tag{15.11}$$

Equations (15.9) and (15.11) indicate that to obtain the expected application reliability it is necessary to obtain the visit statistics (the mean and the variance of the number of visits) to the components and interfaces. These visit statistics can be obtained by mapping the application architecture represented by its probabilistic control flow graph to a DTMC model and then solving the resulting DTMC model. The mapping can be achieved using the following procedure. First, each application component i is mapped to a corresponding state i in the DTMC model. Additionally, if $w_{i,j} > 0$, indicating that a direct transfer of control from component i to component j is possible, a state (i, j) is added to the model. For each state (i, j) in the DTMC model, two transitions are added. First a transition from state i to state (i, j) is added. The probability of this transition, denoted $p_{i,(i,j)}$, is set to $w_{i,j}$. A second transition is added from state (i, j) to state j. The probability of this transition, denoted $p_{(i,j),j}$, is set to 1.00. Setting the transition probabilities in this manner preserves the behavior of the application architecture as described by its probabilistic control flow graph. The resulting DTMC model can then be solved as discussed in Sect. 15.2 to obtain the mean and the variance of the number of visits to the components and interfaces of the application.

15.3.2 Reliability Optimization

In this section, we describe the optimization methodology which produces a desirable mapping of application components to the nodes of a distributed system to mitigate the impact of interface failures, subject to the following two mapping constraints:

- **Node size constraints:** this specifies an upper bound on the number of components each node of a distributed system can host. An upper bound on the number of components on a single node can be due to the limits on the processing capability of the node.
- **Component location constraints:** this specifies an allowable list of nodes for the deployment of each component. This allowable list usually comprises of a subset of all the possible nodes in the distributed system. Restrictions on component deployment may arise due to several reasons including a mismatch between the hardware and the operating environment of a node and that required by the component.

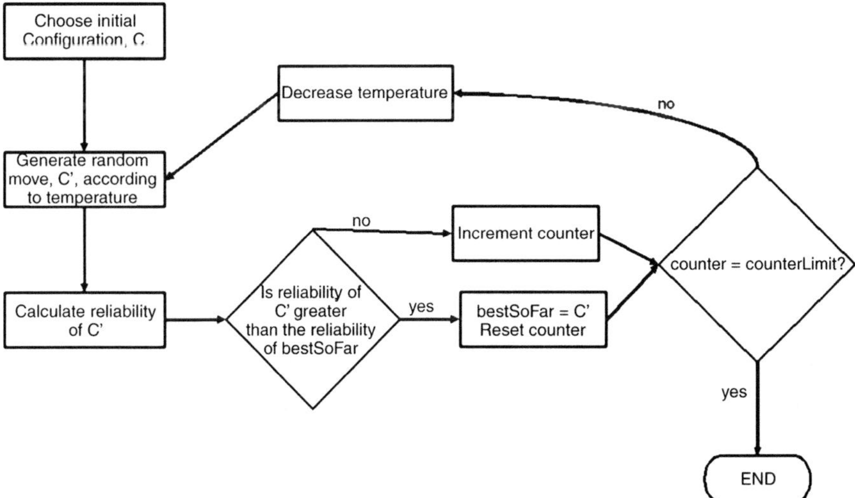

Fig. 15.1 Conceptual flow of the simulated annealing algorithm

A simulated annealing algorithm is defined by its set of potential random moves, employed to shift from one application configuration to another. It makes a series of random moves from the initial application configuration until a proportionate number of scenarios have been investigated. In our particular implementation of the simulated annealing algorithm, we consider only one type of move. Specifically, a move is made by selecting a component and a node separately and at random. If the selected node is different than the current node of the selected component, the move is made. The neighborhood of an application configuration S is then defined as the set of all application configurations obtained by applying a series of zero or more random moves. It should be noted that when a move is made which co-locates

1. find random schedule S;

2. $bestSoFar \leftarrow cost(S)$;

3. $counter \leftarrow 0; phase \leftarrow 0$;

4. **while** $phase \leq maxPhases$ **do**

5. $counter \leftarrow 0$;

6.   **while** $counter \leq counterLimit$ **do**

7.   select a random move m from $neighborhood(S)$;

8.   let S' be the system orientation obtained from S with m;

9.   **if** $reliability(S') > reliability(S)$ **then**

10. $accept \leftarrow$ **true**;

11.   **else**

12. $accept \leftarrow$ **true** with probability exp(-$\Delta R/T$),

13.   **false** otherwise;

14.   **if** $accept$ **then**

15. $S \leftarrow S'$;

16.   **if** $reliability(S') > bestSoFar$ **then**

17. $counter \leftarrow 0; phase \leftarrow 0$;

18. $bestSoFar \leftarrow reliability(S')$;

19.   **else**

20. $counter++$;

21. $phase++$;

22. $T \leftarrow T \cdot \beta$;

Fig. 15.2 Simulated annealing algorithm

two components that are originally placed on different nodes to the same node, the algorithm eliminates only the hardware contribution to the interface reliability. The software contribution to the interface reliability is still considered, since it is a factor regardless of the placement of the components as discussed in Sect. 15.3.

The simulated annealing algorithm, derived directly from [22] and described in complete detail in Fig. 15.2 operates as follows. As in other optimization techniques, the algorithm starts from an initial configuration. The selection of this configuration is trivial, inconsequential, and may be a random mapping of components to nodes. Given a temperature T, the algorithm makes a random move in the neighborhood of the initial configuration. The objective function given by (15.9) is calculated for the new configuration and the variation Δ in the objective function caused by the move is also determined. If $\Delta > 0$ (*i.e.*, the new configuration has a higher reliability than the previous configuration), the move is applied. Otherwise, the non-improving move will be applied with probability $\exp(-\Delta/T)$.

The probability of accepting a non-improving move decreases over time. This behavior is obtained by decreasing the temperature as follows. The algorithm listed in Fig. 15.2 uses a variable *counter* that is incremented for each move that is not applied and reset to zero when the best solution found so far is improved upon. When *counter* reaches a specific limit, the temperature is adjusted to $T \cdot \beta$ (where β is a fixed constant smaller than 1) and the *counter* is reset to zero.

15.4 Illustrations

In this section, we illustrate the analysis and the optimization methodologies. Towards this end, we first describe the characteristics of the applications used for illustration. Subsequently, we demonstrate the analysis methodology followed by the optimization methodology.

15.4.1 Description of Applications

The applications used to illustrate the analysis and the optimization methodologies have the following characteristics.

The first application has seven components and its architecture is shown in Fig. 15.3. The visit statistics to the components and interfaces obtained by solving the DTMC model of the application architecture shown in Fig. 15.4 are summarized in Table 15.1. The small application size for the first case study is chosen deliberately so that the configuration choices can be enumerated exhaustively to determine the optimal configuration which maximizes application reliability. This provides us with comparison points against which the deployment configurations obtained using our optimization framework can be verified.

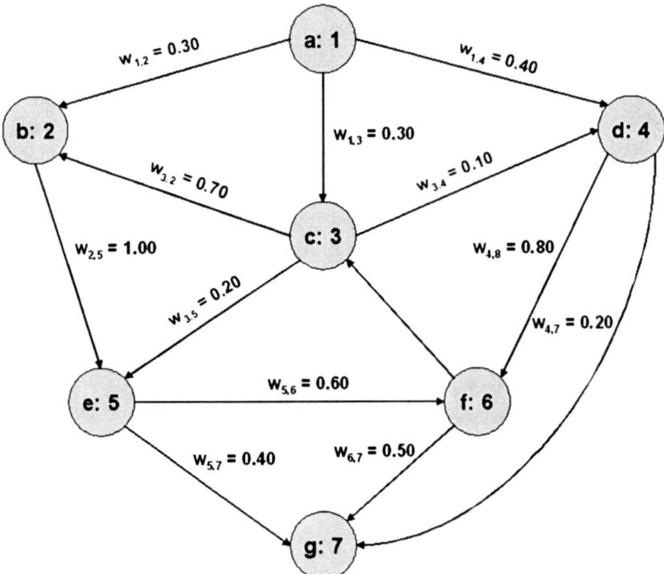

Fig. 15.3 Application architecture for case study I

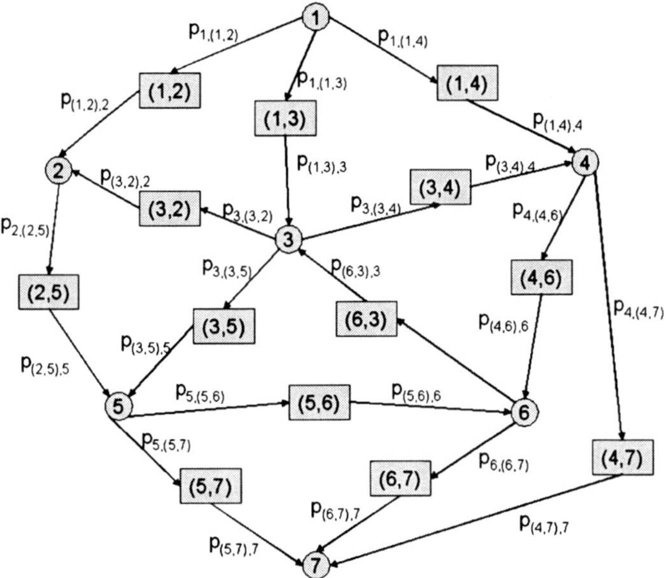

Fig. 15.4 DTMC model for architecture in case study I

In the second case study, we consider an application with eighteen components with the architecture shown in Fig. 15.5. The DTMC resulting from the mapping process is cumbersome to visualize and hence not shown here. The visit statistics

Table 15.1 Visit statistics of components and interfaces (case study I)

Component/Interface	Mean	Variance	Component/Interface	Mean	Variance
1	1.0000	0.0000	2	0.8580	0.6441
3	0.7971	0.8780	4	0.4797	0.3052
5	1.0174	0.7785	6	0.9942	0.8991
7	1.0000	0.0000	$(1,2)$	0.3000	0.2100
$(1,3)$	0.3000	0.2100	$(1,4)$	0.4000	0.2400
$(2,5)$	0.8580	0.6441	$(3,2)$	0.5580	0.5863
$(3,4)$	0.0797	0.0826	$(3,5)$	0.1594	0.1617
$(4,6)$	0.3837	0.2810	$(4,7)$	0.0960	0.0867
$(5,6)$	0.6104	0.7155	$(5,7)$	0.4070	0.2413
$(6,3)$	0.4971	0.6967	$(6,7)$	0.4971	0.2500

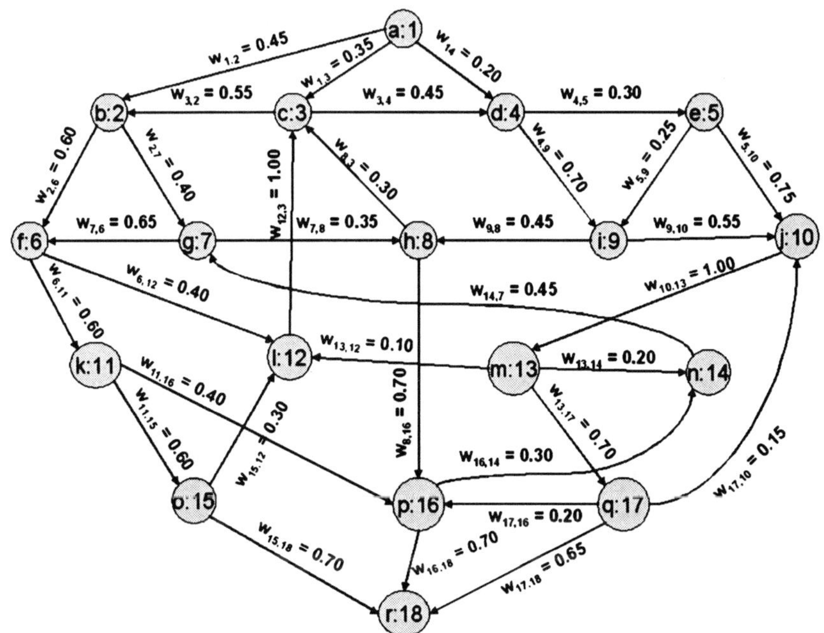

Fig. 15.5 Application architecture for case study II

obtained by solving the DTMC model are listed in Table 15.2. Exhaustive enumeration in the case of the second application is infeasible, since there would be approximately $4 \cdot 10^{22}$ different system configurations. Estimates indicate that a brute-force exhaustive enumeration algorithm executing on a modern computer would take many days to find an optimal system configuration. Thus, the purpose of this case study is to illustrate the value of the optimization framework to find optimal or close to optimal configurations efficiently and expeditiously.

Without loss of generality and for the purpose of illustration, we set the reliability of each component to 0.997, the reliability of the hardware portion of each interface

Table 15.2 Visit statistics of components and interfaces (case study II)

Component/Interface	Mean	Variance	Component/Interface	Mean	Variance
1	1.0000	0.0000	(4, 9)	0.4304	0.3285
2	0.9570	0.8150	(5, 9)	0.0461	0.0449
3	0.9219	1.0485	(5, 10)	0.1383	0.1242
4	0.6148	0.3909	(6, 11)	0.4938	0.3281
5	0.1845	0.1609	(6, 12)	0.3292	0.4114
6	0.8230	0.7522	(7, 6)	0.2488	0.2424
7	0.3828	0.3490	(7, 8)	0.1340	0.1256
8	0.3484	0.3028	(8, 3)	0.1045	0.1160
9	0.4544	0.4112	(9, 8)	0.2144	0.1997
10	0.4922	0.4779	(9, 10)	0.2621	0.2114
11	0.4938	0.3281	(10, 13)	0.4922	0.4779
12	0.4673	0.6436	(11, 15)	0.2963	0.2548
13	0.4922	0.4779	(11, 16)	0.1975	0.1589
14	0.2676	0.2586	(12, 3)	0.4673	0.6436
15	0.2963	0.2548	(13, 12)	0.0492	0.0522
16	0.5638	0.3335	(13, 14)	0.0985	0.0965
17	0.6122	0.5490	(13, 17)	0.3446	0.3206
18	1.0000	0.0000	(14, 7)	0.2676	0.2586
(1, 2)	0.4500	0.2475	(15, 12)	0.0889	0.0949
(1, 3)	0.3500	0.2275	(15, 18)	0.2074	0.1644
(1, 4)	0.2000	0.1600	(16, 14)	0.1692	0.1668
(2, 6)	0.5742	0.5398	(16, 18)	0.3947	0.2389
(2, 7)	0.3828	0.3490	(17, 10)	0.0918	0.1163
(3, 2)	0.5070	0.6599	(17, 16)	0.1224	0.1259
(3, 4)	0.4148	0.3467	(17, 18)	0.3979	0.2396
(4, 5)	0.1845	0.1609			

to 0.993 and the reliability of the software portion of each interface to 0.995. Using these values, the reliability of an interface between two components when they are deployed on the same node is 0.995 and when they are deployed on separate nodes is 0.991.

15.4.2 Analysis Methodology

To demonstrate the importance of factoring interface failures, we compute the application reliability with and without interface failures for both applications and summarize these reliabilities in Table 15.3. Additionally, Table 15.3 displays the reliability of each software application when considering the software portion of the interfaces, but not the hardware portion. These results clearly indicate that the application reliability estimate obtained without considering interface failures is optimistic. This is despite the fact that the reliability of each individual interface itself is significantly high. Furthermore, considering the software and the hardware portions of the interface separately allows an inclusion of interfaces at a finer level of gran-

Table 15.3 Reliabilities with and without interfaces

Case study #	Application reliability		
	without interfaces	with interfaces	with interface software
I	0.98171	0.92288	0.95673
II	0.96761	0.86462	0.92326

ularity in the reliability computation, thereby leading to a more accurate estimate than what can be obtained by using a single value to represent interface reliability.

Next, we seek to analyze the sensitivity of the application reliability to the interface reliabilities through two different sets of experiments for each application. In the first experiment, we set the component reliabilities to 1.00 and vary the reliabilities of both the hardware and software portions of the interfaces from 0.990 to 0.999 in steps of 0.001. In the second experiment, we set the reliabilities of the components and the hardware portion of the interfaces to 1.00 and vary the reliability of the software portion of the interfaces from 0.990 to 0.999 in steps of 0.001. In both the experiments, the reliabilities of all the interfaces were varied simultaneously. In each experiment, the application reliability was computed for each setting of the interface reliabilities using (15.11). It can be observed from Figs. 15.6 and 15.7 that the application reliability decreases as interface unreliability increases. Again, we also see that considering only the software portion leads to an optimistic estimate of

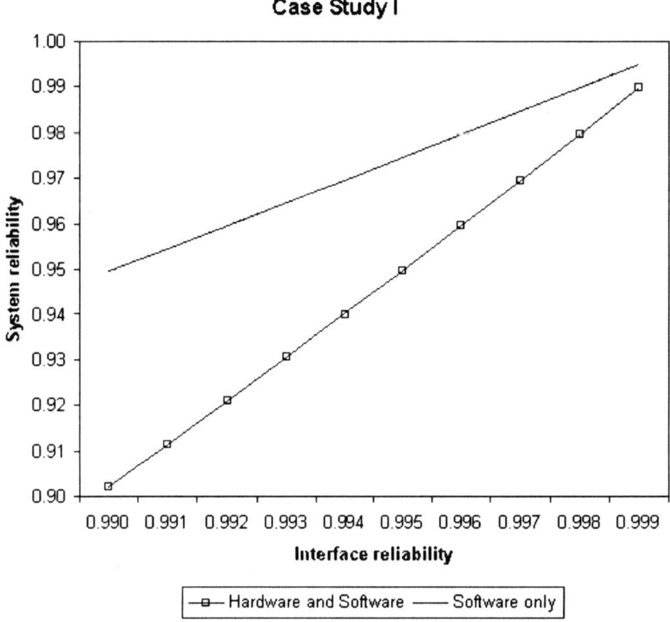

Fig. 15.6 Sensitivity of application reliability to interface reliabilities (case study I)

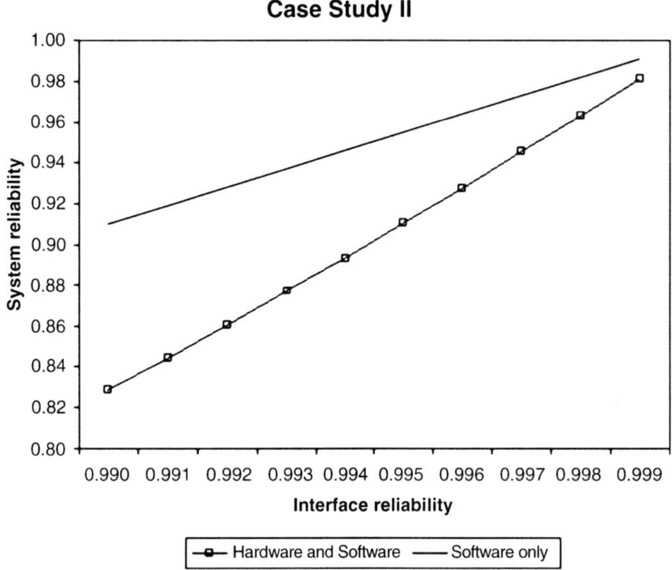

Fig. 15.7 Sensitivity of application reliability to interface reliabilities (case study II)

reliability compared to considering both the hardware and software portions of the interfaces. Although these results are intuitive and expected, the expressions developed in Sect. 15.3 enable a quantification of this influence.

15.4.3 Optimization Methodology

To facilitate a discussion of the results obtained from the optimization methodology, we first describe notation to represent the deployment of components across nodes. We use letters a, b, c, *etc.* to denote nodes, and numbers 1, 2, 3, *etc.* to denote application components. Subsequently, we use the notation, *node*: *component*$_1$, *component*$_2$, ..., *component*$_k$ to denote the deployment of components on a node. For example, the notation a: 1, 2 indicates that components 1 and 2 are mapped to node a. The nominal values for the component and interface reliabilities are assumed to be the same as used to illustrate the analysis methodology.

15.4.3.1 Case Study I

The initial deployment configuration of the application components, using the above notation, is given in Fig. 15.3. We performed a series of different experiments where each experiment is characterized by a unique set of constraints. In each experiment we begin with the initial deployment configuration shown in Fig. 15.3 and explore alternative configurations using both simulated annealing and exhaustive enumeration to arrive at the one which maximizes the application reliability.

Experiment I.1: Boundary Conditions

In this experiment, we consider two extreme scenarios at the two ends of the spectrum. At the one extreme we limit the number of components that can be deployed on each node to one. In this case, each component is placed on a separate node and the configuration which maximizes the application reliability, as determined by both simulated annealing and exhaustive enumeration, is identical to the original mapping shown in Fig. 15.3. The application reliability for this configuration is 0.92288. The other end of the spectrum is the scenario where there is no limit on the number of components that can be deployed on a single node. Additionally, there are no constraints on the assignment of components to nodes. It is intuitive, that given no restrictions, the optimization algorithm will tend to place all the components on one node, thus eliminating the contribution of the hardware portion of the interface. Both the simulated annealing and exhaustive enumeration algorithms produce this configuration, which has a reliability of 0.95673.

Experiment I.2: Location Constraints

In this experiment, the number of components which can be deployed on each individual node is unbounded, however, each component is permitted to reside only on a specific subset of the nodes. The list of allowable nodes for each component is given in Table 15.4. The configuration that maximizes system reliability under these constraints is given by c: 1, 4 and f: 2, 3, 5, 6, 7 and this configuration has a reliability of 0.94898. This configuration is produced both by the simulated annealing and exhaustive enumeration algorithms.

Experiment I.3: Size Constraints

In this experiment, we limit the components which can be allocated to each node. The maximum number of components that can be allocated to node a, b, c, d, e, f and g are 3, 1, 3, 2, 2, 2 and 4 respectively. The application configuration which

Table 15.4 Component location constraints (Experiments I.2 and I.4)

Component #	Node	Component #	Node	Component #	Node
1	a, b, c, d, g	2	b, c, e, f, g	3	a, d, f, g
4	b, c, d, e	5	a, b, c, e, f	6	a, d, e, f, g
7	a, e, f, g				

maximizes the reliability under these constraints is as follows: a: 1, 2, 3; c: 5, 6, 7; and e: 4. This configuration which is obtained both using exhaustive enumeration and simulated annealing has a reliability of 0.94031.

Experiment I.4: Combination of Location and Size Constraints

This experiment combines the restrictions of Experiments II.2 and II.3. Therefore, in this experiment the maximum number of components which can be supported by each node is the same as in the case of Experiment I.3 and the list of allowable nodes for each component is given in Table 15.4. The optimal configuration under these constraints is given by: a: 5, 6, 7; e: 4; and g: 1, 2, 3. This configuration, produced by both exhaustive enumeration and simulated annealing has a reliability is 0.94031.

As expected, the application reliability produced in Experiments I.2, I.3 and I.4 is between the two extreme ends of the reliability spectrum determined in Experiment I.1. Furthermore, as the constraints become more stringent, the optimal reliability of the application decreases. There are two points of interest for each method used for optimization in the above experiments: (i) the ability to maximize the application reliability, and (ii) the execution time of the algorithm. The feasibility of the simulated annealing-based framework is demonstrated by the fact that it obtains the same configurations as those obtained using exhaustive enumeration in all the experiments. The execution times for both these algorithms are listed in Table 15.5. Execution times for the simulated annealing algorithm ran less than 12,610 ms (or 12.6 s), while execution times for exhaustive enumeration ran from 12,468 ms to 216,953 ms (or 12.4 sec to 3.6 min). In each experiment, the execution time of exhaustive enumeration was longer than its simulated annealing counterpart. The simulated annealing algorithm executed between 28% and 1620% faster than exhaustive enumeration. The difference in the execution times of the simulated annealing and exhaustive enumeration is expected to grow dramatically as a function of the number of nodes in the software system. Exhaustive enumeration may take several days to complete even for applications with just eight to nine components, rendering this approach useless for sufficiently large software systems.

Table 15.5 Comparison of execution times (ms) (case study I)

Experiment	Simulated annealing	Exhaustive enumeration
I.1	12,610	216,953
I.2	12,484	123,390
I.3	13,765	20,672
I.4	9719	12,468

15.4.3.2 Case Study II

Following the expanded notation introduced to represent the mapping of components to nodes, Fig. 15.5 depicts the system as a series of hosts (*a* through *r*) and components (1 through 18). Using the deployment shown in Fig. 15.5 as the initial configuration, we conducted a series of four experiments similar to the first case study. The experiments for this application were conducted solely using the simulated annealing algorithm, since exhaustive enumeration is infeasible in this case.

Experiment II.1: Boundary Conditions

As in the first case study, we begin by considering the extreme ends of a spectrum of possible component to node mappings. First, we limit the number of components allowed on each node to one. Consequently, the optimization algorithm has no viable search space to explore and places each component on a separate node. The configuration which maximizes application reliability is therefore identical to the original one given in Fig. 15.5. The application reliability for this configuration is 0.78026. The other end of the spectrum is the scenario in which there is no limit on the number of components which can be co-located on a single node. It is clear that with no restrictions in place, the optimization algorithm will tend to place all the components on one node, thus eliminating the effect of hardware interface failures on the application reliability. The application reliability when all components are placed on the same node is 0.83334.

Experiment II.2: Location Constraints

In this experiment, the number of components which can be supported by each individual node is unbounded. However, each component is permitted to reside only on a specific subset of the nodes. The list of allowable nodes for each component is

Table 15.6 Component location constraints (Experiments II.2 and II.4)

Component #	Allowable hosts	Component #	Allowable hosts
1	*a,b,g,r*	2	*b,c,e*
3	*c,d,e*	4	*a,d,f*
5	*b,c,e,f*	6	*a,b,e,f*
7	*b,e,g,i,j,l*	8	*h,j,n,p,r*
9	*a,b,d,f,i,q*	10	*a,b,f,j,m*
11	*c,d,k,r*	12	*e,f,g,k,l,p*
13	*m,o,p,q*	14	*a,b,l,n,o*
15	*d,l,o,p*	16	*m,n,p*
17	*f,j,q,r*	18	*l,r*

given in Table 15.6. The application configuration which maximizes system reliability under these constraints is given in Fig. 15.8 and has a reliability of 0.80685.

Experiment II.3: Size Constraints

In this experiment we limit the components which can be allocated to each node. The maximum number of components that can be allocated to each node is given in Table 15.7. There are no constraints on mapping of components to nodes. The application configuration which maximizes system reliability under these node size constraints is shown in Fig. 15.8 and has a reliability of 0.81257.

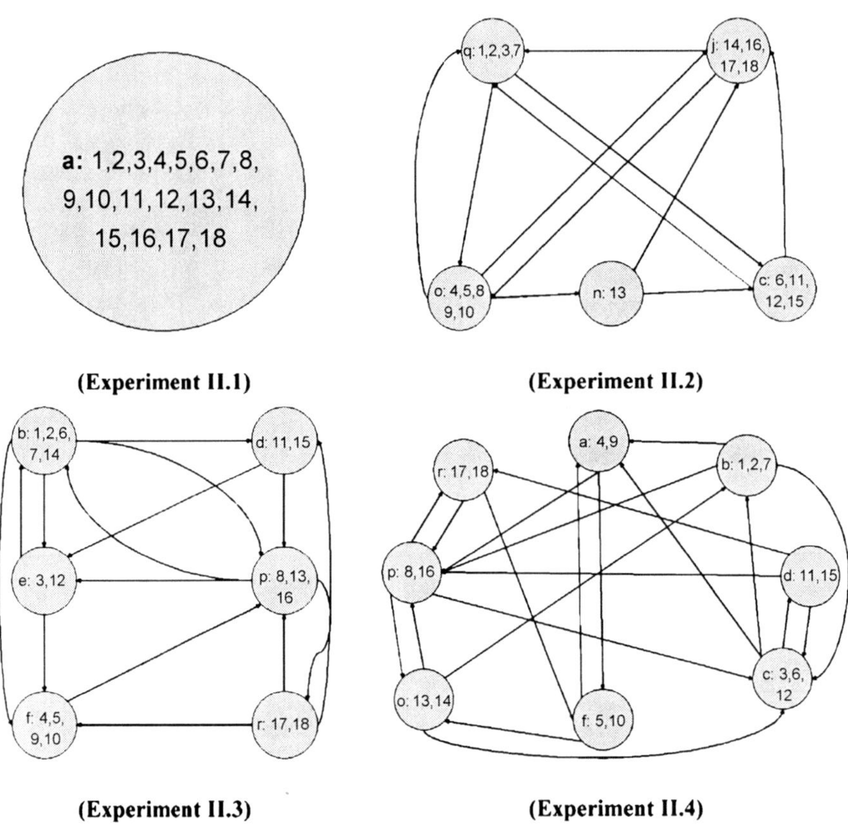

Fig. 15.8 System configurations obtained from simulated annealing (case study II)

Table 15.7 Node size constraints (Experiments II.3 and II.4)

Host #	Max. components	Host #	Max. components
a	2	j	4
b	3	k	2
c	4	l	3
d	2	m	1
e	4	n	2
f	2	o	5
g	2	p	2
h	3	q	4
i	2	r	2

Experiment II.4: Combination of Size and Location Constraints

This experiment combines the restrictions of Experiments II.2 and II.3. Therefore, the maximum number of components which can be supported by each node is given in Table 15.7 and the list of allowable nodes for each component is given in Table 15.6. The application configuration which maximizes the reliability under these constraints is given in Fig. 15.8 and has a reliability of 0.80038.

Similar to the first case study, the application reliabilities obtained in Experiments II.2, II.3, and II.4 lie in between the two extreme reliabilities obtained in Experiment I. This should be expected since each of these experiments consists of some constraints, while the extreme cases consist of either an unconstrained or a maximally constrained system.

The execution time of the simulated annealing algorithm for each of the experiments is listed in Table 15.8. The longest execution time is only 4,255,810 ms. (70.9 min) making the algorithm practical for systems which cannot be optimized using exhaustive enumeration. Furthermore, typical execution times are even lower averaging at only 3,557,488 ms (or 59.3 min).

Table 15.8 Execution time (ms) of simulated annealing – case study II

Experiment	Execution time
II.1	4,255,810
II.2	2,651,953
II.3	3,396,877
II.4	3,925,312

15.5 Related Research

This section summarizes related research and discusses the advantages of our work with respect to the prevalent efforts.

Architecture-based software reliability analysis has been an active area of research in the last few years. However, the majority of the current efforts ignore interface failures. The work closest to the reported research is by Cukic *et al.* [7]. In this research, a UML deployment diagram is used to represent the placement of components across nodes. Execution paths through the control flow graph of the application are enumerated. The reliability of each path is computed as the product of the component and interface reliabilities along the path. The application reliability is then obtained by averaging the path reliabilities. This approach suffers from several disadvantages. First, it cannot consider infinite paths that may be present in the application architecture due to the presence of loops. Second, since the approach is purely computational, it does not easily facilitate sensitivity and predictive analysis and optimization.

Our analysis approach enjoys several advantages compared to the one proposed by Cukic *et al.* [7]. First, it is state-based and hence can consider infinite loops analytically. Second, the analytical reliability function produced facilitates sensitivity and predictive analysis and forms the basis of optimization. Finally, the optimization methodology takes the next step and provides systematic guidance to determine a deployment configuration of application components across the nodes of a distributed system to mitigate the influence of interface failures on application reliability.

15.6 Conclusions and Future Research

This chapter presented a methodology for architecture-based software reliability analysis considering interface failures. A significant advantage of the methodology is that it produces an analytical function which relates the application reliability to the reliabilities and visit statistics of the components and interfaces comprising the application. An optimization methodology, based on the simulated annealing technique, which uses the analytical reliability function as the basis, and provides a deployment configuration of the application components across the nodes of a distributed system for maximal application reliability was also developed. We illustrated the potential of the analysis and optimization methodologies using several examples.

Demonstration of the framework for a practical application where exhaustive enumeration is infeasible, making it necessary to assess the optimality of the deployment configuration using expert judgment is the focus of our future work. Another direction of future research will be the development of an optimization approach which takes into consideration the fractional processing capacities of the nodes. Once these techniques are available, we propose to develop an optimization framework which enables two-way tradeoffs between application performance and reliability. Our future research also focuses on developing techniques to estimate component and interface reliabilities using different software artifacts.

Acknowledgements The research at University of Connecticut was supported by the following grants: (i) Large Grant from the Univ. of Connecticut Research Foundation, and (ii) CAREER award (#CNS-0643971) from the National Science Foundation.

References

1. Cukic B (2005) The virtues of assessing software reliability early. *IEEE Software*, May/June 2005, pp 50–53
2. Gokhale S (2005) Software reliability analysis incorporating second-order architectural statistics. *Intl. Journal of Reliability, Quality and Safety Engineering*, 12(3):267–290
3. Gokhale S, Trivedi KS (2006) Analytical models for architecture–based software reliability prediction: A unifcation framework. *IEEE Trans. on Reliability*, December 2006, 55(4):578–590
4. Goseva-Popstojanova K, Hamill M, Perugupalli R (2005) Large empirical case study of architecture-based software reliability. In: *Proc. of Intl. Symposium on Software Reliability Engineering (ISSRE)*, November 2005, pp 43–52
5. Goseva-Popstojanova K, Kamavaram S (2003) Assessing uncertainty in reliability of component–based software systems. In: *Proc. of Intl. Symposium on Software Reliability Engineering (ISSRE)*, November 2003, pp 307–320
6. Krishnamurthy S, Mathur AP (1997) On the estimation of reliability of a software system using reliabilities of its components. In: *Proc. of Eighth Intl. Symosium on Software Reliability Engineering (ISSRE)*, November 1997, Albuquerque, New Mexico, pp 146–155
7. Yacoub S, Cukic B, Ammar HH (2004) A scenario based reliability analysis approach for component based software. *IEEE Trans. on Reliability*, December 2004, 53(4):465–480
8. Kemeny JG, Snell JL (1960) *Finite Markov Chains*. Van Nostrand Reinhold, New York
9. Trivedi KS (2001) *Probability and Statistics with Reliability, Queuing and Computer Science Applications*. John Wiley, 2nd edition
10. Dennis JE, Schnabel R (1983) *Numerical Methods for Unconstrained Optimization and Nonlinear Equations*. Prentice Hall, Englewood Cliffs, NJ, USA
11. Luenberger DG (1984) *Linear and Nonlinear Programming, Second Edition*. Addison-Wesley, Reading, Massachusetts
12. Cormen T, Leiserson C, Rivest R (1991) *Introduction to algorithms*. McGraw Hill Inc.
13. Greiner R (1992) Probabilistic hill-climbing: Theory and applications. In: *Proc. of the Ninth Canadian Conference on Artifcial Intelligence*, pp 60–67, Vancouver, 1992. Morgan Kaufmann
14. Fogel LJ, Owens A, Walsh MJ (1966) *Artifcial Intelligence Through Simulated Evolution*. Wiley Publishing, New York
15. Holland JH (1975) *Adaptation in Natural and Artifcial Systems*. University of Michigan Press, Ann Arbor
16. Glover F, Laguna F (1997) *Tabu Search*. Kluwer Academic Publishers, Norwell, MA, USA
17. Lee Y, Ellis JH (1996) Comparison of algorithms for nonlinear integer optimization: Application to monitoring network design. *Journal of Environmental Engineering*, pp 524–529
18. Kirkpatrick S, Gelatt CD, Vecchi MP (1983) Optimization by Simulated Annealing. *Science, Number 4598, 13 May 1983*, 220, 4598:671–680
19. Battiti R, Tecchiolli G (1994) Simulated annealing and tabu search in the long run: A comparison on qap tasks. *Computer Math. Applic.*, 28(6):1–8

20. Paulli J (1993) Information utilization in simulated annealing and tabu search. *COAL Bulletin*, 22(28–34)
21. Bain LJ, Engelhardt M (1980) *Introduction to Probability and Mathematical Statistics*. Duxbury Press, Belmont, CA, 1980.
22. Anagnostopoulos A, Michel L, Hentenryck PV, Vergados Y (2003) A simulated annealing approach to the traveling tournament problem. In: *Proc. of Intl. Conference on the Integration of Constraint Programming, Artifcial Intelligence and Operations Research*

Chapter 16
Software Reliability Growth Models Based on Component Characteristics

Takaji Fujiwara[1], Shinji Inoue[2], Shigeru Yamada[2]

[1] Development Department 2, Development Division, Fujitsu Peripherals Limited
35 Saho, Katoh-shi, Hyogo 673-1447, Japan
[2] Department of Social Systems Engineering, Faculty of Engineering, Tottori University
4-101 Minami, Koyama-cho, Tottori-shi, Tottori 680-8552, Japan

16.1 Introduction

In recent highly informative society, computer systems are used in various fields, and play an important role. With regard to computer systems, high reliability has been achieved with hardware systems due to reliability techniques. On the other hand, it is very difficult to obtain high reliability of the software system, so that the reliability of the entire system has been still unsatisfactory in the software field compared with hardware. Once system failure occurs due to latent faults in the software system, the computer system is entirely useless and considerable damage may be sustained. Occasionally, critical faults may develop that pose a serious threat to human life. Therefore, one important issue is to develop highly reliable software systems.

As one of the solutions to this problem, we have used software reliability assessment technologies of software reliability growth models (abbreviated as SRGMs) [1–4]. In these technologies, stochastic models based on non-homogeneous Poisson processes (abbreviated as NHPPs) have been often used. These SRGMs describe the time-dependent behavior of the cumulative number of faults detected in the testing phase of software development and the operation phase, which is regarded as the software reliability growth process. Thus, they are intended for reliability assessment of the overall software system. This is because conventional SRGMs have been built based on the following three basic assumptions: [5–7]

(1)   The software system consists of one domain.
(2)   The testing-domain expands with progress of the testing.
(3)   The testing-domain is extended to the overall software system.

The testing domain means the set of testing paths in the software system influenced by executed test cases in the testing. These basic assumptions can be changed as follows by considering the component-based structure of the software system:

(1)   Many software systems consist of multiple sub-components (abbreviated as SCs) and a main-component (abbreviated as MC) to call them. In this case,

the set of testing paths influenced by execution of test cases exists within each component.

(2) Unless an MC is tested, the SC called from it is not tested. Therefore, the testing domain may not be extended with testing progress.

(3) Because the testing domain exists in each component, the testing domain should eventually be expanded to the overall component, but may not be extended to the whole software system.

Moreover, because the development size, the degree of complexity, and the importance of each implemented component differ from each other, these testing efforts are also different. Therefore, to describe the software fault-detection phenomenon more precisely, we need to develop and apply a suitable SRGM to each component.

In this chapter, we propose SRGMs with the characteristics of each component to overcome the above issues. Then, in order to take these properties of each component into consideration, we discuss our notion of the module composition (MC and SCs) of a software system in Sect. 16.2. In Sect. 16.3, our SRGMs are formulated based on NHPPs by considering the component characteristics discussed in Sect. 16.2, and the estimation method of model parameters is also mentioned. Finally, in Sect. 16.4, by applying fault-count data observed in the actual software development projects, we show numerical examples of software reliability assessment and results of goodness-of-fit comparison of our SRGMs with conventional SRGMs based on certain criteria.

16.2 Module Composition

In recent years, the internal structure of software systems has been changing with various development methods. We can give some examples, such as the Windows applications accompanying the spread of Windows OS, the Web applications accompanying rapid progress of the Internet technique, and so on. However, even if the style of these software system has been changing, the situation of the fundamental module composition is almost the same. A processing requirement unit has shifted to the component instead of the function. That is, an SC is called from a MC, and then the processing expected in the SC is performed.

Now, let us take Microsoft WORD® to explain the notion of module composition in this chapter. An MC indicates the fundamental function of the software system, as shown in Fig. 16.1. That is, we define the field of text input, which is the main function of a word processor and the field of menu bar, which calls each function as an MC. We also define an SC as the function displayed by clicking the alternative of the menu bar. For example, Fig. 16.2 shows the dialog of the SC functions displayed by clicking the "Page Setup", which is the alternatives in the "File" menu bar. We note that we can divide the alternatives in the "File" menu of Microsoft WORD® into the following functions:

- The functions which can be processed only by an SC (Print, Save, and so on),
- The functions in which MC and SC influence each other (Page Setup, and so on).

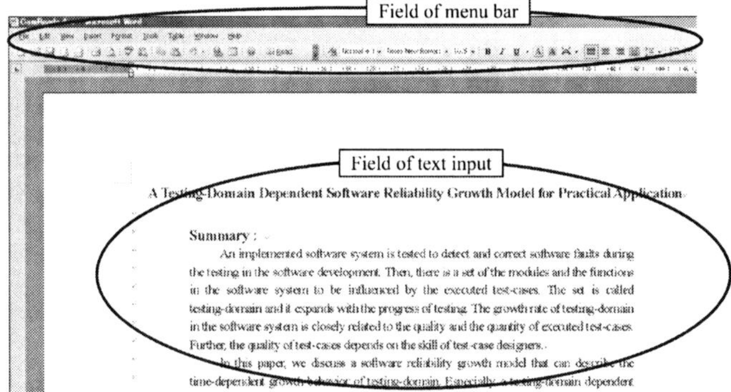

Fig. 16.1 The main function of Microsoft WORD

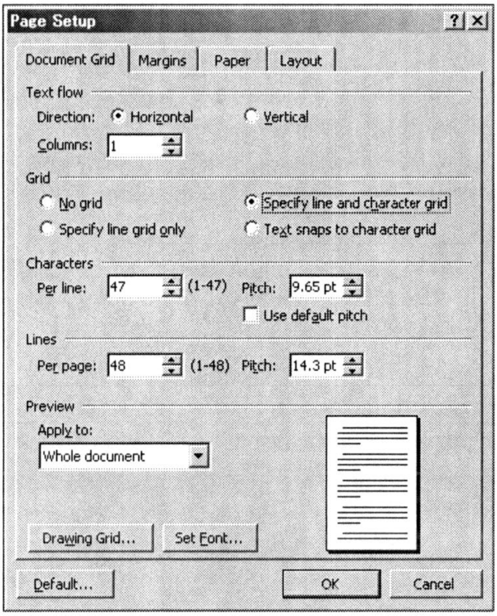

Fig. 16.2 The displayed SC functions ("Page Setup" dialog)

Since a general software product comprises the module composition shown in Fig. 16.3, we can consider that each SC has structures that are called from an MC, and performs selected requirement processing. Therefore, the latent faults in such a software system are classified into the following:

- **Surface fault**: The faults latent in a MC, which can always be detected,
- **Concealment fault**: The faults latent in SCs. And if the SCs are not called from a MC, the faults cannot be decteded.

Then, we can assume that the degree of testing-coverage for an MC affects the realization of high quality of a SC from Fig. 16.3.

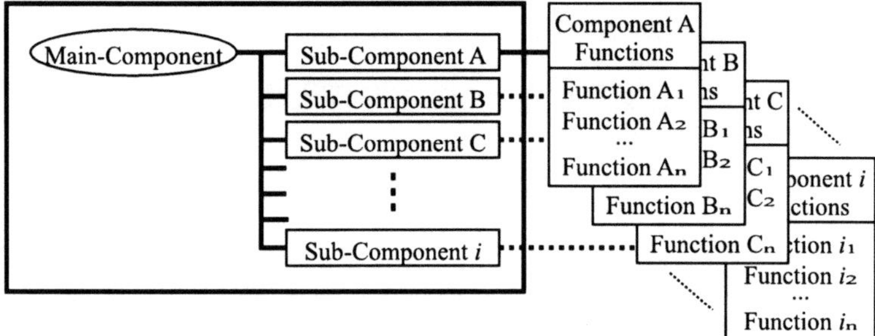

Fig. 16.3 Summary of module composition

16.3 Software Reliability Growth Modeling

In this section, we discuss software reliability growth modeling based on an NHPP with the characteristics of module composition discussed in Sect. 16.2. In order to describe the fault-detection phenomenon, we now define $\{N(t), t \geq 0\}$ as a counting process representing the cumulative number of faults detected up to testing time t. In this chapter, we assume that $\{N(t), t \geq 0\}$ follows an NHPP formulated as:

$$\Pr\{N(t) = n\} = \frac{\{H(t)\}^n}{n!} \exp[-H(t)] \ (n = 0, 1, 2, \cdots),$$

$$H(t) = \int_0^t h(x) \, \mathrm{d}x, \tag{16.1}$$

where $\Pr\{A\}$ is defined as the probability of event A, and $H(t)$ is a mean value function of an NHPP that represents the expectation of $N(t)$. Furthermore, $h(t)$ in (16.1) is called an intensity function of an NHPP, which represents the instantaneous fault detection rate at testing time t.

16.3.1 Basic SRGM Based on Component Characteristics

We propose an SRGM with the characteristics of module composition discussed in Sect. 16.2. To start with, we set the following assumptions in order to formulate the time-dependent behavior of the fault-detection phenomenon of each component (MC and SCs):

(1) The detected faults are corrected and removed without introduction of new faults.
(2) The latent faults in a component are distributed uniformly over the component.
(3) The number of faults detected within a component during $(t, t + \Delta t]$ is proportional to the number of remaining faults in the component at testing time t.

(4)   The SC is tested when it is called from MC.

(5)   The fault-detection rate in the SC depends on the degree of testing-progress indicated by the testing path coverage, and so forth, of MC.

From these assumptions, we can obtain the following differential equations for MC and SC_i $(i = 1, 2, \cdots, j)$ respectively:

$$\frac{dn_m(t)}{dt} = \beta\,[N_m - n_m(t)] \quad (N_m > 0,\ 0 < \beta < 1), \tag{16.2}$$

$$\frac{dn_{s_i}(t)}{dt} = b_{s_i}(t)\,[N_{s_i} - n_{s_i}(t)] \quad (N_{s_i} > 0,\ 0 < b_{s_i}(t) < 1), \tag{16.3}$$

where

$$
\begin{aligned}
i &= \text{the index number of SCs,} \\
N_m &= \text{the initial fault content in MC,} \\
N_{s_i} &= \text{the initial fault content in } SC_i, \\
\beta &= \text{the proportionality constant, which represents the fault detection rate} \\
&\quad\ \text{per fault remaining in MC,} \\
b_{s_i}(t) &= \text{the fault detection rate per fault remaining in } SC_i \text{ at the testing time } t, \\
n_m(t) &= \text{the cumulative number of faults detected up to testing time } t \text{ within MC,} \\
n_{s_i}(t) &= \text{the cumulative number of faults detected by testing time } t \text{ in } SC_i.
\end{aligned}
$$

From assumption (5), the fault detection rate per fault remaining in SC_i at the testing time t, $b_{s_i}(t)$, can be defined as follows [6]:

$$b_{s_i}(t) = \frac{\beta}{N_m} n_m(t). \tag{16.4}$$

Thus, by solving the differential equations (16.2) and (16.3) with respect to $n_m(t)$ and $n_{s_i}(t)$ under the initial conditions $n_m(0) = 0$, $n_{s_i}(0) = 0$, and (16.4), we can obtain the following mean value functions for MC and SC_i respectively:

$$n_m(t) = N_m\,\{1 - \exp[-\beta t]\}, \tag{16.5}$$

$$n_{s_i}(t) = N_{s_i}\,\{1 - \exp[1 - \beta t - \exp(-\beta t)]\}. \tag{16.6}$$

Consequently, the expected cumulative number of faults detected up to testing time t for the whole software system, $H(t)$, can be represented as follows:

$$H(t) \equiv H_b(t) = n_s(t) + \sum_{i=1}^{j} n_{ci}(t). \tag{16.7}$$

An SRGM in (16.7) is called BCC-SRGM and represented as $H_b(t)$. BCC-SRGM in (16.7) is reflecting the components' characteristics since this SRGM can estimate the inherent parameters for each component which are MC and SCs.

16.3.2 Generalization of BCC-SRGM

In this section, we discuss generalized software reliability growth modeling by improving the BCC-SRGM discussed in the previous section. Before the modeling, we have to consider the difference of the testing periods of MC and SCs. That is, it should be noted that the testing period of MC is longer than that of each SC. Then, considering the fault detection rate of each SC is not same, we can change (16.4) into:

$$b_{s_i}(t) = \frac{b(t)}{N_m} n_m(t) \cdot p_i \quad (i = 1, 2, \cdots, j \,;\, p_i > 0), \tag{16.8}$$

by incorporating the reaction factor p_i $(i = 1, 2, \cdots, j)$. In this chapter, we define the reaction factor, p_i, as the ratio of the number of test cases corresponding to the component among the total number of test-cases as follows:

$$p_i = \frac{\text{Total number of test-cases for SC}i}{\text{Total number of test-cases of whole system}} \times (\text{Newly developed rate}). \tag{16.9}$$

Furthermore, the fault detection rate is not necessarily constant throughout the testing phase. That is, since it is known that the fault detection rate is related to the learning phenomenon, we have to consider the function of the time-dependent behavior of the fault detection rate, $b(t)$, instead of β in (16.4) of the BCC-SRGM. In this chapter, $b(t)$ means the fault detection rate per fault remaining in MC at the testing time t [1–4]. Therefore, we employ the following three kinds of typical fault detection rate functions into $b(t)$:

- **CFDR** (Constant Fault Detection Rate): The fault detection rate is constant throughout the testing.

$$b(t) \equiv b_c(t) = \beta \ (\beta > 0). \tag{16.10}$$

- **DFDR** (Decreasing Fault Detection Rate): The fault detection rate is a monotonically decreasing function with respect to the testing time t.

$$b(t) \equiv b_d(t) = \beta \cdot \exp[-\beta t] \ (\beta > 0). \tag{16.11}$$

- **IFDR** (Increasing Fault Detection Rate): The fault detection rate is a monotonically increasing function with respect to the testing time t.

$$b(t) \equiv b_i(t) = \frac{\beta^2 t}{1 + \beta t} \ (\beta > 0). \tag{16.12}$$

In the above equations (16.10)–(16.12), β means the fault detection rate per fault remaining in the software system.

Thus, we describe more realistically the time-dependent behavior of the fault detection phenomenon in MC by introducing the following assumption:

(6)   The fault detection rate is not necessarily constant throughout the testing.

From this assumption, we propose the following differential equation for MC:

$$\frac{\mathrm{d}n_m(t)}{\mathrm{d}t} = b(t)\left[N_m - n_m(t)\right]. \tag{16.13}$$

Consequently, by substituting (16.10)–(16.12) for $b(t)$ into the differential equation in (16.13), and solving with respect to $n_m(t)$ under the initial condition $n_m(0) = 0$, we can obtain the following mean value functions for MC, respectively:

$$n_m(t) \equiv n_{m_c}(t) = N_m \left\{ 1 - \exp\left[-\int_0^t b_c(x)\,\mathrm{d}x \right] \right\}$$
$$= N_m \left\{ 1 - \exp[-\beta t] \right\}, \tag{16.14}$$

$$n_m(t) \equiv n_{m_d}(t) = N_m \left\{ 1 - \exp\left[-\int_0^t b_d(x)\,\mathrm{d}x \right] \right\}$$
$$= N_m \left\{ 1 - \exp[-(1 - \exp[-\beta t])] \right\}, \tag{16.15}$$

$$n_m(t) \equiv n_{m_i}(t) = N_m \left\{ 1 - \exp\left[-\int_0^t b_i(x)\,\mathrm{d}x \right] \right\}$$
$$= N_m \left\{ 1 - (1 + \beta t)\exp[-\beta t] \right\}. \tag{16.16}$$

Then, by substituting (16.14)–(16.16) for the mean value function of MC, $n_m(t)$, into the fault detection rate function, $b_{s_i}(t)$, in (16.8), and solving the differential equation in (16.3) with respect to $n_{s_i}(t)$ under the initial condition $n_{s_i}(0) = 0$, we can obtain the following mean value functions for SC_i according to each fault detection rate function respectively:

$$n_{s_i}(t) \equiv n_{sc_i} = N_{s_i} \left\{ 1 - \exp\left[-p_i \left(e^{-\beta t} - \beta[1 - t] \right) \right] \right\}, \tag{16.17}$$

$$n_{s_i}(t) \equiv n_{sd_i} - b_d(t) \cdot N_{s_i} \left\{ 1 - \exp\left[-p_i \left(e^{-(1 - e^{-\beta t})} - e^{-\beta t} \right) \right] \right\}, \tag{16.18}$$

$$n_{s_i}(t) \equiv n_{si_i} = N_{s_i} \left\{ 1 - \exp\left[-p_i \left\{ (2 + \beta t)\, e^{-\beta t} - (2 - \beta t) \right\} \right] \right\}. \tag{16.19}$$

However, when calculating the mean value function for SC_i, we have to use the estimated value based on each SC_i's testing time t_{i_l} $(l = 1, 2, \cdots, t)$. Consequently, the expected cumulative number of faults detected up to testing time t for the whole software system, $H(t)$, can be represented as follows:

$$H(t) = n_m(t) + \sum_{i=1}^{j} n_{s_i}\left(t_{i_l}\right), \quad (i = 1, 2, \cdots, j,\ l = 1, 2, \cdots, t). \tag{16.20}$$

Henceforth, SRGMs applied CFDR, DFDR, and IFDR are represented as $H_c(t)$, $H_d(t)$, and $H_i(t)$, respectively, in this chapter.

16.3.3 Parameter Estimation

The reliability growth parameters, N_m, N_{si}, and β, in mean value function, $H(t)$, in (16.7) and (16.20) can be estimated by the method of maximum likelihood. We assume that data pairs, (t_k, y_k) $(k = 1, 2, \cdots, n; 0 < t_1 < t_2 < \cdots < t_n)$, has been observed in the testing phase of software development, where y_k is the cumulative number of faults detected in a given testing time interval $(0, t_k]$. Then, based on the NHPP formulated by (16.1), the joint probability mass function, *i.e.*, the likelihood function, L, is given by

$$
\begin{aligned}
L &= \Pr\{N(t_1) = y_1, N(t_2) = y_2, \cdots, N(t_n) = y_n\} \\
&= \prod_{k=1}^{n} \frac{\{H(t_k) - H(t_{k-1})\}^{(y_k - y_{k-1})}}{(y_k - y_{k-1})!} \exp[-H(t_n)],
\end{aligned}
\tag{16.21}
$$

where $t_0 \equiv 0$ and $y_0 \equiv 0$. Equation (16.21) is transformed into the log-likelihood function. The reliability growth parameters, N_m, N_{si}, and β, can be estimated by solving the simultaneous likelihood equation of each system numerically. The reliability growth parameters, N_m and β, can be estimated with the fault-count data observed in the MC testing, and N_{si} estimated by using the fault-count data obtained by each SC testing.

16.4 Numerical Examples for Software Reliability Analysis

16.4.1 Estimation of Model Parameters

In this section, we apply the proposed SRGMs to the fault-count data observed in the testing-phase of actual software development projects in order to show the applicability of our models. We adopt two data sets called DS1 and DS2 for numerical examples in this chapter. These data consist of 33 and 70 data pairs for the whole software system respectively. These data sets have been observed in the actual testing phases, and shown in the following:

DS1: (t_k, y_k) $(k = 1, 2, \cdots, 33 \; ; \; t_k \text{ (days)})$,
DS2: (t_k, y_k) $(k = 1, 2, \cdots, 70 \; ; \; t_k \text{ (days)})$,

where the measurement unit of testing time t_k represents calendar days respectively. DS1 has been observed from the WEB application development, which manages PCs working within the system. In addition, this WEB application has been programmed by JAVA using Eclipse development environment, consists of an MC and 8 SCs, and about 180 faults have been detected so far. On the other hand, DS2 has been observed from the WEB application development, which supports system engineers. This WEB application has also been programmed by Visual Basic using the .NET development environment, consisting of an MC and 5 SCs, and about 170

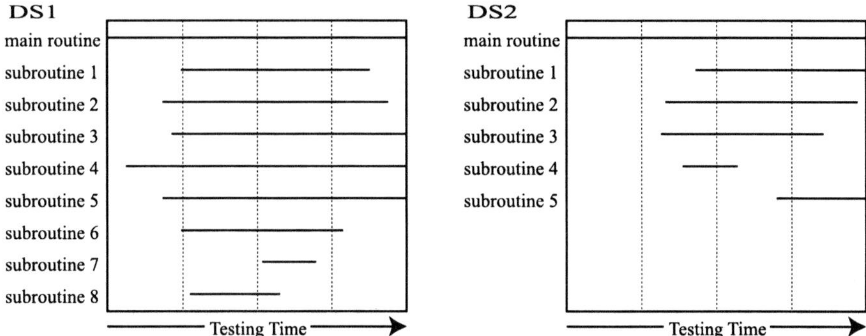

Fig. 16.4 The testing periods for MC and each SC in DS1 and DS2

Table 16.1 Results of model comparisons based on the MSE

| | | Proposed SRGMs | | | Exponen- | Delayed | T-DS |
	$H_b(t)$	$H_c(t)$	$H_d(t)$	$H_i(t)$	tial	S-shaped	SRGM
DS1	18.01	17.37	**6.31**	12.11	245.57	53.39	25.77
DS2	**10.31**	16.77	31.70	69.65	34.17	101.99	13.89

faults have already been detected. Furthermore, testing processes for a MC and each SC in DS1 and DS2 are shown in Fig. 16.4.

16.4.2 Goodness-of-fit Comparisons

Now, we compare the proposed SRGMs with three conventional NHPP models, an exponential SRGM [8], a delayed S-shaped SRGM [9], and a testing-domain dependent SRGM with skill factor (abbreviated as T-DS SRGM) [6,7]. As a criterion for these goodness-of-fit comparisons, we use the <u>m</u>ean <u>s</u>quared <u>e</u>rrors (abbreviated as MSE) [2, 3], and the <u>A</u>kaike's <u>I</u>nformation <u>C</u>riterion (abbreviated as AIC) [10]. Letting n be the observed number of data, M the number of parameters in a SRGM, and $\widehat{H}(t)$ the estimated mean value function, we can calculate the value of the MSE as follows:

$$\text{MSE} = \frac{1}{n-M} \sum_{k=1}^{n} \left[y_k - \widehat{H}(t_k) \right]^2. \tag{16.22}$$

Accordingly, we can conclude that the SRGM having the smallest value of the MSE fits best to the observed data set. The comparison results based on the MSE are shown in Table 16.1.

Next, the AIC can be calculated by the following equation:

$$\text{AIC} = 2M - 2 \cdot (\text{the logarithmic maximum likelihood of an SRGM}). \tag{16.23}$$

Table 16.2 Results of model comparisons based on the AIC

		Proposed SRGMs			Exponen-	Delayed	T-DS
	$H_b(t)$	$H_c(t)$	$H_d(t)$	$H_i(t)$	tial	S-shaped	SRGM
DS1	136.45	136.29	**130.79**	135.50	220.62	192.26	188.96
DS2	**249.66**	258.65	260.93	299.85	281.35	315.94	290.57

When the values of AIC for compared SRGMs are calculated, if the difference among the calculated values is 1–2 or more, we can judge that the difference is significant and the SRGM with the small value has good suitability. If the difference in the AIC is 1 or less, the superiority or inferiority of compared SRGMs cannot be judged, and the same degree of goodness-of-fit is meant. The results of model comparisons based on the AIC are shown in Table 16.2.

From Tables 16.1 and 16.2, we can say that our SRGMs fit better to the actual fault-count data than conventional SRGMs in terms of two comparison criteria. We can also see that proposed SRGM based on DFDR, $\widehat{H_d}(t)$, has the best goodness-of-fit, especially to the DS1 indicating an S-shaped growth curve. On the other hand, in DS2 in which the software reliability is not sufficiently high, the BCC-SRGM, $\widehat{H_b}(t)$, indicates the best goodness-of-fit for the both comparison criteria.

Based on the estimation results for DS1 and DS2, the estimated mean value functions $\widehat{H_d}(t)$ and $\widehat{H_b}(t)$ that represents the best suitability to the arranged actual fault-count data and its 90% confidence limits are shown in Figs. 16.5 and 16.6 respectively. We can also observe a good fit to DS1 and DS2 from Figs. 16.5 and 16.6 respectively.

In this way, we find that the reliability growth curve that fully reflected each component's characteristics can be represented by the proposed SRGM, $\widehat{H_c}(t)$, in

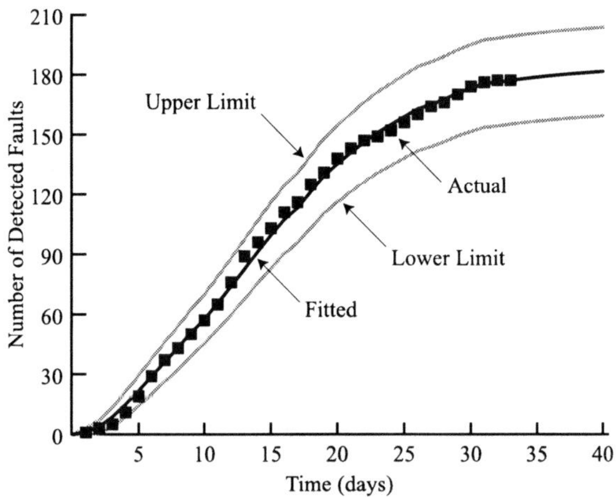

Fig. 16.5 Estimated mean value function $\widehat{H_d}(t)$ for DS1

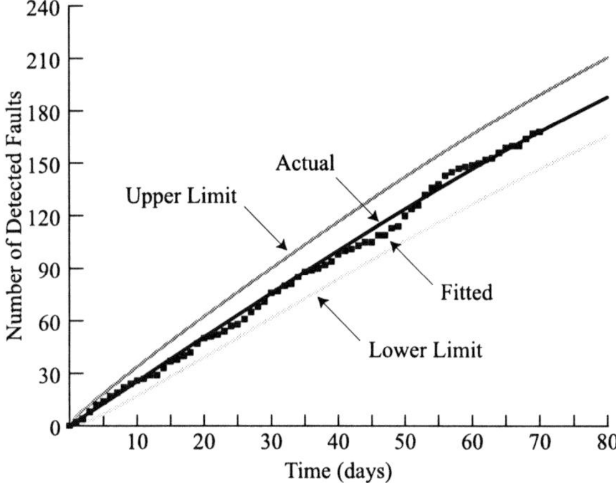

Fig. 16.6 Estimated mean value function $\widehat{H_b}(t)$ for DS2

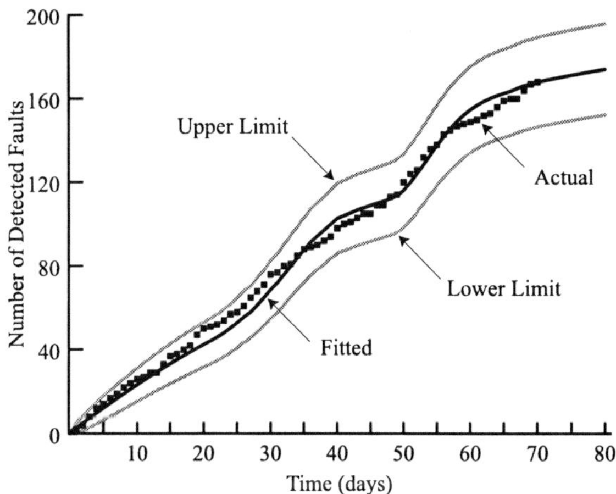

Fig. 16.7 Estimated mean value function $\widehat{H_c}(t)$ for DS2

Fig. 16.7. That is, in Fig. 16.7, we can say that the characteristics of the fault detection phenomenon of SC_3 and SC_5 that cannot be expressed by conventional SRGMs are reflected. Thus, from the above-mentioned results, without increasing model parameters in an SRGM, we can consider that highly accurate estimation is possible based on the proposed models.

16.5 Concluding Remarks

In this chapter, we have proposed a BCC-SRGM, and three kinds of generalized SRGMs incorporating three kinds of fault detection rate functions, such as CFDR,

DFDR, and IFDR, reflecting the time-dependent behavior of a fault detection rate in order to extend BCC-SRGM, based on an NHPP in consideration of characteristics of module composition of the software system. In particular, these SRGMs have been built with the relationship between an MC and SCs. Furthermore, the generalized SRGMs have been incorporated into the reality that each SC's testing time is not equivalent to the software system's testing time. In addition, two data sets observed in the actual development projects of our company have been applied to these SRGMs. Next, numerical illustrations on software reliability analysis of the actual data have been presented. As a result of these model comparisons with the existing SRGMs, we have seen that the proposed SRGMs had better goodness-of-fit to the actual fault-count data. Among those, we have obtained the result that our SRGM based on the DFDR represents the best suitability to the actual fault-count data, DS1, which indicates the S-shaped reliability growth curve, and the BCC-SRGM shows the best suitability to the actual fault data, DS2, which still lies in the middle of growth. We have also shown that our SRGMs proposed in this chapter could describe the reliability growth curve reflecting each SC's characteristics.

In future studies, we need to apply many actual fault data sets to these SRGMs, and to investigate the effectiveness of these SRGMs more. Moreover, in order to improve the accuracy of the proposed SRGMs, we must apply the testing domain theory [4–7].

References

1. Yamada S, Takahashi M (1993) Introduction to Software Management Model. Kyoritsu-Shuppan, Tokyo
2. Yamada S (1994) Software Reliability Models – Fundamentals and Applications. JUSE Press, Tokyo
3. Pham H (2000) Software Reliability. Springer-Verlag, Singapore
4. Yamada S, Fujiwara T (2004) Software Reliability: Model, Tool, Management. Society of Project Management, Chiba
5. Ohtera H, Yamada S, Narihisa H (1990) Software reliability growth model for testing-domain. Transactions of the IEICE J73-D-I:170–174
6. Fujiwara T, Yamada S (2000) Software reliability growth modeling based on testing-skill characteristics: model and application. Transactions of the IEICE-A J83-A:188–195
7. Fujiwara T, Yamada S (2001) Testing-domain dependent software reliability growth models and their comparisons of goodness-of-fit. Proceedings of the Seventh ISSAT International Conference on Reliability and Quality in Design, pp 36–40, August
8. Goel AL, Okumoto K (1978) A time dependent error detection rate model for a large scale software system. Proceedings of the 3rd USA-Japan Computer Conference, pp 35–40
9. Yamada S, Osaki S (1985) Software reliability growth modeling: models and applications. IEEE Trans Software Engineering SE-11:1431–1437
10. Akaike H (1976) What is the information criterion AIC? (in Japanese) Suri Kagaku 14:5–11

Part IV
Quality Engineering in Design

Chapter 17
Statistical Analysis of Appearance Quality for Automotive Rubber Products

Shigeru Yamada, Kenji Takahashi

Department of Social Systems Engineering,
Faculty of Engineering, Tottori University,
Tottori-shi, 680-8552 Japan

17.1 Introduction

In recent years, customer needs for industrial products have become diverse, and traditional mass production has evolved into manufacturing of a wide variety of products in small quantities. Even in such a situation, manufacturers have to produce their products at low cost and high quality. Multiobjective production creates a great number of product specifications. It is difficult to produce at strict specification but of the high quality which consumer and customer require. The main objective of manufacturers is to reduce defects of products for reduction in quality cost while continuing to produce products of high quality. Under such a background, quality control operation is an effective method of problem solving.

Sponge corner materials for automotive products are generally called weatherstrip. A weatherstrip is required to fit a cross-sectional shape to the structure of doors and maintain the appearance quality. As for quality problems of weatherstrip products, surface appearance quality is very important, as well as performance. Higher performance and appearance quality are required for producing automative parts with the increasing diversification of design of cars.

The problem for rubber products companies is that the surface of products becomes bloomed, which means poor appearance. The problem makes the surface of products turn white. We believe that the causes of the problem lie in the production process, specifically the conditions of vulcanization and compounding agent. At the same time, even if we are able to solve the problem of bloom, the performance of products may fall. Therefore, we need to fulfil both performance and quality of appearance.

The appearance quality problem in a rubber product company is that the surface of products becomes bloomed. We consider that the cause of the bloom phenomenon in the production process results from conditions of cure and rubber chemicals. Then, we have not only to solve quality problem of blooming, but ensure the performance of products, *i.e.*, we must fulfil both performance and appearance quality required for the rubber products.

In this paper, we conduct several kinds of the design of experiment based on a quality engineering approach [1–3] to identify the causes of a bloom phenomenon to improve the appearance quality and the performance of rubber products. Applying the design of experiment based on orthogonal array $L_{16}(2^{15})$, we analyze two kinds of performance data of the rubber product, *i.e.*, Hari and *CS*, which were observed from the factory experiments. Then, we can derive the process average and 95% confidence limits for these two performance measures. Further, executing multivariate analysis [4] such as multiple regression and discriminant analyses, we propose several schemes to eliminate the bloom phenomenon.

17.2 Description of Product and Defect Phenomenon

Sponge corner materials for automotive products are generally called weatherstrips. A weatherstrip is required to fit a cross-sectional shape to the structure of doors and maintain the appearance quality. The weatherstrips seal up the door and the body of cars. In fact, the weatherstrips prevent rain and dust from entering the car interior. They are rubber strips, black in color.

Generally, there are two kind of rubber products based on the state of rubber, sponge and dense. Rubber products for automotive products are sponge corner materials. Recently, sponge corner materials for automotive products have caused many appearance quality problems in the production process under manufacturing environment of a wide variety of products in small quantities.

Specifically, a bloom means a problem in appearance quality, whereby the surface of rubber products turns white. The causes of the problem are considered as follows:

1. Insoluble vulcanization agent and accelerator for overcompounding emerge from the surface of rubber products.
2. Compounding agent cannot be dissipated uniformly in the process of mixing, due to the high temperature of mixing, the long time of mixing, under-curing and so forth.
3. Compounding agent itself has an effect on the bloom phenomenon (2).

When we solve the problem of the bloom phenomenon, we cannot desregard performance as an automotive rubber product. That is, sponge corner materials for automotive products have to fulfil combined required quality characteristics of appearance and performance.

We focus on the following factors in this paper:

* Resolving the bloom phenomenon $\cdots$ Swell, Blem of gas
* Maintenance and enhancement of appearance quality $\cdots$ *CS*
* Solid state properties $\cdots$ Vulcanizing time
* Cost

17.3 Identification of Bloom Phenomenon

First, we performed qualitative analysis of the occurrence frequency of bloom phenomenon and the investigation of methods for reproducing them. Next, we derived the compounding condition from the results of the qualitative analysis. Such qualitative analysis meant checking what constituents were included in certain samples. We also had to derive the conditions of mixing and the molding factors from the production process flow investigation, where the compounding conditions were types and quantities of rubber chemical, the conditions of mixing are thermal histories and the degree of dispersion, and the molding factors were vulcanizing temperature and time.

As compounding conditions, we set traditional sponge corner materials against newly developed ones and estimated the compounding agent causing the bloom phenomenon. Next, we investigated the methods for reproducing them. Then, we could identify the cause as the derivative of a bloom phenomenon with the qualitative analysis. Therefore, selecting the compounding agent as the measures of bloom, we considered the behavior of rubber chemicals.

As the conditions of mixing, we reproduced actual manufacturing conditions by a laboratory experiment to check the bloom phenomenon. As a result, it was shown that it was very difficult to reproduce the bloom phenomenon in a laboratory experiment because of the mass production line. The capability of the machines and dispersion difference between the actual mass production line and laboratory experiment were also considered as the causes.

As the molding factors, dividing the type of products according to the occurrence frequency of bloom phenomena, we checked the conditions of bloom in three periods. As a result, the vulcanized condition was found to have an impact on bloom phenomena. At this time, by changing the conditions of the vulcanizing temperature and time, and the volume of product, the molding temperature and volume of product were shown to have an impact.

Finally, identifying the derivation factor of bloom from the compounding conditions, we selected the compounding agent by applying quantitative measures of bloom. For the molding factors, we specified the temperature of vulcanized rubber as a derivation factor of bloom. The efficiency of machines and dispersant difference were cited as reasons why it was difficult to replicate the bloom phenomenon. In addition, because we found that bloom could not be reproduced in a laboratory experiment, we decided to optimize performance in the laboratory experiment and to check the bloom phenomenon on mass production line.

17.4 Orthogonal Arrays

Selecting four factors from the analysis of bloom phenomena, we set two levels on each factor as shown in Table 17.1. Next, we had to optimize the factors in turns of performance measures.

Table 17.1 Controllable factors in the design of experiment

	Factor A	Factor B	Factor C	Factor D
Level 1	1.00 *phr*	Compounding agent X	0.00 *phr*	1.15 *phr*
Level 2	0.80 *phr*	Compounding agent Y	0.25 *phr*	0.80 *phr*

Applying orthogonal array $L_{16}(2^{15})$ to the experiment for the four control factors with two levels (see Table 17.2), we conducted a laboratory experiment with interactions among the selected factors and analyzed them.

The performance measures to be optimized were Swell and CS, where Swell is the part where the expanded rubber is crooked when removed from the die assembly after vulcanizing, and CS the rate of change of height, often called compression permanent deformation. The former measure of Swell is evaluated by sight, and the latter measures of CS by the no return rate often compressing the product of 50% of the height. We can define the latter measures of CS (17.1):

$$CS = \frac{T_0 - T_1}{T_2}(T_2 = T_0/2)\,, \tag{17.1}$$

where T_0 and T_1 are the untested height and the tested height respectively.

In the following, we analyze the experiment data on Swell and CS with a cumulative method called cumulative χ^2-test, and with a logit transform respectively.

Table 17.2 Orthogonal arrays

No	1	2	3	4	5	6	7	8	9	10	11	12	13	14	15
1	1	1	1	1	1	1	1	1	1	1	1	1	1	1	1
2	1	1	1	1	1	1	1	2	2	2	2	2	2	2	2
3	1	1	1	2	2	2	2	1	1	1	1	2	2	2	2
4	1	1	1	2	2	2	2	2	2	2	2	1	1	1	1
5	1	2	2	1	1	2	2	1	1	2	2	1	1	2	2
6	1	2	2	1	1	2	2	2	2	1	1	2	2	1	1
7	1	2	2	2	2	1	1	1	1	2	2	2	2	1	1
8	1	2	2	2	2	1	1	2	2	1	1	1	1	2	2
9	2	1	2	1	2	1	2	1	2	1	2	1	2	1	2
10	2	1	2	1	2	1	2	2	1	2	1	2	1	2	1
11	2	1	2	2	1	2	1	1	2	1	2	2	1	2	1
12	2	1	2	2	1	2	1	2	1	2	1	1	2	1	2
13	2	2	1	1	2	2	1	1	2	2	1	1	2	2	1
14	2	2	1	1	2	2	1	2	1	1	2	2	1	1	2
15	2	2	1	2	1	1	2	1	2	2	1	2	1	1	2
16	2	2	1	2	1	1	2	2	1	1	2	1	2	2	1
Factor	B	$B \times C$	C	D	$B \times D$	$C \times D$	e_1	e_1	e_1	e_1	e_1	e_1	A	e_1	e_1

(e_1 :first order error)

17.5 Analysis of Swell

17.5.1 Cumulative Method

Using the experiment data based on the orthogonal array as shown in Table 17.2, we converted them into the cumulative frequency and performed the analysis of variance as follows:

1. Computing the cumulative frequency from observed data:
   We computed the cumulative frequency from the evaluation data of Swell, categorized as $\bigcirc$ (better), $\triangle$ (average), and $\times$ (worse) observations. This experiment is performed 3 times consecutively. These groups are divided into groups I, II, and III. The cumulative frequency is shown in Table 17.3.
2. Calculation of weights:
   We derived the weight W_1 for group I and the weight W_2 for group II. P_I and P_II indicate the count percentages of groups I and II respectively. Then, we had

$$W_1 = \frac{1}{P_\mathrm{I}(1 - P_\mathrm{I})} = \frac{48^2}{17(48 - 17)}$$
$$= 4.37 , \tag{17.2}$$
$$W_2 = \frac{1}{P_\mathrm{II}(1 - P_\mathrm{II})}$$
$$= 5.66 . \tag{17.3}$$

Table 17.3 Experiment results of Swell and cumulative frequency

No	Experiment Conditions	Observations of Swell Evalution $\bigcirc$	$\triangle$	$\times$	Cumulative frequency $\bigcirc$	$\triangle$	$\times$
1	$A_1B_1C_1D_1$	0	3	0	0	3	3
2	$A_2B_1C_1D_1$	1	2	0	1	3	3
3	$A_2B_1C_1D_2$	3	0	0	3	3	3
4	$A_1B_1C_1D_2$	0	3	0	0	3	3
5	$A_1B_1C_2D_1$	3	0	0	3	3	3
6	$A_2B_1C_2D_1$	1	2	0	1	3	3
7	$A_2B_1C_2D_2$	0	3	0	0	3	3
8	$A_1B_1C_2D_2$	3	0	0	3	3	3
9	$A_2B_2C_1D_1$	0	1	2	0	1	3
10	$A_1B_2C_1D_1$	0	0	3	0	0	3
11	$A_1B_2C_1D_2$	0	0	3	0	0	3
12	$A_2B_2C_1D_2$	0	0	3	0	0	3
13	$A_2B_2C_2D_1$	3	0	0	3	3	3
14	$A_1B_2C_2D_1$	1	2	0	1	3	3
15	$A_1B_2C_2D_2$	1	2	0	1	3	3
16	$A_2B_2C_2D_2$	1	2	0	1	3	3

Table 17.4 Analysis of variance for Swell

Source	Degree of freedom	Sum of sqaures	V_n biased variance	F-value	F_0'	Pure sum of guares	Contribution ratio
A	2	0.21	0.11	○			
B	2	16.55	8.27	6.42**	8.15**	15.58	16.23
C	2	21.65	10.82	8.40**	10.66**	20.68	21.54
D	2	0.21	0.11	○			
$B \times C$	2	15.09	7.55	5.86**	7.44**	14.12	14.71
$B \times D$	2	0.94	0.47	○			
$C \times D$	2	2.40	1.20	○			
e_1	16	20.62	1.29	4.50**			
(e_1')	24	24.37	1.02				
e_2	64	18.35	0.287				
e	88	42.15	0.48			44.66	47.53
T	94	96				96	100.00

(e_2: second order error, ○: pooled in the error factor,
**:significant of 1% level, *: significant of 5% level)

Table 17.5 Selection of the optimal level

	B_1			B_2		
	○	△	×	○	△	×
C_1	4	8	0	0	1	11
C_2	7	5	0	6	6	0

3. Analysis of variance (ANOVA):
   Calculating the statistical quantities for the *analysis of variance*, we obtained
   the summary of ANOVA as shown in Table 17.4.
4. Derivation of optimal conditions of Swell:
   From Table 17.4, it was found that Factors B and C and their interaction $B \times C$
   are significant at 1% level. Therefore, we selected combination $B_1 C_1$ as optimal
   conditions for Swell measure, as shown in Table 17.5.

17.5.2 Process Average and 95% Confidence Limits

1. Process average of Swell:
   (Group I)

$$\hat{\mu}_1(B_1, C_2) = \overline{B_1 C_2}$$
$$= \frac{7}{12}$$
$$= 0.58 . \tag{17.4}$$

(Group II)

$$\hat{\mu}_{\text{II}}(B_1, C_2) = \overline{B_1 C_2}$$
$$= \frac{(7+5)}{12}$$
$$= 1.00 . \qquad (17.5)$$

2. 95% confidence limits:

We obtained the following 95% confidence limits for groups I and II where V_e and n_e represent the total error variance and the effective number of replications, respectively, and F-value ($F_{88}^1(0.05)$) is referred to the table of F-distribution [1,2]:

(Group I)

$$\hat{\mu}(B_1, C_2) = \overline{B_1 C_2} \pm \sqrt{F_{88}^1(0.05) \times V_e \times \hat{\mu}(1 - \hat{\mu}) \times \frac{1}{n_e}}$$
$$= [0.39, 0.77] . \qquad (17.6)$$

(Group II)

$$\hat{\mu}(B_1, C_2) = \overline{B_1 C_2} \pm \sqrt{F_{88}^1(0.05) \times V_e \times \hat{\mu}(1 - \hat{\mu}) \times \frac{1}{n_e}}$$
$$= [0.83, 1.00] . \qquad (17.7)$$

17.6 Analysis of *CS*

17.6.1 Logit Transformation and Data Analysis

In a similar method to Sect. 17.5.1, using the experiment data of *CS* after performing the logit transformation, we conducted data analysis based on a quality engineering approach. Calculating the SN ratios of the original data x_{ij} for i-th experiment number and j-th repetition by using

$$\eta_i = -10 \log_{10} \left(\frac{1}{m} \sum_{j=1}^{m} x_{ij}^2 \right) . \qquad (17.8)$$

We have the transformed data as shown in Table 17.6. In (17.8), η_i is the SN ratio for i-th experiment, and m means the number of repetitions. Then, we obtained the summary of ANOVA as shown in Table 17.7 by performing the analysis of variance based on Table 17.6. From Table 17.7, it was found that factors C, D are significant at 5% level, and their interaction $C \times D$ is significant at 1% level in terms of *CS* measures. Therefore, we selected combination $C_1 D_2$ as the optimal condition

Table 17.6 Experiment data of CS and logit ratio

	Experiment	CS (3 repetitions)			Logit transformation
No	Condition	$n=1$	$n=2$	$n=3$	SN ratio
1	$A_1B_1C_1D_1$	44	44	40	7.39
2	$A_2B_1C_1D_1$	41	42	44	7.46
3	$A_2B_1C_1D_2$	48	47	43	6.74
4	$A_1B_1C_1D_2$	47	46	49	6.49
5	$A_1B_1C_2D_1$	43	44	42	7.33
6	$A_2B_1C_2D_1$	43	44	43	7.26
7	$A_2B_1C_2D_2$	44	43	41	7.39
8	$A_1B_1C_2D_2$	43	41	44	7.39
9	$A_2B_2C_1D_1$	46	44	46	6.87
10	$A_1B_2C_1D_1$	42	43	42	7.47
11	$A_1B_2C_1D_2$	50	49	49	6.14
12	$A_2B_2C_1D_2$	47	45	44	6.87
13	$A_2B_2C_2D_1$	45	45	41	7.19
14	$A_1B_2C_2D_1$	46	44	43	7.06
15	$A_1B_2C_2D_2$	43	44	44	7.20
16	$A_2B_2C_2D_2$	44	44	44	7.13

Table 17.7 Analysis of variance for logit transformed data of CS

Source	f	S	V	F_0	F_0'	S'	ρ
A	1	0.01	0.01	◯			
B	1	0.15	0.15	◯			
C	1	0.40	0.40	6.60*	7.46*	0.35	16.21
D	1	0.45	0.45	7.36*	8.32*	0.40	18.38
$B \times C$	1	0.00	0.00	◯			
$B \times D$	1	0.00	0.00	◯			
$C \times D$	1	0.65	0.65	10.67*	12.07**	0.60	27.77
e	8	0.48	0.06				
(e')	12	0.65	0.05			0.81	37.64
T	15	2.15				2.15	100.00

(◯: pooling error, **:the 1% level, *: the 5% level)

for CS measure as shown in Table 17.8. Comparing the results based on between logit-transformed and original data, we obtained the same results in the analysis of variance as shown in Table 17.7.

Table 17.8 Select optimal level

C_1D_1	C_1D_2	C_2D_1	C_2D_2
43.17	47.00	43.58	43.25

17.6.2 Process Average and 95% Confidence Limits

1. Process average for SN ratio data:

$$\hat{\mu}(C_1, D_1) = \overline{C_1 D_1} = \frac{29.14}{4}$$

$$= 7.30 . \tag{17.9}$$

2. 95% confidence limits for SN ratio data:

$$\hat{\mu}(C_1, D_1) = \overline{C_1 D_1} \pm \sqrt{F_{12}^1(0.05) \times V_{e'} \times \frac{1}{n_e}}$$

$$= 7.30 \pm 0.23 . \tag{17.10}$$

3. Inverse transformation into original data:
   By using the relationship as

$$\mu' = \sqrt{10^{-\frac{\mu}{10}} \times 100} , \tag{17.11}$$

we obtained the following 95% confidence limits transformed into original data:

$$\hat{\mu}'(C_1, D_1) = [42.02, 44.30] , \tag{17.12}$$

where μ and μ' represent the estimated values in the SN-ratio and original data measures respectively.

17.6.3 Process Average and 95% Confidence Limits Under Simultaneous Optimal Conditions of Swell and CS

Considering the simultaneous optimal conditions for Swell and CS measures based on Tables 17.5 and 17.8, we obtained simultaneous control factor conditions to B_1, C_2 and D_2. Then, we have the following process average and 95% confidence limits:

1. Process average:

$$\hat{\mu}(B_1, C_2, D_2) = \overline{B_1 C_2} + \overline{D_2} - \overline{T} . \tag{17.13}$$

2. 95% confidence limits of Swell:
   (Group I)

$$\hat{\mu}(B_1, C_2, D_2) = \overline{B_1 C_2} + \overline{D_2} - \overline{T} \pm \sqrt{F_{88}^1(0.05) \times V_e \times \hat{\mu}(1 - \hat{\mu}) \times \frac{1}{n_e}}$$

$$= [0.36, 0.80] . \tag{17.14}$$

(Group II)

$$\hat{\mu}(B_1, C_2, D_2) = \overline{B_1 C_2} + \overline{D_2} - \overline{T} \pm \sqrt{F_{88}^1(0.05) \times V_e \times \hat{\mu}(1 - \hat{\mu}) \times \frac{1}{n_e}}$$

$$= [0.81, 1.00].$$ (17.15)

3. 95% confidence limits of *CS*:

$$\hat{\mu}(B_1, C_2, D_2) = \overline{B_1 C_2} + \overline{D_2} - \overline{T} \pm \sqrt{F_{12}^1(0.05) \times V_{e'} \times \frac{1}{n_e}}$$

$$= 7.37 \pm 0.26.$$ (17.16)

4. Inverse transformation into original *CS* data:

$$\mu'(B_1, C_2, D_2) = [41.54, 44.11].$$ (17.17)

17.7 Discriminant Analysis for Swell Measures

Based on the result of the cumulative method in Chap. 5 and regarding significant factors *B* and *C* as explanatory variables, we executed discriminant analysis. In this analysis, the following three cases were treated as the objective variables:

(i) We define the appearance quality evaluations ($\bigcirc$ and $\triangle$) except for $\times$ as 1 and the evaluations ($\times$) as 2 (see Table 17.10).

(ii) We define the appearance quality evaluations ($\bigcirc$ and $\times$) except for $\triangle$ as 1 and the evaluations ($\triangle$) as 2 (see Table 17.11).

(iii) We define the appearance quality evaluations ($\bigcirc$) as 1 and the evaluations ($\triangle$ and $\times$) except for $\bigcirc$ as 2 (see Table 17.12).

Then, we have the following discriminant equations for cases (i)–(iii).

$$\text{(case (i)): } z = -7.00B_i + 28.00C_i + 2.33 \qquad (i = 1, 2).$$ (17.18)

$$\text{(case (ii)): } z = 1.00B_i - 4.00C_i \qquad (i = 1, 2).$$ (17.19)

$$\text{(case (iii)): } z = -1.40B_i + 5.60C_i - 0.23 \qquad (i = 1, 2).$$ (17.20)

Using these discriminant equations and the observed data in Tables 17.10–17.12, it is possible to compute the scores as shown in Tables 17.13–17.15 respectively. In Tables 17.13–17.15, if the score calculated from Eqs. 17.18–17.20 becomes more than 0, the predicted value is given as 1, otherwise the predicted value is given as 2. Therefore, we found that cases (ii) and (iii) except for case (i) were not acceptable results because the experiment data for factors *B* and *C* were not originally observed for performing the discriminant analysis.

Table 17.9 Analysis of variance table of raw data

Source	f	S	V	F_0	F_0'	S'	ρ
A	1	1.33	1.33	○			
B	1	12.00	12.00	○			
C	1	33.33	33.33	6.54*	7.38*	30.43	12.32
D	1	36.75	36.75	7.21*	8.14*	33.84	13.70
$B \times C$	1	0.00	0.00	○			
$B \times D$	1	0.08	0.08	○			
$C \times D$	1	52.08	52.08	10.22*	11.54**	49.18	19.91
e_1	8	40.75	5.09	2.31*			
(e_1')	12	54.17	4.51				
e_2	32	70.67	2.21				
e	43	124.83	2.90			133.54	54.66
T	47	247.00				247.00	100.00

(○: pooling error, **: the 1% level, *: the 5% level)

Table 17.10 Discriminant analysis data (case(i))

No	Factor B (B_i)	Factor C (C_i)	Objective variable (Z_i)
1	0.00	0.00	1
2	0.00	0.00	1
3	0.00	0.00	1
4	0.00	0.00	1
5	0.00	0.25	1
6	0.00	0.25	1
7	0.00	0.25	1
8	0.00	0.25	1
9	1.00	0.00	2
10	1.00	0.00	2
11	1.00	0.00	2
12	1.00	0.00	2
13	1.00	0.25	1
14	1.00	0.25	1
15	1.00	0.25	1
16	1.00	0.25	1

17.8 Multiple Regression Analysis for CS Measures

First, the correlation among four independent variables and objective one was analyzed by using the SN ratio data of CS (see Table 17.16). Then, the correlation matrics was obtained as shown in Table 17.17. Factors A and B had no strong correlation to the SN ratio. Factors C and D had a stronger correlation to SN ratio than factors A and B.

Table 17.11 Discriminant analysis data (case (ii))

No	Factor B (B_i)	Factor C (C_i)	Objective variable (Z_i)
1	0.00	0.00	2
2	0.00	0.00	2
3	0.00	0.00	1
4	0.00	0.00	2
5	0.00	0.25	1
6	0.00	0.25	2
7	0.00	0.25	2
8	0.00	0.25	1
9	1.00	0.00	1
10	1.00	0.00	1
11	1.00	0.00	1
12	1.00	0.00	1
13	1.00	0.25	1
14	1.00	0.25	2
15	1.00	0.25	2
16	1.00	0.25	2

Table 17.12 Discriminant analysis data (case (iii))

No	Factor B (B_i)	Factor C (C_i)	Objective variable (Z_i)
1	0.00	0.00	2
2	0.00	0.00	2
3	0.00	0.00	1
4	0.00	0.00	2
5	0.00	0.25	1
6	0.00	0.25	2
7	0.00	0.25	2
8	0.00	0.25	1
9	1.00	0.00	2
10	1.00	0.00	2
11	1.00	0.00	2
12	1.00	0.00	2
13	1.00	0.25	1
14	1.00	0.25	2
15	1.00	0.25	2
16	1.00	0.25	2

A multiple regression analysis is conducted by using the experiment data for significant factors C and D as shown in Table 17.16. Then, we obtained the estimated multiple-regression equation as

$$y = 1.26 \times C_i + 0.96 \times D_i + 6.00 \qquad (i = 1, 2, \cdots, 16) , \qquad (17.21)$$

Table 17.13 Discriminant table (case (i))

No	Observed value	Predicted value	Score
1	1	1	2.333
2	1	1	2.333
3	1	1	2.333
4	1	1	2.333
5	1	1	9.333
6	1	1	9.333
7	1	1	9.333
8	1	1	9.333
9	2	2	−4.667
10	2	2	−4.667
11	2	2	−4.667
12	2	2	−4.667
13	1	1	2.333
14	1	1	2.333
15	1	1	2.333
16	1	1	2.333

Table 17.14 Discriminant table (case (ii))

No	Observed value	Predicted value	Score
1	2	−	0.000
2	2	−	0.000
3	1	−	0.000
4	2	−	0.000
5	1	2	−1.000
6	2	2	−1.000
7	2	2	−1.000
8	1	2	−1.000
9	1	1	1.000
10	1	1	1.000
11	1	1	1.000
12	1	1	1.000
13	1	−	0.000
14	2	−	0.000
15	2	−	0.000
16	2	−	0.000

where the adequacy of analysis was checked as shown in Table 17.18 of ANOVA. The actual measurement and predicted values computed by using (17.21) for original data are shown in Fig. 17.1 and Table 17.19.

The estimated multiple regression equation is significant at 5% of F-test. However the precision of analysis is not so high, because the multiple correlation one and determination coefficient are low. One reason is that we used the orthogonal array data of prespecified level values for factors C and D.

Table 17.15 Discriminant table (case (iii))

No	Observed value	Predicted value	Score
1	2	2	−0.233
2	2	2	−0.233
3	1	2	−0.233
4	2	2	−0.233
5	1	1	1.167
6	2	1	1.167
7	2	1	1.167
8	1	1	1.167
9	2	2	−1.163
10	2	2	−1.163
11	2	2	−1.163
12	2	2	−1.163
13	1	2	−0.233
14	2	2	−0.233
15	2	2	−0.233
16	2	2	−0.233

Table 17.16 Analyzed data of CS

No	A	B	C	D	SN ratio
1	1.00	0.00	0.00	1.15	7.39
2	0.80	0.00	0.00	1.15	7.46
3	0.80	0.00	0.00	0.80	6.74
4	1.00	0.00	0.00	0.80	6.49
5	1.00	0.00	0.25	1.15	7.33
6	0.80	0.00	0.25	1.15	7.26
7	0.80	0.00	0.25	0.80	7.39
8	1.00	0.00	0.25	0.80	7.39
9	0.80	1.00	0.00	1.15	6.87
10	1.00	1.00	0.00	1.15	7.47
11	1.00	1.00	0.00	0.80	6.14
12	0.80	1.00	0.00	0.80	6.87
13	0.80	1.00	0.25	1.15	7.19
14	1.00	1.00	0.25	1.15	7.06
15	1.00	1.00	0.25	0.80	7.20
16	0.80	1.00	0.25	0.80	7.13

Table 17.17 Correlation matrix

	A	B	C	D	SN ratio
A	1	0	0	0	− 0.075
B	0	1	0	0	0.26
C	0	0	1	0	0.43
D	0	0	0	1	0.458
SN ratio	− 0.075	− 0.26	0.43	0.458	1

Table 17.18 Analysis of variance for multiple regression analysis

	f	S	V	F_o
Due to regression	2	0.846	0.423	4.235*
Error	12	1.154	0.096	
Total	15	2.144		

Table 17.19 Actual measurement and predicted values for original data of CS

| Logit transformation | | Original data | |
SN ratio	Predicted value 1	Average of actual measurement values	Predicted value 2
7.39	7.10	42.67	44.17
7.46	7.10	42.33	44.17
6.74	6.76	46.00	45.91
6.49	6.76	47.33	45.91
7.33	7.41	43.00	42.60
7.26	7.41	43.33	42.60
7.39	7.07	42.67	44.28
7.39	7.07	42.67	44.28
6.87	7.10	45.33	44.17
7.47	7.10	42.33	44.17
6.14	6.76	49.33	45.91
6.87	6.76	45.33	45.91
7.19	7.41	43.67	42.60
7.06	7.41	44.33	42.60
7.20	7.07	43.67	44.27
7.13	7.07	44.00	44.27

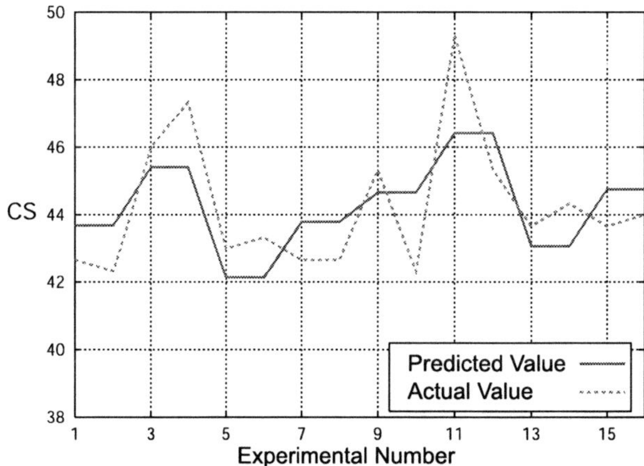

Fig. 17.1 Results of multiple regression analysis

17.9 Concluding Remarks

From the analysis results above, we have concluded that optimal conditions for the bloom problem are combination of $B_1C_2D_2$, and factor A is irrelevant to this problem. Performance of Swell and CS for there optimal conditions in the laboratory experiment has been checked on the actual mass production line in terms of appearance quality. At the result, there have been no bloom problems.

In this paper, we conducted statistical data analysis for quality improvement in actual production lines based on the design of the experiment. Here, we applied two methods of cumulative method and logit transformation in a quality engineering approach. Further, we conducted multivariate analysis such as multiple regression and discriminant analysises for prediction of the quality characteristic. Next, we obtained some reasonable analysis results. In this, improved appearance quality of weatherstrip products has been achieved.

References

1. Taguchi G (1976) Design of experiment (I), 3rd edn (in Japanese). Maruzen, Tokyo
2. Taguchi G (1977) Design of experiment (II), 3rd edn (in Japanese). Maruzen, Tokyo
3. Yamada S, Kimura M, Takahashi M (1998) Statistical quality control for TQM (in Japanese). Corona Publishing, Tokyo
4. IO JUSE (2004) User's manual for JUSE-StatWorks/ V4.0, JUSE, Tokyo

Chapter 18
Present Worth Design of Engineering Systems with Degrading Components

Young Kap Son, Gordon J. Savage

Systems Design Engineering, University of Waterloo,
Waterloo, Ontario, Canada N2L 3G1

18.1 Introduction

The ability of a manufacturer to design and produce a reliable and robust product that meets the customer's short- and long-term expectations with low cost and short product development time is the key for success in today's market. Customers' expectations include quality at the start of a product's life and both functionality and performance over a planned lifetime (*e.g.*, warranty time). Quality may be defined as conformance of performance measures to specifications [1]. Functionality is related to hard failures of components, meaning that the system ceases to function completely. Performance over time considers so-called soft failures wherein the system operates but performance measures do not meet their limit specifications. Often the design addresses only quality and it is hoped that performance and functionality will be acceptable [1–4]. However, performance and functionality over time are important to customers and must be ensured.

A fall-off of performance and functionality arises from material and dimensional changes in components that occur because of, for example, temperature, time and wear. These changes are referred to as degradation. Performance reliability has been invoked to mitigate component degradation. Design methods for performance reliability improvement include both parameter (*e.g.*, mean values) and tolerance design, such that the influence of environmental conditions and operating conditions on degradation is minimized. The traditional performance reliability design methods are based on sampling approaches (*i.e.*, Monte Carlo simulation). To present, they have mainly focused on determining mean values of components that maximize the Mean Time To Soft Failure (MTTSF) and minimize its variance [5, 6].

Such time-related measures may not be meaningful to both engineers and managers, and thus monetary measures are more inclusive. For example, a decrease in cost of 10% is quite clear to everyone compared to a decrease in MTTSF of 10% that may be clear only to quality and reliability engineers. Present worth of expected quality losses has been invoked as a monetary measure. There are a few research activities to determine initial means and tolerances of responses using this approach. In brief, they consider a response $Z(t)$ and a quadratic loss function $L(Z(t)) = k(Z(t) - m)^2$, where m is the target at an arbitrary time, and then define present worth in terms of the expected loss as

$$PW_L = \int_0^T E[L(Z(t))] \, e^{-rt} \, dt$$

where r is the user's discount rate (*i.e.*, interest rate in economic analysis). For a normally distributed response we have the present worth formula

$$PW_L = \int_0^T k \left[(\mu_Z(t) - m)^2 + \sigma_Z^2(t) \right] e^{-rt} \, dt \, .$$

The term k is a deviation cost coefficient defined as the ratio of the unit cost associated with quality loss divided by $(USL - LSL)^2/4$ where USL and LSL are the upper and lower specification limits respectively for the response.

Teran *et al.* considered the expected quality loss to be a continuous cash flow stream to be minimized over a planning horizon $(0, t_L)$ [7]. They invoked and then extended the quality loss function for product response degradation. For different time-variant profiles of mean and variance they provided the best design parameter values (*i.e.*, initial mean and variance of a system response, $\mu_Z(0)$ and $\sigma_Z(0)$ by minimizing present worth. Chou *et al.* augmented the PW_L term above with the production cost and then performed tolerance allocation formulated as the optimization problem

$$\text{Minimize } C_T(\textbf{tol}, r, t_L) = C_P(\textbf{tol}) + PW_L(\textbf{tol}, r, t_L)$$

Subject to

$$\textbf{tol}_L \leq \textbf{tol} \leq \textbf{tol}_U$$

where C_T represents the total expected cost up to time t_L, $C_P(\textbf{tol})$ is the production cost and $\textbf{tol}_L$ and $\textbf{tol}_U$ indicate lower and upper limits of the tolerances [8, 9]. The production cost involved reciprocal models of the tolerances [10]. In the two above activities, the mean and variance of the product performances over time were assumed to be in general quadratic polynomial functions of time.

There are three main concerns in the Taguchi-based loss function design methods mentioned above. First, the application of the expected quality loss function is valid only for known statistical distributions (*e.g.*, normal distribution) of system performance measures at each time. Second, the loss function used is limited since a) its extension to *smaller/larger-is-best* performance metrics is not always obvious, and b) the calculation method used for the coefficient k in the loss function is not convincing, especially for multi-response systems. Third, the approach cannot be extended to systems wherein multiple time-variant responses come from degradation in multiple components.

The allocation of means and tolerances to provide quality, performance reliability and functional reliability in engineering systems is a challenging problem. Traditional measures to help select the best means and tolerances include probabilistic measures and mean time to failure and its variance; however, they have some shortcomings. In this paper, a monetary measure based on present worth is invoked as a more inclusive metric. We consider the sum of the production cost and the expected loss of quality cost over a planned horizon at the customer's discount rate. Key to the approach is a probabilistic-based loss of quality cost that incorporates the cumulative distribution function that arises from time-variant distributions of system performance measures due to degrading components. The proposed framework greatly extends classical reliability approaches and Taguchi-based loss function methods that use assumed degradation profiles of system responses. As a case study, the design of an automobile overrun clutch assembly shows the practicality and promise of the approach. More specifically, in Sect. 18.2, we present modeling of time-variant system performances from component degradation processes in terms of limit-state functions so that no response distributions need be assumed and all performance metrics (*i.e.*, target/smaller/larger-is-best) are permitted. In Sect. 18.3, we provide a non-sample-based probability approach using set theoretical formulation, in terms of unions of non-conformance regions, to approximate the cumulative distribution function of soft failure [11]. In Sect. 18.4, economic design formulations based on present worth for general multi-response systems are discussed. A monetary measure in terms of the sum of a probabilistic loss of quality cost and a production cost is developed. In Sect. 18.5, implementation of various economic design formulations for the design of an automotive overrun clutch is explained as a case study.

18.2 Modeling of Time-variant Systems

18.2.1 Component Degradation

Component degradation due to the effects of environmental and operating conditions produces time-variant characteristics in a component (*i.e.*, degradation) with the consequence that the system performance varies over time as well. Component degradation processes can be expressed in terms of the way random variables change over time [12]. Let us extract from the components a vector of random design variables denoted as $\mathbf{V} = [V_1, V_2, \ldots, V_m]$: these may be dimensions, resistances, spring constants and so forth. Let p be the design parameter vector comprising, for example, means and standard deviations that characterize $\mathbf{V}$, then $\mathbf{p} = [\mu_1, \mu_2, \ldots, \mu_m, \sigma_1, \sigma_2, \ldots, \sigma_m]$. In general, samples of the m arbitrarily distributed design variables (now denoted as $\mathbf{v}$) can be mapped to a vector $\mathbf{u}$ comprising m uncorrelated standard normal variables using the Rosenblatt transformation [13]. The transformation is denoted by the general implicit form $\Gamma(\mathbf{v}, \mathbf{p}, \mathbf{u}) = 0$, although in many cases the transformation is explicit. Next consider the random variable degradation models for the components denoted as $\mathbf{X}(t) = [X_1(t), X_2(t), \ldots X_m(t)]$. After a system has been placed in operation, design parameter $\mathbf{p}$ drifts or degrades with time [12]. Samples of component degradation distributions (denoted by the vector $\mathbf{x}(t)$) are functions of both $\mathbf{p}$ and time (t) and the standard normal vector $\mathbf{u}$. Since parameter degradation over time is in general evaluated from the initial design set $\mathbf{p}$, we let $\mathbf{p}(t) = \mathbf{w}(\mathbf{p}, t)$ and write conveniently

$$x_i(t) = f_i(\mathbf{p}(t), \mathbf{u}) \quad \text{for} \quad i = 1, 2, \ldots, m \,. \tag{18.1}$$

For example, if a design variable V is normally distributed with the initial parameters μ and σ, then $\mathbf{p} = [\mu, \sigma]$ and the $u - v$ transformation is simply $v = \mu + \sigma u$. A common degradation function has the form $x(t) = (\mu + \sigma u)(1 + d(t))$, where $d(t)$ is the normalized change in $\mathbf{p}$.

18.2.2 Time-variant Limit-state Functions

A system model that relates outputs to inputs may be formed by either a mechanistic approach using the interactions of components, or, an empirical approach using response surface methodology. In both approaches, the q uncertain performance measures (*e.g.*, responses) $\mathbf{Z}$ are written as functions of the m degradation variables $\mathbf{X}$, in the particular explicit form,

$$Z_i(t) = z_i(\mathbf{X}(t)) \,. \tag{18.2}$$

We relate responses to their specification limits by limit-state functions of the form

$$g_i(\mathbf{x}(t)) = \{ z_i(\mathbf{p}(t), \mathbf{u}) - \zeta \} \,, \tag{18.3}$$

where z_i is a response and ζ is either a lower or upper specification. (Note: For upper specifications the negative of the right side of (18.3) is used.) For any limit-state function, we define

$$g(\mathbf{x}(t)) > 0, \mathbf{x}(t) \in \text{Conformance region (Success region, } S)$$
$$g(\mathbf{x}(t)) = 0, \mathbf{x}(t) \in \text{Limit-state surface (LSS)}$$
$$g(\mathbf{x}(t)) < 0, \mathbf{x}(t) \in \text{Non-conformance region (Failure region, } F)$$

A limit-state function has one non-conforming region. For n limit-state functions, the union of all such regions defines the non-conformance region of the system. Using known distribution functions of $\mathbf{V}$ and a design vector $\mathbf{p}$, a limit-state function at time zero, say $g(\mathbf{v})$ in $\mathbf{v}$-space is mapped to $g(\mathbf{u})$ in $\mathbf{u}$-space. This mapping is the same for arbitrary time when $\mathbf{x}(t)$ is used. The limit-state surface drifts due to the combination of the design set $\mathbf{p}$ and time, and thus, the success and fail regions change accordingly.

18.3 Cumulative Distribution Function Modeling

In order to evaluate the cumulative distribution function of time to soft failure (CDF) numerically, it is necessary to break time into discrete steps. Consider a fixed time step h and a time index denoted as l where $l = 0, 1, \ldots L$ then $t_l = l \times h$ is the time at the lth step, and $t_L = L \times h$ is the life time. The cumulative distribution function at time t_L for design parameter vector $\mathbf{p}$ can be approximated using a series system reliability concept as

$$F(\mathbf{p}, t_L) \approx \Pr\left\{ \cup_{l=0}^{L} [\cup_{i=1}^{n} (g_i(\mathbf{p}(t), \mathbf{u}) \leq 0)] \right\} . \tag{18.4}$$

In order to help us evaluate (18.4), let us define an instantaneous failure (*i.e.*, non-conformance) region of the ith limit-state function at any selected discrete time t_l as

$$E_{l,i} = \{\mathbf{u} \in \mathbf{U} : g_i(\mathbf{p}(t_l), \mathbf{u}) \leq 0\} .$$

Then, the system instantaneous failure region up to time t_l, denoted as $\mathbf{E}_l$ is expressed as

$$\mathbf{E}_l = E_{l,1} \cup E_{l,2} \cup \cdots \cup E_{l,n} = \bigcup_{i=1}^{n} E_{l,i} . \tag{18.5}$$

From [11], the incremental failure probability, from time t_l during time interval h, is written as

$$\Delta F(\mathbf{p}, t_l) = \Pr(\mathbf{E}_{l+1} \cup \mathbf{E}_l) - \Pr(\mathbf{E}_l) . \tag{18.6}$$

The probability in (18.6) is easily evaluated by Monte Carlo simulation using the complete limit-state function or for a good second-order approximation, wherein only pairs of intersections are invoked, we order probabilities in decreasing order, and denote these individual failure sets as $\Pr(E_{l,i}^{o})$. Now, the first term on the right side of Eq. (18.6) is rewritten as an upper bound

$$\Pr_{\mathrm{U}}(\mathbf{E}_{l+1} \cup \mathbf{E}_l) = \sum_{i=1}^{2n} \Pr(E_{l,i}^{o}) - \sum_{i=2, j<i}^{2n} \max\left(\Pr(E_{l,i}^{o} \cap E_{l,j}^{o}) \right) \tag{18.7a}$$

and the second term as a lower bound

$$\Pr_{\mathrm{L}}(\mathbf{E}_l) = \Pr(E_{l,1}^{o}) + \sum_{i=2}^{n} \max\left(\left[\Pr(E_{l,i}^{o}) - \sum_{j=1}^{i-1} \Pr(E_{l,i}^{o} \cap E_{l,j}^{o}) \right], 0 \right) . \tag{18.7b}$$

We now have the conservative approximation of (18.6)

$$\Delta F(\mathbf{p}, t_l) \cong \mathrm{Pr}_{\mathrm{U}}(\mathbf{E}_{l+1} \cup \mathbf{E}_l) - \mathrm{Pr}_{\mathrm{L}}(\mathbf{E}_l) \,.$$

The cumulative distribution function at time t_L is evaluated as

$$F(\mathbf{p}, t_L) = \mathrm{Pr}(\mathbf{E}_0) + \sum_{l=0}^{L-1} (\Delta F(\mathbf{p}, t_l)) \,, \tag{18.8}$$

where the first term on the right represents the non-conformance (*e.g.*, quality) at time zero. In this paper, (18.8) is evaluated using FORM (First-Order Reliability Method) and second-order bounds on union probability whose detailed explanations are shown in [11].

18.4 Formulation of Economic Design Problems

18.4.1 Present Worth Evaluation of Design

Herein we adapt the work of Faber [14] and Aktas *et al.* [15] to help define the expected capitalized loss of quality cost of a system, denoted as C_{LQ}^E, up to the planned time t_L all brought to the same reference time. The general form is

$$C_{\mathrm{LQ}}^E(t_L) = \int_0^{t_L} c_{\mathrm{F}}(t) f_T(t) \mathrm{e}^{-rt} \, \mathrm{d}t \,, \tag{18.9}$$

where $c_{\mathrm{F}}(t)$ is the cost of failure at time t, r is the real rate of interest, e^{-rt} is the discount factor and $f_T(t)$ is the probability density function, conditional on the event that the system is in a conforming state at time $t = 0$. In the case when $f_T(t)$ is unknown and we have a constant failure cost denoted as c_{F}, an approximation using discrete time events and evaluations of $F(t)$ provide the expected loss of quality cost expressed in (18.9) as

$$C_{\mathrm{LQ}}^E(\mathbf{p}, t_L, r) = c_{\mathrm{F}} \sum_{l=1}^{L} (F(\mathbf{p}, t_l) - F(\mathbf{p}, t_{l-1})) \mathrm{e}^{-rt_l} \,. \tag{18.10}$$

For a mass-produced product, it is necessary to take into account the fact that some non-conformance may be present at time zero. The non-conforming products may be eliminated by inspection for a cost of c_S, or accepted and introduced to the market so that $c_S = c_{\mathrm{F}}$. In either case, we introduce an additional cost term $c_S F(\mathbf{p}, t_0)$. Further, let the production cost be $C_{\mathrm{P}}(\mathbf{p})$ and now the total cost is

$$C_{\mathrm{T}}(\mathbf{p}, t_L, r) = C_{\mathrm{P}}(\mathbf{p}) + c_S F(\mathbf{p}, t_0) + C_{\mathrm{LQ}}^E(\mathbf{p}, t_L, r) \,. \tag{18.11}$$

The total cost expressed as (18.11) provides a monetary metric for choosing design parameters $\mathbf{p}$ under a fixed interest rate r for given costs c_S and c_F, and planned time t_L. It should be noted that if we use only the first two terms in Eq. (18.11), we are essentially performing so-called design for quality.

18.4.2 Formulation of Economic Design Problems

We propose the three following design problems.

Design (a): suppose we wish to perform only design for quality, then the design problem (*i.e.*, allocation of means and tolerances of design variables) can be formulated into the following optimization problem:

$$\text{Minimize } C_P(\mathbf{p}) + c_S F(\mathbf{p}, t_0) \tag{18.12}$$

Subject to

$$\mathbf{p}_L \leq \mathbf{p} \leq \mathbf{p}_U$$

Design (b): suppose we wish to minimize the total expected cost up to a planned time t_L under the interest rate r, then the design problem can be formulated into the following optimization problem:

$$\text{Minimize } C_T(\mathbf{p}, t_L, r) \tag{18.13}$$

Subject to

$$\mathbf{p}_L \leq \mathbf{p} \leq \mathbf{p}_U$$

Design (c): suppose we augment design (b) above with i) a quality policy at shelf time t_0 and/or ii) a performance reliability policy at a specified later time (*e.g.*, maintenance time) t_M, then the design problem can be formulated into the following constrained optimization problem:

$$\text{Minimize } C_T(\mathbf{p}, t_L, r) \tag{18.14}$$

Subject to

$$F(\mathbf{p}, t_0) \leq Y_0$$
$$F(\mathbf{p}, t_M) \leq Y_M$$
$$\mathbf{p}_L \leq \mathbf{p} \leq \mathbf{p}_U$$

For example of (c), suppose we select a six-sigma quality policy (*i.e.*, 3.4 defects per million), then we set $Y_0 = 3.4 \times 10^{-6}$. If we augment this with a B10 for 10 years life performance reliability policy [16], (*i.e.*, 10% of the systems will have failed by 10 years of operation), we set $Y_M = 0.1$ and $t_M = 10$ years.

18.5 Case Study

An automotive overrun clutch discussed in detail [3,4,17–19] is shown in Fig. 18.1.

The clutch assembly comprises four different types of parts: one hub, one cage, four rollers and four springs. The springs push the rollers out so that they remain in contact with both the cage and the hub: their forces are small and neglected herein. If the hub is turning counter-clockwise, relative to the cage, the rollers bind, supplying a torque so that the cage turns with the hub. If the hub is turning clock-wise, relative to the cage, the rollers slip and there is no torque transmission. In order to provide the proper operation over time, the clutch designer must select: a) dimensions that include the hub diameter D, the roller diameter d, the cage outer diameter B, the cage inner diameter A and the roller length L; b) the number of rollers N; and c) a material with the modulus of elasticity E and Poisson's ratio v.

The choice of nominal values and tolerances at the design stage affects operation of the clutch over time as surfaces wear to produce dimensional changes [17–19]. In this clutch design, we assume that for the hub, roller and cage the material is 1020 steel.

There are three performance measures of interest. The first is the contact angle, shown as α, and its value is important to ensure proper binding and slipping. If the value of the angle is greater than the upper specification limit or lesser than the lower specification limit, the clutch does not work correctly and it must be reworked

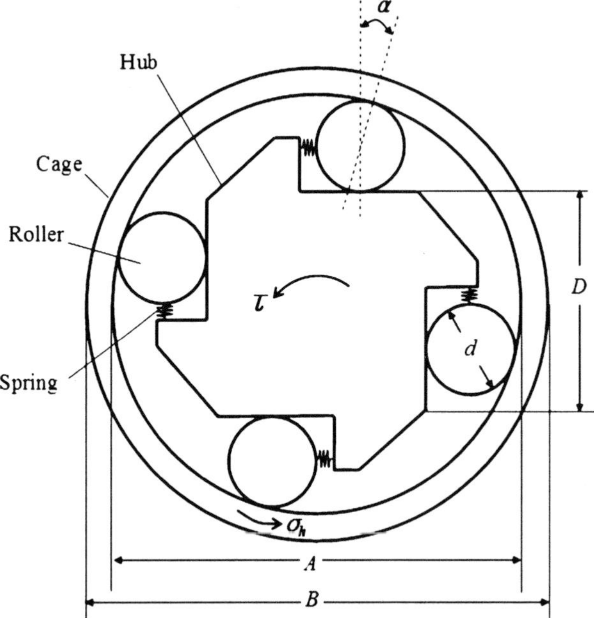

Fig. 18.1 Overrun clutch assembly

or scrapped. The second performance measure is the torque capacity τ arising form the friction between the hub, roller and cage. The design must provide sufficient torque without exceeding material compressive strengths. The third measure is the hoop stress σ_h arising from the circumferential roller friction force on the cage. This stress must be kept below a maximum to prevent fatigue failure in the pipe-like cage.

The various equations that relate performance measures to design variables come from [4, 17–19]. First, we define the ratio

$$S = \frac{D+d}{A-d} \, , \tag{18.15}$$

then we write

$$\alpha = \cos^{-1}(S) \, , \tag{18.16}$$

$$\tau = \frac{NL}{4} \left(\frac{\sigma_c}{c_1}\right)^2 \frac{D^2 d}{4(D+d)} \sin(\alpha) \, , \tag{18.17}$$

$$\sigma_h = \frac{N}{2\pi} \left(\frac{\sigma_c}{c_1}\right)^2 \frac{Dd}{D+d} \frac{\cos(\alpha)}{A} \frac{B^2 + A^2}{B^2 - A^2} \, , \tag{18.18}$$

where σ_c is the average permissible contact stress and

$$c_1 = \frac{1}{4} \sqrt{\frac{\pi}{\frac{2(1-v^2)}{E}}} \, .$$

The control design variables $\mathbf{V}$ are selected to be those involved in each performance measure, and thus they are the hub diameter $V_1(D)$ the roller diameter $V_2(d)$, and the inner cage diameter $V_3(A)$. All three design variables are assumed to be normal and independent. We let the standard deviation be written by the statistical tolerance relation $\sigma_i = \text{tol}_i/3$ so that the design parameters are the means and tolerances of design variables and hence $\mathbf{p} = [\mu_1, \ldots, \mu_3, \text{tol}_1, \ldots, \text{tol}_3]$. The remaining design variables in the performance measure relations are considered to be constant and deterministic. Thus, the roller length L and the cage outer diameter B are set as 80 mm and 120 mm respectively. Moreover, we have $v = 0.29$ and $E = 207$ GPa, and $\sigma_c = 3.79$ for 1020 steel [19].

The three control design variables degrade over time and now are denoted as X_1, X_2 and X_3 respectively. More specifically, the surface wear on the clutch parts causes the inner dimension of the cage (*i.e.*, x_3) to increase over time while the diameters of the hub (*i.e.*, x_1) and roller (*i.e.*, x_2) decrease over time. We let the wear degradation rate over time be constant [19] and choose a linear degradation model of the form $x_i(t) = v_i(1 + d(t))$, where $d(t)$ indicates the normalized change in v_i during operating time. Herein $d(t) = k \cdot t$ and based on the average yearly mileage, we assume the wear rate $k = 2.5 \times 10^{-4}$ mm/year.

A sensitivity analysis shows that the angle and hoop stress are most affected by a change in the cage diameter (*i.e.*, x_3) and thus we consider herein only the degradation of the cage. Now, for the $\mathbf{u} - \mathbf{v}$ space relation $v_3 = \mu_3 + \sigma_3 u_3$ the degradation model in terms of the distribution parameters and time can be written as

$$x_3(t) = \mu_3(1 + 2.5 \times 10^{-4}t) + \frac{\mathrm{tol}_3}{3}(1 + 2.5 \times 10^{-4}t)u_3 \, ,$$

where $\mu_3(t)$ and $\sigma_3(t)$ are evident. In this clutch design, the degradation rates of the hub and rollers are considered to be zero and thus following the above derivation

$$x_i = v_i = \mu_i + \frac{\mathrm{tol}_i}{3}u_i \quad \text{for } i = 1 \text{ and } 2 \, .$$

The three performance measures, α, τ and σ_h become time-variant random variables and are denoted as Z_1, Z_2 and Z_3, and now we have three time-variant models $Z_i = z_i(X_1, X_2, X_3)$ for $i = 1$, 2 and 3.

These measures combined with their performance specifications provide limit-state functions. More specifically, the angle has a target-is-best quality metric with upper and lower specifications 0.122 ± 0.035 radians respectively, and we have the two limit-state functions

$$g_1 = 0.157 - z_1(\mathbf{p}, \mathbf{u}, t)$$
$$g_2 = z_1(\mathbf{p}, \mathbf{u}, t) - 0.087$$

The torque has a larger-is-best quality metric with lower value $3000 \, N - m$ and we have

$$g_3 = z_2(\mathbf{p}, \mathbf{u}, t) - 3000 \, .$$

Finally, the hoop stress has a small-is-best quality metric with upper value 400 MPa thus

$$g_4 = 400 \times 10^6 - z_3(\mathbf{p}, \mathbf{u}, t) \, .$$

The tree time-variant limit-state surfaces are readily formed by writing $g_i(\mathbf{p}, \mathbf{u}, t) = 0$.

18.5.1 Initial Design and CDF

Consider an initial design taken from [3, 4] where the three means and three tolerances of the design variables are $\mathbf{p}_i = [55.29, 22.86, 101.69, 0.25, 0.3, 0.4]$. The plot of the cumulative distribution function using Eq. (18.8) is shown in Fig. 18.2.

In order to get some idea of how the CDF for our three responses is constructed, consider Fig. 18.3 where the four limit-state surfaces for the two design variables x_1 and x_3, with x_2 held constant, are drawn at times $t = 0$ and $t = 10$ years in $\mathbf{u}$-space.

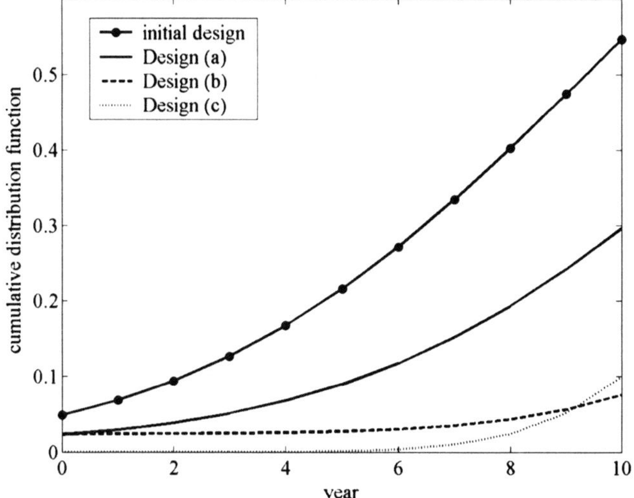

Fig. 18.2 Cumulative distribution functions for each design

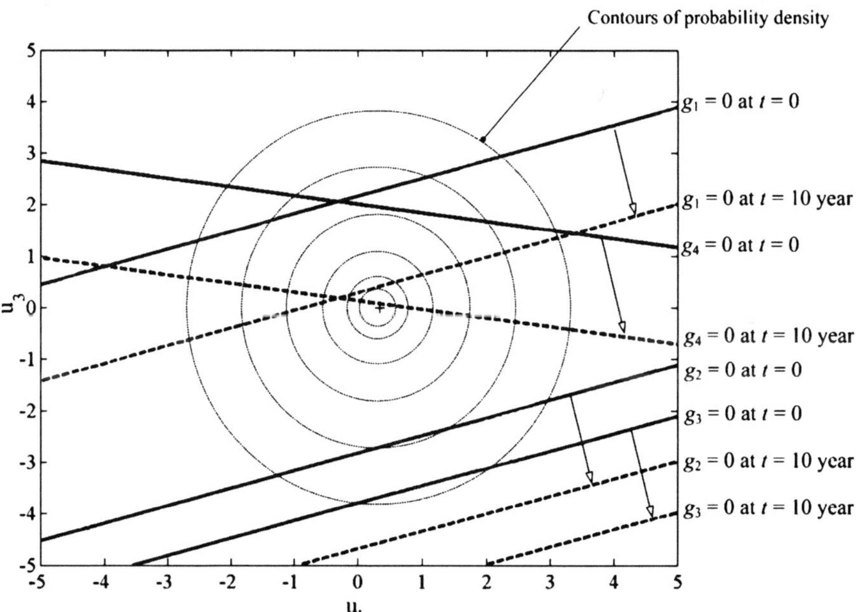

Fig. 18.3 Limit-state surfaces at times $t = 0$ and $t = 10$ years

Clearly, the time-shifts in the limit-state surfaces of the angle and hoop stress are the primary contributors to the shape of the CDF. That is, both $g_1 = 0$ and $g_4 = 0$ (the limit-state surfaces associated with the angle's upper specification and the hoop stress) shift towards the origin over time and contribute an increase in the probability of non-conformance, while $g_2 = 0$ and $g_3 = 0$ (the limit-state surfaces associated with the angle's lower specification and the torque) shift away from the origin over time and contribute a decrease in the probability of non-conformance. Since the limit-state surfaces are close to planar, FORM provides excellent probability evaluations. The plan now is to improve the performance by re-shaping the CDF.

18.5.2 New Designs

Consider new designs for a planned time $t_L = 10$ years and an interest rate $r = 3\%$. Cost factors for scrap and rework are $c_S = \$20$ and $c_F = \$20$. For a design that requires tolerance allocation, we introduce the production cost (in dollars) for the clutch assembly adopted from [4] as

$$C_p\left(\mathbf{p}\right) = \left(3.5 + \frac{0.75}{\text{tol}_1}\right) + \left(3.0 + \frac{0.65}{\text{tol}_2}\right) + \left(0.5 + \frac{0.88}{\text{tol}_3}\right).$$

The upper and lower bounds for the means (in mm) are

$$55.0973 \le \mu_1 \le 55.4973,$$
$$22.6600 \le \mu_2 \le 23.0600, \text{ and}$$
$$101.4900 \le \mu_3 \le 101.8900.$$

The upper and lower bounds for the tolerances (in mm) are

$$0.12 \le \text{tol}_1 \le 0.25,$$
$$0.08 \le \text{tol}_2 \le 0.3, \text{ and}$$
$$0.2 \le \text{tol}_3 \le 0.4.$$

We consider the three design cases presented earlier and denoted as

a) Design for quality (*i.e.*, no expected loss of quality),
b) Design for total cost, and
c) Design for total cost with the quality constraint $Y_0 < 3.4$ defects/million, and, the performance reliability constraint at time $t_M = 10$ years such that $Y_M < 0.1$.

The three cumulative distribution functions for the respective design are shown in Fig. 18.2. The design parameters and the corresponding costs for the three different new designs are shown in Table 18.1.

Let us compare design (a) with design (b) with reference to Table 18.1 and Fig. 18.2. Design (b) provides a reduction in the total cost by \$2.7645 per clutch. This cost reduction is made up by a small investment of \$0.8411 to reduce the tolerance of the cage which in turn provides a large decrease in the expected loss of quality cost of \$3.6056. Hence, 100,000 clutch assemblies manufactured using design (b) instead of design (a) could provide about \$276,450 profit. Next, consider design (c) which is marginally feasible since only the roller tolerance is above its low limit. We see in Table 18.1 that the introduction of both the high quality policy and the high performance reliability policy contributes to a considerable increase in production cost with no return in lowering the loss of quality cost and thus design (c) presents the highest total cost of all designs.

Table 18.1 Optimal design parameters and the corresponding costs in dollars for the three different designs

Parameters and costs	Design (a)	Design (b)	Design (c)
μ_1 (mm)	55.4973	55.4973	55.4973
μ_2 (mm)	22.6789	22.7322	22.6830
μ_3 (mm)	101.4900	101.4900	101.4900
tol_1 (mm)	0.2500	0.2500	0.1200
tol_2 (mm)	0.3000	0.3000	0.0839
tol_3 (mm)	0.3960	0.2873	0.2000
C_T ($)	19.2726	16.5081	26.9241
C_p ($)	14.3890	15.2301	25.4008
C_{LQ}^E ($)	4.8836	1.2780	1.5233

18.6 Conclusions

In this paper, we have presented an economic-based design of engineering systems that allocates the means and tolerances of degrading components using production costs, quality costs and loss of quality over time costs in terms of present worth. The loss of quality over time for multi-response systems is obtained from the system CDF which is built numerically using set-theoretic concepts in terms of time-variant limit-state functions. Probabilities have been evaluated using FORM, although any other approach, such as Monte Carlo sampling, would work. Arbitrary system performance metrics (smaller/target/larger-is-best) may be used due to the generality of limit-state functions as shown in our example. Constrained optimization problems have been formulated to help allocate simultaneously means and tolerances of design variables. A comparison of three design philosophies shows the weakness of using only design for quality or for the introduction of overly stringent levels of quality and performance reliability. Indeed, the strength of the approach is that designs may be investigated via the costs for any level of quality and performance reliability with a given time horizon and discount rate. The proposed design method could be used as an economical means of building in robustness to environmental and operating conditions (plus manufacturing variability) that designers and users are unable to control. Work is ongoing to apply the proposed method to the determination of optimum warranty time.

Acknowledgements This work was partially funded by the National Science and Engineering Research Council of Canada (NSERC).

References

1. Savage GJ, Carr SB (2001) *Interrelating quality and reliability in engineering systems*, Quality Engineering 14:137–152
2. Spence R, Soin SR (1997) *Tolerance Design of Electronic Circuits*. Imperial College Publishing, London
3. Seshadri R, Savage GJ (2002) *Integrated robust design with probability of conformance metric*. International Journal of Materials and Product Technology 17:319–337
4. Choi HR, Park M, Salisbury E (2000) *Optimal tolerance allocation with loss functions*. ASME Journal of Manufacturing Science and Engineering 122:529–535
5. Styblinski MA (1991) *Formulation of the drift reliability optimization problem*, Microelectronics Reliability 31:159–171
6. Van den Bogaard JA, Shreeram J, Brombacher AC (2003) *A method for reliability optimization through degradation analysis and robust design*. IEEE Reliability and Maintainability Symposium, Tampa, Fla., pp. 55–62
7. Teran A, Pratt DB, Case KE (1996) *Present worth of external quality losses for symmetric nominal-is-better quality characteristics*. Engineering Economist 42:39–52
8. Chou C, Chang C (2000) *Bivariate tolerance design for lock wheels by considering quality loss*. Quality and Reliability Engineering International 16:129–138
9. Chou C, Chen C (2001) *On the present worth of multivariate quality loss*. International Journal of Production Economics 70:279–288
10. Hauglund FH, Chase KW, Greenwood WH, Loosli BG (1990) *Least cost tolerance allocation for mechanical assemblies with automated process selection*. ASME Manufacturing Review 3:49–59
11. Son YK, Savage GJ (2006) *Set theoretic formulation of performance reliability of multiple response time-variant systems due to degradations in system components*. Quality and Reliability Engineering International Online 23:171–188
12. Jones JA (1999) *A toolkit for parametric drift modeling of electronic components*. Reliability Engineering and System Safety 63:99–106
13. Rosenblatt M (1952) *Remarks on a multivariate transformation*. Annual of Mathematical Statistics 23:470–472
14. Faber MH (1997) *Risk based structural maintenance planning*. In: Guedes SC (ed) Probabilistic Methods for Structural Design, Solid Mechanics and Its Applications. Kluwer Academic Publisher, Dordrecht, pp. 377–402
15. Aktas E, Moses F, Ghosn M (2001) *Cost and safety optimization of structural design specifications*. Reliability Engineering and System Safety 73:205–212
16. Ryu D, Chang S (2005) *Novel Concepts for Reliability Technology*. Microelectronics Reliability 45:611–622
17. South DW, Mancuso JR (1994) *Mechanical Power Transmission Components*. Marcel Dekker, New York
18. Xue W, Pyle R (2004) *Optimal Design of Roller One Way Clutch for Starter Drives*. SAE World Congress, Detroit, Michigan
19. Chesney DR, Kremer JM (1997) *Generalized equations for roller one-way clutch analysis and design*. SAE paper 970682, pp. 27–37

Chapter 19
Economic-statistical Design of a Logarithmic Transformed S^2 EWMA Chart

P. Castagliola[1], G. Celano[2], S. Fichera[2]

[1]IRCCyN & Université de Nantes, Institut Universitaire de Technologie de Nantes, Avenue du Professeur Jean Rouxel, BP 539, 44475 Carquefou Cedex, France
[2]Dipartimento di Ingegneria Industriale e Meccanica, University of Catania, Viale Andrea Doria 6, 95125 Catania, Italy

19.1 Introduction

Exponentially weighted moving average (EWMA) control charts are an efficient means in detecting small process shifts, both in position and dispersion of the collected data. Implementing an EWMA chart to control a manufacturing process requires the computation and plotting of a random variable which is a function of the current sample statistic and of the past samples collected from the process. This allows the EWMA to prevail over the traditional Shewhart chart in terms of statistical sensitivity when small shifts in the process position and/or dispersion are expected. The aim of this chapter is to present the economic-statistical design of a S^2 EWMA control chart for the on-line control of the process dispersion. The investigated chart operates through a control statistic based on a logarithmic transformation of the sample variance to make possible working on an approximately standard normally distributed random variable. Since the implementation of control charts to monitor process stability has become normal practice within an industrial manufacturing environment, designing economically the control chart is an important managerial aspect of SPC that should be carefully taken into account by practitioners. However, as widely suggested in the literature, the statistical reliability of the chart with respect to the probability of signaling false alarms should not be neglected. For this reason, in this chapter the economic design of the S^2 EWMA chart is discussed and achieved, respecting a statistical constraint related to the in-control Average run length (ARL_0), which allows the expected number of false alarms to be controlled and maintained at an acceptable level. An extensive data set of examples is presented to show how process costs and parameters affect the design of the S^2 EWMA chart. The numerical analysis has been organized as a 2^7 unreplicated factorial plan. The first step of the analysis presents a comparison with a S^2 Shewhart chart to quantify the cost savings achievable through the EWMA; then a sensitivity analysis on the EWMA chart design variables is performed to show how the input parameters derived from the process affect the decisions concerning the chart design.

Hoang Pham, *Recent Advances in Reliability and Quality in Design*
©Springer 2008

19.2 Literature Review

Manufacturing processes are characterized by a certain amount of variability which in some cases can strongly affect the quality of the outcome. Generally, process variability consists of an "inherent" or "natural" variability, which cannot be eliminated and can be considered as a background noise. Sometimes a further source of variability due to the occurrence of a special cause is present in the process: this variability sums to the natural variability and can lead to an unacceptable level of process performance. Several causes can give raise to an excessive amount of process variability: defective raw materials, incorrect methods of working, erroneous machine setups or operating conditions, operator errors, and so on The increase in process variability is generally reflected on the process mean and/or process dispersion: a shift in process mean causes the outcome be far from the target with a consequent production of an excessive number of nonconformities; an increase in process dispersion leads to a larger spread of data, which corresponds to a lower uniformity in process output: the result is once again an excessive number of produced nonconforming items. Processes characterized exclusively by the presence of the natural variability are said to operate in the "in-control" condition; when a special cause occurs, the process state is said to be "out-of-control".

Statistical process control SPC provides a large set of techniques designed to help practitioners in monitoring the state of control of a manufacturing process and quickly detecting special causes when they occur: control charts are a widely used on-line control technique and can be implemented on sample statistics related to both process mean and dispersion. An efficient implementation procedure for control charts should include the contemporary implementation of two charts for monitoring both the process mean and dispersion: to do this, the sample mean Shewhart $\bar{X}$ and R, S or S^2 control charts are widely described in SPC manuals and used by practitioners, (Montgomery, 2004). However, since the early 1980s, the use of the EWMA (exponentially weighted moving average) statistic as a process monitoring tool has become more and more popular in the statistical process control field. Implementing an EWMA chart to control a manufacturing process requires the computation and plotting of a random variable, which is a function of the current sample statistic and of the past samples collected from the process. This allows the EWMA to prevail over the traditional Shewhart chart in terms of statistical sensitivity when small shifts in the process position and/or dispersion are expected. The properties and design strategies of the EWMA control chart for the mean (introduced by Roberts (1959)) have been thoroughly investigated by Robinson and Ho (1978), Crowder (1987, 1989), Lucas and Saccucci (1990) and Steiner (1999). The use of the EWMA as a tool for monitoring the process variability has received attention by Wortham and Ringer (1971), Sweet (1986), Ng and Case (1989), Crowder and Hamilton (1992), Hamilton and Crowder (1992), and MacGregor and Harris (1993), Gan (1995), Amin *et al.* (1999), Lu and Reynolds (1999), and Acosta-Mejia *et al.* (1999), and more recently by Castagliola (2005), who developed a new two-sided S^2 EWMA control chart as an extension of the Crowder and Hamilton (1992),

Hamilton and Crowder (1992) initial approach, but based on a three-parameter logarithmic transformation. The S^2 EWMA control chart suggested by Castagliola (2005) has the following advantages: (a) the easy-to-use scheme for computing the control limits and the parameters a, b and c of the logarithmic transformation; (b), the improved performance in terms of normality/symmetry of the transformed sample variances; and (c) the possibility of detecting increases and decreases of the nominal standard deviation in a more similar way: in fact, the logarithmic transformation allows an approximately standard normal variable to be managed thus avoiding the possibility of "biased *ARLs*"; a "biased *ARL*" chart is one characterized by equal tails probability limits and the possibility of having larger out-of-control *ARLs* than the in-control *ARL*. A comparison with Crowder and Hamilton (1992), Hamilton and Crowder (1992) papers shown that the S^2 EWMA control chart proposed by Castagliola (2005) gives smaller out-of-control *ARL* than the other procedures.

Before implementing a chart on a process, its design parameters should be selected, whichever the observed variable. For a Shewhart chart the selection of the sample size n, the width of control limits L and the sampling interval h is required to start the on-line process control. For an EWMA control chart, a fourth parameter λ_E called the smoothing parameter and related to the weight of the past samplings on the actual plotted point, should be selected.

Among the literature studies concerning control charts, the development of criteria for the selection of the chart design parameters has received a great deal of attention. In particular, the chart design performed to meet the economic objective has been extensively investigated. Since the early contribution to the economic design of a Shewhart chart for monitoring the sample mean (Duncan, 1956), the design problem has focused on the selection of the sample size n and/or the width of control limits L and/or sampling interval h aimed to minimize a process expected cost per time unit. Among the several contributions in this field of research, the economic model from Lorenzen and Vance (1986) is worth mentioning: it extends the model proposed by Duncan and can be applied to several charts, both for quantitative and qualitative variables; this model represents the starting point of the largest part of papers dealing with the economic design of a control chart, whichever the controlled sample statistic and the chart scheme. Here, due to the limited availability of space the references to all these contributions are not reported.

Generally, designing economically a control chart yields design parameters that do not lead to a high performing statistical behavior of the chart: very often, the unconstrained economic optimization leads to chart schemes characterized by an excessive number of false alarms, that is several points outside the control limits without any process shift; thus, even if the overall cost is minimized, the high probability of signaling false out-of-control conditions results in a low reliable chart, which can lead the decision maker to misleading conclusions: as a consequence, one can prescribe excessive unnecessary process adjustments, which often result in an unnatural variability increase of the monitored quality characteristic. Furthermore, although the false alarm costs can be quantified accurately, it is very difficult to model and evaluate numerically the cost associated to the loss of practitioner's confidence on the chart when too many false alarms are signaled. For these reasons, even if the addition of an upper constraint on the probability of signaling a false alarm from the chart causes an increase of expected process costs, in literature it is strongly advised to select the economic design of the chart parameters under a statistical constraint.

In an economic design constrained statistically, the optimization of the chart design parameters is achieved by minimizing the expected cost per time unit under the condition of respecting a pre-assigned expected number of samples to be taken between two false alarms, *i.e.*, a fixed in-control average run length ARL_0 for the chart. One of the earlier contributions proposed in the field of statistically constrained economic design of a control chart is discussed in Saniga (1989).

In the field of $\overline{X}$ Shewhart-type charts, Celano and Fichera (1999) proposed two schemes of multi-objective optimization based on properly designed satisfaction functions optimized by means of a genetic algorithm. The economic, statistical and economic-statistical designs of Shewhart $\overline{X}$ and R, $\overline{X}$ and S charts were compared each other in McWilliams *et al.* (2001). Economic-statistical models for the design of a sample mean EWMA chart have been investigated in literature by Torng *et al.* (1995), who achieved an optimal design through a Hooke and Jeeves optimization procedure, and Tolley and English (2001), who compared an EWMA and a combined EWMA-$\overline{X}$ control scheme and demonstrated how very often it is superfluous to manage both charts within the same application. Up to now, in the field of the statistically constrained economic design of an EWMA chart no contribution was devoted to charts monitoring sampling statistics related to the process dispersion.

The aim of this chapter is to present the statistically constrained economic design of a S^2 EWMA control chart. The investigated EWMA chart monitors a statistic that depends on the sample variance through a logarithmic transformation. The model from Lorenzen and Vance was assumed as the objective function corresponding to the economic goal which minimizes the hourly costs of the process operating condition. The statistical goal was expressed through a constraint on the in-control ARL_0. To test the efficiency of the proposed S^2 EWMA in terms of cost savings a numerical analysis has been performed on several process scenarios. A benchmark of 128 datasets of costs, times and process parameters has been designed by starting from the illustrative example provided in Lorenzen and Vance (1986): the hourly costs related to the S^2 EWMA chart implementation have been computed and compared with those corresponding to the implementation of a S^2 Shewhart chart; finally, a sensitivity analysis on the EWMA chart design variables has been performed to show how the input parameters derived from the process affect the decisions concerning the chart design.

The rest of the paper is organized as follows: the next paragraph presents the S^2 EWMA chart scheme, summarizing the statistical assumptions which are at the basis of the monitored transformed statistic; then, the economic model of Lorenzen and Vance is briefly recalled and the Markov chain developed to evaluate the $ARLs$ is presented; finally, the computational results are reported and discussed.

19.3 The Logarithmic Transformed S^2 EWMA Chart

Let $X_{k,1}, \ldots X_{k,n}$ be a sample of n independent normal, $N(\mu, \sigma_0)$, random variables, where μ is the process mean, σ_0 is the nominal process standard deviation, and k is the subgroup number. The investigated chart is used to monitor the process dispersion, therefore the out-of-control condition for the process corresponds to the occurrence of a special cause, which leaves unchanged the process mean μ and shifts the standard deviation from σ_0 to σ_1: the entity of this shift is quantified through the parameter $\tau = \sigma_1/\sigma_0$. For the development of the economic model, σ_0 is assumed as known. Let S_k^2 be the sample variance of subgroup k, $i.e.$,

$$S_k^2 = \frac{1}{n-1} \sum_{j=1}^{n} \left(X_{k,j} - \overline{X_k} \right)^2$$

where $\bar{X}_k$ is the sample mean of subgroup k. Following a recommendation by Box $et\ al.$ (1978), Crowder and Hamilton (1992) suggested the application of the classical EWMA approach to the logarithm of the successive sample variances, $i.e.$, $T_k = \ln S_k$. The main motivation for this approach is that $T_k = \ln S_k$, which has a log-gamma distribution (Johnson $et\ al.$, 1994), tends to be more normally distributed than the sample variance. More recently, Castagliola (2005) suggested applying the following transformation to S_k^2

$$T_k = a + b \cdot \ln \left(S_k^2 + c \right)$$

where a, b and $c > 0$ are three constants, and then, to use the classical EWMA approach on the T_k

$$Z_k = (1 - \lambda) \cdot Z_{k-1} + \lambda T_k$$

The main motivation of this transformation is that if the constants a, b and c are judiciously selected, then the distribution of T_k will results in a better standard normality than the original transformation by Crowder and Hamilton (1992). The control limits of the S^2 EWMA control chart (corresponding to the Z_k) are

$$LCL = E\left(T_k\right) - L \times \left(\frac{\lambda_E}{2 - \lambda_E} \right)^{1/2} \times \sigma\left(T_k\right)$$

$$UCL = E\left(T_k\right) + L \times \left(\frac{\lambda_E}{2 - \lambda_E} \right)^{1/2} \times \sigma\left(T_k\right)$$

where L is a positive constant (for example $L = 3$), and where $E(T_k)$ and $\sigma(T_k)$ are the theoretical expectance and standard deviation of T_k. The central line CL of the chart equals $E(T_k)$. Castagliola (2005) proved that the constants a, b and c are necessarily equal to

$$b = B(n)$$

$$c = C(n)\,\sigma_0^2$$

$$a = A(n) - 2B(n)\ln(\sigma_0)$$

Table 19.1 Constants $A(n)$, $B(n)$, $C(n)$, $E(T_k)$, $\sigma(T_k)$ and Z_0 for $n = 2, \ldots, 15$

n	$A(n)$	$B(n)$	$C(n)$	$E(T_k)$	$\sigma(T_k)$	Z_0
2	−0.5096	1.4464	0.8064	0.08205	0.8195	0.346
3	−0.6627	1.8136	0.6777	0.02472	0.9165	0.276
4	−0.7882	2.1089	0.6261	0.01266	0.9502	0.237
5	−0.8969	2.3647	0.5979	0.00748	0.9670	0.211
6	−0.9940	2.5941	0.5801	0.00485	0.9765	0.193
7	−1.0827	2.8042	0.5678	0.00335	0.9825	0.178
8	−1.1647	2.9992	0.5588	0.00243	0.9864	0.167
9	−1.2413	3.1820	0.5519	0.00182	0.9892	0.157
10	−1.3135	3.3548	0.5465	0.00141	0.9912	0.149
11	−1.3820	3.5189	0.5421	0.00112	0.9927	0.142
12	−1.4473	3.6757	0.5384	0.0009	0.9938	0.136
13	−1.5097	3.8260	0.5354	0.00074	0.9947	0.131
14	−1.5697	3.9705	0.5327	0.00062	0.9955	0.126
15	−1.6275	4.1100	0.5305	0.00052	0.9960	0.122

where $A(n)$, $B(n)$, and $C(n)$ are three functions depending only on the sample size n. The values for $A(n)$, $B(n)$, and $C(n)$ are determined by assuming that the following log-normal distribution:

$$f_L\{x|A(n), B(n), C(n)\} = \frac{B(n)}{x}\varphi\{A(n) + B(n)\ln(x + C(n))\}$$

is the closest to the distribution of the population variance S^2 according to a criterion based on the first three moments $E(S^2)$, $V(S^2)$ and $\gamma_3(S^2)$: a detailed description of how $A(n)$, $B(n)$, and $C(n)$ are computed is reported in Castagliola (2005) and Castagliola *et al.* (2006). Table 19.1 reports the values of $A(n)$, $B(n)$, and $C(n)$ for $n = 2, \ldots, 15$.

If $\tau = 1$, the process is assumed to operate in the "in-control" condition; the cumulative distribution function (cdf) $F_{T_k}(t|n)$ of T_k depends only on n:

$$F_{T_k}(t|n) = F_G\left\{\exp\frac{t - A(n)}{B(n)} - C(n)\left|\frac{n-1}{2}, \frac{2}{n-1}\right.\right\}$$

The cdf F_{T_k} is defined for $t \geq A(n) + B(n)\ln\{C(n)\}$. $F_G(x|u, v)$ is the cdf of the following gamma distribution:

$$f_G(x|u, v) = \begin{cases} 0 & (x \leq 0) \\ \dfrac{x^{u-1}\exp(-x/v)}{v^u\Gamma(u)} & (x > 0) \end{cases}$$

which for $u = (n-1)/2$ and $v = 2/(n-1)$ coincides with the distribution of the variance S^2.

The computation of $E(T_k)$ and $\sigma(T_k)$ has been achieved by numerical quadrature. Finally, a reasonable value for Z_0 is

$$Z_0 = A(n) + B(n)\ln\{1 + C(n)\}$$

As can be seen, this value depends only on n and not on σ_0. Whichever the value for n, Z_0 can be replaced by 0 in practice with little effects. Table 19.1 reports the values of $E(T_k)$, $\sigma(T_k)$ and Z_0 for $n = 2, \ldots, 15$.

When the process shifts to the out-of-control condition, $i.e.$ $\tau \neq 1$, the variable T_k will depend both on n and τ:

$$F_{T_k}(t\,|n, \tau) = F_G\left\{\exp\frac{t - A(n)}{B(n)} - C(n)\left|\frac{n-1}{2}, \frac{2\tau^2}{n-1}\right.\right\}$$

$F_G(x|u, v)$ is still the cdf of the gamma distribution, which for $u = (n-1)/2$ and $v = 2\tau^2/(n-1)$ coincides with the distribution of the shifted variance $\tau^2 S^2$. The interested reader can get more details in Castagliola (2005).

Example 1: The goal of this example is to show how the S^2 EWMA control chart behaves in the case of an increase and a decrease in the nominal process variability. The first 100 data points plotted in Fig. 19.1 (top and bottom) consist of 20 identical subgroups of $n = 5$ observations randomly generated from a normal (20, 0.1) distribution (corresponding to an in-control process), while the last 50 data points of Fig. 19.1 (top) consist of 10 subgroups of $n = 5$ observations randomly generated from a normal (20, 0.2) distribution (the nominal process standard deviation σ_0 has increased by a factor of 2), and the last 50 data points of Fig. 19.1 (bottom) consist of 10 subgroups of $n = 5$ observations randomly generated from a normal (20, 0.05) distribution (the nominal process standard deviation σ_0 has decreased by a factor of 2). The corresponding 30 sample variances are plotted in Fig. 19.2, for the two cases (increasing and decreasing).

At this step, the asymmetry between increasing and decreasing sample variances is particularly noticeable. If $n = 5$ and $\sigma_0 = 0.1$, then $b = 2.3647$, $c = 0.5979 \times 0.1^2 = 0.005979$ and $a = -0.8969 - 2 \times 2.3647 \times \ln(0.1) = 9.9929$. The 30 transformed sample variances T_k are plotted in Fig. 19.3 and the S^2 EWMA sequence along with the S^2 EWMA control limits $LCL = 0.0075 - 3\sqrt{0.1}/1.9 \times 0.967 = -0.658$ and $UCL = 0.0075 + 3\sqrt{0.1}/1.9 \times 0.967 = 0.673$ ($\lambda = 0.1$ and $L = 3$) are plotted in Fig. 19.4. The S^2 EWMA control chart clearly detects an out-of-control signal at the 25-th subgroup (in the increasing case) and at the 26-th subgroup (in the decreasing case), pointing out that an increase/decrease of the process variability occurred.

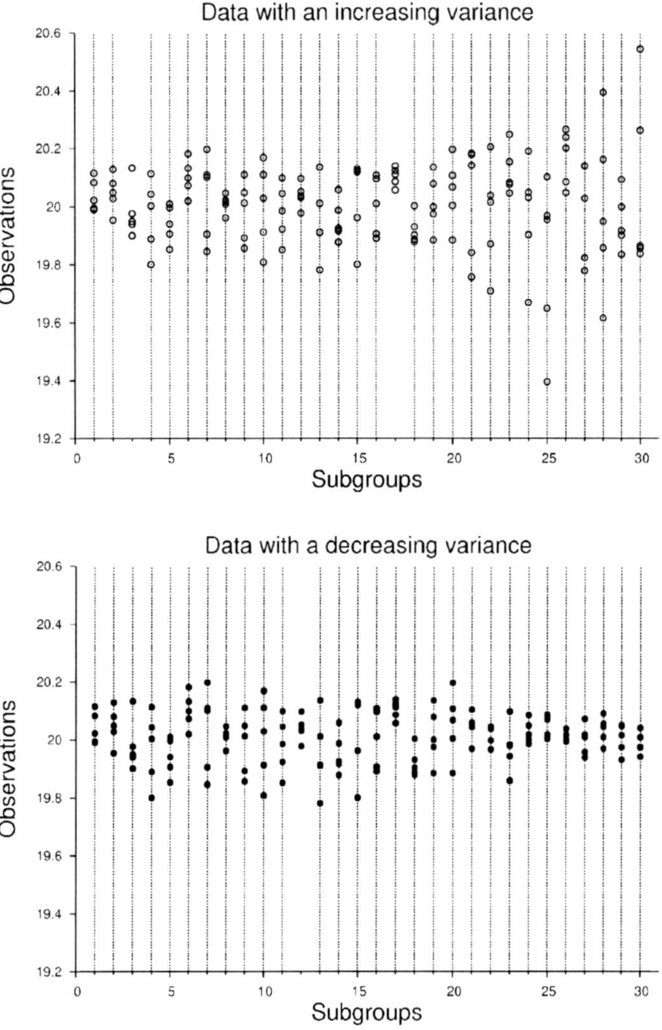

Fig. 19.1 Data with an increasing variance (*top*), and with a decreasing variance (*bottom*)

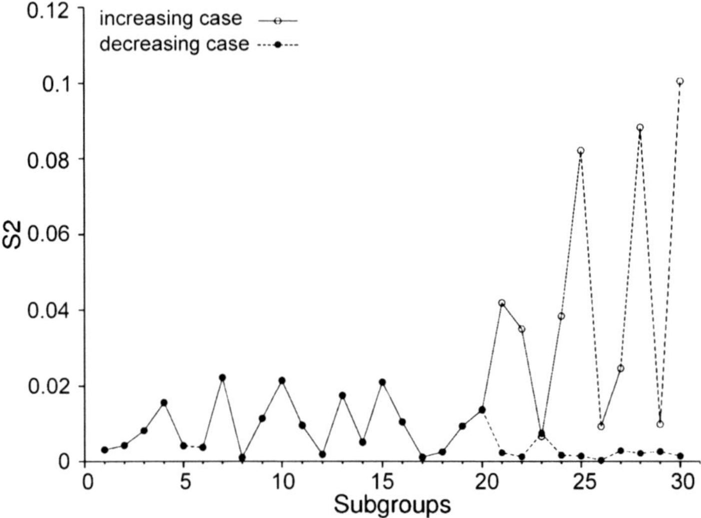

Fig. 19.2 Sample variances S_k^2 corresponding to the data of Fig. 19.1

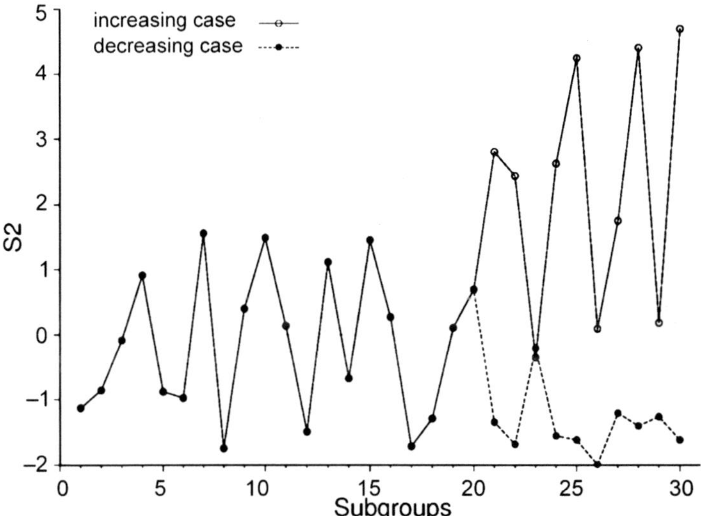

Fig. 19.3 Transformed sample variances $T_k = a + b \cdot \ln\left(S_k^2 + c\right)$ corresponding to the data of Fig. 19.1

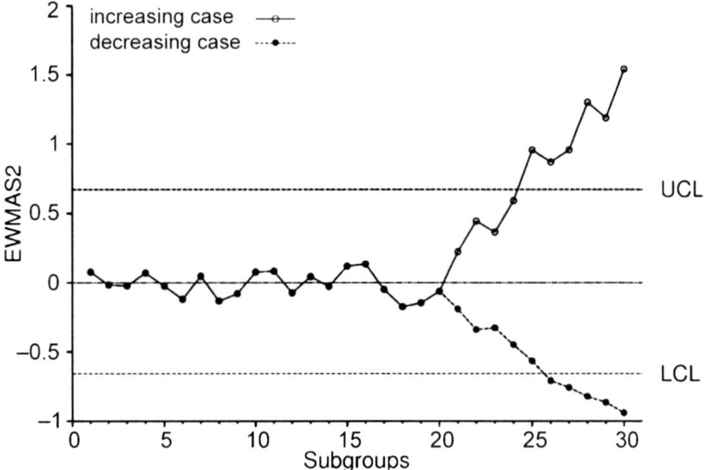

Fig. 19.4 S^2 EWMA control chart ($\lambda_E = 0.1, L = 3$) corresponding to the data of Fig. 19.1

19.4 The Economic Design of the S^2 EWMA

19.4.1 Introduction

Before implementing a control chart the selection of its design parameters is needed: for example, to implement an efficient Shewhart chart, the sample size n, sampling interval h and width of control limits L should be judiciously selected by practitioners; to design an EWMA chart one more parameter should be fixed: the EWMA smoothing parameter λ_E which weights the effect of the past process history.

Generally, the selection of the chart parameters is based on an empirical approach: for example, when a Shewhart control chart should be implemented to monitor the sample mean from a critical process parameter a sample size $n = 5$, a sampling frequency $h = 1$ time unit and a width of control limits $L = 3$ are suggested by SPC manuals; these values stem from considerations of statistical efficiency: in fact, for a large shift in the process mean they allow a low expected number of false alarms and a fast signal of an out-of-control condition. However, these values are not optimal for other operating conditions such as small shifts of the process mean or shifts in process variance. Thus, in the literature many strategies have been proposed to select the optimal chart parameters for each process scenario. A widely investigated approach takes into account a process hourly cost function $Hc = C/T$ made up of costs associated to the process operating conditions and the chart implementation; the total hourly cost function Hc can be modeled as a function depending on the chart design parameters: the values of the chart design parameters which lead to the minimization of the total cost function can be assumed as the optimal economic chart design.

However, as widely reported in literature, selecting the chart design parameters to achieve a pure economic goal can lead to serious problems concerning the chart statistical efficiency: very often, the economically selected chart parameters provide charts having an extremely high probability of false alarms. Thus, even if the hourly cost is mathematically minimized, the loss in chart reliability due to the high frequency of false alarms represents a very important "cost" that cannot be numerically quantified. To overcome this problem, the economic design of the chart can be constrained statistically by introducing a lower bound on the expected number of false alarms. In this chapter, this kind of constrained optimization will be pursued.

19.4.2 Formulation of the Mathematical Model

For the proposed S^2 EWMA chart, the total hourly cost Hc has been determined by doing reference to the economic model by Lorenzen and Vance (1986). According to this model, the process production horizon can be divided into cycles made up of the following phases: production, monitoring and adjustment following the occurrence of a special cause. Each cycle begins with the production process in the "in-control" state and continues operating until an "out-of-control" signal is plotted on the chart and the following corrective actions are completed. The corrections are considered as perfect and restore the process to a condition "as good as new". Following an adjustment in which the process is returned to the in-control state, a new cycle begins. The repeated cycle of the process defined under this structure corresponds to a renewal process. A process considering costs incurred in such renewal process corresponds to a renewal-reward process, see Ross (1970).

The failure mechanism is modeled through an exponential random shock model: due to the occurrence of a special cause the process suddenly shifts to the out-of-control condition. This assumption restricts the implementation of the proposed model to all those processes characterized by a constant failure rate: as a consequence, this model is not suitable for progressive aging processes. The working of metals by means of cutting tools or forming dies is an example of such processes: if a preventive maintenance policy excluding the wear phenomenon by early replacement of the tool has not been designed, due to wear the probability of occurrence of a special cause increases with time and the failure rate increases during the process cycle. In this case, the exponential random shock mechanism should be replaced by a Weibull distribution: for major details, the interested readers can refer to Hu (1984), Banerjee and Rahim (1988), Rahim (1993).

Finally, it is worth noting that the implementation of an economic model requires a full knowledge of the process operating conditions. This can happen only after a preliminary period of chart implementation during which the charts have been designed in accordance to practical considerations and practitioners take confidence with the procedures related to their use: several interesting suggestions can be found in the SPC manuals, see for example Montgomery (2004). After this "warm up" period, before starting to take into account the economic aspects related to the control chart, the practitioners should have collected:

- the parameters, *i.e.*, mean and standard deviation, and the shape of the statistical distribution modeling the controlled parameter measures;
- the failure rate of the process affecting the quality of the monitored parameter, which should be constant;
- the fraction of non-conforming items with process both in-control and out-of-control conditions;
- some cost and time quantities, which will be discussed below.

To define the economic model for the investigated S^2 EWMA chart the following notation has been adopted:

δ_1 is a parameter set equal to 1 if production continues during searches, equal to 0 if production is stopped, (in this chapter $\delta_1 = 1$);

δ_2 is a parameter set equal to 1 if production continues during repair, equal to 0 if production is stopped, (in this chapter $\delta_2 = 0$).

λ is the number of occurrences per hour of an assignable cause (the failure rate): it represents the parameter of the exponential distribution that models the length of the "in-control" period:

$$pdf: f(t) = \lambda\, e^{-\lambda t}$$

τ represents the entity of the shift of the process standard deviation from σ_0 to σ_1. Models considering a multiple special cause structure will not be taken into account here. To gain more insight about this concern, the interested reader can refer to Duncan (1971) and Wu *et al.* (2004). Once the "in-control" standard deviation is known, one can optimize the chart design with respect to a shift τ of the standard deviation:

$$\tau = \frac{\sigma_1}{\sigma_0}$$

C_0 is the cost per hour of operation due to nonconformities produced while the process operates in the "in-control" state. This cost can be computed by taking into account the hourly rate of production r [parts/h], the in-control fraction non-conforming p and the unit cost of producing a non-conforming part C_{nc} [\$/part]:

$$C_0 = r \cdot p \cdot C_{nc} \ [\$/h]$$

C_1 is the cost per hour of operation due to non-conformities produced while the process operates in the "out-of-control" state. This cost can be computed by taking into account the hourly rate of production r [parts/h], the out-of-control fraction non-conforming p' and the unit cost of producing a non-conforming part C_{nc} [\$/part]:

$$C_1 = r \cdot p' \cdot C_{nc} \ [\$/h]$$

c_f is the fixed component of sampling cost. It includes costs related to the repair crew and the out-of-pocket expenses of inspectors' salaries and wages.

c_v is the variable component of sampling cost. It includes the costs associated to the single measure sampling. In the case of destructive testing it should include the cost of the single item sampled and eliminated.

T_0 is the expected search time when a false alarm occurs. It includes the times required to assemble the repair crew and to search for a special cause when it has not been occurred, respectively denoted as T_{rc} [h] and T_{fa} [h]:

$$T_0 = T_{rc} + T_{fa} \ [\text{h}]$$

T_1 is the expected time required to find an assignable cause after a plotted point outside the control limits. It includes the times required to assemble the repair crew T_{rc} [h] and to search for a special cause T_{sc} [h] when it has been occurred. Generally, T_1 coincides with T_0, i.e. $T_{sc} = T_{fa}$:

$$T_1 = T_{rc} + T_{sc} \ [\text{h}]$$

T_2 is the expected time required to repair the process.

W is the cost for locating and repairing an assignable cause when it exists. To compute this cost the practitioners need to know the hourly process downtime cost C_d $[\$/\text{h}]$, the hourly cost C_{rc} $[\$/\text{h}]$ of the repair crew and the times needed to find the assignable cause and repair the process:

$$W = C_{rc} \cdot (T_1 + T_2) + C_d \cdot [(1 - \delta_1) \cdot T_1 + (1 - \delta_2) \cdot T_2] \ [\$]$$

Y is the cost per false alarm. This cost derives from the time required to assemble the repair crew and to maintain the process in the downtime condition to search for a special cause:

$$Y = C_{rc} \cdot T_1 + C_d \cdot [(1 - \delta_1) \cdot T_1] \ [\$]$$

E is the time required to sample and measure a single part. It depends on the operator's skill and on the complexity of the measuring apparatus.

Once the cost and time input data have been collected, the minimization of costs can be pursued. To do this, the expected length T of a cycle and the costs C incurred during this cycle should be computed. For an easier understanding of the terms contributing to the cycle length T, the reader can refer to Fig. 19.5. The expected length of the cycle T can be calculated as the sum of the "in-control" period plus the "out-of-control" period:

$$T = \frac{1}{\lambda} + \frac{(1 - \delta_1) \cdot s \cdot T_0}{ARL_0} + h \cdot ARL_1 - \zeta + n \cdot E + T_1 + T_2$$

the first term in the above expression represents the expected time corresponding to the "in-control" period of the cycle, expressed as the inverse of the process failure rate, i.e., the mean time between failures (MTBF).

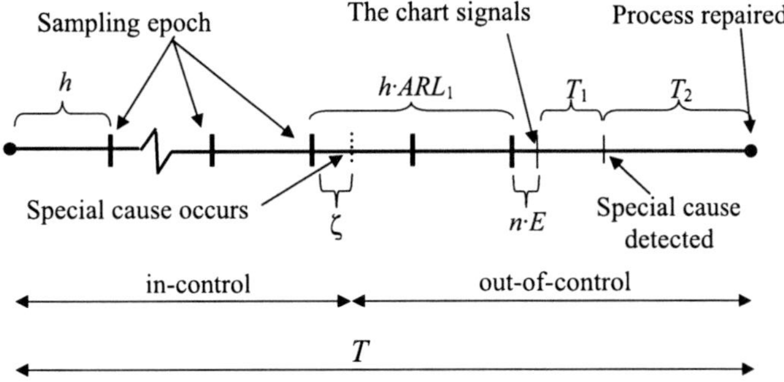

Fig. 19.5 The length T of the cycle time of the process

The second term of this expression takes into account the effect of false alarms on the length of the cycle: it is equal to the expected time T_0 required to search for a false alarm times the expected number of false alarms within a cycle. The variable s represents the expected number of samples taken while the process is in the "in-control" condition:

$$s = \sum_{i=0}^{\infty} i \cdot (\text{assignable cause occurs between the } i\text{th and } (i+1)\text{st samples})$$

$$s = \frac{e^{-\lambda h}}{1 - e^{-\lambda h}}$$

The sum of the remaining terms in the right hand side of cycle time length expression equals the expected length of the "out-of-control" period of the cycle. The term $h \cdot ARL_1$ represents the average time to signal of the chart when a special cause has occurred. The procedure to compute the average run length ARL of the S^2 EWMA chart will be presented below. The variable ζ represents, given the occurrence of an assignable cause between the j and $j+1$ sample, the expected time of occurrence of the cause within this interval:

$$\zeta = \frac{\int_{hi}^{h(i+1)} \lambda'(x - hi)\, e^{-\lambda'x} dx}{\int_{hi}^{h(i+1)} \lambda' e^{-\lambda'x} dx} = \frac{1 - (1 + \lambda h) \cdot e^{-\lambda h}}{\lambda \cdot \left(1 - e^{-\lambda h}\right)}$$

The term $n \cdot E$ gives the time required to take a sample, to collect the n data and to plot the point on the chart; the terms T_1 and T_2 sum over the time needed to find and repairing the special cause.

The expected cost per cycle C includes the costs due to nonconformities with process in the "in-control" and "out-of-control" conditions, here denoted as C_a and C_b respectively; the expected cost C_c associated to false alarms; the cost C_d corresponding to the detection and removal of a special cause; the expected cost C_e required for sampling during the cycle:

$$C = C_a + C_b + C_c + C_d + C_e \; [\$]$$

The expected cost C_a due to nonconformities while the process is operating in the "in-control" condition is a function of the Mean Time Between Failures $MTBF = 1/\lambda$ and the hourly cost C_0:

$$C_a = \frac{C_0}{\lambda} \ [\$]$$

The expected cost C_b due to nonconformities while the process is operating in the "out-of-control" condition is equal to the hourly cost C_1 times the length of the "out-of-control" period of production during the cycle:

$$C_b = C_1 \left(h \cdot ARL_1 - \zeta + n \cdot E + \delta_1 \cdot T_1 + \delta_2 \cdot T_2 \right) \ [\$]$$

The expected cost C_c due to false alarms is equal to the cost per false alarm Y times the expected number of false alarms during the "in-control" period:

$$C_c = \frac{s \cdot Y}{ARL_0} \ [\$]$$

The cost C_d required to detect and remove a special cause is equal to W:

$$C_d = W \ [\$]$$

The sampling cost C_e can be computed as the cost incurred to collect a sample times the number of samples taken during the entire cycle length T:

$$C_e = (c_f + c_v \cdot n) \left(\frac{1}{\lambda} + \frac{(1 - \delta_1) \cdot s \cdot T_0}{ARL_0} + h \cdot ARL_1 - \zeta + n \cdot E + \delta_1 \cdot T_1 + \delta_2 \cdot T_2 \right) / h$$

Assuming that the sequence of production, monitoring and adjustment can be represented as a renewal-reward stochastic process, (Ross, 1970), the expected cost Hc per hour can be determined by dividing C to T:

$$Hc = \frac{C}{T} \ [\$/h]$$

19.4.3 Computation of the ARLs for the S^2 EWMA

The statistical properties of a control chart are usually evaluated by means of its average run length (ARL). Depending on process operating conditions the interpretation of the meaning of ARL for a control chart is twofold: if the process is "in-control", the ARL equals the expected number of samples to be taken between two successive false alarms and is denoted as ARL_0; when the process is "out-of-control" the ARL is denoted as ARL_1 and equals the expected number of samples to be taken between the occurrence of a special cause and its signal from the chart. When designing a chart, a practitioner should try to limit as more as possible the number of

Fig. 19.6 The control interval of the S^2 EWMA chart divided into p subintervals representing the states of the Markov Chain

false alarms and to have a chart very quick in signalling the occurrence of a special cause. This means that: i) the ARL_0 should be settled sufficiently high (generally, $ARL_0 > 200$) to avoid a loss in confidence on interpreting the points falling outside the control limits of the chart; ii) the ARL_1 should be very small to get the out-of-control signal after a very short time, since the special cause has occurred. It is common practice to design control charts having a selected value for the ARL_0 and trying to minimize the ARL_1.

The ARL of the S^2 EWMA control chart can be numerically evaluated by using an approach based on a discrete time Markov chain approximation. This Markov chain approach, originally proposed by Brook and Evans (1972) for the CUSUM chart has been adapted by Lucas and Saccucci (1990). To get the Markov chain model, the interval between the upper and lower control limits LCL and UCL is divided into $p = 2m + 1$ subintervals of width 2δ, see Fig. 19.6. The number p of subintervals has been set equal to 121 to achieve a good compromise between the approximation due to the finite approach and the computational effort depending on the size of the transition probability matrix of the Markov chain. Each interval represents a feasible *state* of the process numerically determined by means of the value of the monitored statistic. When the number of subintervals p is sufficiently large, the finite approach provides an effective method that allows the ARL of a control scheme to be accurately evaluated.

The control statistic:

$$Z_k = (1 - \lambda)Z_{k-1} + \lambda T_k$$

is said to be in *transient state* j at time k if:

$$H_j - \delta < Z_k < H_j + \delta \quad \text{for } j = -m, -m+1, \ldots, 0, \ldots, m,$$

where H_j represents the midpoint of the jth interval. The control statistic is in the absorbing state if Z_k falls outside the control limits. The process is assumed to be "in-control" whenever Z_k is in a transient state, and is assumed to be "out-of-control" whenever Z_k falls in the absorbing state. Based on this mathematical modeling, given the initial state of the Markov chain the ARL can be interpreted as the expected number of transitions before the process falls into the absorbing state. Referring to the formulas characterizing the behavior of the discrete time absorbing Markov chains it holds, (Winston, 2004):

$$ARL = \mathbf{d}^{\mathbf{T}} \cdot (\mathbf{I} - \mathbf{P})^{-1} \cdot \mathbf{1}$$

where $\mathbf{d}$ is the initial probability $(p \times 1)$vector, $\mathbf{I}$ is the $(p \times p)$ identity matrix, $\mathbf{P}$ is the transition $(p \times p)$ probabilities matrix, and $\mathbf{1}$ is a $(p \times 1)$ vector of ones.

The initial probability vector $\mathbf{d}$ contains the probabilities that the control statistic starts in a given state. This vector is such that for $i = -m, -m+1, \ldots, 0, \ldots, m$:

$$d_i = \begin{cases} 1 & \text{if } H_i - \delta < Z_0 < H_i + \delta \\ 0 & \text{otherwise} \end{cases}$$

Consequently, this vector contains only a single element equal to 1, being the remaining $2m$ entries equal to 0. For both the ARL_0 and ARL_1 computation it will be assumed that the chain starts from state 0, $i.e.$, the Z_i corresponding to the initial state of the chain falls within the state containing the central line CL: as a consequence, $d_0 = 1$.

The transition probability matrix $\mathbf{P}$ contains the one step transition probabilities of the Markov chain. The generic element $p_{i,j}$ of $\mathbf{P}$ represents the probability that the control statistic goes from state i to state j in one step. To determine this probability, it is assumed that the control statistic is equal to H_j whenever it falls within the state j, $i.e.$,

$$p_{i,j} \cong \Pr\left[\lambda^{-1}\left\{(H_j - \delta) - (1 - \lambda)H_i\right\} < T_k \leq \lambda^{-1}\left\{(H_j + \delta) - (1 - \lambda)H_i\right\}\right]$$

Introducing the cdf of the random variable T_k, the expression above can be rewritten as:

$$p_{i,j} \cong F_{T_k}\left[\lambda^{-1}\left\{(H_j - \delta) - (1 - \lambda)H_i\right\}\right] - F_{T_k}\left[\lambda^{-1}\left\{(H_j + \delta) - (1 - \lambda)H_i\right\}\right]$$

19.4.4 Formulation of the Constrained Optimization Problem

The economic design of the S^2 EWMA control chart achieves the minimization of the hourly cost Hc by properly selecting the parameters n, h, L and λ_E. Furthermore, a low expected number of false alarms should be assured by the chart. Therefore, the problem of the economic-statistical design of the S^2 EWMA control chart can be formalized as the minimization of the non-linear function Hc under a constraint on the expected number of false alarms signaled by the chart, which can be fixed through the assignment of a proper value for the "in-control" ARL:

$$Hc(n^*, h^*, L^*, \lambda_E^*) = \min_{n,h,L,\lambda_E} [Hc(n, h, L, \lambda_E)]$$

subject to: $ARL_0 = ARL_0^*$

To obtain the optimal parameters $(n^*, h^*, L^*, \lambda_E^*)$ the following strategy can be followed:

1. For $\tau = 1$ and each sample size n extracted out from the range $n_{\min} \leq n \leq n_{\max}$, find the set of couples (L, λ_E) within the ranges $L_{\min} \leq L \leq L_{\max}$ and $\lambda_{E_{\min}} \leq \lambda_E \leq \lambda_{E_{\max}}$ satisfying the statistical constraint on the ARL_0:

$$ARL(1, n, L, \lambda_E) = ARL_0^*$$

At the end of this step, n sets of couples (L, λ_E) satisfying the statistical constraint will be available.

2. For each sample size n and couple (L, λ_E) select the sampling interval h^* out from the range $h_{\min} \leq h \leq h_{\max}$ which minimizes the hourly cost Hc:

$$Hc(n, h^*, L, \lambda_E) = \min_{n,h,L,\lambda_E} [Hc(n, h, L, \lambda_E)]$$

3. Among the minimum expected costs Hc found for each sample size n, sampling interval h^* and couple (L, λ_E), the optimum will correspond to the smallest value of Hc:

$$Hc(n^*, h^*, L^*, \lambda_E^*) = \min_{n,h^*,L,\lambda_E} [Hc(n, h^*, L, \lambda_E)]$$

19.5 The Economic Statistical Design of the S^2 EWMA: a Numerical Analysis

19.5.1 Methodology

The hourly costs for the S^2 EWMA have been determined by means of the procedure presented above and have been compared with those obtained for a S^2 Shewhart chart. A benchmark of 128 examples was considered; costs and process inputs for the economic model were adapted from the numerical example proposed in Lorenzen and Vance (1986): 32 datasets have been determined by assuming two levels for different cost and process parameters; furthermore, two levels for the shift τ have been taken into account: $\tau = 0.5$ (reduction in the variance of the controlled quality characteristic-process improvement) and $\tau = 1.5$ (increase in the variance of the controlled quality characteristic-process deterioration). Although the condition corresponding to the low level for τ is less critical for the quality of process output than the other, a chart able to quickly find reductions in process dispersion allows a rapid signal of process improvement to be detected and the procedure of reformulation of new process control limits to be started. Even if in practice it is not correct at all, the same "out-of-control" costs were considered for both the investigated shifts in process dispersion: this allowed to model the benchmark of examples as a two level full factorial design. Finally, two levels for the "in-control" average run length ARL_0 were assumed as statistical constraints: $ARL_0 = 200, 370$; these values represent a good compromise between the need of a statistically reliable chart and the search for a design solution which allows costs to be maintained at an acceptable level. Thus, the benchmark of examples was structured as a 2^7 unreplicated factorial design. Table 19.2 shows the selected values for process costs and parameters, and the fixed ranges for the design variables: each example results from the combination of the factors fixed at either the high or low level; all the possible alternatives were considered and 128 datasets were obtained.

Table 19.2 Process parameters and design variables ranges for the investigated benchmark

Factors	(−)	(+)	Fixed parameters	
λ [fail/h]	0.02	0.05	E [h]	0.0833
C_0 [\$/h]	57.12	114.24	T_0 [h]	0.0833
C_1 [\$/h]	600	949.2	T_1 []	0.0833
W [\$]	488.7	977.4	T_2 []	0.75
Y [\$]	488.7	977.4	c_f [\$]	0
τ(shift)	0.5	1.5	c_v [\$]	4.22
ARL_0	200	370	δ_1	1
			δ_2	0

Ranges for variables

n	$[3, 4, \ldots, 15]$
h	$[0.2, 0.25, \ldots, 5]$
L	$[2.5, 2.501, \ldots, 2.9]$
λ_E	$[0.05, 0.06, \ldots, 0.7]$

19.5.2 Evaluation of the Cost Savings vs. the S^2 Shewhart

Tables 19.3–19.6 show the optimal designs for the S^2 EWMA and the hourly cost Hc for the S^2 Shewhart chart when $ARL_0 = 200, 370$ respectively. For the entire benchmark of problems, the averaged percentage cost savings $Cs = (Hc_{\text{Shew}} - Hc_{\text{EWMA}})/Hc_{\text{Shew}}$ achievable by means of EWMA with respect to the Shewhart chart were computed: when $ARL_0 = 370$ and $\tau = 0.5$ (1.5) the averaged cost savings are equal to 4.22% (20.17%); similarly, when $ARL_0 = 200$ and $\tau = 0.5$ (1.5) they are equal to 1.21% (19.72%). All the datasets characterized by $\tau < 1$ show a percentage cost saving achievable by the S^2 EWMA chart implementation less than 10%; on the other hand, when $\tau > 1$ the savings vary between 14% and 38%. Four datasets indicate that the Shewhart chart outperforms the EWMA, (with an average saving less than 0.3%); all these examples have been optimized by assuming $ARL_0 = 200$ as statistical constraint, the failure rate λ and the out-of-control cost C_1 set at the low level (−), the false alarm cost Y fixed at the high level (+). All the remaining 124 examples show that the proposed EWMA approach significantly outperforms the Shewhart chart. Figure 19.7 reports an histogram relative

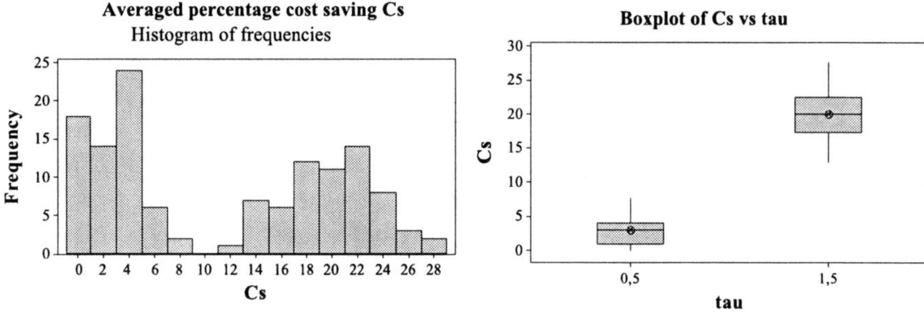

Fig. 19.7 Distribution of the average percentage cost saving Cs: histogram of frequencies and box plot vs. tau

to the frequencies of Cs and a box plot vs. the entity of the shift τ and clearly shows a bimodal behavior of data; when the process dispersion increases, *i.e.* $\tau > 1$, the cost savings achievable by means of the S^2 EWMA vs. the S^2 Shewhart improve: this result is very positive because it confirms the convenience in adopting the proposed chart when shifts which lead to a poor process performance are expected. An analysis of costs contributing to the total expected hourly cost Hc allows to affirm that the trend of Cs vs. the entity of the shift τ mainly depends on the expected sampling cost C_e: when $\tau > 1$ the sample size n required by the S^2 EWMA is significantly smaller than the S^2 Shewhart leading to significant cost savings on C_e; conversely, the higher sampling frequency h designed for S^2 EWMA brings to quite larger costs C_c due to false alarms than the S^2 Shewhart: however, this effect is not so strong as to counterbalance the saving on the sampling costs.

Considering the economic importance of the sample size n on the average cost saving Cs, some other important findings are worthy of note:

- The S^2 EWMA optimal sample size corresponds to $7 < n^* < 10$ when $\tau = 0.5$ and $n^* = 3$ when $\tau = 1.5$: this suggests not only that the selected range $[3, 15]$ for the sample size n is sufficient to find the optimum but also that the S^2 EWMA is very sensitive to dispersion increases in data ($\tau > 1$) and does not need too large samples to quickly detect increases in process dispersion. For the data corresponding to the example of Lorenzen and Vance (1986), Fig. 19.8 shows the optimal hourly costs Hc of the EWMA vs. fixed values of the sample size n for the different combinations for τ and ARL_0.
- The Shewhart chart is always designed with an optimal sample size n^* larger than the S^2 EWMA: in particular, when $\tau = 1.5$, it results $n^* \geq 6$ whichever the selected example; this explains the significantly higher values for the percentage cost savings achieved through the EWMA chart when $\tau = 1.5$: thanks to the logarithmic transformation of the sample variance the proposed EWMA becomes very sensitive to process dispersion increases; in fact, the possibility of dealing with a symmetric quasi-normal distribution avoids the presence of a high type II error β for the chart when $\tau > 1$; on the other hand, high values for β can occur for the S^2 Shewhart chart working on the asymmetrical distributions of the

sample variance. Hence, the higher sensitivity of the EWMA chart with respect to the Shewhart allows a lower optimal sample size to be fixed without affecting the ARL_1 value: significant reductions in the expected sampling costs C_E within a cycle derive, and an acceptable level the out-of-control cost C_B can be maintained. These results demonstrate how the implementation of the S^2 EWMA chart is economically advantageous when increases in process dispersion ($\tau > 1$) are expected. For process improvements in terms of variability, ($\tau < 1$), the over-performance of the S^2 EWMA vs. the Shewhart chart is smaller because the difference between the optimal sample sizes for the two charts reduces.

To further investigate other effects perhaps having influence on the difference Cs between the hourly costs of the two charts, the set of examples was studied as a 2^7 unreplicated factorial plan with the percentage cost saving Cs assumed as response

Table 19.3 Optimal S^2 EWMA designs for $ARL_0 = 200$, $\tau = 0.5$

S^2 Shew.	S^2 EWMA					
$Hc\,[\$/h]$	n^*	L^*	h^*	λ_{E^*}	ARL_1	$Hc\,[\$/h]$
134.72	9	2.683	1.35	0.46	2.85	134.43
190.20	8	2.674	0.80	0.43	3.23	186.58
186.49	9	2.683	1.45	0.46	2.85	186.45
237.40	8	2.674	0.85	0.43	3.23	234.60
134.72	9	2.683	1.35	0.46	2.85	134.43
235.69	6	2.644	0.45	0.31	4.40	225.89
212.34	9	2.683	1.10	0.46	2.85	210.53
283.95	7	2.662	0.55	0.38	3.72	275.24
143.60	9	2.683	1.40	0.46	2.85	143.34
210.43	8	2.674	0.85	0.43	3.23	207.16
195.33	9	2.683	1.45	0.46	2.85	195.33
257.51	9	2.683	1.00	0.46	2.85	255.01
168.85	9	2.683	1.05	0.46	2.85	166.77
256.36	7	2.662	0.55	0.38	3.72	247.05
221.32	9	2.683	1.10	0.46	2.85	219.55
304.57	7	2.662	0.55	0.38	3.72	296.31
135.61	15	2.717	2.45	0.70	1.75	135.89
191.37	9	2.683	0.95	0.46	2.85	188.8
187.33	15	2.717	2.60	0.70	1.75	187.59
238.49	9	2.683	1.00	0.46	2.85	236.59
161.05	9	2.683	1.10	0.46	2.85	159.84
237.29	8	2.674	0.65	0.43	3.23	229.64
213.48	9	2.683	1.10	0.46	2.85	212.56
285.48	8	2.674	0.65	0.43	3.23	278.73
144.48	15	2.717	2.45	0.70	1.75	144.75
211.56	9	2.683	1.00	0.46	2.85	209.29
196.16	15	2.717	2.60	0.70	1.75	196.42
258.57	9	2.683	1.05	0.46	2.85	256.93
170.02	9	2.683	1.10	0.46	2.85	168.87
257.91	8	2.674	0.65	0.43	3.23	250.64
222.44	9	2.683	1.15	0.46	2.85	221.58
306.06	8	2.674	0.65	0.43	3.23	299.73

Table 19.4 Optimal S^2 EWMA designs for $ARL_0 = 200$, $\tau = 1.5$

S^2 Shew.	S^2 EWMA					
$Hc\,[\$/h]$	n^*	L^*	h^*	λ_{E^*}	ARL_1	$Hc\,[\$/h]$
150.19	3	2.326	0.60	0.05	4.58	118.29
210.21	3	2.326	0.40	0.05	4.58	159.72
201.11	3	2.326	0.65	0.05	4.58	171.45
256.69	3	2.326	0.45	0.05	4.58	209.76
150.19	3	2.326	0.60	0.05	4.58	118.29
257.49	3	2.326	0.30	0.05	4.58	187.50
231.13	3	2.326	0.50	0.05	4.58	188.72
305.69	3	2.326	0.35	0.05	4.58	238.67
158.92	3	2.326	0.65	0.05	4.58	127.40
230.15	3	2.326	0.40	0.05	4.58	181.23
209.80	3	2.326	0.65	0.05	4.58	180.54
276.43	3	2.326	0.45	0.05	4.58	231.09
188.04	3	2.326	0.50	0.05	4.58	144.16
278.20	3	2.326	0.30	0.05	4.58	209.41
240.02	3	2.326	0.50	0.05	4.58	197.92
326.24	3	2.326	0.35	0.05	4.58	260.38
151.39	3	2.326	0.65	0.05	4.58	121.81
212.54	3	2.326	0.45	0.05	4.58	164.65
202.18	3	2.326	0.70	0.05	4.58	174.77
258.72	3	2.326	0.50	0.05	4.58	214.48
181.19	3	2.326	0.50	0.05	4.58	139.58
261.87	3	2.326	0.35	0.05	4.58	194.13
233.05	3	2.326	0.55	0.05	4.58	193.22
309.78	3	2.326	0.35	0.05	4.58	244.88
160.10	3	2.326	0.7	0.05	4.58	130.88
232.36	3	2.326	0.45	0.05	4.58	185.98
210.86	3	2.326	0.70	0.05	4.58	183.83
278.33	3	2.326	0.50	0.05	4.58	235.61
190.07	3	2.326	0.50	0.05	4.58	148.78
282.45	3	2.326	0.35	0.05	4.58	215.84
241.91	3	2.326	0.55	0.05	4.58	202.39
330.21	3	2.326	0.35	0.05	4.58	266.59

variable and the parameters reported in Table 19.2 as factors; even if a statistical analysis of the effects influence cannot be performed for a 2^k unreplicated plan by means of an ANOVA table, due to the lack of the experimental error estimate, the entity of the effects can be graphically studied by a normal probability plot on the standardized effects, for further details see Montgomery (2005). In this analysis only the first order terms were taken into account.

Figure 19.9 shows the normal probability plot for the investigated plan; the entity τ of the shift in process dispersion is the parameter having the largest effect on the response variable Cs: as discussed above, the cost savings achievable through the S^2 EWMA are larger when τ is greater than 1, that is when process

Table 19.5 Optimal S^2 EWMA designs for $ARL_0 = 370$, $\tau = 0.5$

S^2 Shew.	S^2 EWMA					
$Hc\,[\$/h]$	n^*	L^*	h^*	λ_E^*	ARL_1	$Hc\,[\$/h]$
142.87	9	2.866	1.25	0.43	3.20	137.73
201.84	7	2.849	0.65	0.33	4.18	191.30
194.16	10	2.873	1.50	0.46	2.88	189.55
248.34	7	2.849	0.70	0.33	4.18	239.13
142.87	9	2.866	1.25	0.43	3.20	137.73
251.06	7	2.849	0.50	0.33	4.18	231.87
222.59	9	2.866	1.00	0.43	3.20	214.68
298.79	7	2.849	0.50	0.33	4.18	280.99
151.66	9	2.866	1.30	0.43	3.20	146.62
221.77	7	2.849	0.65	0.33	4.18	211.83
202.92	10	2.873	1.50	0.46	2.88	198.40
268.14	9	2.866	0.95	0.43	3.20	259.43
179.40	9	2.866	0.95	0.43	3.20	171.07
271.48	7	2.849	0.50	0.33	4.18	252.88
231.50	9	2.866	1.00	0.43	3.20	223.68
319.16	7	2.849	0.50	0.33	4.18	302.01
143.42	10	2.873	1.45	0.46	2.88	138.59
202.55	9	2.866	0.90	0.43	3.20	192.78
194.69	11	2.88	1.70	0.48	2.62	190.29
249.01	9	2.866	0.95	0.43	3.20	240.32
171.22	9	2.866	1.00	0.43	3.20	163.28
252.03	7	2.849	0.50	0.33	4.18	234.12
223.31	9	2.866	1.00	0.43	3.20	215.89
299.74	7	2.849	0.50	0.33	4.18	283.24
152.21	10	2.873	1.45	0.46	2.88	147.45
222.47	9	2.866	0.90	0.43	3.20	213.13
203.44	11	2.88	1.70	0.48	2.62	199.12
268.79	9	2.866	0.95	0.43	3.20	260.56
180.14	9	2.866	1.00	0.43	3.20	172.28
272.45	7	2.849	0.50	0.33	4.18	255.13
232.22	9	2.866	1.05	0.43	3.20	224.88
320.09	7	2.849	0.55	0.33	4.18	304.17

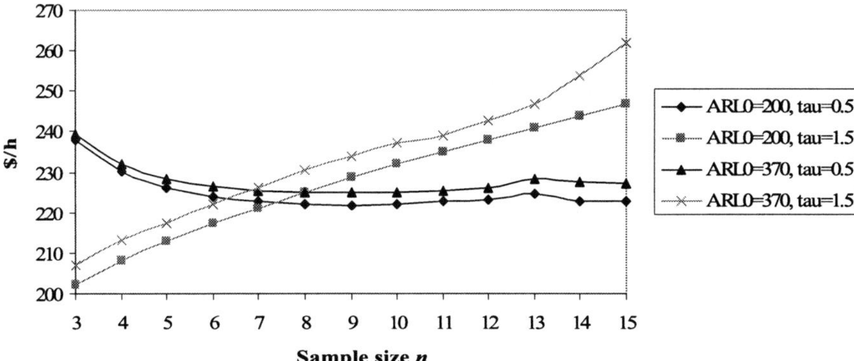

Fig. 19.8 Expected optimal hourly cost Hc vs. the sample size n for the S^2 EWMA, $\lambda = 0.02$, $C_0 = 114.24\ \$/\text{hours}$, $C_1 = 949.2\ \$/\text{hours}$, $W = 977.4\ \$/\text{hours}$, $Y = 977.4\ \$/\text{hours}$

Table 19.6 Optimal S^2 EWMA designs for $ARL_0 = 370$, $\tau = 1.5$

S^2 Shew.	S^2 EWMA					
$Hc\,[\$/h]$	n^*	L^*	h^*	λ_E^*	ARL_1	$Hc\,[\$/h]$
156.18	3	2.548	0.50	0.05	5.87	123.32
220.27	3	2.548	0.35	0.05	5.87	167.14
206.70	3	2.548	0.55	0.05	5.87	176.18
265.83	3	2.548	0.35	0.05	5.87	216.89
156.18	3	2.548	0.50	0.05	5.87	123.32
272.57	3	2.548	0.25	0.05	5.87	197.34
239.37	3	2.548	0.40	0.05	5.87	195.06
320.15	3	2.548	0.30	0.05	5.87	248.24
164.84	3	2.548	0.55	0.05	5.87	132.37
239.80	3	2.548	0.35	0.05	5.87	188.42
215.32	3	2.548	0.55	0.05	5.87	185.22
285.19	3	2.548	0.40	0.05	5.87	237.99
196.65	3	2.548	0.40	0.05	5.87	150.58
292.93	3	2.548	0.25	0.05	5.87	219.13
248.19	3	2.548	0.40	0.05	5.87	204.24
340.37	3	2.548	0.30	0.05	5.87	269.78
156.84	3	2.548	0.55	0.05	5.87	125.54
221.29	3	2.548	0.35	0.05	5.87	170.40
207.32	3	2.548	0.60	0.05	5.87	178.33
266.72	3	2.548	0.40	0.05	5.87	219.77
188.77	3	2.548	0.40	0.05	5.87	144.50
274.74	3	2.548	0.30	0.05	5.87	201.76
240.27	3	2.548	0.45	0.05	5.87	197.93
321.98	3	2.548	0.30	0.05	5.87	252.10
165.49	3	2.548	0.55	0.05	5.87	134.58
240.78	3	2.548	0.40	0.05	5.87	191.64
215.94	3	2.548	0.60	0.05	5.87	187.33
286.03	3	2.548	0.40	0.05	5.87	240.81
197.59	3	2.548	0.40	0.05	5.87	153.68
294.98	3	2.548	0.30	0.05	5.87	223.29
249.06	3	2.548	0.45	0.05	5.87	207.06
342.16	3	2.548	0.3	0.05	5.87	273.63

deterioration is expected and a very sensitive tool is required to monitor the process. Equally significant is the improvement in cost savings due to the variation of failure rate λ and the "out-of-control" cost per hour C_1 due to producing nonconformities: a variation of these parameters from the low to the high value brings to a 2% improvement in the Cs response. All the mentioned effects positively influence the behavior of the S^2 EWMA with respect to the S^2 Shewhart, that is varying them from low to high level corresponds to an increase in cost savings. The statistical constraint ARL_0 also plays an influencing role on the percentage cost saving: in particular, the utilization of the S^2 EWMA scheme becomes more and more effective than the Shewhart chart when larger values for the ARL_0 are assumed as constraints: that is, when the occurrence rate of false alarms should be maintained low, the cost surplus due to the statistical constraint is less heavy for the EWMA than the Shewhart chart. On the other hand, the "in-control" cost per hour C_0 due to produc-

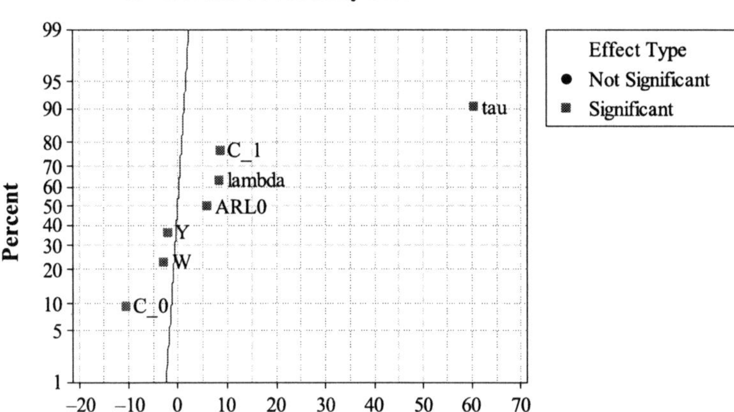

Fig. 19.9 2^7 factorial plan: normal probability plot of the standardized effects for the percentage cost saving Cs response variable

ing non-conformities has a negative influence on cost savings; however, the average cost saving Cs achievable by the S^2 EWMA implementation when C_0 is at the high (+) level still remains equal to about 10%. The remaining main effects Y and W have a less significant influence on Cs. All the interactions have a negligible effect on Cs, see the interaction plot reported in Fig. 19.10.

19.5.3 A Sensitivity Analysis on the Design Parameters of the S^2 EWMA

A further investigation on the design parameters of the S^2 EWMA chart completed the study. To understand if the process costs and parameters have a significant effect on the design variables of the S^2 EWMA, a sensitivity analysis was performed by assuming at each step one of the four design variables n, L, h, λ_E and the hourly cost Hc as response of the experiment; then, the effects influence was evaluated by means of normal probability plots, see Fig. 19.11. Only the first order effects were taken into account. Table 19.7 summaries the obtained results: each row indicates the effect of a factor on the response; the responses are listed in the table columns. A positive sign entry in the table means that varying the effect from low to high level increases the response; a negative sign means that varying the effect from low to high level decreases the response. The process failure rate λ has a negative influence on the design parameters altogether: thus, for processes characterized by more frequent occurrences of special causes or increases in process dispersion, the S^2 EWMA chart is designed with smaller sample size n, width of control limits L, sampling interval h and higher dependence λ_E on past process history: due to the

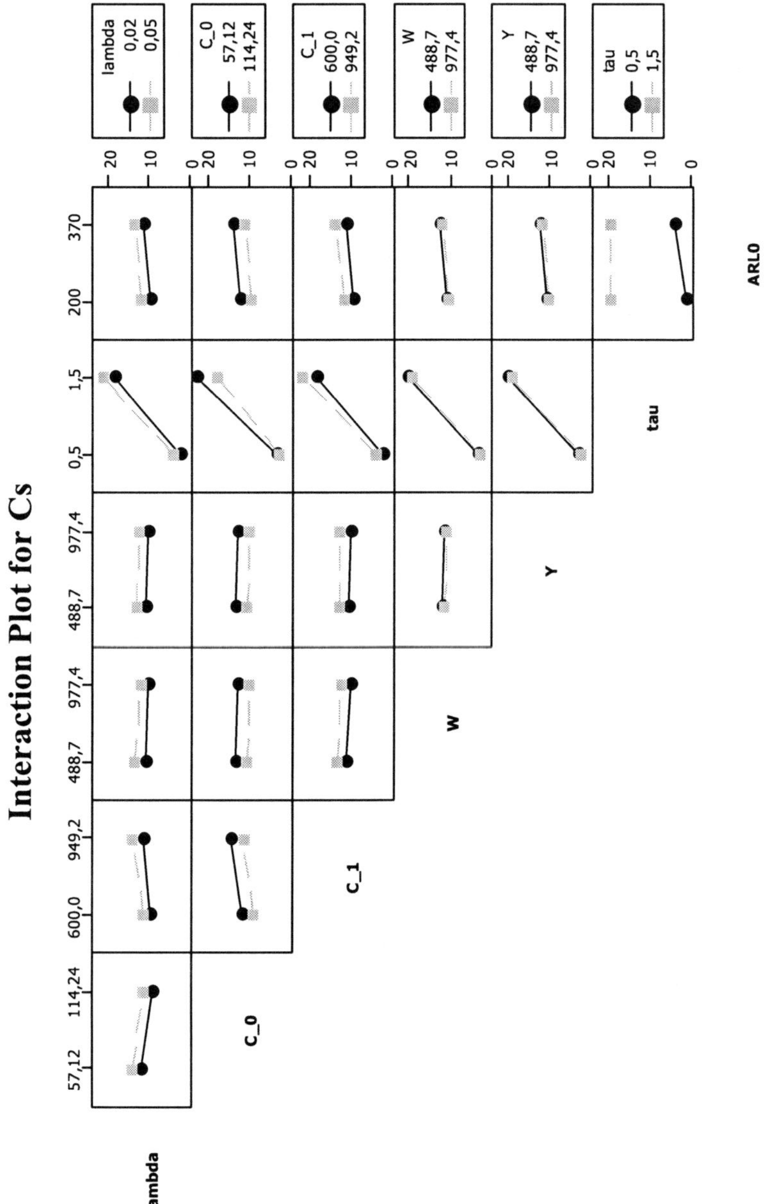

Fig. 19.10 2^7 factorial plan: interaction plot for the response variable Cs

shorter length of the "in-control" period within the process cycle, there is the need of a chart able to quickly detect the "out-of-control" condition through a narrower control interval, higher sampling frequency and stronger correlation among the plotted

Table 19.7 Sensitivity analysis on the S^2 EWMA design variables and on the hourly expected cost Hc

Effect [#]	n	L	h	λ_E	$Hc\,[\$/h]$
Main					
λ [failures/h]	-	-	- -	-	+++
C_0 [\$/h]					+++
C_1 [\$/h]	-		- -	-	+++
W [\$]					+ +
Y [\$]			+		+
τ (shift)	- - -	- -	- -	- - -	- -
ARL_0		+ +	-		+

[#] (+++): high positive influence (+ +): medium positive influence (+): low positive influence;
(- - -): high negative influence (- -): medium negative influence (-): low negative influence

points. Finally, the failure rate λ is the process parameter having the highest positive effect on the process hourly costs Hc, with an average increase of 61 \$/h. The cost C_0 due to non-conformities produced by the process in the "in-control" condition has not statistical effect on the design parameters of the S^2 EWMA, whereas it shows a strong positive influence on Hc. The cost C_1 due to non-conformities produced by the process in the "out-of-control" condition shows a negative effect on all the design parameters; in particular, it has influence on the sampling interval: this result derives from the fact that when high "out-of-control" costs C_1 are expected the sampling frequency should be increased to limit as more as possible the duration of the "out-of-control" condition; furthermore, as expected, C_1 has an high influence on Hc. The cost W to detect and eliminate a special cause does not affect the decision concerning the design parameters of the S^2 EWMA chart; on the other hand, it has a positive influence on the hourly cost Hc. The cost per false alarm Y shows a slight positive influence on h; the influence on the other design parameters n, L and λ_E can be considered negligible. Moreover, the influence of Y on hourly cost Hc is very slight with respect to the other parameters discussed above. The entity of the process shift τ has a negative influence on the chart design parameters: as discussed above this influence is high for the sample size n; the smoothing parameter λ_E is negatively affected by τ: for positive shifts in the process variance the weight of the past process history is improved to quickly detect the "out-of-control" condition. Considering the approximate assumption of equal cost C_1 regardless of the "out-of-control" process condition ($\tau < 1$ or $\tau > 1$), the hourly cost Hc is negatively affected by the entity of the shift τ: once again, this result reflects the high efficiency of the proposed S^2 EWMA in detecting "out-of-control" conditions corresponding to process deterioration: in few words increases in the process dispersion are detected faster than decreases. The statistical constraint ARL_0 has a positive influence on L: this fact depends on the relation between L and ARL_0 by means of the Markov chain model; the ARL_0 only slightly affects negatively the sampling frequency, whereas it has not a practical influence on the other parameters. Finally, it has a very light influence on the hourly cost Hc: this is a very interesting finding be-

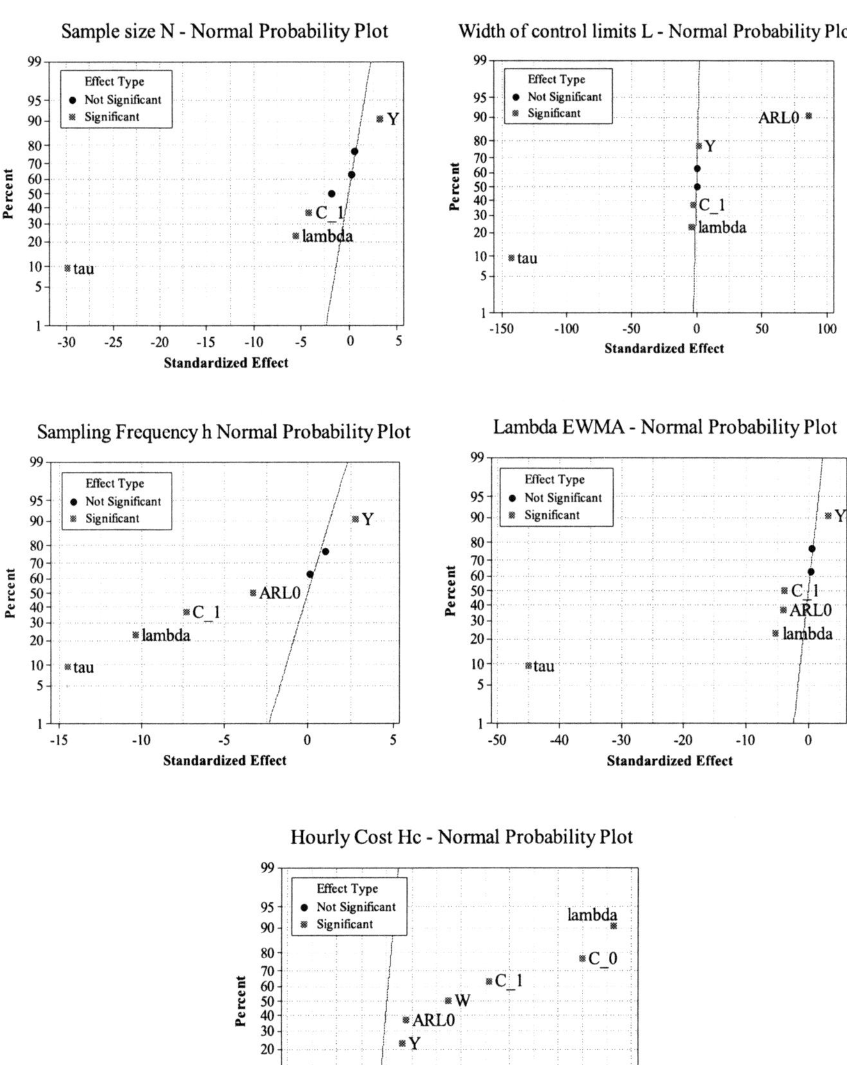

Fig. 19.11 Sensitivity analysis on the design variables n, L, h, λ_E and on the hourly cost Hc: normal probability plots of the standardized effects

cause it means that the S^2 EWMA allows one to achieve an high statistical reliability without excessively improving the related costs.

19.6 Conclusions

Monitoring the dispersion of sample data related to a parameter which affects the output quality of a manufacturing process is a key activity for a correct implementation of Statistical Process Control. A logarithmic transformed S^2 EWMA optimally designed with respect to economic aspects and statistical properties was proposed in this chapter for the on-line control of the sample variance. The selected designs for the chart parameters allow practitioners to implement a statistical tool which has a minimum hourly cost and a low probability of false alarms signals. The proposed S^2 EWMA chart was compared with a S^2 Shewhart chart by considering an extensive benchmark of problems corresponding to several process cost and operating parameters. The examples were organized as a 2^7 factorial design to quantify the influence of each process cost and operating parameter on the overperformance of the EWMA against the Shewhart chart. In particular, significant cost savings can be achieved when the process dispersion is expected to increase, $i.e.$, for process operating conditions corresponding to a quality deterioration of the process output, and for strict constraints on the expected number of false alarms: this result can be achieved thanks to the fact that the sample size n required by the S^2 EWMA is significantly smaller than the S^2 Shewhart and allows significant savings on the sampling cost. A sensitivity analysis on process costs and operating parameters was performed assuming as responses of the factorial plan all the design variables and the hourly process expected cost Hc corresponding to the S^2 EWMA implementation. It was demonstrated that:

- The failure rate λ negatively affects all the design variables and strongly influences the hourly process expected cost Hc;
- The shift in process dispersion τ negatively affects all the design variables and has a medium negative influence on Hc: this means that the proposed S^2 EWMA is particularly suited to detect special causes leading to process deterioration;
- The cost per false alarm Y only affects sampling frequency h and has a slight influence on Hc;
- The "out-of-control" cost C_1 due to non-conformities negatively affect h: that is, when C_1 improves higher sampling frequencies are required to minimize costs; furthermore, C_1 strongly influences the hourly process expected cost Hc;
- The cost factors W and C_0 have no effect on the design variables, whereas they affect positively the hourly cost Hc;
- The statistical constraint ARL_0 has a positive influence on L. Furthermore, it slightly influences the hourly cost Hc: this finding confirms the possibility of achieving an high statistical reliability for the S^2 EWMA without excessively improving the related costs.

Acknowledgements This work is partially funded by the Projet International de Coopération Scientifique PICS-3753 entitled "Méthodes statistiques adaptatives pour la surveillance de la variabilité de procédés" of the CNRS (Centre National de la Recherche Scientifique) and by the Project "Carte di controllo avanzate per il monitoraggio della variabilità di processo" funded in the year 2005 by the University of Catania

References

1. Acosta-Mejia CA, Pignatiello JJ Jr, Rao BV (1999) *A comparison of control charting procedures for monitoring process dispersion.* IIE Transactions 31:569–579
2. Amin RW, Wolff H, Besenfelder W, Baxley R Jr (1999) *EWMA control charts for the smallest and largest observations.* Journal of Quality Technology 31:189–206
3. Banerjee PK, Rahim MA (1988) *Economic design of $\bar{X}$-control charts under Weibull shock models.* Technometrics 30:407–414
4. Box GE, Hunter WG, Hunter JS (1978) *Statistics for Experimenters.* Wiley, New York, ISBN: 0-471-09315-7
5. Brook D, Evans DA (1972) *An approach to the probability distribution of CUSUM run length.* Biometrika 59:539–549
6. Castagliola P (2005) *A New S^2-EWMA control chart for monitoring the process variance.* Quality and Reliability Engineering International 21:1–14
7. Castagliola P, Celano G, Fichera S (2006) *Monitoring process variability using EWMA.* In: Pham H (ed.) Springer Handbook of Engineering Statistics. Springer, New York, ISBN: 1-85233-806-7
8. Celano G, Fichera S (1999) *Multiobjective economic design of a X control chart.* Computers and Industrial Engineering 37:129–132
9. Crowder SV (1987) *A Simple method for studying run-length distributions of exponentially weighted moving average charts.* Technometrics 29:401–407
10. Crowder SV (1989) *Design of exponentially weighted moving average schemes.* Journal of Quality Technology 21:155–162
11. Crowder SV, Hamilton MD (1992) *An EWMA for monitoring a process standard deviation.* Journal of Quality Technology 24:12–21
12. Duncan AJ (1956) *The economic design of x charts used to maintain current control of a process.* Journal of American Statistical Association 51:228–242
13. Duncan AJ (1971) *The economic design of $\bar{X}$ control charts when there is a multiplicity of assignable causes.* Journal of the American Statistical Association 66:39–53
14. Gan FF (1995) *Joint monitoring of process mean and variance using exponentially weighted moving average control charts.* Technometrics 37:446–453
15. Hamilton MD, Crowder SV (1992) *Average run lengths of EWMA control charts for monitoring a process standard deviation.* Journal of Quality Technology 24:44–50
16. Hu PW (1984) *Economic design of an X-control chart under non-poisson process shift.* TIMS/ORSA Joint National Meeting, San Francisco, Calif., p. 87
17. Johnson NL, Kotz S, Balakrishnan N (1994) *Continuous Univariate Distributions.* Wiley, New York, ISBN: 0-471-58495-9
18. Lorenzen TJ, Vance LC (1986) *The economic design of control charts: a unified approach.* Technometrics 28:3–10
19. Lucas JM, Saccucci MS (1990) *Exponentially weighted moving average control schemes: properties and enhancements.* Technometrics 32:1–12
20. MacGregor JF, Harris TJ (1993) *The exponentially weighted moving variance.* Journal of Quality Technology 25:106–118
21. McWilliams TP, Saniga EM, Davies DJ, (2001) *Economic, statistical and economic-statistical design of X-bar and R charts and Xbar and S charts.* Journal of Quality Technology 33:234–241
22. Montgomery DC (2004) *Introduction to Statistical Quality Control.* 5th edn. Wiley, New York, ISBN: 0-471-65631-3
23. Montgomery DC (2005) *Design and Analysis of Experiments.* 5th edn. Wiley, New York, ISBN: 0-471-31649-0
24. Ng CH, Case KE (1989) *Development and evaluation of control charts using exponentially weighted moving averages.* Journal of Quality Technology 21:242–250
25. Rahim MA (1993) *Economic design of $\bar{X}$ control charts assuming weibull in-control times.* Journal of Quality Technology 25:296–305

26. Roberts SW (1959) *Control chart tests based on geometric moving averages.* Technometrics 1:239–250
27. Robinson PB, Ho TY (1978) *Average run lengths of geometric moving average charts by numerical methods.* Technometrics 20:85–93
28. Ross SM (1970) *Applied probability models with optimization applications.* Holden-Day, San Francisco, ISBN: 0-318-74688-3
29. Saniga EM (1989) *Economic statistical control-chart designs with an application to $\bar{X}$ and R charts.* Technometrics 31:313–320
30. Steiner SH (1999) *Exponentially weighted moving average control charts with time varying control limits and fast initial response.* Journal of Quality Technology 31:75–86
31. Sweet AL (1986) *Control charts using coupled exponentially weighted moving averages.* IIE Transactions 18:26–33
32. Tolley GO, English JR (2001) *Economic designs of constrained EWMA and combined EWMA control schemes.* IIE Transactions 33:429–436
33. Torng JCC, Cochran JK, Montgomery DC, Lawrence FP (1995) *Statistically constrained economic design of the EWMA control chart.* Journal of Quality Technology 27:250–256
34. Winston WL (2004) *Operations Research.* Applications and Algorithms, 4th edn. Brooks/Cole–Thomson Learning, Belmont Calif., USA, ISBN: 0-534-42362-0
35. Wortham AW, Ringer LJ (1971) *Control via exponential smoothing.* Logistics Review 7:33–40
36. Wu Z, Shamsuzzaman M, Pan ES (2004) *Optimization design of control charts based on Taguchi's loss function and random process shifts.* International Journal of Production Research 42:379–390

Chapter 20
Risk Management Techniques
for Quality Software Development

Toshihiko Fukushima[1], Shigeru Yamada[2]

[1] Department of Quality Assurance, Nissin Systems Co., Ltd, 293-1, Ayahorikawa-cho Ayanokouji Sagaru, Horikawa Street, Shimogyo-ku, Kyoto-shi, Kyoto, 600-8482 Japan
[2] Department of Social System Engineering, Tottori University, 4-101 Minami, Koyama-cho, Tottori-shi, Tottori, 680-8552, Japan

20.1 Introduction

Due to the rapid growth of the IT community, users are demanding both very specific requirements and quick delivery time. Under this pressure, it is often difficult for project managers to meet these high expectations. Consequently, many of risks remain latent in almost all software development projects. The use of countermeasures after failures shows a certain degree of "management failure". Therefore, we have to manage such risks at an early stage for projects to be successful. Figure 20.1 shows the relationship between these risks and QCD result of project.

In the past, the development division of Nissin Systems Co. Ltd had been determining the delivery date and budget even before the requirements were completed. These projects were unable to be accurately planned, which led to QCD failures and ultimately, project failures. Figure 20.2 shows the main causes of our company's three worst project failures. These causes were failures in the project plan, requirement analysis, outsourcing management, and progress management. Performing

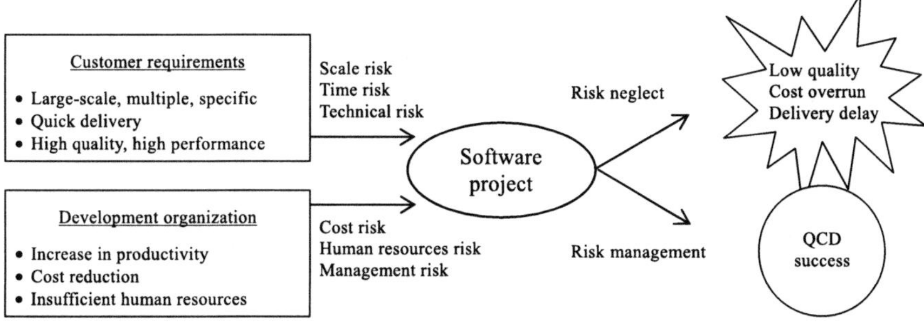

Fig. 20.1 Relationship between risks and QCD result

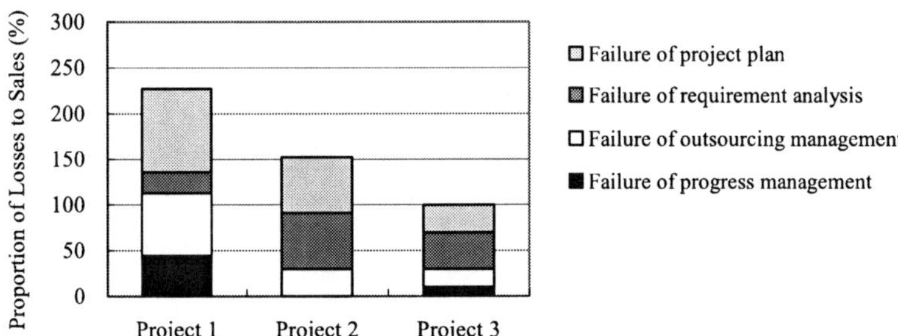

Fig. 20.2 Cause of big failure projects

failure-causal analysis on 1 year's worth of failed projects showed that the cost excess ratio (excess costs/total sales) were 2.9% for requirements management, 4.3% for project planning, 1.8% for progress management, 2.6% for subcontract management, and 0.9% for quality assurance. It became evident that project management was needed at the onset of the project, as it was found that vague requirements and the unfamiliarity of risks were the biggest causes of project failures.

To prevent recurrence, the Quality Management Division introduced risk management activities to identify and mitigate such project risks. Risk mitigation activities were implemented within the development division for the failed projects resulting in an overall decrease of 12.4%. Figure 20.3 shows the reduction of excess-cost after risk management was implemented.

Recently, the Quality Management Division has implemented risk management, process quality assurance (QA), and product QA activities, to manage software-development projects (see Fig. 20.4).

(1)   Risk management activities encompass the activities that identify, evaluate and track risks, and develop risk management plans.

(2)   Process QA activities support project management by using activities such as project plan review and earned value management (EVM) [1, 2] to monitor project progress.

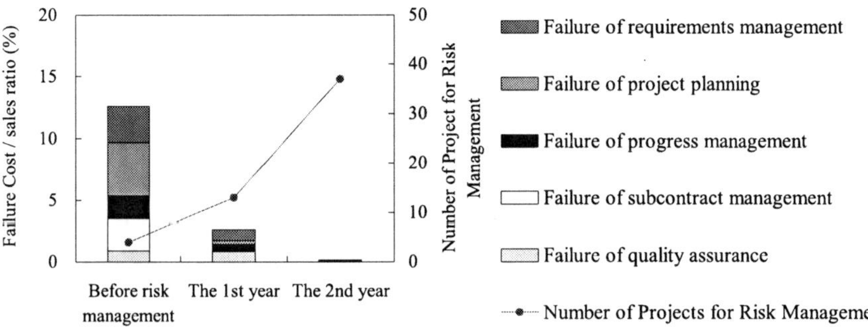

Fig. 20.3 Reduction of failure cost

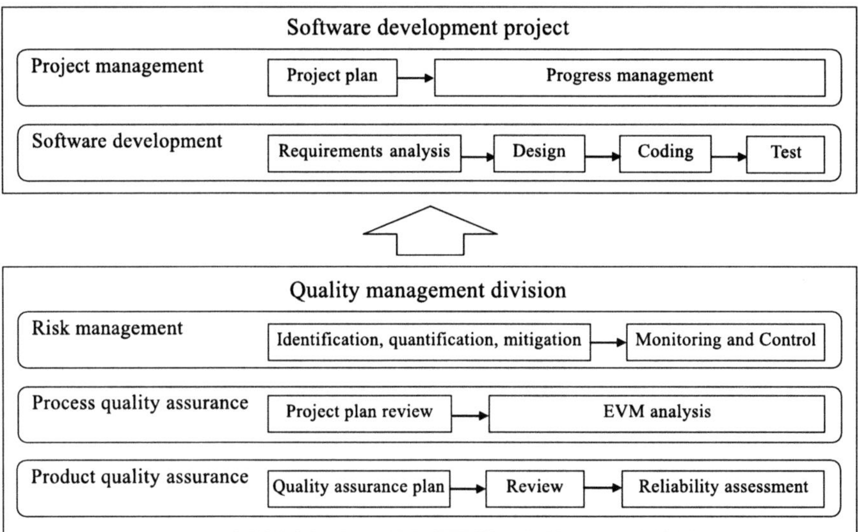

Fig. 20.4 Activities of quality management division

(3)   Product QA activities cover those activities which guarantee software product quality by evaluating design-reviews and tests. Moreover, it uses software reliability assessment techniques [3] to analyze the quality of software.

Through these activities, project failures due to risk are prevented.

This chapter analyzes the effect of the above activities on the project result by performing a multiple regression analysis. Using this analysis, the effects of project management factors on the project result is examined, and the method to manage a successful project is discussed.

20.2 Project Risk Management

20.2.1 Project Risks in Practice

Project risk is anything that might jeopardize the successfulness of a project. Many risks are latent in software projects. For example, when developing a new system, the risk that the schedule becomes less clear is latent. In our experience, the risks are considered to emanate from three sources: arbitrary scheduling, ambiguous requirement definition, and the problem of human factors. Main risks that were experienced in actual projects are shown below:

(1)   Schedule risk

- Arbitrary schedule which does not take into account the size of the project
- Unreasonable schedule

(2)   Requirement definition risk

- New development or new technology
- Ambiguous quality specification or software products
- Unclear scope
- Insufficient consideration of client's needs

(3)   Human factor risks

- Project manager's lack of management skill
- Insufficient developers with appropriate knowledge of technologies
- Insufficient communication within project team

20.2.2 Risk Management Activities

The main risk management activity used in our company is to review projects that may fall below QCD requirements. For this activity, interviews were conducted with the project manager during weekly risk management meetings regarding risk identification, risk quantification, and risk mitigation. During development, risks are tracked to estimate whether there were any signs of more risk. The following procedures are performed (see Fig. 20.5):

(1)   Risk identification:
       At the start of a project, risk is identified by interviewing the project manager using a risk checklist (see Table 20.1) which is then analyzed.
(2)   Risk quantification:
       The identified risks are quantified as a risk ratio using the risk checklist. Moreover, evaluation uses three areas of interest; danger, pursuit, and safety.
(3)   Risk mitigation:
       Risk mitigation is planned based on the result of the evaluation from step 2.
(4)   Risk monitoring and control:
       During development, regular risk management meetings are held to monitor and control risk. Projects are then prioritized based on risk evaluation and risk ratio.

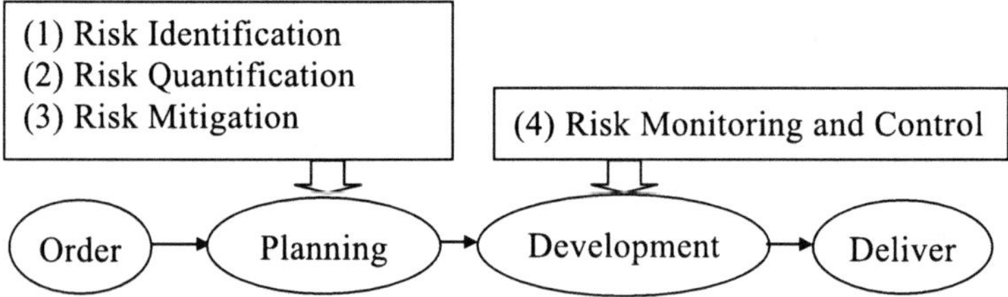

Fig. 20.5 Outline of the risk management

Table 20.1 Risk checklist

Development System			weight(i)
Development scale	(1)	There are many functions or its size is undefined	3
	(2)	System architecture is complicated or complexity is unknown	3
Functional requirement	(1)	Can't understand for which purpose the system is used	4
	(2)	End user function or operator function is ambiguous	6
	(3)	Can't know the most important quality requirement demanded by the user	4
Efficiency requirement	(1)	Can't agree with efficiency requirements, or can't grasp efficiency	4
	(2)	There is no track record of the efficiency requirement	4
System extensibility	(1)	The document does not describ the system's maximum size	2

Organization			
Basic technology	(1)	Project member has no experience with the application	4
	(2)	Member does not have sufficient skill	2
	(3)	Member has no experience with the technology	2
Project management	(1)	The role and responsibility of project manager are undefined	2
	(2)	The project manager has little experience	4
	(3)	There is no time to validate quality assurance	4
	(4)	Organization doesn't have a member's margin	2
Schedule	(1)	Schedule is unexplained	2
	(2)	The size of software outputs has not been estimated	2
Resource	(1)	Development environment is insufficient	2
	(2)	Test environment is insufficient	2
Outsourcing	(1)	Subcontractor is not resident	2
	(2)	Output of subcontractor is not clear	4
	(3)	A great portion of development software is outsourcing	4
	(4)	Skill of subcontractor is insufficient	2

Customer			
Promised delivery date	(1)	The delivery date is earlier than the estimated completion date	6
Customer	(1)	Customer's specification is insufficient, or often changed	1
	(2)	Customer often interrupts during development	1
	(3)	Customer is a new client	2
	(4)	There is a deficiency of communication within the customer's company	1
	(5)	The delay resulting from customer's inputs impacts the project	1

Contract			
Estimation	(1)	Order has not been received, or extent of contract is not clear	4
	(2)	The deliverables or the due date does not decided	4
	(3)	The estimate is too soft or the budget is suppressed	4
	(4)	Field-testing cost is undefined	2
Acceptance testing	(1)	The customer acceptance-testing period is not determined	4

Risk ratio			100

20.2.3 Risk Identification and Quantification

Risk is identified by interviewing the project manager using a risk checklist. A risk interview is conducted by a specialist with development and management skills of software development projects. Moreover, the checklist can detect the usual risks easily since actual failure cases are included.

The checklist shown in Table 20.1 lists the causes of failure from past projects by classifying them into system characteristics, organization of development, customer characteristics and contract details. Using these identified risks, the risk ratio of the project is calculated using the following mathematical formula:

$$\text{Risk ratio} = \sum (\text{risk factor (i) in the checklist} \times \text{weight (i)}), \qquad (20.1)$$

where each item on the checklist has weight (i), and the risk ratio can reach a maximum of 100 points.

Furthermore, with regards to damage control, we focus on three areas: whether the risk is (i) dangerous, (ii) pursuable, and (iii) what the safe conditions are. The dangers of projects are frequently assessed at regular risk management meetings.

20.2.4 Risk Mitigation

Projects plan to mitigate identified risks. Risk mitigation strengthens the processes of the project based on our standard software process. In many cases, the requirement analysis and the project planning process are strengthened by risk mitigation. The standard software process defines the organization's role and processes in the software development of our company by using a guideline "Organization process definition" which is KPA (Key process area) of Capability Maturity Model (CMM) level 3 [4]. In addition, the standard software process is concerned with defining a contract, development, management, and quality assurance process based on the CMM.

Examples of risk mitigation to prevent ambiguous requirements are shown below:

(1) Ensure that the requirements specification's review covers the following points: client's purpose of using the system, checks all system functions for loopholes, it is clearly defined for testing.
(2) Obtain client's approval of the requirements document early in the project.
(3) Ensure that the requirements written in the requirements specification are covered in system test cases.
(4) Use software reliability growth model [3] to evaluate software quality. The software reliability growth model is a well known model used as a method of presuming the number of faults that remains in software using the number of faults detected during testing. Figure 20.6 is an example of a project that has

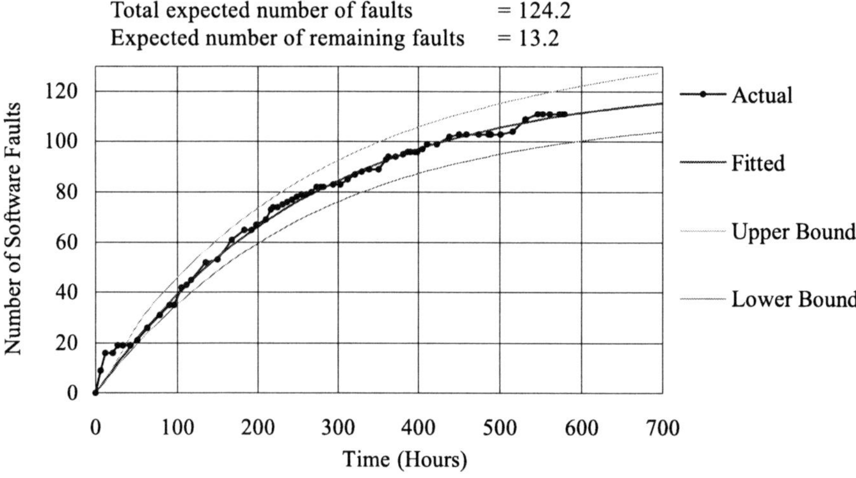

Fig. 20.6 Applying the exponential software reliability growth model

applied the software reliability growth model. With this figure, we are able to determine shipment quality using the ratio of "expected number of remaining faults" to "the total expected number of faults". This example resulted in a successful project because there were no customer complaints.

20.2.5 Risk Monitoring and Control

During software development, the risk ratio is reevaluated using the risk checklist for identification of existing risks and any new risks. With regard to risk monitoring and control, it is evaluated based on the result of progress management using EVM. EVM uses the relationship between cost and schedule performance to determine the progress of the project. Using the EVM, we can readily compare how much work has actually been completed against the amount of work planned. Therefore, EVM can be used as a means to track the influence of schedule risks. By analyzing the difference between plan and fact, the accuracy of re-planning is improved and risk can be mitigated.

Figure 20.7 shows an example of progress management using the EVM technique. The horizontal axis is time transition and the vertical axis is the cost. Planned value (PV) is the value of the physical work scheduled, per the authorized budget to accomplish the scheduled work. Actual cost (AC) is total costs incurred that must relate to whatever cost was budgeted within the planned value and earned value in accomplishing work during a given time period. Earned value (EV) is the sum of the approved cost estimates for activities completed during a given period.

The risk management activity judges whether risk exists or not based on the following QCD control limit:

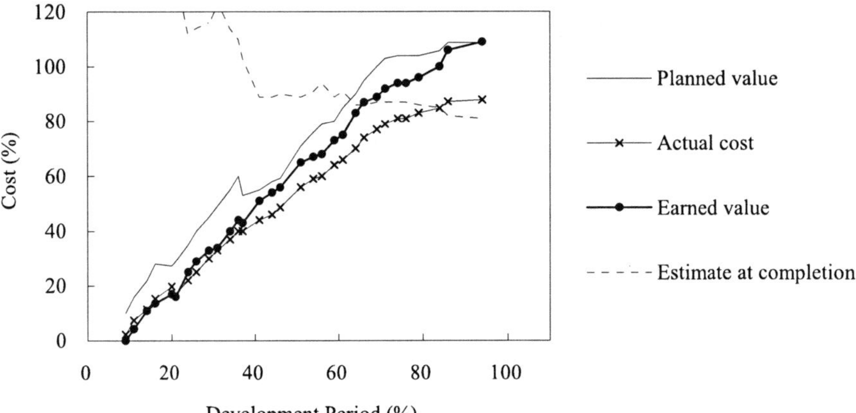

Fig. 20.7 An example of EVM results (PV, AC, and EV)

(1)   Quality:

- Did the design-review pass?
- Is the design-review on time, within 10% of the plan?

(2)   Cost:

- Is the EVM cost variance (CV) less than 10% of the planned value?

(3)   Delivery:

- Is the EVM time variance (TV) less than 10% of the planned value?

where 10% is the QCD control limit. CV is the difference between EV and AC, *i.e.*, CV = EV − AC. TV is the difference between EV and PV, *i.e.*, TV = EV − PV (see Fig. 20.8).

When a project exceeds the above-mentioned QCD control limit, the level of risk becomes serious. As a result the risk management team has to (i) analyze the causes, (ii) plan risk mitigation, and (iii) pursue the validity of the mitigation. If control

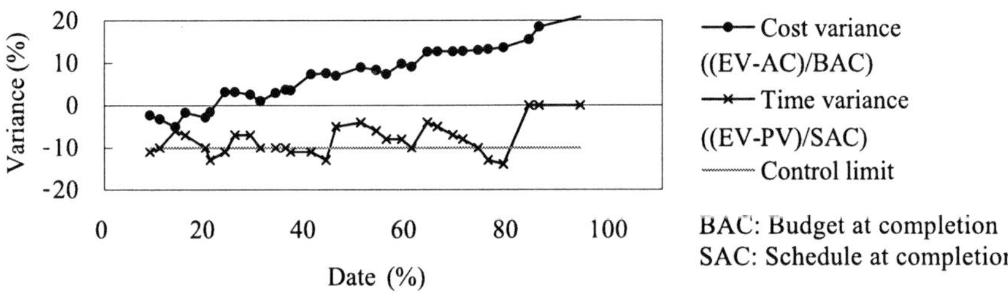

Fig. 20.8 An example of EVM results (CV, TV). BAC budget at completion, SAC schedule at completion

limits have been surpassed for 2 weeks or more, a "risk alarm" is activated and the senior executive is made aware of the problem. Then risk mitigation is performed at the company level.

20.3 Project Effect Analysis

It is important to continuously perform statistical analysis of measurement data collected from software development projects to improve project management techniques. This will improve the forecasting and controlling of projects and lead to more QCD success.

This section quantitatively evaluates projects by using measurement data from actual projects, and discusses the effectiveness of the project management to achieve high software product quality and to meet development costs [5].

20.3.1 Assumptions

Our activities in risk management, process QA, and product QA have been changing the results of our projects. A multiple linear regression analysis was applied by using data collected from these activities to obtain relationships between these activities and the project's result. These relationships can predict the results of a project at an early stage as shown in (20.2) and (20.3). Using these equations, the effects of various factors on software products quality and development cost can be examined.

Software product quality
$$= F(\text{risk management, process QA, product QA}) , \qquad (20.2)$$
Software development cost
$$= G(\text{risk management, process QA, product QA}) , \qquad (20.3)$$

where mathematical formulas $F(\ )$ and $G(\ )$ are given by linear regression equations.

Two variables were used as the dependent variables: Number of faults (Y_1) as a measure of software product quality and budget variation (Y_2) as a measure of software development cost. Five variables were used as independent variables: risk ratio of project initiation (X_1) and speed of risk mitigation (X_2) as measures of risk management, frequency of EVM (X_3) as a measure of process QA, frequency of review (X_4) and pass rate of review (X_5) as measures of product QA. These variables are explained below:

Y_1: Number of faults $=$ (the number of faults detected during acceptance testing and in operation).
Y_2: Budget variation $=$ (actual cost at completion/budget at completion).
X_1: Risk ratio of project initiation $= \sum$ (risk factor (i) in the checklist $\times$ weight (i)).

X_2: Speed of risk mitigation = (the date when the risk ratio reaches 30 points or less − the start date)/(expected completion date − the start date). The reason we use 30 points as a benchmark is because almost all projects that reached their QCD target had a risk ratio of 30 points or less.

X_3: Frequency of EVM = the frequency of EVM analysis per project size. There were several examples of where the project risks could be mitigated with more frequent EVM analysis as the managers were able to deal with problems early on.

X_4: Frequency of review = the frequency of design-review per project size. There were several examples where frequent reviews increased the quality of the product.

X_5: Pass rate of review = the pass rate of the first design-review. This measures the level of judgment of the reviewer.

20.3.2 Correlation Analysis

To estimate the parameters of the model, a regression analysis was performed by using the data from ten real projects (see Table 20.2). The correlation of the five independent variables (X_1, X_2, X_3, X_4 and X_5) and two dependent variables (Y_1, Y_2) was analyzed by using the real data (see Table 20.3). The following correlation characteristics were obtained from Table 20.3:

(1) X_1 and X_2 shows a strong positive correlation with Y_1 and Y_2, so multiple linear regression analysis can be applied.
(2) X_1 shows a strong correlation with X_2 and X_3.
(3) X_5 has no correlation with Y_1.

Based on the correlation analysis, X_2, X_3, X_4 and X_5 were selected as the important factors because X_1 is multicollinear with X_2 and X_3.

20.3.3 Multiple Linear Regression

A multiple linear regression analysis with Y_1 was conducted by using the data shown in Table 20.2. The multiple-regression equation for software product quality given by (20.4) was obtained, as well as the normalized multiple regression expression is given by (20.5). In order to check the goodness-of-fit adequacy of the model, the coefficient of multiple determination (R^2) was checked, and given as 0.5624. Furthermore, the squared multiple correlation coefficient, adjusted for the degrees of freedom (adjusted R^2), was given as 0.3438. The output of the multiple linear regression analysis is shown in Tables 20.4 and 20.5.

Table 20.2 Analyzed data

Project No.	Risk management		Process QA	Product QA			Software products quality	Software development cost
	Risk ratio of project initiation	*Speed of risk mitigation*	*Frequency of EVM*	*Frequency of review*	*Pass rate of review*		*Number of faults*	*Budget variation*
	X_1 (0–100)	X_2 (0–1.00)	X_3 (time/scale)	X_4 (time/scale)	X_5 (0–1.00)		Y_1	Y_2
1	73	1.00	0.21	0.05	0.50		12	1.13
2	47	0.33	0.78	0.10	0.50		0	0.92
3	38	0.87	1.18	0.14	0.50		13	1.12
4	35	0.95	0.74	0.12	1.00		3	1.04
5	34	0.51	2.70	0.22	0.50		0	0.93
6	32	1.00	1.70	1.71	0.67		1	1.19
7	32	0.28	0.79	0.07	0.00		0	0.92
8	24	0.00	0.99	0.62	0.74		0	1.03
9	23	0.00	1.84	0.31	0.50		0	0.82
10	13	0.00	2.42	0.54	1.00		0	1.00

Table 20.3 Correlation matrix

	X_1	X_2	X_3	X_4	X_5	Y_1	Y_2
X_1	1						
X_2	0.6328	1					
X_3	-0.6563	-0.3538	1				
X_4	-0.3376	0.1248	0.3321	1			
X_5	-0.3018	0.0495	0.2377	0.2903	1		
Y_1	0.6507	0.6432	-0.4551	-0.2891	-0.0722	1	
Y_2	0.3517	0.7377	-0.2746	0.4750	0.3019	0.5889	1

Table 20.4 Estimated parameters for software product quality

Term	Coefficient	SE	p	95% CI of Coefficient
Intercept	1.4190	3.7329	0.7169	-7.7152 to 10.5532
X_2	7.6944	3.5950	0.0761	-1.1022 to 16.4911
X_3	-0.7582	2.0276	0.7213	-5.7194 to 4.2031
X_4	-3.3709	3.0405	0.3100	-10.8108 to 4.0691

$$Y_1 = 7.6944X_2 - 0.7582X_3 - 3.3709X_4 + 1.4190 , \qquad (20.4)$$

$$Y_1^N = 0.6426X_2 - 0.1181X_3 - 0.3301X_4 , \qquad (20.5)$$

where Y_1^N is normalized equation (20.4).

A multiple linear regression analysis with Y_2 was also conducted. As a result, the multiple-regression equation of software development cost has been given by (20.6), and the normalized multiple regression expression by normalized data equation is given by (20.7). To check the goodness-of-fit adequacy of this model, R^2 has been

Table 20.5 Table of analysis of variance for software product quality

Source of variation	SSq	DF	MSq	F	p
Due to regression	134.4	3	44.8	2.57	0.1499
Error	104.5	6	17.4		
Total	238.9	9			

given as 0.7281 and adjusted R^2 has been given as 0.5922. The output of the multiple linear regression analysis is shown in Tables 20.6 and 20.7.

$$Y_2 = 0.1607X_2 - 0.0314X_3 + 0.1077X_4 + 0.9307 , \qquad (20.6)$$

$$Y_2^N = 0.6012X_2 - 0.2189X_3 + 0.4726X_4 , \qquad (20.7)$$

where Y_2^N is normalized equation (20.6).

From (20.5) and (20.7), the order of degree affecting the dependent variables is $X_3 < X_4 < X_2$. It has been found that the "speed of risk mitigation" is an important impact factor on determining the number of software product's faults and the software development cost.

Accuracy for predicting the number of faults is shown in Fig. 20.9. Figure 20.9 indicates that the correlativity between the actual number of faults and predicted number of faults is not so high, as well as the coefficient of multiple determination. As for cause, the number of faults for 6 of the 10 projects analyzed was 0. Nevertheless, (20.4) is helpful in predicting many projects, which have many actual faults. From Fig. 20.9, we can understand that quality is improved by early risk mitigation and many reviews.

Accuracy for predicting cost is shown in Fig. 20.10. Figure 20.10 shows that the budget variation can be predicted with high accuracy, as well as the coefficient of

Table 20.6 Estimated parameters for software development cost

Term	Coefficient	SE	p	95% CI of Coefficient
Intercept	0.9307	0.0657	< 0.0001	0.7700 to 1.0914
X_2	0.1607	0.0632	0.0441	0.0059 to 0.3154
X_3	-0.0314	0.0357	0.4132	-0.1186 to 0.0559
X_4	0.1077	0.0535	0.0907	-0.0232 to 0.2386

Table 20.7 Table of analysis of variance for software development cost

Source of variation	SSq	DF	MSq	F	p
Due to regression	0.087	3	0.029	5.36	0.0392
Error	0.032	6	0.005		
Total	0.119	9			

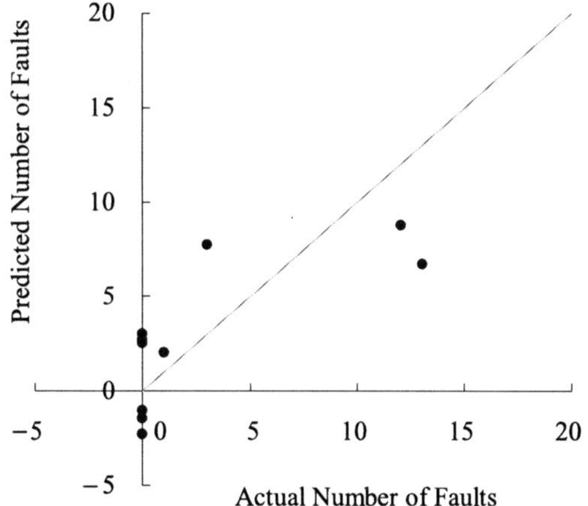

Fig. 20.9 Accuracy for predicting the number of faults

multiple determination. From Fig. 20.10, we can understand that cost is improved by early risk mitigation, although it is increased by review.

20.3.4 Effectiveness Evaluation of Management Factor

(1) Evaluation of risk management
When risk management, process QA, and product QA results were compared, risk management was evaluated to have the greatest positive influence on the

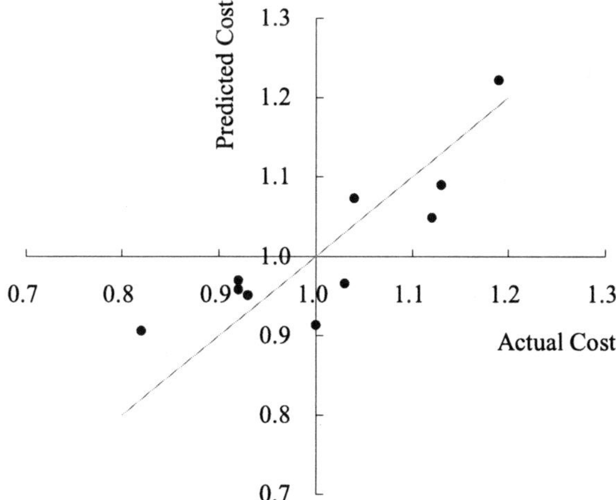

Fig. 20.10 Accuracy for predicting cost

project result. The reason is that only risk management activities directly support the project managers.

(2) Evaluation of product QA

Because of the design-review activity's poor judgment, pass rate of review (X_5) was deemed to not have any influence on the product quality. Therefore, we determined that accuracy of design-review was one of the risks. After that, risk management activities supported the design-review activities.

20.4 Conclusions

This chapter discussed a method to implement a risk management system within an organization, and proposed a method to quantitatively evaluate the effect of management factors by using prediction equations based on the measured data from real projects.

By devising multivariate factors and normalized variables of software development using a multiple linear regression analysis, the prediction equations were able to obtain accurate results. As a result, these showed that risk management activities that support the project manager were very effective in producing good project results. Therefore, we think that this quantitative evaluation method is effective in promoting the improvement of project management techniques that software development organizations can implement.

In the future, by using the quantitative evaluation method, the Plan-Do-Check-Act cycle of our improvement will be established. We also aim to continue to improve project management techniques to develop high-quality software.

References

1. Nohzawa T (1999) International Standard Project Management: PMBOK and EVMS (in Japanese). JUSE Press, Tokyo
2. Tominaga A (2003) Explanation: Earned Value Management (in Japanese). The Society of Project Management
3. Yamada S (1994) Software Reliability Modeling: Fundamentals and Applications (in Japanese). JUSE Press, Tokyo
4. Carnegie Mellon University Software Engineering Institute (CMU/SEI) (1995) The Capability Maturity Model: Guidelines for Improving the Software Process (Sei Series in Software Engineering), Addison-Wesley, New York
5. Fukushima T, Fukuta A, Yamada S (2005) Early-stage product quality prediction by using software process data," Proceedings of the 11th ISSAT Internnational Conference on Reliability and Quality in Design, St Louis, Missouri, USA, pp. 261–265

Part V
Application in Engineering Design

Chapter 21
Recent Advances in Data Mining
for Categorizing Text Records

W. Chaovalitwongse[1], H. Pham[1], S. Hwang[1], Z. Liang[1], C.H. Pham[2,3]

[1] Department of Industrial and Systems Engineering, Rutgers University, USA
[2] Cisco Systems Inc., San Jose, California, USA
[3] Department of Electrical Engineering, San Jose State University, USA

21.1 Introduction

In a world with highly competitive markets, there is a great need in almost all business organizations to develop a highly effective coordination and decision support tool that can be used to become a daily life predictive enterprise to direct, optimize and automate specific decision-making processes. The improved decision-making support can help people to examine data on the past circumstances and present events, as well as project future actions, which will continually improve the quality of products or services. Such improvement has been driven by recent advances in digital data collection and storage technology. The new technology in data collection has resulted in the growth of massive databases, also known as data avalanches. These rapidly growing databases occur in various applications including service industry, global supply chain organizations, air traffic control, nuclear reactors, aircraft fly-by-wire, real time sensor networks, industrial process control, hospital healthcare, and security systems. The massive data, especially text records, on one hand, may contain a great wealth of knowledge and information, but on the other hand, contain other information that may not be reliable due to many uncertainty reasons in our changing environments. However, manually classifying thousands of text records according to their contents can be demanding and overwhelming. Data mining has gained a lot of attention from researchers and practitioners over the past decade as an emerging research area in finding meaningful patterns to make sense out of massive data sets. It can be viewed as a process to extract hidden predictive information or pattern from data in large collections. Data mining has been successfully applied to correlating text records in many industries. It has been used to extract and integrate information from multiple sources to provide value-added services, such as customizable Web information gathering, comparative shopping, grouping medical records, and product diagnosis.

During the past few years, there has been an explosion of interest in data mining research applied to correlation/categorization problems of text records, which arise in many different industries (*e.g.*, telecommunication, pharmaceutical and life sciences, media, finance and banking, insurance, retail, public sector, and manufacturing) [1]. However, the real challenge in this line of research is that a typical text record only represents factual information (communicative intentions) in a complex, rich, and opaque manner [2]. Consequently, unlike numerical and fixed field data, it cannot be analyzed by traditional data mining methods and existing software products and tools due to (1) the format of the data source and its temporal nature, (2) high sensitivity and variation of the entry data, (3) impact of data representation on appropriate mining functions, and (4) high degree of dependency between the attributes in a pattern. In addition, a text record has no standard or consistent attributes embedded in it. There are several ways to describe and store the information distributed in a text record. The size of accessible text data can be very large although they may contain a great wealth of knowledge. Extracting useful information and representing it as data attributes require a tremendous amount of work in scanning all the documents and recognizing their importance in a text record. Thus, the increase in accessible retrieval information in a text record from database systems has caused an information flood in spite of hope of becoming knowledgeable about various topics [3]. In order to overcome this challenge, a special branch of data mining, called text mining, has been widely studied over the past few years to acquire useful knowledge from large amounts of text records such as internal reports, technical documents, e-mail messages, medical records, defect records, and so on.

Similar to data mining, linguistic knowledge for Natural Language Processing (NLP) is among the most widely used methods for knowledge extraction from text records. However, the information extraction process is typically focused on domain-specific lexical and semantic information stored in a database. For instance, Message Understanding Conferences (MUCs) is used to find a specified class of events (*e.g.*, company mergers) and to fill in a template for each instance of such an event. Therefore, this technology cannot solve the problem of finding novel patterns (or new knowledge) rather than predefined patterns (or existing knowledge) in a specified class and extracting information that may be used to discover trends in the domain described in the text record database.

For all the reasons mentioned above, research in text mining is still in its infancy. Data mining researchers have also been exploring new data mining frameworks that can be tailored specifically to encounter the problem of extracting and representing information from massive text records. This chapter presents an overview practice of correlating and categorizing text records as well as recent advances in data mining research for mining text records. The chapter is structured as follows. An overview of real life text mining problems is provided in Sect. 21.2. Section 21.3 presents a brief background in data mining research for readers to get familiar with existing standard approaches in text mining. Section 21.4 discusses the state-of-the-art approaches in data mining that may be applicable to analyzing complex categorizing text records. Section 21.5 provides a discussion on several research challenges in analyzing text records and future directions of text mining. Here, we give a list of acronyms that will be used throughout the paper: ANOVA = analysis of variance, CART = classification and regression tree, CHAID = Chi-Square automatic interaction detector, NLP = natural language processing, MDS = multidimensional scaling, MUC = message understanding conference, PDP = product development process, SVM = support vector machine.

21.2 Text Mining in Practice

In general, correlating/categorizing text records is the task of quickly and accurately identifying text records corresponding to the same entity or group from the database. The entity or group can be varied depending on the criterion of correlation or categorization such as individuals, products, geographical regions. The text record correlation/categorization can also be referred to many different tasks such as entity heterogeneity, entity identification, object isomerism, and instance identification. The applications of text record correlation/categorization are many. In customer-service systems for example, it can be used for marketing, customer relationship management, fraud detection, customer service improvements, and data warehousing [4]. The most important issue for text mining is how to represent the contents of textual data in a text record. Such representation should be standard or uniform so that, along with many other purposes, one can apply on-line monitoring statistical process or data mining approaches to obtain useful results, feedbacks and findings. As we know that the content in text records could be massive and varies greatly, there will be sizable information in the text records that cannot be retained within the representation and cannot appear in the final output. The representation system of text mining is known as a pre-processing step for data transformation. The transformation and the sized of transformed data can vary significantly depending on specific domains [5]. For example, text record databases such as patent documents and Medline data contain much larger size of keywords or vocabularies than the databases in contact centers or product development centers. Once an appropriate representation function is determined, the next step is the selection of statistical analysis or data mining approaches, which can be adapted to the new text records' data formats. Next we discuss some real life operations where the text mining research has played such a critical role.

21.2.1 Product Development Process

A product development process (PDP) is the sequence of steps, processes or activities which an enterprise employs to conceive, design and commercialize a product [6]. The PDP needs to be well organized and coherent to ensure the efficient delivery and the quality of a final product. Although every organization may follow a slightly different PDP, the basic elements are usually the same. The major PDP steps are as follows: planning, design, production, service, and support [7]. Each of these PDP steps requires much effort to ensure product quality and reliability.

A wide variety of text mining tools is currently being used in the industry for many purposes. They can be used in fault diagnosis, process and quality control and machine maintenance tasks. Advances in text mining can give great impacts on product design and process planning, which have been shown to constitute 60–80% of the total manufacturing costs [7]. Another application of text mining tools in product development process is the use of probabilistic networks and decision trees for failure diagnosis of process units [8].

21.2.2 Customer Service and Product Diagnosis

The application of text mining in customer service and product diagnosis is focused on finding valuable patterns and rules in text (problem) records that indicate trends and significant features about specific topics. The text record database for this process may be a collection of warranty repair information from service centers, which contains records of the repair actions, customer complaints and individual product details carried out at the service centers [6]. This text database is used mainly to maintain transaction records for repair. The database is usually in a form of hybrid fixed-format and free-form text fields. Fixed-format fields are those that have strict formatting criteria for the type, range and precision of its contents such as record IDs, components, products and versions. Free-text fields, on the other hand, are unstructured with no formatting requirements. The real challenge in mining this type of text database is the high flexibility of the text contents in the free-text field. In addition, the database size can grow at a rate of several thousand records per month. Most importantly, the size of a problem record, itself, can be dynamic and keeps on growing until the problem is solved. In most cases, for control reasons, problem records cannot be edited after being entered. Because of this limitation, problem records will include any incorrect or misleading information that was entered. It will also contain any possible fixes that were unsuccessful [9].

Correlating/categorizing text records takes into account two very important issues: functional and technical. The functional issue may arise when a single product problem may actually (1) straddle or involve other products, (2) contain or encompass more than one product issue, (3) be the result of third party interactions or dependencies [9]. The technical issue arises when the amount of accessible problem records data has been increasing. Problem records are in a form of series of texts, which are intrinsically multi-dimensional. When taking the temporal property (the order of texts) into account, the mining process has to face with the exponential increase in computational requirements of analyzing sequential data as opposed to other types of data.

An example of the use of text mining in customer service is the association analysis between products. If a customer detects an issue using a product, he/she can report it to an engineer at customer service by opening a "Problem Record". A problem record is then stored in a database as a hybrid record (fixed-format and free-form text). The record usually contains all of the information about a problem from its first reporting, up to and including its resolution. When a new customer detects the same issue as previously described, another service engineer will try to look for associations of the current issue and the past records in the database. The service priority might be biased, since the new engineer might look for associations that he/she, personally, deems important and some important associations might be overlooked in the process. Therefore, the association analysis in data mining can be used to generate all association rules present in the database, subject to statistical confidence and support constraints. Another example of text mining application is to provide access to classified problem records while reducing the effort of processing the problem records. Instead of a manual process to go through every record according to their content, advances in text mining may be used to configure the desired categories and to train the data query process.

21.2.3 *Improved Healthcare Quality with Electronic Medical Records*

The introduction of the electronic patient medical record has allowed physicians and patients to examine the cost and quality in healthcare. In addition, patient medical records contain a wealth of information that can prove invaluable for the conduct of clinical research. Large numbers of patient medical records can now be examined in a hospital or private practice setting, and patients can be tracked through the healthcare system [10]. In general, these clinical records are largely maintained in free-text form. Text mining research can be applied to clinical records by constructing a reliable and efficient method to extract structured information and examine the totality of patient records. Then text mining analysis can provide physicians a quantitative decision-making tool to determine whether there is a general consensus as to the treatment of patients. This tool is very important in the emergency unit (EU) at a hospital for initial diagnosis to determine if those patients have emergent conditions. Generally, the triage values of emergent, urgent, and non-urgent are assigned based upon the initial complaint that will be stored in the clinical records. The use of data mining has been successfully applied in a study of mining variety of patient information with breast complaints to determine if the patients have breast cancer [11].

21.3 Background in Data Mining

During the last decade, there have been an explosive growth and advances in digital data collection and storage technology. Advances in data collection, widespread use of bar codes for most commercial products, and the computerization of many business and government transactions have flooded us with information embedded in massive databases, also known as data avalanches, and generated an urgent need for new techniques and tools that can intelligently and automatically assist us in transforming this data into useful knowledge [2]. Many such new techniques and tools are studied and described in the emerging field of data mining. Data mining can be broadly defined as the analysis of large observational data sets to find hidden and unsuspected relationships and to summarize the data in novel ways that are both understandable and useful to the data owner [12]. The term data mining is also often known as knowledge discovery, which is referred to the process of extracting useful information from databases [9]. Data mining can be viewed as a powerful technology with great potential to help researchers and practitioners focus on the most important information in their databases. Data mining can play a very crucial role in the product development process. The information from an initial analysis of the strengths and weaknesses of products can be used to improve the products' superiority in today's highly competitive markets. The status of the products' predecessors encrypted in problem text records (*e.g.*, defect records, customer relationship management, fraud detection, data warehousing) can be used to obtain the needed information about reported product problems. This operation lends itself to research challenges in information retrieval/matching. Data mining tools can also be used to predict future trends and behaviors, allowing businesses to make proactive, knowledge-driven decisions. The automated, prospective analyses offered by data mining move beyond the analyses of past events provided by retrospective tools typical of decision support systems. In practice, there are two main goals in data mining studies: prediction and description [13]. Prediction has to deal with analyzing some variables (with attributes or features) in the data set to estimate or predict unknown or future values of other variables of interest. Description has to deal with extracting patterns embedded in the data that can be interpreted or explained by any field experts.

21.3.1 Data

Data can be defined as observations, numbers, or text records that can be processed by a computer. Today, data are being accumulated in vast and growing amounts in different formats and different databases. Examples of data that we can see in our everyday lives include operational or transactional data (*e.g.*, sales, cost, inventory, payroll, and accounting), non-operational data (*e.g.*, industry sales, forecast data, and macro economic data), data about the data itself (*e.g.*, logical database design or data dictionary definitions).

21.3.2 Information and Knowledge

The patterns, associations, or relationships among all the recorded data can provide information. For example, analysis of retail point of sale transaction data can yield information on which products are usually purchased together. Generally, information can be converted into knowledge about historical patterns and future trends. For example, summary information on retail supermarket sales can be analyzed in light of promotional efforts to provide knowledge of consumer buying behavior. Thus, a manufacturer or retailer could determine which items are most susceptible to promotional efforts. One famous example of supermarket transactions is the association rule between diapers and beers. In practice, the information that one wants to discover for making a better decision are association rules, classification rules, clustering and segmentation, similarity search, and sequential patterns.

21.3.3 Data Mining Process

The data mining process starts from understanding the problem to translating the knowledge extracted from the data to the users [4]. In data mining, domain experts are very critical as they are the ones who understand the meaning of the data. They can also identify quality problems of the data. In this section, we briefly describe the data mining process as follows. Data sampling/selection is the first step where the data are selected by domain experts for the modeling process. Data cleaning/preprocessing is the second step where the domain experts cleanse and reformat the data, as the database can accept data only in a certain format. From the database, the users can present the data in a useful format, such as a graph or table. Data transformation/exploration is the third step where the data are extracted, transformed, and analyzed by data analysis tools (*e.g.*, statistics) to explore the useful information in the data. Pattern extraction and discovery is the forth step where modeling tools are used by selecting tables, records, and attributes for typical tasks and they also create new derived attributes that might represent useful patterns. In this step, the modeling and evaluation tasks are coupled. They can be repeated several times to change parameters until optimal performances are achieved. Reporting/visualization is the last step where the data mining experts provide data access to the end users and decide how to use the data mining results. A generic data mining process is illustrated in Fig. 21.1.

Nevertheless, for research purposes, the data mining analysis may involve the following processes [6]. The first step is summarizing the data, which is a process to reduce or transform the data for interpretation and analysis without losing any important information. The second step is finding hidden relationships, which is a process to identify important, yet non-obvious, facts, patterns, relationships, anomalies or trends in the data. The last step is making predictions or decisions, which is a process to estimate or calculate for something that is unknown or the decision to be made.

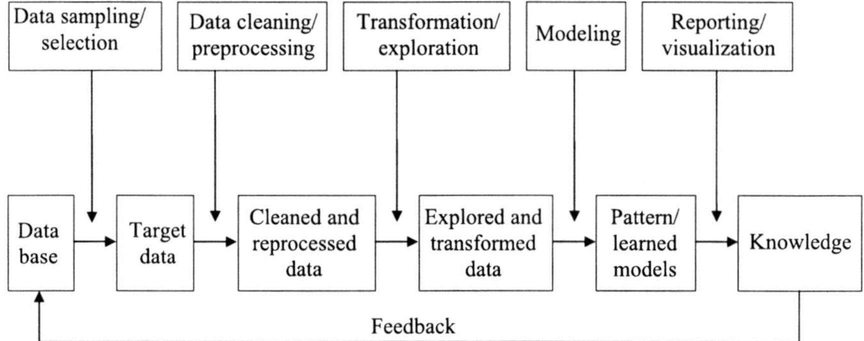

Fig. 21.1 A generic data mining process

21.4 State-of-the-art in Data Mining

In this section, we give a review of the state-of-the-art approaches in data mining and discuss how they can be applied to text mining. Recalling the data mining process, once appropriate concepts are extracted from the data, each piece of text contents for text mining, we can apply various statistical analysis methods in data mining to structured text records. The methods in data mining we discuss here include basic statistics, clustering, nearest neighbor, decision trees, induction rules, neural networks, factor analysis, log-linear analysis, and multidimensional scaling. These data mining methods may enable us to analyze the content of text records from the viewpoints of various semantic functions or textual patterns.

It is worth noting that before one applies data mining techniques to the data in the form of text records, appropriate concepts used to extract information from each piece of text content in a text record are required. These concepts are very crucial in text mining as much knowledge can be extracted from text contents, such as linguistic knowledge for Natural Language Processing (NLP) and domain-specific lexical and semantic information that may be stored in a text record. These important concepts in text mining can be viewed as a data pre-processing step and are often referred to the *document-handling technologies* [3]. In document search, the emphasis is put on the extraction of text record related to some specific topics, in which the data can be represented by character strings or keywords. In document organization, as data mining experts focus on the clustering and classification of text records, the data can be characterized by a vector space model representing a set of keywords. Analysis of keyword distribution is then performed to generate sets (clusters) of documents. In knowledge discovery from text records, the objective is to extract interesting information or messages from content. The text records can be transformed by semantic concepts and investigated using NLP, generic data mining, or visualization techniques to discover embedded information (*e.g.*, trend patterns, association rules). For example, a simple function that examines the increase and decrease of occurrences of each concept in a certain period may allow us to analyze trends in topics. In addition, the semantic classification of concepts may allow us to study the content of text records from the viewpoints of various semantic categories.

21.4.1 Basic Statistics

In data mining, statistical analyses are usually used to discover patterns and build predictive models which are driven by the data. Many of today's data mining techniques grew out of the statistical analyses and the basic concepts of probability, independence and causality, and even over fitting. The knowledge in basic statistics is itself the foundation on which data mining is built. Histogram is a very obvious example, as a prediction can be expressed as a histogram through multiple possible prediction values, each accompanied by a probability and other statistics. Specifically, a histogram is a collection of possible values of data that can be used to estimate event probabilities or constitute a prediction model. Another example is the regression analysis. As we all learned in our basic statistics courses, regression model can be constructed from input data to provide an excellent way to visualize two-dimensional data and provide descriptive techniques for classification.

Bayesian decision theory is one of the most widely used statistical approaches to the problem in data mining. The basic idea of this approach is based on an estimation of probabilistic decisions [2], which can be described by the Bayes formula given by:

$$\text{posterior} = \frac{\text{likelihood} \times \text{prior}}{\text{evidence}}.$$

We can estimate the prior and likelihood probabilities from the value of input data and convert it to a *posteriori* probability. Then the Bayes decision rule for minimizing the probability of error can be expressed in terms of the posterior probability by calculating conditional and prior probabilities [2]. This Bayesian approach has been widely used in several data mining tasks (*e.g.*, classification [14], anomaly detection [1]) to solve many emerging problems in diverse domains (*e.g.*, Internet [3], epidemiology [8], pharmacology [13]).

21.4.2 Clustering

Clustering is a method used to group records together (or simply, divide a database into groups). Clustering sometimes can be viewed as segmentation. In general, clustering is used for the exploration of inter-relationships among a collection of patterns, by organizing them into disjoint and homogeneous clusters. Intra-connectivity is a measure of the density of connections between the instances (data points) of a single cluster. A high intra-connectivity indicates a good clustering arrangement because the data points grouped within the same cluster are highly dependent on each other. Inter-connectivity is a measure of the connectivity between distinct clusters. A low degree of interconnectivity is desirable because it indicates that individual clusters are largely independent of each other. Given a set of data points, the goal of clustering is to find data clusters or segments such that (1) data points in one cluster are more similar to one another; (2) data points in separate clusters are less similar to one another. In other words, the ultimate objective of clustering is to maximize the intra-connectivity and minimize the inter-connectivity. The choice of similarity measure is a key property in any clustering method as it is used to calculate the degrees of intra-connectivity and inter-connectivity.

k-means clustering, first introduced in the mid-1970s [13], is among the most well studied clustering algorithms. The goal of k-means clustering is to find $k$cluster centers that minimize a squared-error criterion function [2]. In other words, the k-means method partitions the data set into k clusters (subsets) such that all points in a given cluster are closest to the same center. A simple solution procedure for k-means algorithm can be described as follows. It first randomly selects k of the data points to represent initial clusters, all remaining data points are assigned to their closer center. The algorithm then computes the new centers by taking the mean of all data points belonging to the same cluster. This procedure is iterated until there is no change in the cluster centers. If k is not known a priori, various values of k can be evaluated until the most suitable one is found. k-means clustering can also be modeled as an optimization problem, whose formulation is given by

$$\min_{c} \sum_{i=1}^{m} \min_{l} \left\| x^i - c^l \right\|_n ,$$

where there is given m points, represented by x, in an n-dimensional space, and a fixed number of cluster, k. The objective of this optimization model is to determine the centers of the cluster, c, such that the sum of the distances of each point to a nearest cluster center is minimized. It is important to note that the effectiveness of this method, as well as of others, relies heavily on the objective function used in measuring the distance between data points. The difficulty is in finding a distance measure that works well with all types of data.

The applications of clustering are many. Market segmentation is a good example [15]. Clustering is used in market segmentation to subdivide a market into distinct subsets of customers where any subset may conceivably be selected as a market target to be reached with a distinct marketing strategy. The clustering quality may be measured by observing buying patterns of customers in same cluster versus those from different clusters. Bioinformatics is another application, where clustering has played such a critical role over the past decade. Clustering has been used to build groups of genes with related expression patterns based on DNA microarrays. These gene groups usually contain functionally related proteins, such as enzymes for a specific pathway, or genes that are co-regulated [16]. Clustering techniques for high throughput experiments on biological data can be a powerful tool for genome annotation, a general aspect of genomics. In sequence analysis, clustering is used to group homologous sequences into gene families, which is a very important concept in bioinformatics, and evolutionary biology. In high-throughput genotyping platforms, clustering algorithms are used to automatically assign genotypes. For text records, clustering may be used by transforming a text record into points in a high-dimensional space, where each dimension corresponds to one possible keyword. The position of a text record in a dimension is the number of times the keyword appears in the text record. Clusters of text records in this space often correspond to groups of documents on the same topic.

21.4.3 Nearest Neighbor

A very important element in clustering is the use of similarity measures. The definition of similarity is very intuitive as it lies on our innate sense of ordering a variety of different objects. Similarity seems to be ubiquitous also allows us to make predictions. The key idea of the nearest neighbor algorithm is that assumption that objects (or data points) that are near (similar) to each other should have similar predicted values (or properties). The well-known k-nearest neighbor is a prediction technique that classifies each record in a dataset based on a combination of the classes of the k record(s) most similar to it in a database. An enhanced version of the k-nearest neighbor is to take a vote from the k nearest neighbors rather than just relying on the average predicted value of all k nearest records, as the prediction may be wrongly influenced by outliers.

The concept of nearest neighbor has been applied to many diverse disciplines, as it is very easy to understand and implement. It has been used deal with computational geometry to make it practicable in learning and vision [15]. The nearest neighbor model has been studied in computational chemistry to investigate the ability of a genotyping assay based on hybridization through expected melting points of nucleic acid duplex stability [17]. However, for text mining, the nearest neighbor rule may become very inefficient when the database grows very large, especially text records. Text records are usually represented as a data element which is high dimensional, or more generally, are represented by a point in large metric space. Therefore, the similarity (or closeness) calculations are computationally expensive.

21.4.4 Decision Tree

A decision tree is a hierarchical predictive model that has a tree structure. Each branch of the tree represents a classification rule and the leaves are partitions or segmentations of the dataset with their predicted values. In addition to the use for building a predictive model, the decision tree method can be used for exploratory analysis. The basic concept of the decision tree is to grow the tree that groups all the data as perfectly as possible. However, one can always construct a perfect tree where there are as many leaves as the data points. Therefore, most decision tree algorithms stop growing the tree when one of the following three criteria is met: The leave contains only one record. All the records in the current branch have identical characteristics. More specifically, decision trees algorithms start with a training dataset which represents the pre-classified records, and each record contains independent and dependent variables. Then they try to find the independent variable that best splits the records into groups where one single class predominates. This step is repeated recursively to generate lower levels of tree until one of the three criteria is met.

Classification and Regression Trees (CART) is among the most well-known classification tree techniques. It was first introduced in the 1980s [18]. The idea behind the CART tree is that each predictor (attribute) is selected based on how well it separates out the records with different predicted values. It also uses an entropy measure to determine whether a given split point (branching) for a given predictor is better than. One of the great advantages of CART is that it has the validation of the model and the discovery of the optimally generalized model. It first builds a very complex tree and then uses the results of cross validation or test set validation to prune the tree back to the optimally generalized tree. The generalization is a very important concept to avoid over fitting the training data, which is most likely to perform well on new, unseen data can be chosen. Chi-Square Automatic Interaction Detector (CHAID) is another well-known version of decision tree. CHAID is similar to CART in the sense that it builds a decision tree, but it is different in the way that it branches the tree (chooses its splits). Instead of using the entropy measure for choosing optimal splits, CHAID employs a chi-square test to determine which predictor (attribute) is most dependent with the predicted values.

Because of their easy-to-interpret tree structure and the ability to easily generate rules, decision trees are among the most widely used techniques for building understandable decision models. For example, the decision trees have been used as a predictive model for complex profit and return-of-investment (ROI) analysis [11]. In spatial data mining, the decision tree concept has been applied to the study of spatial risk analysis such as epidemic risk or traffic accident risk in the road network by taking account implicit spatial relationships in addition to other object attributes [19].

The main advantage of decision tree for mining text records is that it is very intuitive, and its classification rules are easy to interpret, which can lead to direct sequential decision procedures. Another important aspect is that decision tree can effectively deal with missing attributes, referred text contents in text mining. When decision tree techniques are used to categorize text records, missing values (keywords in metric space) can be handled via surrogates. Surrogates are split values and predictors that mimic the actual split in the tree and can be used when the data for the preferred predictor is missing. However, the decision tree might not perform well with text records containing semantic information. Keywords in text records may be correlated and treating them as independent variables may not always be the best separators.

21.4.5 Neural Network

A neural network, inspired by the structure of biological neurons, is a non-linear predictive model that has been well studied over the past few decades [18]. It is an interconnected group of artificial neurons (represented by data attributes) that uses a mathematical or computational model for information processing. In most cases, a neural network is an adaptive system that changes its structure based on a prediction task, which is usually referred to external or internal information that flows through the network. Specifically, a neural network is a non-linear statistical data modeling or decision making tool for extracting complex relationships between inputs and outputs or to find patterns in data. Similar to the decision tree, one can train the structure of the neural network based on the training data and generate classification rules. However, both techniques may suffer from the same drawbacks when dealing with semantic text. Each neuron (or keywords in text mining) may be correlated and the relationship may be more complex than a network can explain. Most importantly, the classification rules generated by a neural network are very difficult to interpret.

Neural networks are used in a wide variety of applications. They have been used in all facets of business from detecting the fraudulent use of credit cards and credit risk prediction to increasing the hit rate of targeted mailings. In a study in [20], the neural networks have been used as a decision support tool for credit card risk assessment within a major bank. This neural networks tool was used to emulate the decisions of the current risk assessment system and to predict the performance of credit card accounts based on the accounts historical data. Neural networks also have a long list of applications in other areas such as the military, from the automated driving of an unmanned vehicle to cognitive science and speech recognition, such as learning the correct pronunciation of the Arabic language. The study in [21] investigates a neural network module for autonomous vehicle following, defined as a vehicle changing its own steering and speed while following a lead vehicle. Two time-delay back propagation neural networks, one for speed control and the other for steering, were studied under manual control and tested using live vehicle following runs. A new type of recurrent neural network (RNN) architecture for speech recognition was proposed in [20]. It was shown that the people's knowledge and understanding of the Arabic alphabet and words has been improved by this new architecture and the learning algorithm.

21.4.6 Support Vector Machine

Support Vector Machine (SVM) is another data mining technique that has been very well studied over the past few decades. The key concept of SVM is to project input data into a higher dimensional space and divide the space with a continuous separation hyperplane by iteratively minimizing the distance of misclassified data points from the line. In other words, SVM generally tries to construct a hyperplane that minimizes the upper bound on the out-of-sample error. There have been many variations of SVM models. One of the most successful models uses the idea that once a data set is transformed to points in a high dimensional space, which is called kernel transformation, every point can be classified by the separating plane. The fundamental concept of SVM is very similar to the neural network approach. When applied to text records, SVM may also suffer from the same downfalls as the neural network does.

This optimization formalism in the SVM framework incorporates the concept of structural risk minimization by determining a separating hyperplane that maximizes not only a quantity measuring the misclassification error, but also the margin separating the two classes. This can be achieved by augmenting the objective of the robust linear programming formulation given by [22]

$$\min_{\omega,\gamma,y,z} \quad \frac{e^T y}{m} + \frac{e^T z}{k}$$
$$\text{s.t.} \quad A\omega - e\gamma - e \geq y$$
$$-B\omega + e\gamma - e \geq z$$
$$y \geq 0, \ z \geq 0,$$

where A and B are matrices of two data groups to be separated by a hyperplane (defined by $A\omega \geq e\gamma$, $B\omega \leq e\gamma$), and y and z are $\{0, 1\}$ decision variables indicating there are points in group A and B violating the hyperplane constraint respectively. The objective function is therefore minimizing the minimum average misclassifications subject to the hyperplane constraint for separating data points from A from data points from B.

Recent applications and extensions of support vector machines have been mainly focused on pattern recognition and classification [23]. SVMs have been applied to many real life problems, including handwritten digit recognition [24], object recognition [25], speaker identification [15], face detection in images [8], and text categorization [26].

21.4.7 Rule Induction

Rule induction is another commonly used data mining technique in unsupervised data mining, where all possible patterns (if-then rules) are systematically extracted from the data. For each pattern, rule induction will statistically determine its accuracy and significance so that an end user knows how strong the pattern is and how likely it is to occur again. In other words, the rules are presented to the user based on the percentage of times that they are correct, known as accuracy, and how often they apply, known as coverage. In text mining, rule induction appears to fit very well with text record database, as the attributes of text records could be of high cardinality (many different values) or many columns of binary fields. For semantic text records, there are exponential combinations of semantic text contents. For keyword-based text records, the number of distinct keywords could be massive. However, the user may have to keep in mind that the text patterns in the database, expressed as the rules, are not always accurate. In other words, it is important to recognize and make explicit the uncertainty in the rule depending on the rule's accuracy and coverage. The rule induction can potentially give a highly automated way to generate the decision rules in several applications (*e.g.*, business and finance). However, it may also suffer from an overabundance of interesting patterns that can complicate the rules, which might not be practical.

21.4.8 Log-linear Analysis

Log-linear analysis has to deal with a study to distinguish between independent and dependent variables in multiple regression or analysis of variance (ANOVA) by analyzing multi-way frequency tables. The multi-way frequency table is usually used to capture various main and interaction effects that add together in a linear fashion to bring about the observed table of frequencies of the data. The frequencies in each entry of the table proportionately reflect the marginal frequencies. The idea of log-linear analysis is as follows. Given the marginal frequencies for two (or more) factors, one can compute the entry (cell) frequencies that would be expected if the two (or more) factors are unrelated. Significant deviations of the observed frequencies from those expected frequencies reflect a relationship between the two (or more) variables. In text mining, one can use this log-linear analysis to validate a keyword model, which can be constructed by postulating independence between all keywords. If any significant deviations occur, we would reject this keyword model. One difficulty is that there could be massive number of keywords and calculating all the marginal frequencies in the table is very tedious. Therefore, the log-linear analysis may rather be used as a post-analysis to improve the performance of other data mining techniques.

21.4.9 Multidimensional Scaling

Multidimensional Scaling (MDS) is a statistical dimensionality reduction technique for mapping high-dimensional data into a low-dimensional target space for data visualization in exploring similarities or dissimilarities in data. Generally, it analyzes proximity data to reveal hidden structure underlying the data. It starts with a matrix of pairwise similarities, then assigns a location of each data point in a low-dimensional space. Then MDS tries to find an optimal configuration of points (mappings) in some multidimensional space such that the interpoint distances are related to the experimentally obtained similarities by some transformation function. MDS appears to fit very well with visualization of groupings of text records. However, for large text records or highly semantic records, MDS might not be computationally efficient.

21.5 Research Challenges

Mining and categorizing text records are an emerging research area in text mining that has an enormous potential for use and benefits in many applications. In the service industry, advances in text mining can help to understand customer preferences in detail. It can potentially improve the ability to predict and take appropriate action to prevent it. In product development process, text mining may help to predict when product components may fail or production equipment need maintenance, and better control both product quality and operating costs.

In spite of its potential benefits, text mining has raised many open research problems as it has to deal with many implementation issues as well as analytical and computational challenges. One challenge is the high percentage of problem records that could not be categorized or grouped. A huge variance of the text record's text length has added some complexity to it as well. While there is usually a very small of text content giving very little problem information, the longer text record may be desirable. However, a lengthy text record will also add a computational challenge. Although text records are massive, a lot can be done to effectively use this information in future research. For example, knowledge of the problem-area field may be incorporated in the text mining process, as it would allow the company to find out the general predictive text analytics used to transform and integrate unstructured text records, and enable traditional data mining analyses.

References

1. Cerrito P, Cerrito JC (2006) Data and text mining the electronic medical record to improve care and to lower costs. SAS SUGI Proceedings paper 077–31
2. Duda RO, Hart PE, Stork DG (2001) Pattern Classification, 2nd edn. Wiley, New York
3. Myllymaki P, Silander T, Tirri H, Uronen P (2001) Bayesian data mining on the web with B-Course. Proceedings of the 1st IEEE International Conference on Data Mining (ICDM-2001), pp. 626–629
4. Frand J (1996) Data mining: what is data mining? www.anderson.ucla.edu/faculty/jason.frand/teacher/technologies/palace/datamining.htm
5. Liu B, Grossman R, Zhai Y (2003) Mining data records in web pages. Proceedings of the 9th ACM SIGKDD International Conference on Knowledge Discovery & Data Mining (KDD-2003), pp. 601–606
6. Myatt GJ (2006) Making Sense of Data: a Practical Guide to Exploratory Data Analysis and Data Mining. Wiley, New York
7. Dagli CH, Lee H-C (1997) Impacts of data mining technology on product design and planning. In: Plonka F, Olling G (eds) Computer applications in production and engineering. Chapman and Hall, Detroit, Michigan, pp. 58–7

8. Osuna E, Freund R, Girosi F (1997) Training support vector machines: an application to face detection. Proceedings of IEEE Conference on Computer Vision and Pattern Recognition, pp. 130–136

9. Han J, Kamber M (2006) Data Mining: Concepts and Techniques, 2nd edn. Morgan Kaufmann/Elsevier, USA

10. Berson A, Smith S, Thearling K (1999) Building Data Mining Applications for CRM. McGraw-Hill, New York

11. Yuan Y, Shaw MJ (1995) Induction of fuzzy decision trees. Fuzzy Sets and Systems 69:125–139

12. Hand DJ, Mannila H, Smyth P (2000) Principles of Data Mining. MIT Press, Mass., USA

13. Hartigan J (1975) Clustering algorithms. Wiley, New York

14. Fan H, Ramamohanarao K (2003) A Bayesian approach to use emerging patterns for classification. Proceedings of the 14th Australasian Database Conference, Adelaide, Australia, pp. 39–48

15. Schmidt M (1996) Identifying Speaker with Support Vector Networks. Proceedings of Interface, Sydney

16. Heyer LJ, Kruglyak S, Yooseph S (1999) Exploring expression data: identification and analysis of coexpressed genes. Genome Research 9:1106–1115

17. von Ahsen N, Oellerich M, Armstrong VW, Schütz E (1999) Application of a thermodynamic nearest-neighbor model to estimate nucleic acid stability and optimize probe design: prediction of melting points of multiple mutations of apolipoprotein B-3500 and factor V with a hybridization probe genotyping assay on the LightCycler. Clinical Chemistry 45:2094–2101

18. Bishop CM (1995) Neural Networks for Pattern Recognition. Clarendon Press, Oxford

19. Zeitouni K, Chelghoum N (2001) Spatial decision tree-application to traffic risk analysis. Computer Systems and Applications, ACS/IEEE International Conference, pp. 203–207

20. Ismail S, Manan bin Ahmad A (2004) Recurrent neural network with backpropagation through time algorithm for arabic recognition. IEEE International Symposium on Communications and Information Technology (ISCIT-2004), pp. 98–102

21. Kehtarnavaz N, Griswold N, Miller K, Lescoe P (1998) A transportable neural-network approach to autonomous vehicle following. IEEE Transactions on Vehicular Technology 47:694–702

22. Bennett KP, Mangasarian OL (1992) Robust linear programming discrimination of two linearly inseparable sets. Optimization Methods and Software 1:23–34

23. Burges CJC (1998) A tutorial on support vector machines for pattern recognition. Data Mining and Knowledge Discovery 2:121–167

24. Scholkopf B, Burges C, Vapnik V (1995) Extracting support data for a given task. Proceedings of the 1st International Conference on Knowledge Discovery and Data Mining 1995, AAAI Press, Mass., USA, pp. 252–257

25. Blanz V, Scholkopf B, Bulthoff H et al. (1996) Comparison of view-based object recognition algorithms using realistic 3d models. Springer Lecture Notes in Computer Science 1112:251–256

26. Joachims T (1997) Text categorization with support vector machines. Technical report, LS VIII Number 23, University of Dortmund, ftp://ftp-ai.informatik.uni-dortmund.de/pub/Reports/report23.ps.Z

Further Reading

Fayyad U, Piatetsky-Shapiro G, Smyth P, Uthurusamy R (1996) Advances in knowledge discovery and data mining. AAAI/MIT Press, Mass., USA

Feldman R, Sanger J (2006) The Text Mining Handbook: Advanced Approaches in Analyzing Unstructured Data. Cambridge University Press, Cambridge, UK

Gu L, Baxter R, Vickers D, Rainsford C (2003) Record linkage: current practice and future directions. Technical Report 03/83, CSIRO Mathematical and Information Sciences, Canberra, Australia, April

Hauben M (2004) Application of an empiric Bayesian data mining algorithm to reports of pancreatitis associated with atypical antipsychotics. Pharmacotherapy 24:1122–1129

Jagielska I, Jaworski J (2001) Neural network for predicting the performance of credit card accounts. Journal of Computational Economics 9:77–82

Kantardzic M (2002) Data Mining: Concepts, Models, Methods, and Algorithms. Wiley-IEEE Press, UK

Kreyss J, Selvaggio S, White M, Zakharian Z (2003) Text mining for a clear picture of defect reports: a praxis report. Proceedings of the 3rd IEEE International Conference on Data Mining (ICDM-2003), pp. 727–730

Agarwal D (2005) An empirical Bayes approach to detect anomalies in dynamic multidimensional arrays. Proceedings of the 5th IEEE International Conference on Data Mining (ICDM-2005), Houston, Texas, pp. 26–33

Liu B, Grossman R, Zhai Y (2004) Mining web pages for data records. IEEE Intelligent Systems 19:49–55

Menon R, Tong LH, Sathiyakeerthi S et al. (2004) The needs and benefits of applying textual data mining within the product development process. Qual Reliab Engng Int 20:1–15

Nasukawa T, Nagano T (2001) Text analysis and knowledge mining system. IBM Systems Journal 40:967–984

Rodin A, Mosley Jr TH, Clark AG et al. (2005) Mining genetic epidemiology data with Bayesian networks application to APOE gene variation and plasma lipid levels. Journal of Computational Biology 12:1–11

Shakhnarovich G, Darrell T, Indyk P (2006) Nearest-neighbor Methods in Learning and Vision: Theory and Practice. MIT Press, Mass., USA

Silver M (1995) Scales of measurement and cluster analysis: an application concerning market segments in the babyfood market. Statistician 44:101–112

Wang XZ, Chen BH, McGreavy C (1997) Data mining for failure diagnosis of process units by learning probabilistic network. Transactions of Institute of Chemical Engineers 75:210–216

Weiss S, Indurkhya N, Zhang T, Damerau F (2005) Text Mining: Predictive Methods for Analyzing Unstructured Information. Springer, New York

Zhou X, Han H, Chankai I et al. (2006) Approaches to text mining for clinical medical records. Proceedings of ACM Symposium on Applied Computing, pp. 235–239

Berry MW (2003) Survey of Text Mining: Clustering, Classification, and Retrieval. Springer-Verlag, New York

Breiman L, Friedman JH, Olshen RA, Stone CJ (1984) Classification and regression trees. Wadsworth & Brooks/Cole Advanced Books & Software, Pacific Grove, Calif

Chapter 22
Quality in Design: User-oriented Design of Public Toilets for Visually Impaired People

Kin Wai Michael Siu

School of Design, The Hong Kong Polytechnic University

22.1 Difficulties and Consequences for VIP in Accessing Public Environments

According to United Nations statistics, about 1/30th of the world's population is visually impaired with different types, levels, and degrees of visual impairment. It is not difficult to realize that visually impaired people (VIP) face different kinds of difficulties and limitations in their daily lives, in particular when they need to interact with public environments and facilities with which they may not be familiar [1].

In recent years, policymakers and researchers in different disciplines such as sociology, architecture, design, and engineering, have conducted more discussions and made increasing efforts to improve this situation. Among the various projects with different perspectives, directions, objectives, and targeted beneficiaries, the key approach is generally to apply technologies to provide convenience, or overcome existing "barriers" to use, for VIP. However, as many of the studies in Europe and America have shown, it remains clear that the situation is still unsatisfactory that complaints by VIP have frequently heard. Mass media also frequently report on accidents or unsatisfactory and unfair environments for VIP. This is not due to the "almighty" nature of the technologies or inventions, but to the fact that these technologies cannot fit the wants and needs of VIP (*i.e.*, actual users) and functions as they were originally planned and intended.

There are three general consequences of the failure to solve these difficulties. First, VIP minimize the number of times that they go out, in particular to places with which they are not familiar. This situation also implies that the so-called "not-so-serious" situation often declared by governments actually does not reflect the seriousness of the real situation. In addition, the inactive participation in public activities and exploring new places more or less reflects the unwillingness, helplessness, and disappointment of VIP. They are compulsorily forced to change their normal and expected routines (as non-VIP) due to the unsatisfactory provision of public environments and facilities. It is the reason, for example, why we always see VIP only gather in some community and activity centers particularly for them. Such a situation more or less also reflects that VIP are excluded from other public spaces as well as public daily life.

Second, if it is absolutely necessary for VIP to go to unfamiliar places, they need to spend quite a lot of time (sometimes an unreasonable amount of time) preparing to go out. This is because both information for VIP and facilities for searching for information are very limited. One will feel surprised if he or she seriously looks at how limited information and information searching facilities are available in the current so-called metropolitan and high living-standard cities. Today, without the initiation and commitment of governments, information providers know that preparing appropriate information does not generate attractive income. Hence, except for some very limited information for visually impaired tourists, whereby this kind of information and facilities are supposed to attract economic benefits and better international image to others, most information and facilities for local VIP are only prepared by the communities of VIP, and the NGOs servicing VIP, although these groups often lack resources to do it.

Third, VIP need to go to new places with the assistance of other people who are not visually impaired or who are VIP but are already familiar with the places. In fact, this consequence in recent years has caused a worsening situation for VIP, while in general family sizes in modern society are getting smaller, a higher ratio of family members go out to work, and relatives are living comparatively far apart away, including in different countries. Seeking other people, in particular family members or relatives, to give help in turn becomes more difficult. However, government and NGO support has not been seen to create significant improvement in this area. Nevertheless, if this third consequence is common, then the situation at least implies or reminds us that VIP today are still unable to live in a society where the quality of designs of facilities are up to a standard that allows them ease of use as they expect. At the very least, VIP are still necessarily and heavily reliant on others.

According to United Nations statistics, about 1/30th of the world's population is visually impaired. Visually impaired people (VIP) face a variety of difficulties in their daily lives. This is the case not only in countries with a shortage of resources or with relatively lower living standards, but also in developed countries which claim to have a good, modern living environment. Most of the time, such difficulties in daily life come from the misunderstanding, inexperience, and disregard of VIP's wants and needs by other members of society, which in turn relate to poor design quality of facilities and products. To increase the public's awareness of this situation, and to promote equal opportunity in society and improve the design quality of public environments and facilities, a research and design project on how VIP access public toilets has been under way since 2004. To maintain better design quality in public toilets, the "FISH" concept has been initiated. This design concept includes the design considerations of friendly, informative, safe, and hygienic. This article reviews the wants and needs of VIP that need to be considered in using public toilets and the help that they need to be given. Based on the findings of the project, this article then identifies several key areas worthy of attention, and discusses how better quality designs for public toilets can been obtained by implementing the FISH concept, which serves the actual wants and needs of users. Through this article, it is expected that further explorations and discussions on the topic will take place and in turn better quality design for VIP in the future.

Therefore, it is necessary and also a priority to improve the design quality of public environments and facilities. To fit the needs and preferences of users (that is, VIP), instead of looking from the perspectives of professionals (for example, planners, designers, engineers), the better way obviously is to look into the issues from the perspectives of VIP. Hence, the study presented in the following paragraphs takes VIP as the core of the investigation, analysis and discussion.

22.2 Deficiencies in Public Toilets for VIP

Today, among all kinds of public environments, when leaving their homes or places with which they are familiar, one of the most difficult places for VIP to go and use is public toilets [2]. (A public toilet is defined as a toilet built and/or managed by a government department.) It is also one of the places VIP are least willing to go and use. It is not hard to imagine how difficult and inconvenient it must be for a VIP to use a public toilet to which he or she has not been to previously. Even if a public toilet is a place with which a VIP is familiar, such facilities are commonly considered, recognized, and understood to be dirty and full of infectious diseases. This situation creates much difficulty and uneasiness for VIP in using public toilets. For example, in 2004, a study was conducted in Hong Kong to allow design students to understand the difficulties of VIP to use public toilets. Fourteen design students were invited to visit a public toilet. They were then asked to leave the toilet, they were blind-folded, and re-entered and used the toilet. The result was so extreme, but not surprising, that all of them agreed that it was nearly impossible for them to use the toilet. They were afraid to touch the environment and facilities [2].

In fact, even in developed countries with more well-established policies and better resources, complaints from VIP regarding the inaccessibility and unpleasantness of public toilet environments and facilities are still frequently heard. Or say, there is no significant improvement in design quality of public toilet environments as well as facilities that can help VIP to access public toilets. The only identified improvement may be an auto-flushing mechanism that gives an advantage to VIP in using the facility, so that they will not to be scared or embarrassed when searching for the buttons or handles of the conventional flushing mechanisms. Even though other automatic facilities claim to give help to VIP, such as automatic water taps, how to approach these kinds of facilities are always in question.

Then, how should we consider and provide better public everyday lives to VIP? How can design quality of public toilets be improved so that VIP can get sufficient support? Or, how can design quality of public toilets be ensured so that VIP can live with respect and dignity?

22.3 Studies on Accessibility of Public Toilets for VIP

To promote and develop public toilets to be more accessible, a research and design project on public toilets in Hong Kong and how VIP access them has been in progress since 2004. The project tries to reach a high design quality by looking from the perspectives of VIP; that is, to promote a high degree of user fitness [3]. The project is considered a long-term investigation on the topic that it does not aim at exploring the situation at a particular time. Instead, it aims at continuous explorations on the changes of the natures and degrees of accessibility of public toilets due to the cultural, social, and economic changes of the society and also technological changes and inventions.

The project is mainly funded by The Hong Kong Polytechnic University. It is under the design theme initiated by the author in 2004: d*hk® (*i.e.*, Designed in, by and for Hong Kong). The project is conducted in collaboration with several non-governmental organizations that are providing services to VIP in Hong Kong. Professional and social organizations from different disciplines have also given help and provided information to the project including the support of the Hong Kong Toilet Association, since 2005. Several government departments have also provided information for the project. Although the seed money for the project was not large at the beginning, more volunteers have joined in the project, including individual professionals from different disciplines, such as sociologists, social workers, designers, architects, medical doctors, and people working in the industry related to health and sanitation products.

Apart from the common quantitative approaches and methods of investigating VIP's needs with regard to public toilets, such as a questionnaire and survey, the project has taken a qualitative approach. The major reason of taking this approach is that reviewing some research and consultancy reports concerning the needs of disabled people in using public toilets and other public facilities, such as transportation, it has been noticed that the findings and comments in these reports are too general to generate significant improvement in these environments and facilities.

In fact, there have been several local and regional studies on the public toilets for disabled people conducted over the past few decades. However, there has been very little improvement, and nearly no named improvement, in the facilities regarding the needs of VIP in recent years. The major reason is that the findings and resultant improved designs do not meet the actual wants and needs of VIP [4]. Therefore, this research project has taken another direction to investigate the wants and needs of VIP in a more in-depth manner. The key objective of the project is, through an understanding of "user experience", to establish a solid foundation related to VIP's needs and wants, and in turn to propose directions and insights for more comprehensive studies in the future [5, 6].

In the project, Hong Kong public toilets have been taken as cases for investigation. The cases cover three territories of Hong Kong: Hong Kong Island, Kowloon, and the New Territories. For example, the cases include a majority of toilets in urban areas with a dense population, while they also include several toilets in country parks.

The term "public toilet" is confined to two major types of public toilets. The first type is the public accessed toilets managed by the government. In Hong Kong, these toilets are planned, designed, constructed, and managed by or contracted out for services by the government. The other type of toilets are those public accessed private toilets, for example, toilets offered in shopping centers.

Due to the limitation of resources, the numbers of each particular case (toilet) are not large. The major aim is to have a wider scope of coverage of the types of cases in Hong Kong, in order to have a more in-depth comparison of different situations.

The research methods include:

(a) Reviews of the existing policies and provisions of public facilities for VIP, including issues related to equal opportunity.
(b) Interviews with the government officers who are responsible for the overall planning, designing, and managing of public toilets.
(c) Interviews with representatives of property management agencies, who are responsible for planning, assigning contracts, and managing public accessed private toilets; for example, property management companies of shopping malls and corporate chain cafes.
(d) In-depth interviews with NGO representatives (for example, social workers, visually impaired volunteers) who provide services to VIP of Hong Kong.
(e) In-depth interviews with VIP with different types, levels, and degrees of vision impairment. The interviewees (VIP) were introduced by the NGO social workers. The major reason is that VIP are sensitive to the outside world as well as strange people. Due to a large number of bad experiences, many of them prefer to keep apart from those with whom they are not familiar with. They trust social workers who have worked with them and act as friends for a long time.
(f) Field observations of how VIP access and use public toilets. VIP were invited to visit and use several common types of toilets for public use, including those specially designed for disabled people (also known as disabled toilets or accessible toilets).
(g) Environmental and product analyses on the designs of existing public toilets and the facilities available in the toilets (*e.g.*, analysis on functions, product semantics, anthropometrics and ergonomics, product safety, adaptability, *etc.*).

22.4 Key Areas of Design Quality for Consideration

According to the findings of the project, there are four key areas related to the design quality of public toilet environments and facilities for VIP, to which it is worth paying attention and which merit further discussion.

First, there have been very few investigations and design developments concerning the particular wants and needs of VIP in using public toilets. The key reasons for this deficiency are the misunderstanding of the real situation and underestimation of the needs of VIP. That is, there is often a wrong concept that visually impairment is not a significant issue in the current society, or further, discrimination in design for VIP is not so serious. And, as commented on by the interviewed VIP, there is a common misunderstanding that the number of VIP is insignificant in our society. People always put blind people as the whole population of VIP. However, the fact is that the meaning of visually impaired is relatively much wider, in particular in some developing countries with poor living environment and insufficient daily nutrition. It is also why the United Nations has identified that about one-thirtieth of the world's population as being visually impaired. For example, a significant number of older people face visually impaired difficulties that they perhaps did not have to when they were younger [7, 8].

Another major reason for the deficiency in investigation and design developments concerning the special needs of VIP is that people consider physical disability as more significant and these people need more help compared to visually impaired people. Then, reviewing the existing public toilets around the world, in particular those special toilets or facilities claimed to be provided for disabled people, most of the designs only narrowly focus on the needs of physically disabled people. In other words, disabled toilets always simply imply toilets for physically disabled people, and even sometimes more narrowly for wheelchair users only (Figs. 22.1 and 22.2).

The second area necessary for consideration is that, except for their disability in vision, VIP in fact have the same needs as people with no visual impairment. The only major difference is that, under the current situation, VIP generally prefer to go only to the places with which they are familiar. Or rather, they prefer not to go to the places with which they are not familiar. However, it does not imply that VIP do not like to go to new places. As pointed out by the interviewed VIP, they do not lack the curiosity, interest or need to go to new places though their interpretation of and expectation on "new" may not be the same as people without visual difficulties.

Fig. 22.1 Although the sign of the disabled toilet commonly seen today is only a symbolic meaning (and graphic design of a sign) for a disabled toilet, it more or less reflects the biased objectives and function of the toilet

Fig. 22.2 The interior setting and facilities of the current disabled toilets are mainly (nearly only) for physically disabled people. People with other disabilities, such as sensory and cognitive disabilities, get less assistance in these toilets

Nevertheless, the interviewed VIP further pointed out that it is not easy for them to overcome the difficulties, or sometimes embarrassment, of trying to access such unfamiliar places. Then, this is how we know that current designs do not meet the needs of VIP, rather than VIP not having needs.

Third, VIP need to spend quite a lot of time preparing if they need to go to a new place. However, as discussed in paragraphs above, it is very difficult for them to collect the necessary information. On the one hand, it is difficult (nearly impossible) to provide very detailed and updated information about a place for public access. For example, it is not hard to imagine how difficult it is to provide detailed and updated interior information about every public toilet. In particular, when there is a slight change in the settings, it can significantly affect a VIP when trying to access such new places. Furthermore, in densely populated cities like Hong Kong, public environments as well as surrounding settings change rapidly and continuously. Such kinds of rapid changes increase the difficulties and limitations for VIP to access updated information for accessing public environments.

On the other hand, accessing information for VIP is not easy; in particular, public information is relatively limited for VIP to access nowadays, even in countries such as the United States, Germany and Japan, which are able to provide public information by using advanced technology. Moreover, many VIP without special carers (assistants) are in the low income sector. They need to access information by themselves and with some tailor-made facilities for VIP. However, these facilities can be quite expensive. This situation also limits these VIP to access information at home. The Hong Kong government, similar to the governments in many countries, today gives very little support to solve this problem. Facilities for searching information are always provided at a small number of locations, for example, public libraries. Then, it is not difficult to imagine how complicated and not making sense that a VIP needs to go to a place (far away from his or her home) to search information in order to go to a new place. The consequence is easy to see, that VIP prefer not to go to unfamiliar places though they may want to go.

Fourth, VIP sometimes need assistance from people without visual impairment or other VIP already familiar with the place in question. However, as mentioned by the VIP and NGO social workers interviewed, it is inconvenient to request family members, relatives, or friends to give help all the time. On the other hand, like most other disabled people, VIP do not like to request or receive too much assistance. They prefer to be independent and use self-help (self-assist) in going about their daily routines [9]. As mentioned above, many VIP are in the low income sector. For those people who are born with visual impairment, they may find it more difficult to learn in the current education systems with little help, and they often having difficulty in getting jobs [10]. It is unrealistic for a VIP to appoint a carer for his or her whole life. Thus, assistance for self-help is a crucial design concept to allow VIP to gain equal opportunity in their daily routines.

VIP also expect to have more freedom. Asking for assistance in some sense implies dependency or lack of independence. As indicated by the social workers interviewed, most VIP prefer to go to the places they are familiar with, not only because of the difficulties and constraints for them to access new places, but also because they want to do the things which "they are able to handle by themselves." Moreover, the VIP interviewed also identified a critical point; that VIP (in particular those who were born with visual impairment) have been learning to be independent because they need to be well prepared for the moment and situation not to have any assistance.

Another critical point is that nearly all disabled people do not expect others to "label" them as disabled. Asking for care and assistance then more or less generates a perception that VIP are disabled, dependent, or worse, a burden to society. This is why many disabled people, not only VIP, refuse to accept assistance; in particular, disabled people hate other "so-called-abled" people showing pity for them when they give help. It is also the reason VIP prefer to stay at home or in familiar places. Or, as stated by the VIP interviewed, they always shorten their time in public places and sometimes drink less water to prevent the need to use public toilets.

Therefore, if design is claimed to solve needs and improve the quality of daily life of human beings, then how can the design quality of public toilets and facilities be improved to meet VIP's actual wants and needs?

22.5 FISH: Better Designs of Public Toilets for VIP

To improve the design quality as well as the assurance of the quality of public toilets and facilities, based on the key areas identified above, the "FISH" concept has been initiated for research and design development. The concept includes four major directions (or, design considerations): friendly, informative, safe, hygienic.

22.5.1 Friendly

In recent years, moving away from designers and designs, "users" have become the focus for high quality design. Taking consideration of user experience is more often recognized as being crucial if a design is to have a high degree of user fitness as well as user friendliness [11–14].

The findings of the abovementioned research project illustrate that a user-friendly public toilet (or, a public toilet by user-oriented design or user-centered design) does not cater only for the physiological factors but also, as suggested by Jordon and Green, for psychological, cultural, social, and ideological factors [3,4,14]. However, many of the current designs are biased on some particular needs of users, mostly on physiological needs. It is one of the reasons why physical disability is always misconstrued as the "total" of all disabilities. It is also the reason why most of the so-called accessible designs appearing in public environments mainly consider the needs of physically disabled people (though they are also very insufficient and unsatisfactory in quality for this segment of the population) [13].

The VIP interviewed complained about designers lack of sufficient knowledge and experience in considering their wants and needs in terms of vision disability. Designers always consider and propose designs from their own experiences and perspectives. Therefore, the so-called quality designs have not fully considered the actual wants and needs of VIP, and sometimes omit their wants and needs. Designers also do not have much sense on the strengths and limitations of VIP. For example, designers may generally know that VIP always need to use their other senses to determine the locations of toilets and facilities, such as VIP using their feet to feel the texture of tactile guide paths or their fingers to read Braille giving directions to toilets and the locations of other facilities. However, without a close working relationship with VIP, not so many designers know that VIP can feel the sound and wind rebounding from walls and changing temperature to determine the distances between themselves and external objects.

Nevertheless, whether the means of searching and the information provided for VIP have high reliability (*i.e.*, easily accessible and understood) is crucial. For example, sound caution signals are commonly used in public environment. But, with today's excessive noises, as well as caution signals in public environments, what kinds of signals and whether such signals can be recognized and not distracting nor misleading by other noises are crucial. Moreover, some standards of tactile guide paths have been generated in recent years. However, quite a lot of designers and

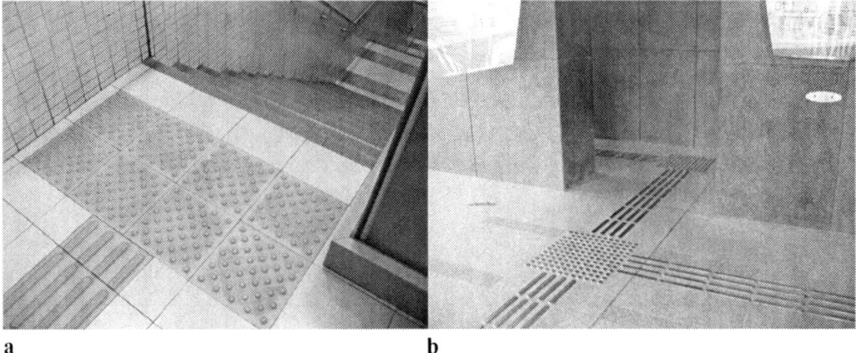

a   b

Fig. 22.3a,b. Many tactile guide paths are wrongly installed. **a** The "positional tiles" are wrongly installed for "warning" purpose. **b** The "warning tiles" are wrongly installed for providing walking direction purpose. One of the reasons is that the raised truncated domes are easier to be arranged in square grid parallel to the sides of the square (*i.e.*, requirements of warning tiles) than in staggered positions (*i.e.*, requirements of positional tile). And, the stainless steel tactile guide paths sometimes are slippery, particularly on rainy days

property management people still wrongly use or mix up different types of guide path surfaces (*e.g.*, patterns of the tiles). For example, not many of these professionals and management people can distinguish the difference between the patterns and functions of the "warning tile" and the "positional tile". (A warning tile consists of raised truncated domes arranged in square grid parallel to the sides of the tile to alert people of potential hazards such as top and bottom of stairs, door openings and at pedestrian crossings. A positional tile consists of raised small dots arranged in staggered positions to indicate change of walking direction.) They seem nearly the same with dots on a flat surface, but in fact different sizes and arrangements of the dots give different meanings to VIP (Fig. 22.3) [15].

Furthermore, whether additional problems will be generated to VIP while searching for information is also important to consider. For example, while searching information at unfamiliar places, VIP are sometimes easily hurt by the sharp edges of facilities, and the gaps between moveable parts such as door jambs and hinges (Figs. 22.4 and 22.5). Poorly maintained and wrongly installed tactile guide paths for VIP also create additional difficulties and confusion. Moreover, to be easy for cleaning, have a longer product lifecycle, and a better appearance (matching or harmonious to the surrounding environment), these days many designers and management companies prefer to use tactile guide paths made of stainless steel. However, the VIP interviewed complained that these kinds of guide paths are more slippery. Additionally, some of the guide paths are claimed to be designed to match and be in harmony with the surrounding environment. Thus, the colors of these guide paths are quite similar to the surrounding environment. However, such kind of designs is contradictory to the original function of guide paths as well as not being distinctive and contrasting enough. Also, these guide paths do not have reflective or florescent color to give help to the VIP with mild grades of visual impairment.

Fig. 22.4 VIP use their fingers and hands to read information or to search for directions. They are easily hurt by the gaps between moveable parts such as door jambs and hinges

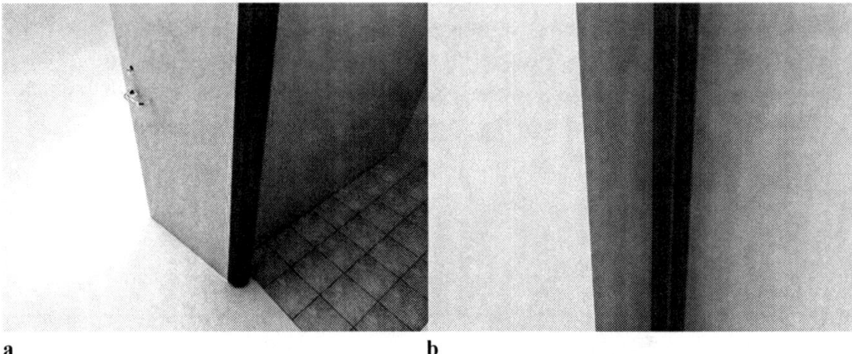

a   b

Fig. 22.5a,b. "Door Gap Protection Covers". A new design for preventing being hurt by door gaps. The design concept guarantees a reliable and quality design not only for VIP but also other people, in particular children and the elderly. The design can be easily installed or attached on different types of doors (Designed by the author. Copyright by The Hong Kong Polytechnic University)

A design successfully applied in particular or even most of the situations and environments does not mean it can be applied to all. For example, while physical contact with the facilities for searching purposes is necessary for VIP, this need is contradictory to the nature and characteristics of public toilets, which are relatively (or at least commonly recognized as) dirtier than other types of public environments. Thus, people (not only VIP) are less willing to touch the facilities of public toilets. Therefore, one of the possible strategies for a user-friendly design is to tackle the problem directly: to minimize contact. That is, automatic facilities are suggested. Although there are also still constraints to implementing the types of fully automatic public toilets that are proposed in some European countries, certain types of automatic facilities intended to minimize contact between VIP and the facilities, for example, flushing mechanisms, water taps, facility position systems, is a good design direction (for example, some models of the designs of "Urilift" are nearly fully automatic; that is, including a self-cleaning process. For details, see www.urilift.nl). As recognized in the recent Beijing, Belfast, and Moscow World Toilet Summits in 2004, 2005 and 2006 respectively, with sufficient resources, public toilets should be moving towards the provision of automatic facilities that provide convenience to different users. Of course, how to implement this idea and what is appropriate in nature, degree, and level of automation are issues that require further investigation, discussion, and tests.

Sometimes, having convenience for a group of users may cause inconvenience to other groups. Nearly all of the interviewed VIP indicated that they did not prefer to use existing disabled toilets, including disabled closets inside toilets. One of the major reasons is that VIP are easily trapped and hurt by the handrails, particularly folding handrails, provided for physically disabled people in such toilets (Fig. 22.6). Another major reason raised by the VIP is that most of them do not have physical disabilities, like wheelchair users. The VIP emphasized that, most of the time, they can take care of themselves without special physical facilities. That is to say, VIP are rather more willing to use common public toilets with common facilities. This expectation is interesting, but not difficult to understand. It more or less reflects VIP's psychological feelings, preferences and needs: they do not want to be labeled as "disabled". Hence, besides giving physiological (functional) advantages, user-

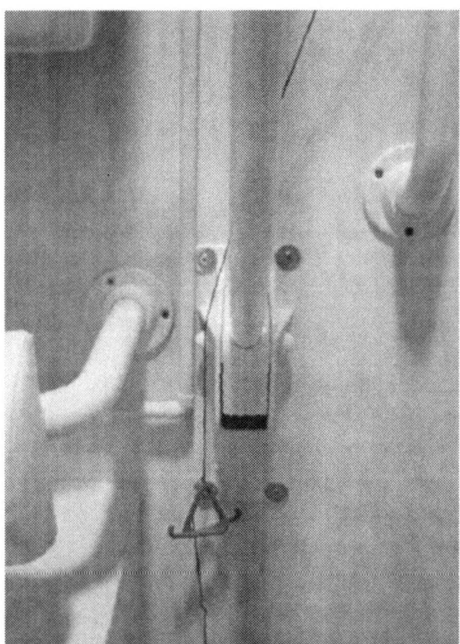

Fig. 22.6 VIP are easily trapped and hurt by the folding handrails and the strings (calling for help) provided for physically disabled people in such toilets

Fig. 22.7 Although the basins are not directly related to the needs of VIP, it is a good illustration of inclusive design. The basins serve not only the needs of people with different heights, but also form a smooth interlocking/transition of different basins that people would not feel strange or embarrassed using no matter which height of basins they are at

friendly design should also consider how to provide psychological as well as other kinds of satisfaction and pleasure to users [14, 16]. Therefore, on the one hand, it is necessary to review the current designs of disabled toilets. On the other hand, inclusive (and universal) design approach must be recommended in the designs of public toilets. The first advantage of this design approach is to promote design quality by developing a design to fit a larger population [17]. The second advantage is that it considers not only the functional and physiological factors but also the psychological factors as discussed above (Fig. 22.7).

Therefore, user-friendly public toilet environment and facility design and provision should not be only for VIP particularly, but also for other users [2, 18]. Further clear evidence obtained from the research and design project to support the inclusive approach in design is that most of the VIP do not want, indeed they refuse and hate, their inconvenience (some of them prefer to call their disability an "inconvenience" instead of disability) to cause any inconvenience to other people. As emphasized by the VIP many times during the interviews, most of them do not want to convey the image to the public that VIP are always asking for more help, more benefits, and more privileges and special care from society. Most importantly, they do not want their convenience to bring them into conflict with other users and distort the image of VIP [19].

User-friendliness further implies the good use of the strengths and specific abilities of particular users [9]. The VIP interviewed agreed that the major difficulty for them in accessing new places is that "new" implies differences, changes, and divergences. They also agreed that they are better in orientation and memory of physical locations and the forms of physical objects than the average person. Therefore, although diversity and choices are important in quality design, "standardization" in some critical elements is recommended [20], in particular in public environments.

Standardization can minimize the difficulties and weaknesses for VIP in new places, and it also makes good use of their strengths.

22.5.2 Informative

Informative means information can be: (i) provided by the right means, (ii) provided at the right time, (iii) provided in the right place, and (iv) understood easily and without any confusion by users [2].

Unlike visually able people, VIP require means other than vision to access public toilets and facilities. Thus, a better quality and more reliable design implies one in which such other means can be effectively used by VIP. For example, VIP are sensitive to the surrounding environment, such as temperature changes, smells, wind directions, and in particular sources of sound. In general, their non-seeing senses are relatively stronger than visually able people. They are also good at orientation and memory of physical locations and the forms of physical objects.

Providing the right means to transmit information further implies a kind of elimination of distracting means and information. Sometimes, VIP have difficulty accessing information if it is distracted or distorted by other information. For example, sound caution signals for VIP are always distracted by other noises, such as continuous announcement or advertisement from loud speakers. Such influence in receiving information is more serious for VIP if a place is new for them. The time for them to receive, interpret, understand, and react to the information takes longer.

Furthermore, although VIP are good in distinguishing different sounds as well as their sources, searching correct indicative information in a new place is still difficult for them, in particular if some signals are not standardized and in turn not easy to understand the meanings behind. Today, it is widely accepted in the design profession that diversity and choices are important in quality design. For example, the bird-sound caution signals for traffic lights in Kyoto create a more countryside-feeling in urban life, while the automatic soft-voice announcements in some of the entrances of elevators and public toilets in shopping centers in Hong Kong give a more relaxed feeling to city dwellers. However, as indicated above, "standardization" in some critical situations and elements is still recommended.

Accuracy and reliability of provided information are therefore crucial. For example, due to the size of the information devices, some information regarding the locations of toilets provided in shopping malls are not to scale. This situation does not create a great problem for visually able people as their eyes are continuously giving them other information (and alternative references, feedback) to find the toilets. However, the similar situation may be relatively more difficult for VIP even when tactile guide paths are provided. Alternative, additional, and supplementary information therefore is necessary. For example, sound signaling devices installed at the entrances of public toilets or information panels may give help for VIP to assure their information received and choices made are correct.

Quality of provided information depends on the accuracy and reliability of information. Yet, such accuracy and reliability sometimes may not be constantly correct or valid. In other words, the effectiveness of the information provided varies in terms of time. For example, the environmental settings and the facilities of a public toilet may change; therefore, updating the information frequently is important. The most common cases that occur in public toilets are some of the closets are under renovation or some of the facilities are out of order. Most of the time, when this kind of situation happens, notices are only provided for visually able people. In fact, VIP need this kind of notice more, since they cannot see the problems, which could pose a danger to them.

In addition, ineffectiveness of information sometimes does not relate to the accuracy and reliability of the content of information or the quality of means. Instead, it relates to the VIP themselves. For example, there are fewer VIP (*i.e.*, younger VIP) who know how to read Braille. Older VIP are also unable to read Braille due to the deterioration of their sensory capabilities. Then, additional and modified systems, such as audio systems and easy-to-read and modified tactile symbols are good alternatives for some VIP [21] (Fig. 22.8).

Providing information by the right means, at the right time, and in the right place also includes the requirement that the information fit the psychological as well as other types of preferences of users. As indicated by the VIP interviewed, not feeling embarrassed while accessing information is one of their major concerns, in particular when other visually able people do not have to do the same things (*i.e.*, search information) as them. For example, VIP do feel embarrassed when they use hands or guiding sticks to search for facilities, such as water taps and urinals, in public toilets that are also full of other users. Thus, designs assisting VIP to approach and receive required information in a minimal amount of time is a critical factor in quality. Moreover, many of the current property management companies like to put a Braille map (tactile map) and words right on the entrance door of public toilets. Such an arrangement was originally intended to give the most direct and convenient

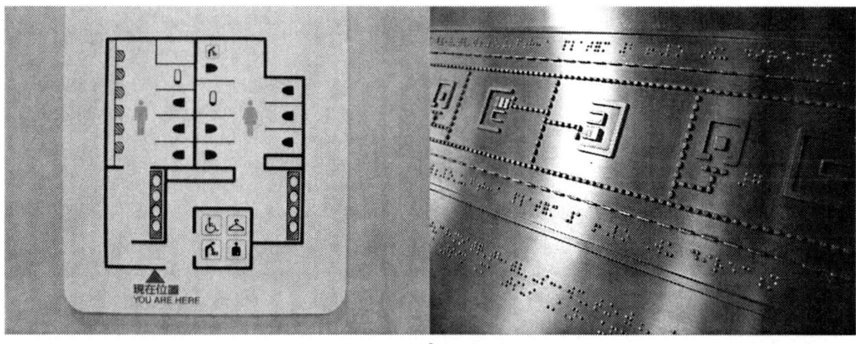

a   b

Fig. 22.8 a,b Orientation information is important for VIP. However, most of this information is not user-friendly or informative for VIP. The information for VIP is only in Braille (text) format. Graphical representation as the figures for VIP is still not common in many public places

position for VIP to search for information to understand the interior settings of the toilets. However, as indicated by the VIP, reading Braille located on an entrance door is extremely embarrassing, especially when people are constantly passing the entrance. Therefore, how to minimize such kind of "strange" behavior or how to change such kind of behavior so as not to be so apparent are critical. For example, Braille and other types and forms of tactile information can be incorporated with the existing handrails (Figs. 22.9 and 22.10). These kinds of simple designs or modifications in designs can give great help to VIP in allowing them to search for required information in their own way without causing any strange, laughable, and irritating feeling to other toilet users.

Fig. 22.9 Braille and other types and forms of tactile information have been incorporated in the handrail. The design provides convenience for VIP and also allows them to search for information in a more natural way that minimizes embarrassment for them

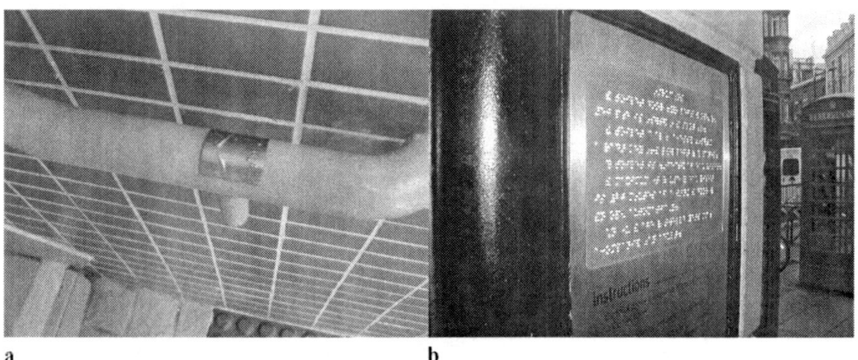

a   b

Fig. 22.10 a,b VIP use their fingers and hands to read information or to search for directions. However, how to direct them to reach the information for reading is a problem that requires consideration. The figures show the tactile information (Braille) on the handrails of a public toilet in Hong Kong and the entrance of a public toilet in England

Informative also implies a high degree of independence to get and understand information. As pointed out by the VIP, sometimes there may be no other user in a public toilet to give them help or for them to ask for information. Therefore, a reliable design of information for VIP must have a high degree of readability (*i.e.*, not by vision) and be understandable without the help of other people. For example, since the 1990s, more tactile guide paths have been paved for VIP to reach the entrances of public toilets. However, up to the present moment, there is still no guide path element (*e.g.*, tile) or other convenient device available for VIP to understand which types of toilets are in front of them when they arrive at the entrances of public facilities. Thus, it is a common scene that in front of the entrance of a public toilet a VIP asks whether the toilet is a male or female toilet. Some people have even suggested that a standard orientation position of male and female toilets can solve this problem. However, this suggestion is impractical, since the architecture and interior settings of most buildings are different, and it is impossible to require them to be the same or standardized. And, as reflected by VIP, they cannot take the risk even if these standards could be established. They would still prefer to ask until they can get (sense) some evidence (information) with a high reliability that allows them to make an informed choice. As they indicated, it is wise for them to make choice on: "What you sense (hear, smell, feel, touch), and then what you do;" instead of "What you guess and remember, and then what you do."

As stated, independence in searching for and understanding information implies that VIP do not need to continually ask other people for assistance. Such situations can prevent embarrassment to VIP, and embarrassment and inconvenience to other people. As the VIP pointed out, continually asking questions and assistance in a toilet is not so convenient, particularly because other people are doing very personal things inside the toilet at the same time and it is not easy for VIP to sense what other people are doing when they want to ask questions. In fact, as reminded by the VIP, even waiting for other people to finish their personal matters and then to ask for assistance is also a very embarrassing situation.

In addition, as discussed above, with regard to the considerations of user-friendliness, information as well as the facilities for providing the information for VIP may sometimes cause inconvenience to other users. For example, continuous sound signals for VIP may cause irritation to other users. It is the same as the existing common disabled toilets which in general occupy more interior space. It is also the reason why some of the small shopping centers do not provide disabled toilets though it is compulsory in many cities. Thus, how to balance benefit of different parties and consideration of different users' wants and needs, and not to over-provide assistance are both important issues [12, 13].

22.5.3 *Safe*

Safety is the most important consideration of all. In particular, since VIP do not have or are weak in their visual capability (*i.e.*, the sense that enables the fastest response), ensuring safe interactions between VIP and toilet environments and facilities are essential for quality design. According to the findings of the project, poor safety in public toilets for VIP comes from three major sources: neglect, unawareness, and inexperience.

Regarding neglect, it is easy to see that quite a lot of the designers and management people only focus their attention on the majority groups. They always have a kind of thinking that serving for the majority groups satisfactorily is equivalent to a good design [22]. On the other hand, some of them are in purpose to disregard the people with special needs, in particular when most of the current ordinances cannot satisfactorily protect the right of accessibility of some minority groups. For example, VIP are always excluded from the consideration list of many shopping centers that toilets in there are always not friendly and even unsafe for VIP to access [2]. In fact, causing difficulties and problems to some users, even a minority group, is a kind of discrimination and also unethical and irresponsible in design [16,23].

Moreover, such deficiency in design quality to a particular group of users sometimes also easily causes negative "ripple effects" to other users; that is, the inconvenience of a group of people may cause further inconvenience to other groups, in particular relating to safety matters [24]. In fact, it is also one of the spirits of inclusive design approach that aims to minimize such negative ripple effects. Besides the issue relating to equal opportunity, caring about the needs of people with special needs by applying inclusive designs can finally minimize the special resource and facilities and the effort of assistance they require, and increase the overall effectiveness of design, and in turn benefit a wider scope of the population.

As stated previously, besides neglect, unsafe public toilets are also caused by unawareness and inexperience. While public toilets, as the name suggests, are used by the public, the settings can and do change all the time. However, without careful consideration, some of changes can easily become traps for VIP. According to some observations, miscellaneous and movable devices are often added to the public toilet environment, such as folding stands for alerting users to the wet floor or cleaning facilities. These kinds of objects and devices may be good for seeing toilet users, but in turn may create problems to some, for instance, VIP (Fig. 22.11). These folding and retractable devices have been used more recently, and include such objects as retractable baby-sitting devices as well as folding handrails for physically disabled people and the elderly. These kinds of retractable and folding devices may fall down accidentally or not be returned to their original positions after use, but without giving any caution notice. These unpredictable and invisible devices are very dangerous for VIP. However, as mentioned by the VIP interviewed, many toilet management people and attendants are unaware of such risks for VIP because of their lack of sufficient training and experience.

To make it safe for VIP to use public toilets (*i.e.*, an unfamiliar place for VIP), one of the possible approaches is to provide standard designs and provisions [4,9]. As discussed before, diversity and choices are important for quality design, since different users have diverse wants and needs, and higher personal satisfaction is expected today. However, it is important to consider a balance in this issue. If the diversity and choices may cause dangerous situations, they should not be encouraged.

Fig. 22.11 Some simple "misplaced" and "unpredictable" objects in public toilets can cause serious inconvenience and even danger to VIP

The second approach to enhance safe public toilet environments is to minimize the "unpredictables" appearing there. In general, unpredictables as well as unexpected objects and matters come from four main sources in public toilets: accidents, deviant behavior, users, and toilet attendants. As its name suggests, "accidents" quite a lot of time are difficult to predict. However, according to the study, many of the accidents (poor quality of toilet environments and unsatisfactory performance of toilet facilities) that happen in public toilets are due to the reason of poor maintenance and management. Slippery floors and sharp edges of broken facilities are two major sources of accidents for VIP. Thus, no matter how good a design original is, continuous high quality management and maintenance is crucial to keep its intended objectives and function operating at an appropriate level. Good management and maintenance also can lower the chance of accidents; that is, minimize the unpredictables [23].

Deviant behavior is always the biggest problem regarding public toilets. As indicated by the VIP interviewed, slight misbehavior may cause serious injury for disabled people. To overcome this, education is essential; and good maintenance and security assurance in public toilets are important, too. Furthermore, sometimes people may not realize that their own ways of using and preferences in a public space may cause problems to others. A careless person forgetting to return a handrail back to its original position or a dropped piece of wet tissue paper may cause serious injury to a VIP. Thus, good civic education for higher awareness and better ethics, designs encouraging good user-practices such as user-friendly designs minimizing inconvenience to users, and good management and maintenance are three key solutions to solve this issue. In addition, according to the observation results, it is very common, because of convenience, for toilet attendants to put things beside walls and at the corners of public toilets. However, such small objects may cause inconve-

nience and danger to VIP. The interviewed cleaners admitted that they do not intend
to do any bad things to VIP as well as other people with special needs. However, the
findings indicate that most of them do not have such kind of awareness.

The third approach is to give instant and effective notice to VIP about the unpredictables to allow them to pay attention and take appropriate action (and reaction).
In other words, it is to make such kinds of invisibles and unpredictables visible (*i.e.*,
sense-able) and predictable by VIP [25]. For VIP, their ears are their eyes, so to
speak. Thus, a sound caution signal is one of the possible solutions. For example,
attaching a sound caution device to a temporary barrier or other obstacle is a good
suggestion. On the other hand, VIP are not only equivalent to blind people. In fact,
quite a lot of VIP can see strong light, flashing light with strong contrast with the surrounding environment, or high-contrast or brightly colored objects or images with
a dark background [26–28] (Fig. 22.12). Thus, using color and light for cautionary
purposes is also a good solution which does not only give caution to VIP but also
for the whole user population.

22.5.4 Hygienic

Although no one would deny safety as the most important consideration in public
toilets, most people would have to suppress their first-impression response, "hygiene", if suddenly asked their key concern with regard to the design of public toi-

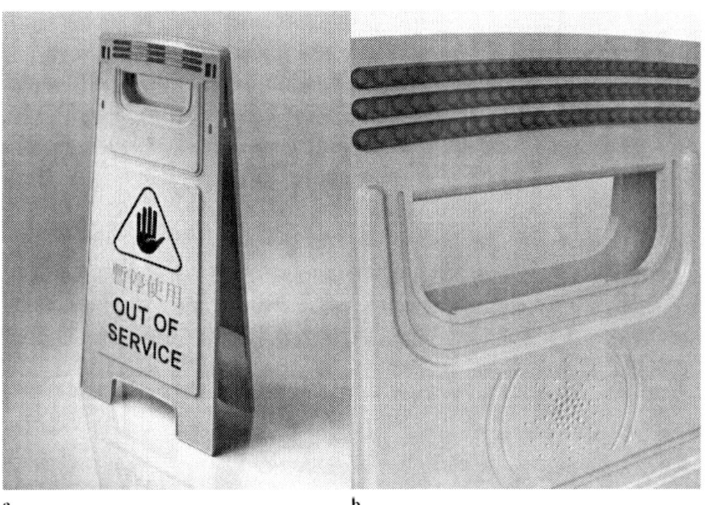

Fig. 22.12 a,b "Light and Sound Caution Devices for Caution Sign". The new inclusive design
for giving visual (flashing LEDs) and sound caution signals to VIP to prevent slipping on the floor
or being hurt by other unexpected objects on the floor (Designed by the author. Copyright by The
Hong Kong Polytechnic University)

lets. It has also been one of the key themes and major topics of the recent World Toilet Summits and international toilet forums and meetings. Moreover, as found in the study conducted by the Hong Kong Toilet Association in 2005, cleanliness and hygiene are two of the major concerns of people [29]. The project was led by Dr Lo Wing-lok, an expert in infectious disease and the Vice President of the Hong Kong Toilet Association. According to the study, 30% of males did not wash their hands after using the toilets. Fear of being infected by the dirty facilities was one of the key factors. Cleanliness and hygiene are also the major factors affecting proper behavior in public toilets, such as flushing toilets and washing hands after using the toilets. In the same way, as the VIP agreed, hygiene is their major concern as well as worry when visiting a public toilet for the first time.

Besides using their ears and guiding sticks, VIP need to use their bodies (hands, fingers, feet, *etc.*) to collect information and interact with the environment and facilities. This situation increases the chance of VIP to have direct contact (or more or less forces them to have direct contact) with the facilities, which are commonly considered full of dirt and hazards, and in turn causes them to be more likely to get infections. Thus, not visiting and using public toilets in a new place is the unwilling but safe choice for many VIP. It is also why there are very few cases reporting VIP getting into trouble in public toilets; one of the major reasons is that VIP seldom go to public toilets with which they are not familiar. Hence, minimizing contact with public toilet facilities which have a high chance of being infected is the design direction fitting VIP' needs.

For the past few decades, some researchers in science and engineering have focused their research and design direction on automatic toilets, such as "smart-toilets". Japanese bathroom manufacturer Toto is now offering a new smart-commode. It can sense when people enter the room and automatically lift the pre-warmed lid. Additional sensors next to the toilet activate an MP3 player with 16 preloaded tunes and a media slot for SD cards so people can load in their own music library. There have been some breakthroughs during the past 10 years. For example, Rentokil Initial has invented a system which can detect the germs on a person's hand and gives a reminder when he or she wants to push the door handle and leave a toilet but without washing or sufficient cleaning of hands. Some companies also launched new toilet facilities that allow people not to have direct physical contact with the facilities but still can operate them, such as paper-towel dispensers. However, under today's technological inventions, it is still difficult and unrealistic in terms of costing and management to aim at a "zero-contact" approach, Therefore, providing appropriate minimal contact with facilities that have a high chance of public contact and infectious disease transmission is an alternative good direction [9,19,21]. Such facilities include water taps, flushing mechanisms, paper-towel dispensers, hand-dryers, door handles, door locks, *etc.* (Fig. 22.13).

Another possible design direction is to separate the facilities that must be contacted by others that have a high potential to be infectious. For example, handrails with Braille and tactile symbols for VIP should be installed in a position and location convenient for VIP but also apart from those facilities that can easily be infected, such as water taps and basins. Moreover, as the interviewed male VIP mentioned,

a   b

Fig. 22.13 a,b An automatic hand-dryer and an automatic water tap. These facilities give convenience to VIP. However, how to assist VIP to locate such facilities (*e.g.*, hand-dryers, water taps) is a crucial question. A good location and orientation system can minimize the human–product contact in public toilets

one of the most unpleasant experiences for them is to search for urinals in public toilets. On the one hand, different from toilets (that is, toilet bowls) which are quite similar in form and position arrangement, there are hundreds of types of urinals in terms of different forms, locations, and even using methods (*e.g.*, vertical and horizontal trenches, bowls, holes). Thus, searching for urinals is a very difficult task, in particular in ensuring the correct position for urinating and whether people are standing in front of the urinals and using them. Hence, VIP needs to use their bodies or other external objects such as guiding sticks to orient themselves to the urinals. Then, more or less VIP are required to touch the urinal during the searching process. Therefore, separating the part of the means for VIP to search information from the part containing or with contact with urine is very essential for VIP.

22.6 Conclusions

The common limitation of a case study is that it is difficult to generalize statistically from such a study, and that it cannot be said to be a "representation of the whole population" of VIP [30]. However, these are not the objectives of the project discussed above. Instead, through the use of in-depth qualitative interviews and observations, the project attempts to identify the difficulties, limitations, and possible directions and areas for improvement in designs for the public. The reviews and discussions above on public toilets aim to generate insights for further investigations and discussions in similar topics.

In considering the VIP's actual wants and needs regarding their uses of public toilets, the "FISH" concept has been initiated and discussed. According the design experience in Hong Kong as well as other Asian cities, the concept is very useful and significant. Moreover, the concept has been verified in other cases related to public environments and facilities, such as street furniture, though the weightings and emphases of the four considerations may be different and sometimes some other considerations may be necessary to be supplemented.

As indicated in the Introduction, to obtain a high design quality, design focus should be shifted from the designers and designs to the users. This shift of focus does not only imply piecemeal and one-time investigation. Instead, while the wants and needs of VIP are continuously changing, along with the practical factors affecting the uses of public toilets, it is necessary to continue investigations and further discussions on the topic. Obviously, enhancing a high quality design of public toilets for the benefit of VIP does not simply imply the prime or narrow considerations of VIP. This would only result in falling into the old trap as the current disabled toilets for physically disabled users that such toilets only narrowly serve physical disabled people, but exclude others. Therefore, it would be better to consider the design of such public environments and facilities by using an inclusive design approach. Only through these kinds of in-depth, careful, inclusive, and continuing action studies, tests, and reviews will high quality designs be generated and assured, and real benefits brought to the public.

Acknowledgements I would like to acknowledge the resources extended by The Hong Kong Polytechnic University to support this study and the presentation of this paper. I would like to thank the Hong Kong Blind Union as well as other NGOs and volunteers (visually impaired persons) for their help in providing information. Special thanks to the Executive Committee Members of the Architectural Services Department and the Hong Kong Toilet Association for their advice and information. I would also like to thank my research assistants and students, Chan Wai-lun, Fu Ho-yin, Lo Chi-hang, and Wan Pak-hong who gave a lot of assistance in collecting the data. Part of the contents of this article has been presented at the ISSAT2006 Conference.

References

1. Frances TJM, Fortuin SWF (2006) Designing universal user interfaces: the application of universal design rules to eliminate information barriers for the visually impaired and the elderly. Retrieved October 31, from http://www.visionconnection.org/Content/Technology/ ForDesignProfessionals/default.htm?cookie%5Ftest=1
2. Siu KWM (2005) Inclusive design: public toilets for blind persons. Conference paper: 2005 World Toilet Summit. World Toilet Organization. Belfast
3. Siu KWM (2005) Pleasurable products: public space furniture with userfitness. Journal of Engineering Design 16:545–555
4. Siu KWM (2005) Userfitness in design: A case study of public toilet facilities. In: Pham H, Yamada S (eds.) Proceedings of the Eleventh ISSAT International Conference: Reliability and Quality in Design. International Society of Science and Applied Technologies, New Brunswick, N.J.
5. Siu KWM (2006) User experience and design. In: Conference Proceedings: 4th Annual Hawaii International Conference on Arts and Humanities [CD publication]. Hawaii International Conference on Arts and Humanities, Honolulu, Hawaii
6. Kuniavsky M (2003) Observing the user experience: a practitioner's guide to user research. Morgan Kaufmann, San Francisco, Calif.

7. Siu KWM (2000) A study of the relationship between the colour vision deficiency and creativity of design and technology students. Korean Journal of Thinking and Problem Solving 10:21–30
8. Meisami E, Brown CM, Emerle HE (2003) Sensory systems: normal aging, disorders, and treatments of vision and hearing in humans. In: Timiras PS (ed.) Physiological basis of aging and geriatrics. CRC Press, Boca Raton, Fla.
9. Willems CG, Vlaskamp FJM, Knops HTP (1997) Use of technology in the support of independent living of people with special needs. In: Anogianakis G, Bühler C, Soede M (eds) Advancement of assistive technology. IOS Press, Amsterdam
10. Crosby FJ, VanDeVeer C (2000) Sex, race, and merit: Debating affirmative action in education and employment. University of Michigan Press, Ann Arbor
11. Wiklund ME (2005) Making medical device interfaces more user-friendly. In: Wiklund ME, Wilcox SB (eds) Designing Usability into Medical Products. Taylor & Francis/CRC Press, Boca Raton, Fla.
12. Siu, KWM (2005) Pleasurable products: public space furniture with userfitness. Journal of Engineering Design 16:545–555
13. Rehabilitation Alliance Hong Kong (2005) Research on the accessibility of public disabled toilets, Rehabilitation Alliance Hong Kong, Hong Kong
14. Jordan PW, Green WS (1999) Human Factors in Product Design: Current Practice and Future Trends, Taylor and Francis, London
15. Universal accessibility: best practices and guidelines (2006) Architectural Services Department, Hong Kong SAR Government, Hong Kong. Retrieved November 12, from http://www.archsd.gov.hk/english/knowledge_sharing/ua/univ_access.htm
16. Greed C (2003) Inclusive Urban Design: Public Toilets. Architectural Press, Oxford
17. Clarkson J (ed.) (2003) Inclusive Design: Design for the Whole Population. Springer, New York
18. Imrie R, Hall P (2001) Inclusive Design: Designing and Developing Accessible Environments. Spon Press, London
19. Anogianakis G, Bühler C, Soede M (eds) (1997) Advancement of Assistive Technology. IOS Press, Amsterdam
20. Stephanidis C, Akoumianakis D, Ziegler J (1997) Universal accessibility and standardisation: new opportunities and prospects. In: Anogianakis G, Bühler C, Soede M (eds) Advancement of Assistive Technology. IOS Press, Amsterdam
21. Placencia-Porrero I, Puig de la Bellacasa Alberola R (eds) (1995) The European context for assistive technology: Proceedings of the 2nd TIDE Congress. IOS Press, Amsterdam
22. Keates S (2003) Countering Design Exclusion: An Introduction to Inclusive Design. Springer, London
23. Goldsmith S (1997) Designing for the Disabled: The New Paradigm. Architectural Press, Oxford
24. Siu KWM (2005) Street furniture design for pedestrian precincts: a balanced approach. In: Conference Proceedings: 3rd Annual Hawaii International Conference on Arts and Humanities [CD publication]. Hawaii International Conference on Arts and Humanities, Honolulu, Hawaii
25. Siu KWM (2006) CASH: criteria for quality toilets in the future. In: Conference Proceedings: World Toilet Summit and Expo: Full Reports of the Summit Participants. World Toilet Organization, and Russian Toilet Association, Moscow, pp. 7–10
26. Electronic Audible Traffic Signal. Retrieved November 12, 2006, from http://www.td.gov.hk/about_us/video/electronic_audible_traffic_signal/index.htm
27. Blind Citizens Australia. Retrieved November 11, 2006, from http://www.bca.org.au/pnws0012.htm
28. General Visual Impairments. Retrieved November 9, 2006, from http://www.brps.org.uk/White/W_Internet_Visual.html
29. Hong Kong Toilet Association (2006) Best Toilet Criteria (Full report). Retrieved October 31, from http://www.hktoilet.org
30. Yin RK (2003) Applications of Case Study Research. SAGE, Thousand Oaks, Calif.

Chapter 23
Assurance Cases for Reliability: Reducing Risks to Strengthen ROI for SCADA Systems

Ann Miller, Rashi Gupta

Department of Electrical & Computer Engineering,
University of Missouri – Rolla, Missouri, USA

23.1 Introduction

Supervisory, Control, and Data Acquisition (SCADA) systems are crucial to many critical infrastructures, typically those that operate in real-time environments. In the past, these systems were based on proprietary protocols and were isolated from other networks. However, as the trend towards standard IT practices has grown, there have been serious concerns regarding the security, reliability and availability of these systems, since they are increasingly connected to an enterprise network. There are many positive business benefits to be gained from this trend; however, this connectivity has increased the risks of exploitation of security vulnerabilities in these non-proprietary systems. These vulnerabilities, if exploited, can result in serious consequences such as degraded performance, loss and/or compromise of critical information, or in the worst case scenario, making these critical systems completely unavailable [1–4].

These attacks not only place the normal operation of these systems at risk but also impact directly on the cost of security, reliability, availability and maintainability of these systems. The immediate economic impact is the cost of repairing or replacing the failed or attacked systems and disruption of business operations. Others include the cost of patching and system recovery, the cost of adding fault tolerance through redundant systems and other maintenance activities. Some of the short-term economic impacts on an organization include the loss of existing customers because of the inability to deliver products or services and a negative impact on the reputation of the organization where as long-term economic impacts include decline in an organization's market valuation and stock prices.

Assurance cases for safety, also known as safety cases, are a common practice in many countries whose regulatory agencies require such cases for critical infrastructures. While reliability is one aspect of safety, this paper proposes assurance cases with a concentration on reliability, specifically for Supervisory, Control, and Data Acquisition (SCADA) systems. It utilizes a goal-based assurance case approach to develop a risk management structure that serves to reduce the impact of these risks in terms of technical and business value [5].The advantage of this approach is that it can begin during the early requirements and analysis phases of a project and continue to be developed throughout a project's life cycle, including operations, maintenance, and de-commissioning.

23.2 Approach

This paper is divided into two parts. In the first part, a business goal aiming to improve the return on investment (ROI) for SCADA system security, reliability, maintainability and availability is set. The choice of these four attributes is made since security risks can result in increased reliability, maintainability and availability (RAM) risks. A risk management structure is developed beginning with risk identification. The technical risks associated with RAM and security are identified along with their impacts. These technical risks are mapped into business risks and their impacts. Risk assessment is done by developing a risk assessment form and a risk severity matrix. The risk assessment form consists of all the technical risk events, their likelihood of occurrence, the impact of the risks and the detection difficultly (how long does it take to detect if a risk has occurred) and when the risks could occur. All of these are categorized under low, medium and high levels of severity. The risk severity matrix is constructed by using the likelihood of occurrence of a risk event and its corresponding impact. Risk events with high severity are represented with red whereas risks with moderate and low risks are shown with yellow and green color respectively. Risk value is then evaluated from the product of impact, likelihood and detection difficulty. This risk value is expressed in terms of ROI. There is an inverse relationship established between the risk value and ROI. Figure 23.1 represents the general risk management structure used in this paper.

In the second part of this paper, assurance cases for each risk event are created using Adelard Safety Case Editor (ASCE) Version 3.5. An attack tree for a SCADA system is developed using Goal Structuring Notation (GSN) [6]. In this attack tree, the security and RAM cases cover all the risks identified for SCADA systems in this report. Industry standards such as the ISO 17799, Critical Infrastructure Protec-

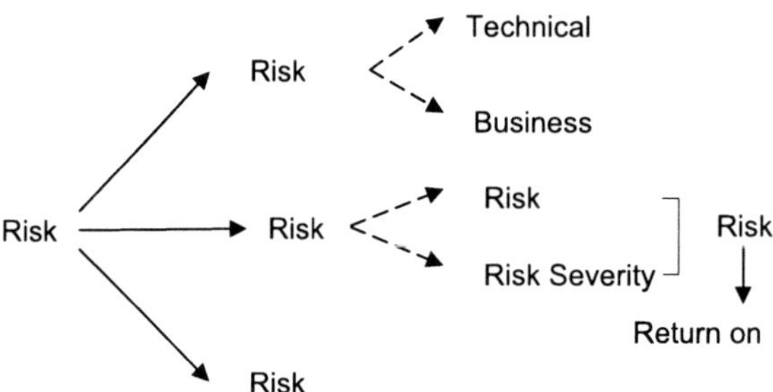

Fig. 23.1 Risk management structure

tion (CIP) are taken into consideration to validate the assurance cases. Further, input from assurance cases is taken to evaluate the reduced impact of the risk and their corresponding detection difficulty. The purpose is to generate and maintain a collection of assurance cases (Security, Reliability, Maintainability and Availability) by taking the view of SCADA systems assurance across all aspects of functionality and operation. The rationale is to analyze the overall impact of assurance cases from a complete system perspective in risk management [7]. The reduced risk value with the help of assurance cases is then calculated. A comparative study between the risk value (ROI) and reduced risk value (enhance ROI) is provided to illustrate how assurance cases are helpful in mitigating risks and thereby improving overall ROI for SCADA systems.

23.3 SCADA Security and RAM Issues – An Overview

SCADA usually refers to a central system that monitors and controls a site or one site within a distributed system. The bulk of the site control is actually performed automatically by a remote terminal unit (RTU) or by a programmable logic controller (PLC) with monitoring by an operator. A SCADA system includes input/output signal hardware, controllers, human machine interface (HMI), networks, communication, and software. SCADA systems are frequently used for critical infrastructures such as oil and gas pipelines, electric utility substations, chemical processing plants, and water and treatment plants. The reliability and availability of SCADA networks and process management systems are critical for utilities to ensure that there is no disruption of service, process redirection, or manipulation of operational data that could result in serious disruption to the critical infrastructure.

In the past, these systems were operated in an isolated or stand-alone environment where computer systems and devices communicated with each other exclusively. These control systems were comprised of proprietary hardware, software and protocols designed specifically for such system operations. Now, with increased connectivity and a move to standard protocols, such as Ethernet and IP, Microsoft Windows and web technologies, these systems are being transformed into an "always connected" real-time environment. This helps to reduce operational costs and improves performance. However, this transition has brought many known vulnerabilities into these systems. Exploitation tools, worms, viruses, Trojans, trapdoors and other malicious software can harm the SCADA systems, network and communication links since significant information about the design, implementation, maintenance and usage of these systems is readily available. The reliability and security concerns for SCADA systems are elevated because these systems are typically not up-to-date with the latest security patches, fixes and best practices. This raises concerns over making system modifications which might affect the time-sensitive operations of SCADA systems. Also, these systems must operate all day, every day, which means that these systems could not be isolated for testing or maintenance. With security risks, availability issues arise. The worst impact of security risks is gaining system access and making a system unavailable for normal operations or modifying the systems to make them untrustworthy and unreliable.

23.4 Risk Identification

23.4.1 Security and RAM Risks Associated with SCADA Systems

The identification and assessment of risks associated with threats that exploit vulnerabilities in SCADA system's design, implementation and configurations are critical to improving the security, reliability, maintainability and availability. Technical risks have direct relation with business risks. With a degraded system performance due to a security attack or system failure, there is direct consequence of disruption or denial in business operations which impacts critical business parameters such as rate of return, annual profit/loss and return on investment. Some of the security and RAM related risks associated with SCADA systems are identified as in Table 23.1.

Reliability, availability and maintainability (RAM) risks and their impacts are outlined in (Table 23.2).

Table 23.1 Security and RAM-related risks associated with SCADA systems

Sample no.	Security risks	Impact
1	Software risks • Buffer overflow • Stack overflow • Heap overflow • Access privilege in program • Format String error • OS related • Software components	Loss of: • Integrity • Data reliability • Data authenticity • Information transparency • System access Other impacts: • Intrusion • Information leak • Information hacking • Misuse of real time critical information • Data modification, interception • Denial of services • Data repudiation • Resource depletion • Accuracy of real time data critical for systems operating in real time environment is compromised

Table 23.1 (continued)

Sample no.	Security risks	Impact
2	Hardware risks • Routers • Switches • Systems • Equipment and machinery • COTS products	• System unavailability • Temporary/permanent disruption of information • Resource depletion • Loss of product/equipment, critical machines
3	Operational/safety risks • VPN • Insufficient patching for 24×7 systems • Operator's error • Modem risks – remote access • Dynamic systems/Evolving systems generally have insufficient security patching	• Unavailability of operational stations • DOS – system freeze • Loss of system access • Loss of information transparency, accuracy, integrity and authenticity • Data reliability lost • Information hacking
4	Network and communication risk • Insecure corporate network • Third party network • Insecure network routing • Network connections • Wireless networking • Network links • Active network attacks • TCP communication protocol • ICCP protocol connections open up SCADA networks to hackers, worms, and other threats • Communication interface • Communication channel	• Unavailability of operational stations • Delay in system response there by disruption in real time information processing • Process disruption – permanent/ temporary • Information/control bypass leading to direct access to control system • Intrusion • Control/information bypass leading to direct access to SCADA systems • Information hacking • Loss of communication system • Data interception • Compromised accuracy of data • Data modification due to man in middle types of attacks
5	Mission critical system (24×7) risk: systems that are crucial for any time sensitive operation and directly relate to a huge financial or human life loss. Since these systems are 24×7 so it is not feasible them for testing and maintenance purposes	• Outage • Loss of system access • System untrustworthy • Disrupted operation • System completely unavailable

Table 23.1 (continued)

Sample no.	Security risks	Impact

6	External risks
	
	1. Cyber attacks:
	  • Man in the middle attack
	  • Worms, viruses, Trojan horses
	  • Back door
	  • Flooding attack
	  • Intrusion
	  • Passive monitoring
	  • Targeted attacks
	  • Unauthorized control
	2. Untrustworthy third party vendors, support organizations/systems' risks, security risks in supply chain
	3. Electronic access
	4. Increased external connectivity and IT standardization: commercial off-the-shelf software replace the propriety process control systems and do not meet the complexities, uniqueness, real time and safety requirements of process control environment
		Loss of:
		
		• Integrity
		• Data reliability
		• Data authenticity
		• Information transparency
		• System access
		
		Other impacts:
		
		• Intrusion
		• Information leak
		• Information hacking
		• Misuse of real time critical information
		• Data modification, interception
		• Denial of services
		• Data repudiation
		• Resource depletion
		• Accuracy of real time data critical for systems operating in real time environment is compromised
		• More security attacks resulting from the known vulnerabilities in standard processes

7	Internal risks • Database attacks • Inappropriate security policy • Sniffing by internal employees • Insecure disposal of systems • Insecure redeployment of systems	• Loss of data accuracy, data availability critical for real time environment • Information leak • Reliability of data is compromised • Data modification • Loss in backup data
8	Insufficient security risks • Passive attacks • Improver intrusion detection system • Improper firewall implementation • Improper/inappropriate antivirus • Single layer of security	• Intrusion • Denial of services • Fault tolerance for mission critical will be lost • Loss in backup data • Increased system vulnerability for all sorts of attack • Other related risks
9	System architecture and interfaces • System interfaces: take advantage of the exploits in the system	• System unavailability • System freeze • Loss of valuable information not from the current system but from all those interfaced with the targeted system

Table 23.2 RAM risks associated to SCADA systems

Sample no.	RAM risks	Impact
1	Software risks For software-based systems, it is difficult to maintain reliability since it is not always feasible to replace them easily; further, "patches" often introduce new errors. • Software systems/components • Operating systems • Software applications	Failure of software applications working independently with high severity will have the following consequences: • Loss of information/data access • Partial failure or complete system failure Dependent software module: Will impact all the other software modules related to it leading to the ceasing of entire real time information process
2	Hardware risk Failure of • Equipment • Machines/systems • PLCs • Radio communication systems • SCADA RTU system	• System unavailability Increases/ Unnecessary resource consumption • Loss of critical machines • Loss of information processing • Increased maintenance
3	Operational risks: dynamic/evolving systems have to have more redundancies since they might be prone to failure when tested under stress or extreme conditions	• System unavailable • Loss of data processing
4	Network/communication risk • Server availability • Radio communication repeaters • Communication links Wire-line methods: Analog P-to-P ATM Mesh networks, frame relay P-to-P T-1 and higher Fiber: SONET Fiber ethernet Radio: DDS (dataphone digital service) Wireless network *etd.*	• Loss of information flow • Delay in system response time • Process disruption, temporary or permanent • Unavailability of operational stations • Complete or partial failure of communication systems

RAM relation:

• Availability is crucial for mission critical operations and real time systems.
• The cost of system and other equipment maintenance should be low.
• The above two are achieved if system reliability is high.
• In case of failure the systems should degrade gracefully.

Table 23.2 (continued)

Sample no.	RAM risks	Impact
5	Mission critical systems Have to be highly available in order to be operational 24×7, *i.e.*, they should be up and running without fail all the time. This calls for their high reliability	In the absence of high RAM, the entire objective of the systems is defeated. There could also be safety related issues concerning with this
6	Remote system maintenance: In absence of system maintenance or with systems that are difficult to manage there are increased risks of system reliability and availability	Less reliable or prone to failure, and therefore, more downtime leading to the disruption of information flow and this is unfavorable for systems working in real time scenarios
7	Internal risks Database RAM: any database system failure will have severe consequences	• Loss of critical data/information availability • Loss of backup data • Loss of information superiority
8	System architecture Complex architecture and interfaces also lead to RAM risks	• System failure under stress/extreme conditions there by making system unavailable • Loss of other systems/subsystems/components interfaced with the main system • Loss of information processing

RAM can be related and quantitatively expressed as:

$$\text{Availability} = \text{MTTF}/(\text{MTTF} + \text{MTTR}) = 1/[1 + (\text{MTTR}/\text{MTTF})]$$

where MTTF = mean time to failure, MTTR = mean time to repair.

23.4.2 Business Risks Associated with SCADA Systems

In critical situations, system availability, data integrity and confidentiality are extremely important. In case of system failure or system downtime, as a result of a security attacks, the value of ROI decreases. While examining the ROI for security and RAM several variables must be considered. The process of determining security and RAM expenditures should take into consideration existing expenditures as well as what might be needed after threat levels are re-evaluated. The cost increases where there are laws, regulations and defense requirements that require certain types of organizations to take extra steps in protecting critical systems from attack. Considering the security and RAM issues of SCADA systems, it is evident that there is a lot at risk. Some of the direct security and RAM related business risks with their impacts, if the technical risks occur, are outlined in Table 23.3.

Table 23.3 Security and RAM related business risks associated with SCADA systems

Sample no.	Business risks	Impacts
1	Operational risks	Disruption in business transactions and business process
2	Financial risks	• Increased cost of reliability by adding redundancies • Increased cost of repair and replacement • Increased cost of security due to patching, clean up, firewall, antivirus, IDS (intrusion detection system) installation • Hindrance in new business • Loss of revenue/business profitability • Loss of business and financial data • Loss of business • Loss of credibility • Cost of redesign/re-engineering the failed systems
3	Market risks	• Loss of customer satisfaction • Hindrance in new business • Decline in stock price • Decline in market value • Loss of revenue/business profitability • Loss of reputation
4	Strategic risks	• Loss of decision superiority • Hindrance in new business
5	Compliance risks	• Violation of company's policies
6	Communication risks	• Loss of interconnection to the whole business network
7	Inability to meet business requirements	• Hindrance in new business

23.5 Mapping Technical Risks into Business Risks

The following sub-sections map the technical risks into business risks identified in the above section.

23.5.1 Mapping Security Risks into Business Risks

The data are given in Table 23.4.

23.5.2 Mapping RAM Risks into Business Risks

The data are given in Table 23.5.

Table 23.4 Technical security risks mapped into business risks and their consequences

Sample no.	Security risk	Impact	Business risk	Business impact
1	Software risks: • Buffer overflow • Stack overflow • Heap overflow • Access privilege in program • Format string error • OS related • Software components	Loss of: • Integrity • Data reliability • Data authenticity • Information transparency • System access Other impacts: • Intrusion • Information leak • Information hacking • Misuse of real time critical information • Data modification, interception • Denial of services • Data repudiation • Resource depletion • Accuracy of real time data critical for systems operating in real time environment is compromised	• Operational risks • Market risks • Financial risks	Loss of: • Revenue • Reputation • Credibility within customers, business partners • Market share • Stock price • Customer data • Business and FI data • Business profitability Other impacts: • Hindrance in new business • Increased cost of security
2	Hardware risks • Routers • Switches • Systems • Equipment and machinery • COTS products	• System unavailability • Temporary/permanent disruption of information • Resource depletion • Loss of product/equipment, critical machines	• Market risks • Financial risks	• Increased cost of security • Loss of market share • Decline in stock price

Table 23.4 (continued)

Sample no.	Security risk	Impact	Business risk	Business impact
3	Operational/safety risks • VPN • Insufficient patching for 24×7 systems • Operator's error • Modem risks – remote access • Dynamic systems/Evolving systems generally have insufficient security patching	• Unavailability of operational stations • DOS – system freeze • Loss of system access • Loss of information transparency, accuracy, integrity and authenticity • Data reliability lost • Information hacking	• FI risk • Compliance risk • Inability to meet business requirements	• FI loss due to security patching • Loss of most personal data
4	Network and communication risk • Insecure Corporate network • Third party network • Insecure network routing • Network connections • Wireless networking • Network links • Active network attacks • TCP communication protocol • ICCP protocol connections open up SCADA networks to hackers, worms, and other threats • Communication interface • Communication channel	1. Unavailability of operational stations 2. Delay in system response there by disruption in real time information processing 3. Process disruption – permanent/temporary 4. Information/control bypass leading to direct access to control system 5. Intrusion 6. Control/information bypass leading to direct access to SCADA systems 7. Information hacking 8. Loss of communication system 9. Data interception 10. Compromised accuracy of data 11. Data modification due to man in middle types of attacks	1. Operational risk 2. Financial risks 3. Communication risk 4. Market risk	1. Delay in business transactions 2. Disruption in business process 3. Cost increase due to security patching 4. Loss of information/decision superiority 5. Loss of interconnection to the whole business network 6. Loss of information quality, information superiority 7. Increased cost due to redundancies

Table 23.4 (continued)

Sample no.	Security risk	Impact	Business risk	Business impact
5	Mission critical system's (24×7) risk: Systems that are crucial for any time sensitive operation and directly relate to a huge financial or human life loss. Since these systems are 24×7 so it is not feasible them for testing and maintenance purposes	• Outage • Loss of system access • System untrustworthy • Disrupted operation • System completely unavailable	• Market risk • Financial risk	Loss of: • Stock price • Reputation • Competence • Customer satisfaction • Business partners • Cost of redesign and re-engineering due to security compromise/flaws • Cost of repairing/replacing system • Disruption on normal operations
6	External risks: 1. Cyber attacks: • Man in middle • Worms, viruses, Trojan infections • Back door • Autorouter • Flooding attack • Intrusion • Passive monitoring • Targeted attacks • Unauthorized control • Untrustworthy third party vendors, support organization/system risks, security risks in supply chain • Electronic access • Increased external connectivity and IT standardization: Commercial off-the-shelf software replace the propriety process control systems and do not meet the complexities, uniqueness, real time and safety requirements of process control environment	Loss of: • Integrity • Data reliability • Data authenticity • Information transparency • System access Other impacts: • Intrusion • Information leak • Information hacking • Misuse of real time critical information • Data modification, interception • Denial of services • Data repudiation • Resource depletion • Accuracy of real time data critical for systems operating in real time environment is compromised • More security attacks resulting from the known vulnerabilities in standard processes	• Market risks • Financial risks • Reputation risk • Operational risk • Compliance risk • Management risk	• Loss of business partner, revenue • Decline in stock price • Loss of contracts • Loss of business profitability • Loss of reputation • Legal action liabilities will increase • Policy violation • Security patches, countermeasures against vulnerability exploits will impact directly on cost • Loss of personal, financial information • Loss of privacy

Table 23.4 (continued)

Sample no.	Security risk	Impact	Business risk	Business impact
7	Internal risks • Database attacks • Inappropriate security policy • Sniffing by internal employees • Insecure disposal of systems • Insecure redeployment of systems	• Loss of data accuracy, data availability critical for real time environment • Information leak • Reliability of data is compromised • Data modification • Loss in backup data	• Market risks • Financial • Reputation • Strategic risk • Operational risk • Compliance risk • Inability to. meet business requirements	• Critical business financial, technical, and most personnel information loss • Loss of business • Regulatory violation • Loss of archival, historical, regulatory, data critical for a business to predict future market trends • Loss of business competence • Loss of revenue • Loss of customer data • Modified financial data • Critical business financial, technical, and most personnel information loss • Security patching • Cost of fixing security flaws • Cost of redesign and re-engineering due to compromised security
8	Insufficient security risks • Passive attacks • Improper intrusion detection system • Improper firewall implementation • Improper/inappropriate antivirus • Single layer of security	• Intrusion • Denial of Services • Fault tolerance for mission critical will be lost • Loss in backup data • Increased system vulnerability for all sorts of attack • Other related risks	• Financial/cost risk • Market risk	• Security patching • Cost of fixing security flaws • Cost of redesign and reengineering due to compromised security
9	System architecture and interfaces • System interfaces: take advantage of the exploits in the system	• System unavailability/system freeze • Loss of valuable information from all those interfaced with the targeted system	• Strategic risks • Financial risks	• Security patching • Cost of fixing security flaws • Cost of redesign and re-engineering due to compromised security

Table 23.5 Technical RAM risks mapped into business risks and their consequences

Sample no	RAM risks/issues	Impact	Business risk	Impact
1	Software risks For s/w systems it is difficult to maintain reliability since its not always feasible to replace them unlike hardware component: • Software systems/components • Operating systems • Software applications	Failure of Software applications working independently with high severity will have the following consequences: • Loss of information/data access • Partial failure or complete system failure Dependent software module: Will impact all the other software modules related to it which can lead to the ceasing of entire real time information process	• Financial risk • Operational risk • Market risk	• Cost of software adding redundancies • Disruption of business process
2	Hardware risk Failure of • Equipment • Machines/systems • PLCs • Radio communication systems • SCADA RTU system	• System unavailability Increases/unnecessary resource consumption • Loss of critical machines • Loss of information processing • Increased maintenance	• Market risk • Financial risk • Operational risk	• Cost of repair/replacement of failed systems/subsystems, components and equipment *etc.* • Cost of adding hardware redundancy • Loss of market share • Cost of redesign and re-engineering the failed systems
3	Operational risks: Dynamic/evolving systems have to have more redundancies since they might be prone to failure when tested under stress or extreme conditions	• System unavailable • Loss of data processing	• Financial • Compliance risk • Inability to meet business requirements	• Delay in business transactions • Disruption in business process • Increase cost due to maintenance

Table 23.5 (continued)

Sample no	RAM risks/issues	Impact	Business risk	Impact
4	Network/communication risk • Server availability • Radio communication repeaters • Communication links Wire-line methods: Analog P-to-P ATM Mesh networks, Frame relay P-to-P T-1 and higher Fiber: SONET Fiber ethernet radio: DDS (dataphone digital service) Wireless network *etc.*	• Loss of information flow • Delay in system response time • Process disruption – temporary or permanent • Unavailability of operational stations • Complete or partial failure of communication systems	• Financial risk • Market risk	• Revenue loss due to loss of information flow • Delay in business transaction • Loss of goodwill • Loss of interconnection to the entire business network • Loss of information and decision superiority
5	Mission critical systems Have to be highly available in order to be operational 24×7 *i.e.*, they should be up and running without fail all the time. This calls for their high reliability	In the absence of high RAM, the entire objective of the systems is defeated. There could also be safety related issues concerning with this	• Severe financial risk • Severe market risk • Personal risk	• Financial loss by redeploying the failed systems • Loss of market trust • Soiled reputation

Table 23.5 (continued)

Sample no	RAM risks/issues	Impact	Business risk	Impact
6	Remote system maintenance: In absence of system maintenance or with systems that are difficult to manage there are increased risks of system reliability and availability	Less reliable or prone to failure, and therefore, more downtime leading to the disruption of information flow and this is unfavorable for systems working in real time scenarios	• Financial risk • Market risk	Loss of: • Stock price • Reputation • Market share *etc.* • Increased cost of maintenance-repairs
7	Internal risks Database RAM-ility: Any database system failure will have severe consequences	• Loss of critical data/information availability • Loss of backup data • Loss of information superiority	• Market risk • Financial risk • Strategic risk • Reputation risk • Inability to meet business requirements	• Revenue loss due to information/data loss • Loss of customer data • Information/decision superiority is lost • Loss of business • Loss of personal data and information
8	System architecture Complex architecture and interfaces also lead to RAM risks	• System failure under stress/extreme conditions there by making system unavailable • Loss of other systems/subsystems/ components interfaced with the main system • Loss of information	• Financial risk • Operational risk	• Increased maintaining loss by repairing the down systems • Loss of business • Cost of redesign and reengineering the failed systems

23.6 Risk Assessment

Risk assessment provides a detailed knowledge base of risk priorities across risk domains, gaps in risk capabilities and areas of under/over-investment. Once risks have been identified and assessed, the steps to properly deal with them become executable. In this paper, two methods for risk assessment are used. Initially, a risk assessment form is constructed for both security and RAM risks. This form contains the risk events, their impact, likelihood, detection difficulty and the event when that particular risk can happen. Using these forms, risk assessment matrices are constructed for both security and RAM related risks.

23.6.1 Risk Assessment Form

23.6.1.1 Security – Risk Assessment Form

Risk likelihood, detection difficulty and impact are classified under three levels:

- High $= 3$
- Medium $= 2$
- Low $= 1$

The categorization between low, medium and high level is done on the basic understanding of the risk categories described above and their impact on the SCADA systems. For detection difficulty, a high level indicates crashing of system without a warning, whereas moderate and low level refer to performance degradation of system at moderate or low intensity.

The risk assessment form also provides risk value which is calculated as:

$$\text{Risk Value} = \text{Likelihood} \times \text{Impact} \times \text{Detection Difficulty}$$

For risk value:

- Low $= 1$–7
- Medium $= 8$–17
- High $= 18$–27

In this report, the risk value is used to denote Return on Investment. The risks are identified from the perspective of cost of investment on the installation, maintenance, testing, redeployment and repair of SCADA systems and subsystems. The relationship between risk value and ROI is inverse, *i.e.*, higher the risk value lower is the return on investment for SCADA systems (Table 23.6).

From Table 23.6, it is evident that 67% of the risk categories come under a high security risk zone and nearly 22% and 11% are under moderate and low security risk zones. This implies that of all the investments made on SCADA systems, a major part of it is invested on countering the impacts of security risks associated with them. This reduces the overall value of ROI.

Table 23.6 Security risk assessment form

Sample no.	Risk event	Likelihood	Impact	Detection difficulty	When	Risk value, ROI
1	Software risks	High	High	Medium	Requirement analysis, designing, programming, testing and maintenance	18 high
2	Hardware risk	Low	Medium	Medium	Post installation	2 low
3	Operational and safety risk	High/low	High	Medium	Remote access, operator's error, information exchange	18 high
4	Network and communication risk	High/medium	High	High/medium	Communication, network links, wireless networks	27 high
5	Risks associated with mission critical systems	High/low	High	High/medium	Post installation. remote system-maintenance, testing, and patching	27 high
6	External risks	High	High	High	Information/data exchange, third party	27 high
7	Internal risks	High	Medium	Medium	Information access, internal employee, disposal	12 medium
8	Insufficient security and passive attacks	Medium	High	Medium	Insufficient firewall installation, improper IDS, security policy	12 medium
9	System architecture	Low	Medium	Medium	System development, testing, maintenance	4 low

Table 23.7 Security risk assessment form

Sample no.	Risk event	Likelihood	Impact	Detection difficulty	When	Risk value
1	Software risk	High	Medium	High	Requirement analysis, designing, programming and installation, testing and maintenance	18 high
2	Hardware risk	Medium	High	Low	Post-installation, under severe/stress conditions	6 low
3	Operational risk	High	High	Medium	Information exchange, system access	18 high
4	Network/communication risk	High/low	High	High/medium	Communication/information exchange	27 high
5	Mission critical system risk	Medium/low	High	Medium	Post-installation, stress conditions	12 high
6	Remote system maintenance risk	Medium	High	High/medium	Post-installation, system access, stress condition	18 high
7	Internal risk	Medium	Medium	High	Information transfer	12 medium
8	System architecture risk	Low	Low	High	System/software development	3 low

23.6.1.2 RAM – Risk Assessment Form

Table 23.7 shows that 63% of the risk categories identified above come under a high RAM risk zone and nearly 25% and 12% come under moderate and low RAM risk zones.

23.6.2 Risk Severity Matrix

Based on the risk assessment form constructed above, risk severity matrices between likelihood of risk event and its impact corresponding to the security and RAM risks are developed below. The zones in dark gray, medium gray and light gray represent high, moderate and low severity risk zone.

23.6.2.1 Security – Risk Severity Matrix

A 4×3 matrix is used to represent security related risk events, their likelihood and severity (Fig. 23.2).

23.6.2.2 RAM – Risk Severity Matrix

A 3×3 matrix is used to represent RAM related risk events, their likelihood and severity (Fig. 23.3).

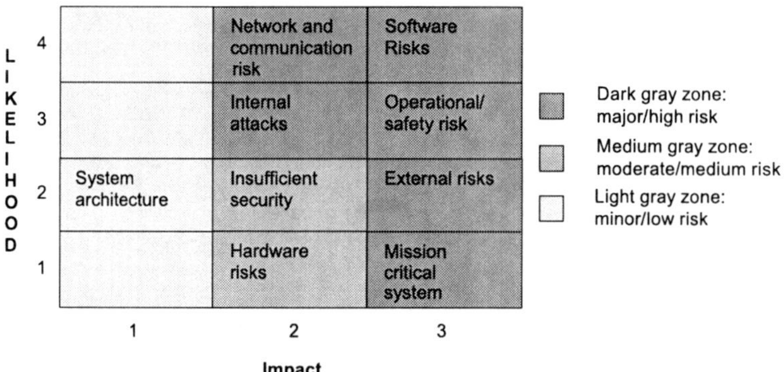

Fig. 23.2 Risk severity matrix for security related risks

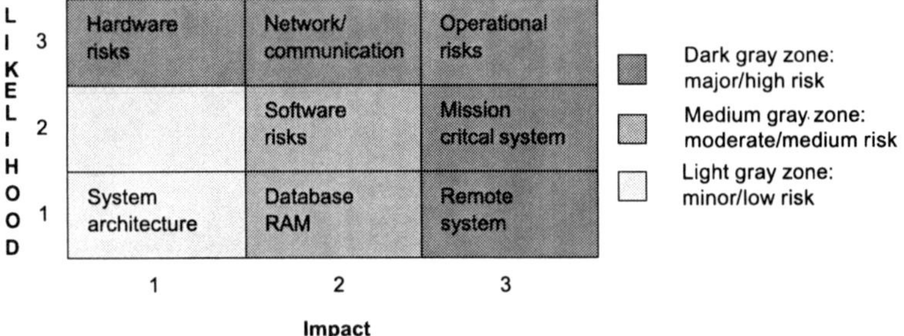

Fig. 23.3 Risk severity matrix RAM related risks

23.7 Goal-based Assurance Case Approach

The goal-based assurance cases developed in [6] ensure that all the risks associated with SCADA systems are covered. In the goal-based approach, specific security and RAM related goals for the SCADA systems are supported by arguments and evidences. The security and RAM cases developed provide a documented body of evidences and valid argument that a SCADA system is adequately secure, reliable, available, maintained for a given application in a given environment. Based on the output of the assurance cases, there is significant reduction in the "likelihood of occurrence of risk event" and its detection difficulty. The new reduced risk value or the enhanced return on investment is represented below.

23.7.1 Security-enhanced ROI

The data are given in Table 23.8.

23.7.2 RAM-enhanced ROI by Using RAM Cases

The data are given in Table 23.9.

With the help of assurance cases the new risk value for security risks is now 67% as low and 33% as moderate. It is noted that there are no more risk levels that are rated high. Similarly, the RAM risks are down to 63% under low and 37% under moderate risk levels. Again, this is based on the stated assumptions; with different assumptions, a new assurance case can be developed.

Table 23.8 Enhanced ROI by using security cases

Sample no.	Risk event	Likelihood	Impact	Detection difficulty	When	Risk value, ROI	Enhanced, ROI
1	Software risks	Medium	High	Medium	Requirement analysis, designing, programming, testing and maintenance	18 high	12 medium
2	Hardware risk	Low	Medium	Low	Post-installation	2 low	2 low
3	Operational and safety risk	Low	High	Medium	Remote access, operator's error, information exchange	18 high	6 low
4	Network and communication risk	Medium	High	Medium	Communication, network links, wireless networks	27 high	12 medium
5	Risks associated with mission critical systems	Low	High	Medium	Post-installation. Remote system maintenance, testing, and patching	27 high	6 low
6	External risks	Medium	High	Medium	Information/data exchange, third party	27 high	12 medium
7	Internal risks	Low	Medium	Low	Information access, internal employee, disposal	12 medium	2 low
8	Insufficient security and passive attacks	Low	High	Low	Insufficient firewall installation, improper IDS, security policy	12 medium	3 low
9	System architecture	Low	Medium	Medium	Development, testing and maintenance	4 low	4 low

Table 23.9 Enhanced ROI by using RAM cases

Sample no.	Risk event	Likelihood	Impact	Detection difficulty	When	Risk value, ROI	Reduced ROI
1	Software risk	Medium	Medium	Medium	Requirement analysis, designing, programming and installation, testing and maintenance	18 high	8 medium
2	Hardware risk	Low	High	Low	Post installation, under severe/stress conditions	6 low	3 low
3	Operational risk	Medium	High	Medium	Information exchange, system access	18 high	12 medium
4	Network/communication risk	Low	High	Medium	Communication/information exchange	27 high	6 low
5	Mission critical system's risk	Low	High	Medium	Post-installation, stress conditions	12 high	6 low
6	Remote system maintenance risk	Medium	High	Medium	Post-installation, system access, stress condition	18 high	12 medium
7	Internal risk	Low	Medium	Low	Information transfer	12 medium	2 low
8	System architecture risk	Low	Low	High	System/software development	3 low	3 low

23.8 Summary and Conclusion

In this paper, an approach to improve return on investment for SACDA systems is described. To realize this, a goal-based assurance case approach, along with a risk management structure, is used. First, all security and RAM-related risks associated with SCADA systems are identified. For security, nine categories of risks were identified, whereas for RAM eight categories of risks were identified. These technical risks were then mapped into business risks affecting the ROI for SCADA system. The risks were prioritized based on their impacts. Later, a risk assessment form was developed. In this form, the likelihood of occurrence of a particular risk event, its impact, detection difficulty and the event when the risk can happen are mapped. This risk assessment form is further utilized to construct risk severity matrices depending on the probability of occurrence and its corresponding impact. The risk severity is

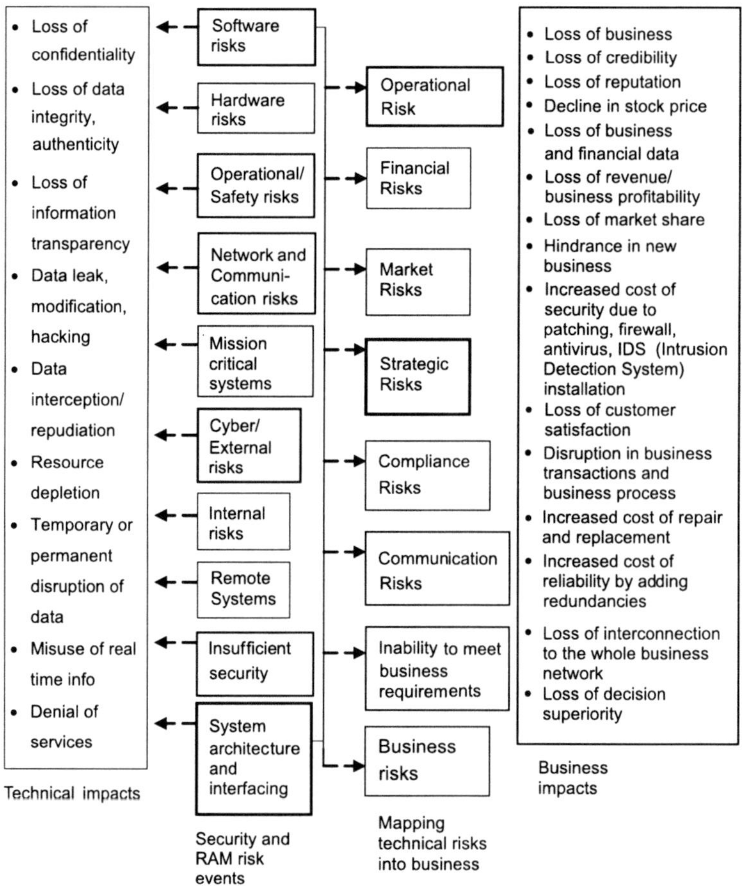

Fig. 23.4 Risk events and associated technical and business impacts

categorized under a level of high, moderate and low risk. These levels are assigned to each of the risk categories. A risk value is calculated by using the formula: Risk value $=$ Likelihood $\times$ Impact $\times$ Detection difficulty. The risk value and ROI are related with an inverse relationship.

Following that, security and RAM cases are developed in such a way that they cover all the identified risks. This helps in reducing the occurrence of risk and their detection difficulty. Based on this, a reduced risk value is recorded. This reduced risk value consequently enhances the ROI (Fig. 23.4).

Assurance cases for reliability, just as assurance cases for safety, are helpful in reducing the risk levels and detection difficulty associated with SCADA systems. There is a significant positive impact on ROI after using security and RAM cases. Thus, it is evident that an enterprise can successfully reduce its operational cost if assurance cases are developed meticulously.

References

1. Miller A (2005) Trends in process control systems security. IEEE Security and Privacy 3:57–60
2. Miller A (2005) Networked process control systems security. Assurance Cases for Security International Workshop, Washington, D.C., June
3. Ramalingam A, Miller A, Erickson KT (2005) SCADA system vulnerability analysis. Proceedings of the Working Together: R&D Partnerships in Homeland Security Conference, Boston, Mass., April 27–28
4. Miller A, Erickson KT (2004) Network vulnerability assessment: a multi-layer approach to adaptivity. NATO Symposium on Adaptive Defence in Unclassified Networks, pp. 13.1–13.8
5. Bloomfield RE, Guerra S, Masera M *et al.* (2006) International Working Group on Assurance Cases (for Security). IEEE Security and Privacy 4:66–68
6. Moleyar KM, Miller A (2007) Formalizing attack trees for a SCADA system. First Annual IFIP WG 11.10 International Conference on Critical Infrastructure Protection, Dartmouth College, Hanover, New Hampshire, March
7. Gray CF, Larson EW (2006) Project Management – the Managerial Process, 3rd edn. McGraw-Hill Series Operations and Decision Sciences, ISBN 0-07-297863-5

Chapter 24
Detecting Driver's Emotion: A Step Toward Emotion-based Reliability Engineering

Shuichi Fukuda

Professor Emeritus, Tokyo Metropolitan Institute of Technology, Japan
Consulting Professor, Stanford University, USA

24.1 Background

In our traditional engineering design, systems are designed so that machines work today in the same way as yesterday, no matter how the situations may change. In short, our traditional goal was to build up a context-independent system. Up to now, situations have not changed appreciably, but today, they change very rapidly and very frequently. Therefore, the context independent approach is no longer effective. To cope with the rapid and frequent changes, a more context-dependent approach is called for.

We could possibly develop many context-dependent approaches. But what should be pointed out is the importance of human cognition and decision making [1]. We could possibly install many sensors and actuators so that a machine could respond to rapid changes. However, it should be stressed that human factors are getting more and more important in securing system reliability. This is not a human factor in traditional reliability engineering, where it is considered as a system element and in a more context-independent framework.

In traditional reliability engineering, an operator is supposed to carry out the designated missions "reasonably enough". If he or she cannot respond "reasonably," then it is called "human error". This may be true when the situations do not change much and the number of decision is significantly limited. However, if the situations change widely and frequently and the number of options in decision making increases, then which option is the best one? Besides, if the changes are very extensive, can a designer predict all the situations at a design stage? No, he or she cannot. There may be no definite "reason" to select one, or if there is, it would take too much time to select one. Then, that means we cannot react in time.

Let us consider aircraft or automobile accidents. How many of them occurred beyond the designer's anticipated scenario?

Let us take the case of China Airlines Flight 140 accident at Nagoya Airport in 1994 in Japan [2]. The first officer made a mistake when the plane was about to land and pressed the take off/go-round button. He did not realize he had done it. The plane suddenly nosed up. The pilot was surprised because if the plane noses up, it will result in a stall. He attempted to nose it down. The fight continued and the plane was stalled and crashed. A total of 264 out of 271 people aboard died.

In this case, we could possibly install a warning system so that if the first officer unintentionally pressed the button, the plane could ask the pilot if it is OK to work the system in such a situation. Then, the pilot would have understood the situation and could have responded better.

Next, let us consider the case of the Iowa DC10 UA Flight 232 accident in Sioux City, Iowa [3]. Three engines blew out and the pilot could not control the plane at first. But through encouraging communication with the air traffic controller, he gradually recovered himself and became aware that he was not flying over the Rockies or over the Atlantic. He realized he was flying over the flat lands of Iowa. He did not have to fly to the nearest airport – an airport was all around him! He could use the highways for landing. Continued communication with the air traffic controller led him to think positively and provided him with more confidence. In the course of conversation, he suddenly became aware that the plane he was flying was a DC10, where all the throttles are designed to operate independently. So if he operated them throttle by throttle, he could possibly land successfully. He did, and 185 out of 285 survived! This accident is known as a textbook case for demonstrating the effectiveness of Crew Resource Management.

Can a warning system work in this case? No. First, such a situation was not anticipated at the design stage. In fact, in the LOFT training, there was no such scenario. The pilot stated at the meeting after the accident that he was very confident of successful landing because he had come through many LOFT programs. However, the fact is, he did not experience such a situational in the LOFT program at all. It was his confidence and his positive thinking that brought him to a successful landing.

Accidents occur when something takes place beyond the expected situations. In such a system as an airplane or an automobile where situations change very frequently and rapidly, human characteristics should be taken more into consideration to secure reliability. Regarding a human as a system element will not work, and human characteristics should be incorporated into design to help an operator in understanding the situation better and in making a better decision. Therefore, a more human centered approach is called for to solve today's reliability issues.

What, then, characterizes humans? It is emotion. It became clear that emotion plays an important role in our cognition and in decision making.

Our traditional approach was such that when an emergency occurs beyond our expectations, then an operator is supposed to understand the situation and control the system. However, in such an emergency, a human's capabilities decrease drastically to about one-third of that of the normal condition. In fact, we know many people die at a single exit in the case of fire or earthquake, although there are many other exits. They panicked, and lost their judgment.

How can we restore their capabilities to their normal level? It is expected that if we could keep an operator emotionally stable, he or she could judge better and act better. That is the main idea behind our current research.

24.2 Emotion

Descartes distinguished reason and emotion. However, Antonio Demasio published a book, *Descartes' Error: Emotion, Reason and the Human Brain*, in which he pointed out reason and emotion cannot be separated [4]. Recent neuroscience tells us that reason and emotion are very closely associated. Daniel Goleman popularized the idea of Emotional Intelligence, in which he insists that our intelligence is very much associated with emotion [5]. Rosalind Picard published *Affective Computing*, in which she stressed the importance of considering emotion in computing [6].

Although reliability engineering has made such remarkable progress, emotion is not really taken into consideration. A high proportion of our efforts was devoted to developing reason-based reliability. However, if we consider flight accidents or car accidents, they seem to indicate that more emotion-based reliability engineering should be explored. This motives us to develop emotional-based reliability engineering.

And this work is a first step toward the goal.

24.2.1 Primary or Basic Emotions

Robert Plutchik proposed eight primary emotions; anger, sadness, joy, disgust, curiosity, acceptance [7]. Paul Eckman proposed five basic emotions anger, fear, sadness, happiness and disgust [8]. The definitions of primary or basic emotions vary from researcher to researcher.

24.2.2 Detecting Driver's Emotion

Our current research is aimed at detecting driver's emotion for safer driving. Our future scenario would be such that if a driver gets angry, a machine detects his or her emotion and a virtual agent or a car navigation system talk to him or her to calm down by saying something like "What a crazy driver that guy is! But I know you are not that sort of a driver. You always drive safely" or something like that, or by turning on a driver's favorite music to distract his attention from the scene to let him or her recover him- or herself.

24.3 Observation of Actual Driving

We installed a camera in a car and observed a driver under real driving conditions. It was soon made clear that lighting changes very often and widely, so that it is very difficult to record a driver's face and process the images in a real situation. So we decided to carry out experiments indoors.

One by-product from this observation is that a driver comes to move more as he gets tired. So if we watch his movements, we could possibly detect his fatigue. Although fatigue is usually not included in emotion, we considered fatigue as well, because it affects emotion and safe driving.

24.4 Experiments Using Driving Simulator

As a first step to carry out experiments indoors, we conducted our experiments using a driving simulator. However, to our surprise, the operator did not show any appreciable facial changes. His eyes were very much fixed, and he almost lost his facial expression. He was too much absorbed in a virtual world. This may be because his field of vision is very narrow in a driving simulator and because a scenario is already built into the system, so the operator is always watching this small field and ready to take action. In a real driving situation, a driver is not aware of the imminent danger and danger comes to him very unexpectedly. In fact, such unexpected events drive him into emotions such as anger. Thus, experiments using a driving simulator completely betrayed our expectations, and did not serve for our purpose. Therefore, we carried out our experiments in a setup shown in Fig. 24.1, with primary emphasis upon detecting emotion in a much simpler manner so that the system can be used while driving.

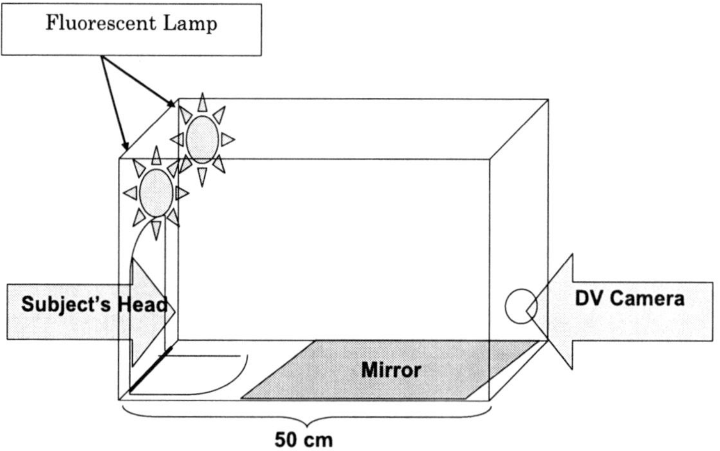

Fig. 24.1 Experimental setup

24.5 Facial Emotional Expression

There are many works which points out that face plays a crucial role in expressing our emotions such as [9]–[33]. Eckman found that emotional face expression do not differ from nation to nation. He developed FACS with Friesen [20]. This connects emotional facial expression with facial muscles.

However, most of these methods are very much complicated and some are useful only for creating face images. Since our system has to be installed in a car, we need a much simpler technique. What is important is simplicity and responsiveness, not accuracy.

24.5.1 Developing a Simpler Technique to Detect Facial Emotion

Consider a cartoon. Although it is simple black and white, it expresses emotions very well [24]. As a first step to extract face feature parameters related to emotion, we prepared real emotional faces and conducted the survey and classified the results, using a principal components analysis and Mahalanobis distance method.

Figure 24.2 shows the basic procedures for this work. Figure 24.3 shows face emotional parameters, which were derived using Principal Components Analysis. Figure 24.4 shows the cartoon face samples.

Based upon preliminary studies, we have developed a simple image processing method. Its procedures are given below.

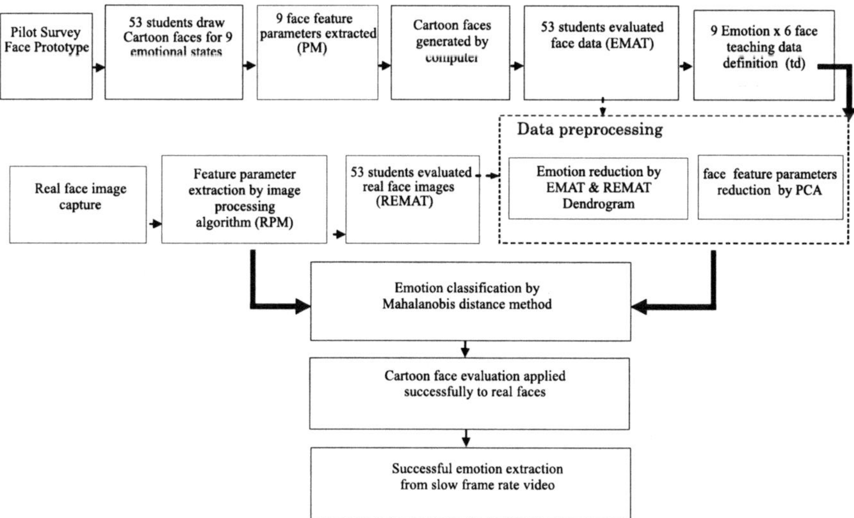

Fig. 24.2 Flow of research

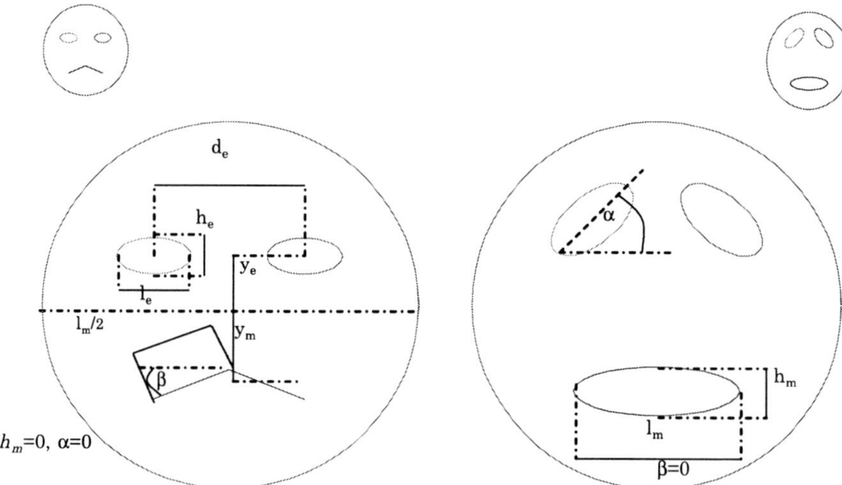

Fig. 24.3 Face feature parameters

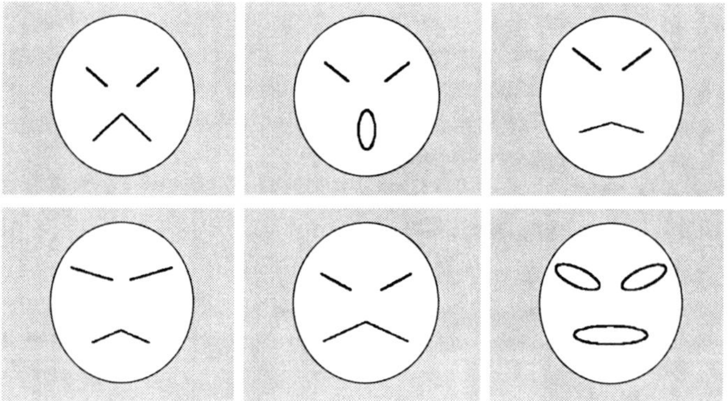

Fig. 24.4 Examples of cartoon angry faces

Segmentation by threshold: A face must be separated from the background. Although there are many works such as [25]–[27], we used a much simpler approach. From an original image, we get an approximation of background by interpolating from the edges of a picture, using linear interpolation from left to right and from up to down. This gives us a picture with a uniform background.

According to Olhausen and Field [28], a histogram is tall in the dark area and small and wide in the lighter region. Using this approach, we can obtain a binary image. Since a face forms an ellipse, it can be extracted by simply taking the pixels closest to the origin in every direction. This is easily done in polar coordinates. The edge of the face remains when it is converted back to Cartesian coordinates.

Ellipse fitting to face: The literature on ellipse fitting is divided roughly into two. One is clustering, such as [29] and [30], and the other is least squares fitting [31,32]. We chose the least squares method developed by Fizgibbon *et al.* [33].

Extraction of face features: Eyes and eyebrows are detected by taking a logical AND between an edge map and its mirror, because eyes and eyebrows are symmetrical with respect to the axis of face. Extraction of the mouth is easy because after fitting the ellipse, we know where the lower part of the face is. The mouth is detected by looking at the first and second order derivative in the x-direction. A high positive value of the first order derivative corresponds to the upper lip and a high positive value of the second order derivative corresponds to the area between the lips.

Emotional classification: Figure 24.5 shows the dendrograms for cartoon faces. Emotions can be reduced from 9 to 7 or to 5 according to these dendrograms.

Figure 24.6 shows the correlation map between emotion evaluation and face feature parameters for cartoon faces. In the cartoon, eye-angle plays a primary role for expressing emotion, while in the real face, it is eyebrows. We omitted eyebrows in the cartoon. When we replace eyebrows in the real face with eye-angle in the cartoon, a very good correspondence can be achieved between a real face and a cartoon face.

Figure 24.7 shows the procedures for extracting emotion from a real face by way of a cartoon model. Our model evaluated emotions 72% correctly and 17% was associated with a similar emotion.

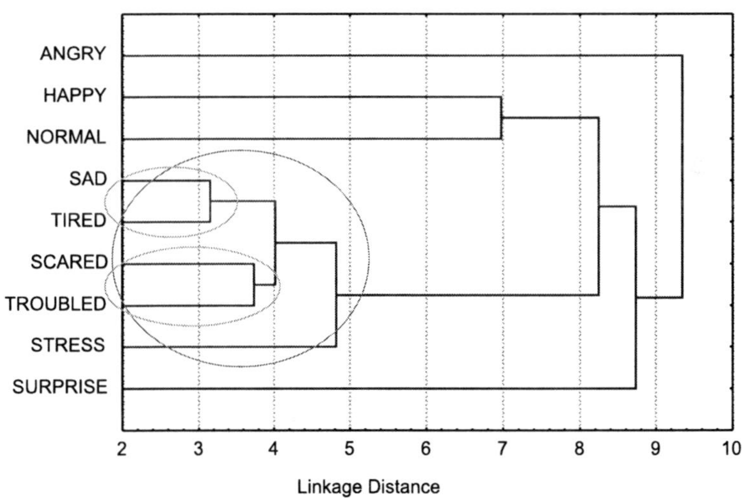

Fig. 24.5 Dendrogram for cartoon faces, single linkage, Euclidean distance

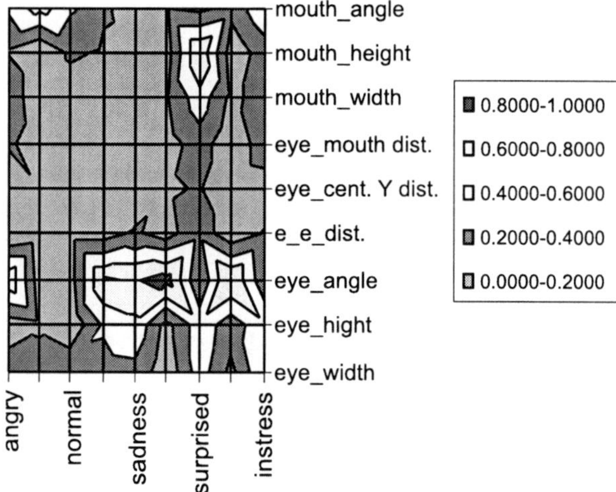

Fig. 24.6 Correlation map between emotion evaluations and feature parameters for cartoon faces

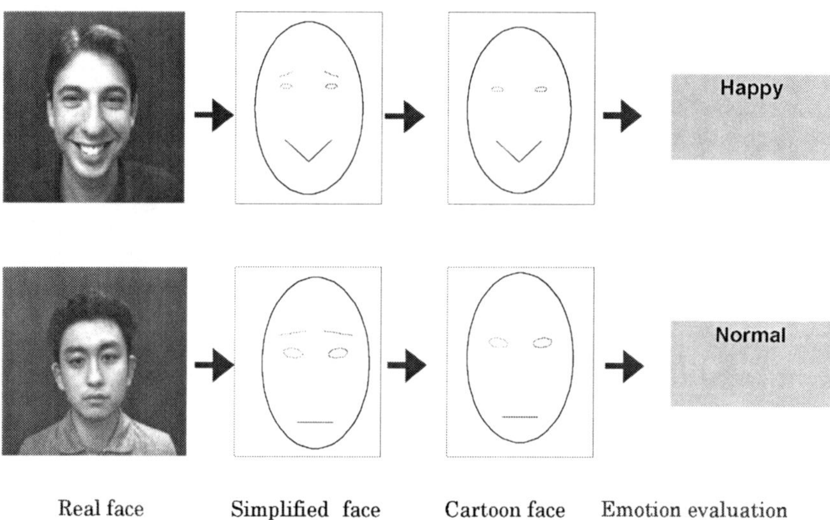

Fig. 24.7 Emotion detection procedures

24.5.2 Detection of Emotion from Real Face

Experiments have been conducted to detect emotion from real face.

If we use a binarization approach, eyebrows are difficult to detect because if we wish to detect eyebrows clearly, shadows appear on the face and it is difficult to process further, and if we set the condition so that the shadow on the face disappear, then eyebrows disappear, too. Therefore, we used color difference approach.

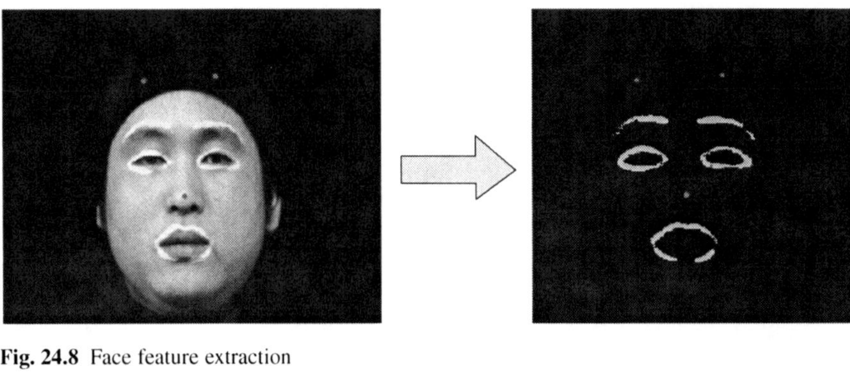

Fig. 24.8 Face feature extraction

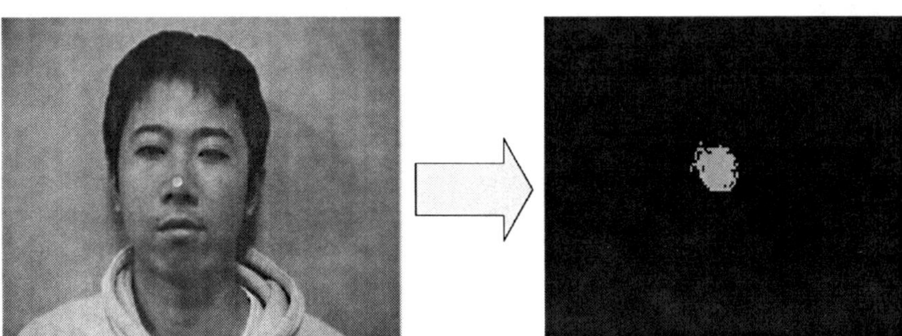

Fig. 24.9 Nose center extraction

However, to observe the changes in face parameters well, the color difference between face parameters and elsewhere should be large. We used pigments that have higher brightness value. Eyebrows were painted red, eyes green and nose center red. Therefore, face parameters were easily detected and identified as shown in Figs. 24.8 and 24.9.

The motion of the top border of the eyes and the top of the eyebrows were compared. These data corresponds very well with more than 85% of the correspondence factor greater than 0.4. Therefore, we did not observe the eyebrow motion in this experiment, and we focused our attention on eye movements alone. In addition, the changes of eye width and eye opening (height) were compared. The change in the opening of the eye (eye height) is far greater than the change of eye width, so we focused our attention on the eye height alone.

Figure 24.10 shows an example of head motion results. The grey line shows surprise and the black line shows anger. The ellipse shows the group by cluster analysis. The grey lines are above the x coordinates, which means that when we are caught by surprise, our face tends to look up. Most of the black lines are below the x-axis. Therefore, when we get angry, our face tends to look down.

Figure 24.11 show examples of an eye opening (eye height) results. These are the normalized results with the threshold value set at 0.2 maximum. It is interesting that emotions are much more expressed in the right eye than in the left eye

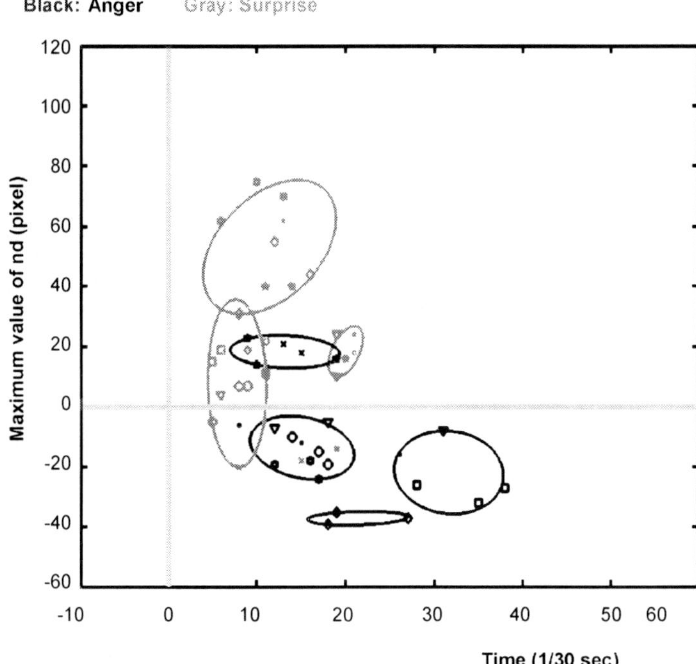

Fig. 24.10 Example of head motion results

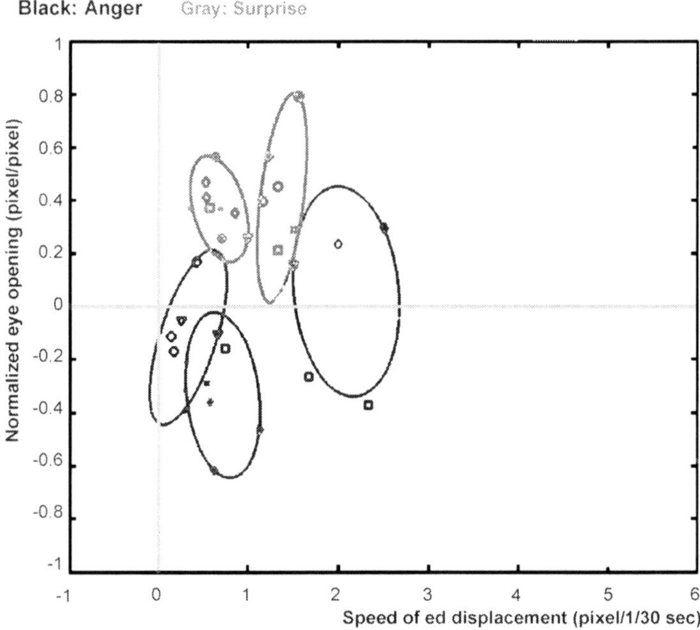

Fig. 24.11 Example of eye opening result (right eye)

in our experiments. The figure shows that we can discriminate anger from surprise.

24.6 Detection of Fatigue

Detection of fatigue was carried out in the same procedures. As the ends of the eyebrows are difficult to identify clearly, the top and bottom of the eyebrows were selected as representative points and their changes were studied. In the case of eyes, both ends and top and bottom of the eyes were selected as representative points. The center of a nose is identified as a crossing point by drawing a line between top and bottom points and drawing a line between left and right points. These top and bottom, and left and right points were where red appears first when scanning from top to down or from left to right.

The representative points and the distances between face parameters were chosen as shown in Fig. 24.12. Further to study the inclination of a head, angles were chosen as shown in Fig. 24.13.

Subjects, all in their early 20s, were tired due to (1) long hours of programming, (2) long hours of programming + lack of sleep, (3) long hours of data analysis, (4) long hours of working on PC + lack of sleep, (5) staying up all night to make a report, (6) making a report + lack of sleep, and (7) programming + making a report. All subjects were tired due to heavy brain work. They were told to say the same sentence. A still image was obtained by 30 frames/s from a moving image.

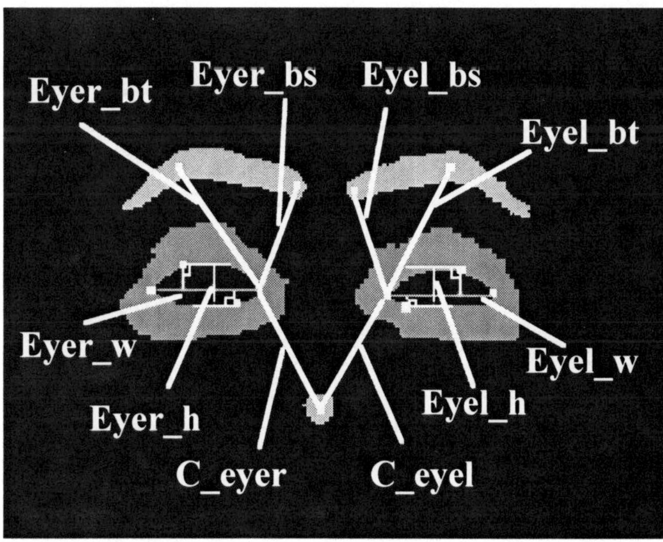

Fig. 24.12 Face parameters

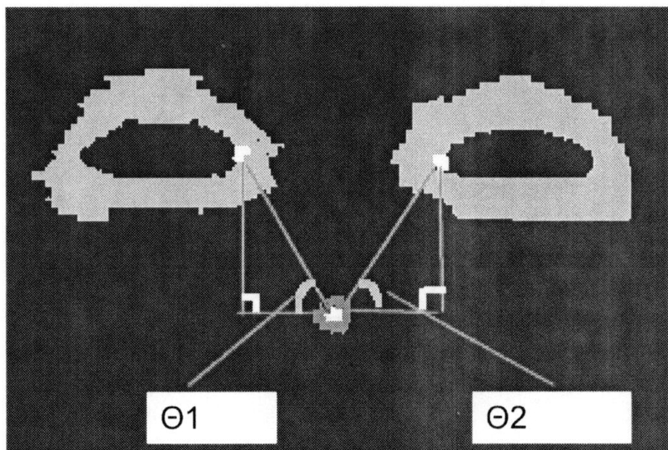

Fig. 24.13 Angle parameter to study face inclination

Figure 24.14 shows left eye changes of height/width of one subject as an example. Other subjects showed similar results. This indicates that whether the subject is alert or tired can be easily detected. Further, even if the subject pretends to be not alert, even though he is really physically alert, the data show clearly he is physically well. Eyes open wide when subjects are alert but less so when they are tired. These results indicate that if we keep observing the changes of face parameters with time, especially eye height/eye width change with time, we can detect fatigue.

However, our preliminary observation of a driver in a real driving condition shows that a driver moves more when he gets tired. If a driver moves more, taking images for observing changes of face parameters would become difficult. Therefore, we observed the overall motion of a head. The results show that when we get tired, our faces tend to look down. This indicates the possibility of detecting fatigue

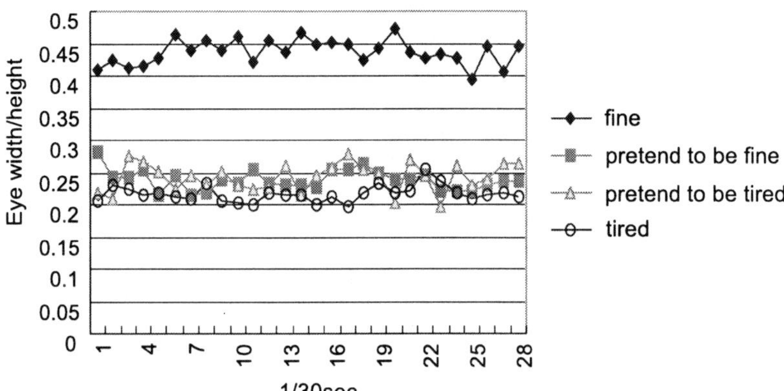

Fig. 24.14 Left eye variation

by a simpler technique other than the image processing used here, if we observe the motion of a head. Therefore, we developed a much simpler technique using an infrared LED array.

24.7 Detection of Dangerous Actions

Our primary objective is to secure safe and reliable driving. Emotional detection is a means to an end, and not our goal. Therefore, when we started to develop a technique to detect fatigue by body motion, we attempted to develop a technique which can be used as widely as possible to detect actions which would endanger driving.

In fact, many automobile accidents are reported to occur due to dangerous actions on the part of a driver. It should also be noted that many traffic accidents happen due to the lack of situational awareness. Therefore, a simple method to extract driver's dangerous actions is developed in an effort to provide a driver with awareness of the situation. This is the same principle as our emotional based approach, because what we attempted in both cases were to provide a driver with full awareness and understanding of the situation which will lead him or her to a better decision and a better action. In this sense, detection of emotion and detection of dangerous actions do not differ at all.

Infrared LED arrays were used to detect dangerous actions. A CCD camera is used because it reacts to infrared lights, although our eyes cannot see them. Infrared lights are detected as white lights in the image processing. Therefore, if an array of infrared LEDs is installed, a dangerous action can be detected by observing how the lights of infrared LEDs change (Fig. 24.15).

Five dangerous body motions as described later were studied, and body motion in normal and safe driving was recorded in order to be compared with them. Although the developed system is quite simple, these dangerous body motions could be successfully detected.

From the observation of a driver in a real driving situation, his body motion changes when he gets tired, compared with his normal driving state. For example, if a driver is not tired, he just looks at a traffic signal with his eyes without a large body movement. However, when he gets tired, he appreciably moves or turns his head toward the signal or road signs. Thus the body moves more when he gets tired.

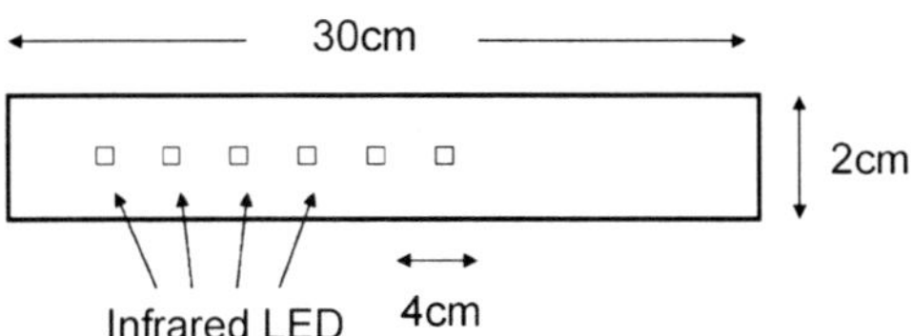

Fig. 24.15 Infrared LED array

Therefore, this technique is expected to be very useful for detecting driver's fatigue in addition to detecting dangerous actions.

The color of an automobile roof is monochromatic. Therefore, if a camera is placed in front of a driver so that the image of his torso and the roof only can be taken, these two images can be separated by means of color. The locations of shoulders can be easily detected because when we scan the image from the edge of a picture, the first point where color other than the background color appears is the location of a shoulder. Head can be detected in the similar manner. If we scan the picture from the left top corner to the bottom right corner, the first point where color other than the background color appears is the location of the top of a head. By noting these three points and their location changes, we can study dangerous actions of a driver: (1) normal driving condition, (2) dozing off, (3) operating a control box, (4) reaching out toward a dashboard to pick up something, and (5) leaning sideways.

The images taken showed that for each of these movements: (1) top of the head can be easily identified, but not shoulders, (2) head drooped, (3) head leans toward right (this is a Japanese car with a steering wheel on the right hand side), (4) the upper portion of the body appears at the top right of a picture, and (5) the upper portion of the body leans toward the right.

These results show when the upper portion of the body leans, the head leans toward the same direction. Therefore, we can estimate the movement of the upper portion of the body simply by noting the movement of the head. This indicates that if we can detect the movement of a driver's head, we can identify the dangerous motions. If images are taken in every 3 min, then our system detects dangerous movements correctly.

24.8 Detection of Emotion from Voice

Voice is another rich expression of emotion. As a preliminary step, the method for detecting emotion from speech was developed. Voice sample data were collected and pitches were extracted using the Cepstrum method. Then, pitches are smoothed and interpolated and power is calculated. After that, chunk intervals are calculated. Then pitches and powers are calculated for each chunk. Next, emotional coefficients are obtained. Based upon these features, voices are classified using the Mahalanobis method and cluster analysis. Angry, disgust, happy, normal, sad, and tired emotions are detected, no matter who the speaker is and no matter what language he speaks. Spanish, Portuguese, French, Italian, Japanese, Korean, German, Flemish, Dutch, English, Macedonian, Russian, Serbo-Croatian voice samples did not show any difference in terms of emotional detection.

24.8.1 Detection of Anger and Fatigue from Voice

The problem with a car driver is that he or she does not speak much. For aircraft, the above technique might be applied to detect emotion when a pilot communicates with an air traffic controller or the first officer, but a car driver speaks only a small number of words or very short sentences. The above technique, however, might be applied to a driver too, in the future when most navigation systems will be voice-manipulated.

To meet today's requirements, we have developed another technique. As anger and fatigue are considered to be most dangerous for driving, an emotional data base for anger and fatigue, together with emotional data in normal driving conditions, was constructed for each driver in a text independent manner. The system identifies whether a driver is angry or tired or in a normal driving condition by noting pitch and pressure.

Sound can be characterized by three elements: pitch, pressure and tempo. Another element which characterizes sound is formant. A formant is a peak in a sound frequency spectrum which results from the resonant frequencies of vocal tracts. The formant with the lowest frequency is called F_1, the second F_2, and the third F_3. Most often, the two first formants, F_1 and F_2, are primarily determined by the position of the tongue. F_1 has a higher frequency when the tongue is lowered, and F_2 has a higher frequency when the tongue is forward. But formants are influenced by the preceding and succeeding phonemes, so that we come up with different results in a text independent system. Therefore, we did not adopt formant as our feature parameter. We did not adopt tempo either, because we were afraid that how we say a sentence might influence the result. In everyday communication, we intentionally use tempo to get our emotional messages through. However, our aim is to detect true emotions rather than intended emotions, so we omitted tempo in our experiments. Therefore, pitch and pressure was taken up to detect emotions.

In this experiment, a system takes the initiative to communicate, because it is expected in real driving that drivers do not say much. The system detects drivers' emotions based upon his response.

1. Emotional data base is constructed for each subject. Three emotional states are recorded: angry, tired and normal. Subjects are told to record the same sentence in this experiment to reduce the variation from subject to subject. However, in real applications, a driver can record anything.
2. Pitch was obtained from time series data and we chose as our primary feature parameter, maximum pitch frequency and as secondary feature parameter, maximum sound pressure.
3. Subjects' voices in different emotions were compared with data in emotional data base and subject's emotional state was determined, using Mahalanobis distance.

Figure 24.16 shows experimental procedures. Emotions can be detected most effectively by sound pressure as shown in Fig. 24.17. Sound pressure increases from tired to normal to angry state. In this experiment, pitch is widely dispersed, so that

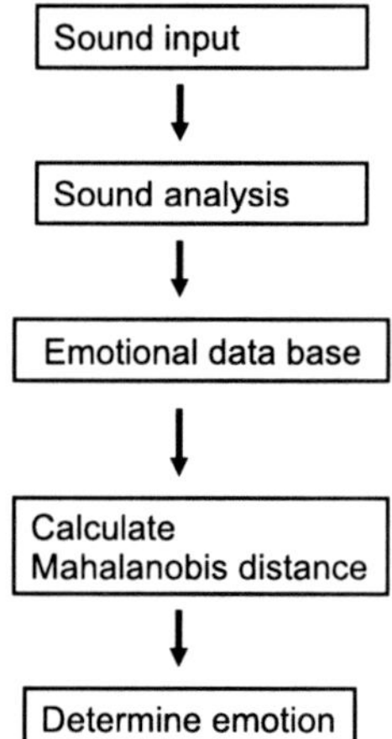

Fig. 24.16 Procedures

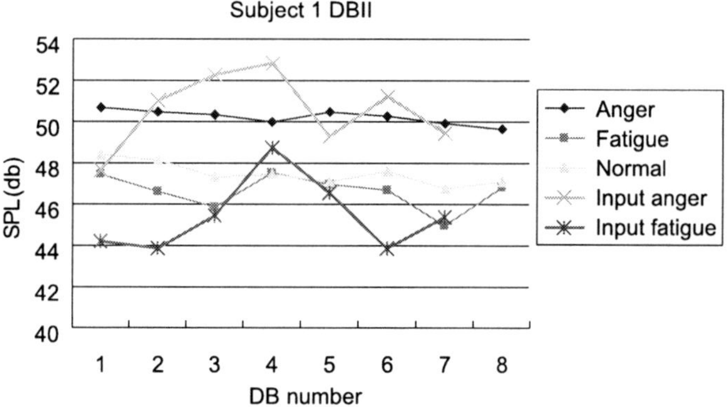

Fig. 24.17 Results

we cannot discriminate tired state from normal one. However, with respect to anger, its pitch value is far greater than the other two. So when it comes to detecting anger alone, pitch is an effective detector.

References

1. Kaiho H (ed) (2002) Psychology of Warm Cognition. Kaneko Shobo (in Japanese)
2. http://en.wikipedia.org/wiki/China_Airlines_Flight_140
3. http://www.clear-prop.org/aviation/haynes.html
4. Demasio AR (2005) Descartes' Error: Emotion, Reason and the Human Brain. Penguin, UK
5. Goleman D (2005) Emotional Intelligence. Bantam Dell Publishing Group, USA
6. Picard RW (2000) Affective Computing. MIT Press, Mass., USA
7. Plutchik R (1980) Emotion: a Psychoevolutionary Synthesis. Harper and Row, New York

8. Eckman P (1972) Emotions in the Human Face. Pergamon Press, UK
9. Walker JH, Sproull L, Subramani R (1994) Using a human face in an interface. Proc Human Factors in Computing Systems, pp. 85–91
10. Walters K, Hofer B (1997) Facial animation: past, present and future. Proceedings of the 24th International Conference on Computer Graphics and Interactive Techniques, pp. 434–436
11. Roth BH (1998) Panel on affect and emotion in the user interface. Proceedings of the International Conference on Intelligent User Interfaces, San Francisco, Calif., pp. 91–94
12. Daugman J (1997) Face and gesture recognition: overview. IEEE Transact Pattern Analysis and Machine Intelligence 19:675–676
13. Eckman P (1994) Facial expression and emotion. American Psychologist 48:384–392
14. Bartlett MS, Hager JC, Eckman P, Sejnowski TJ (1999) Measuring facial expressions by computer image analysis. Psychophysiology 36:253–263
15. Cohn JF, Katz GS (1998) Bimodal expression of emotion by face and voice. Proceedings of the 6th ACM International Multimedia Conference on Face/Gesture Recognition and Their Applications, pp. 41–44
16. Lien JJ, Kanade T, Cohn JF, Li CC (1998) A multi-method approach for discriminating between similar facial expressions, including intensity estimation. Proceedings of the IEEE International Conference on Computer Vision and Pattern Recognition
17. Cohen JF, Lien JJ, Kanade T et al. (1998) Beyond prototypic expressions: discriminating subtle changes in the face. Proceedings of the 7th IEEE Robot and Human Communication Workshop, Takamatsu,Kagawa, Japan
18. Mase K (1991) Recognition of facial expression from optical flow. IEICE Transact E74:3474–3483
19. Parke FI (1991) Techniques for facial animation. In: Thalman NM, Thalmann D (eds) New Trends in Animation and Visualization. Wiley Professional Computing, New York, pp. 229–241
20. Eckman P, Friesen WV (1978) The facial action coding system. Consulting Psychologists Press
21. Gosseline P, Kirouac G, Dore FY (1997) Components and recognition of facial expression in the communication of emotion by actors. What the face reveals. Oxford University Press, Oxford, pp. 243–267
22. Chernoff H (1973) The use of faces to represent points in k dimensional space geographically. J Am Statist Assoc 68:361–368
23. Shuichi F (ed) (2006) HCD Handbook – Human Centered Design. Maruzen (in Japanese)
24. McCloud S (1994) Understanding Comics, the Invisible Art. Harper Perennial, Harper Collins, New York
25. Saber E, Tekalp A (1996) Face detection and facial feature extraction using color, shape and symmetry based cost functions. ICPR '96, pp. 654–658
26. Chow G, Li X (1993) Toward a system for automatic facial feature detection. Pattern Recognition 26:1739–1755
27. Hu J, Yan H, Sakalli M (1999) Locating head and face boundaries for head-shoulder images. Pattern Recognition, 32:1317–1333
28. Olhausen BA, Field DJ (2000) Vision and the coding of natural images. Am Sci 88:238–245
29. Leavers L (1992) Shape Detection in Computer Vision Using the Hough Transform. Springer-Verlag, London, New York
30. Yuen HK, Illingworth J, Kittler J (1989) Detecting partially occluded ellipses using the Hough transform. Image and Vision Computing 7:31–37
31. Rosen PL, West GA (1995) Nonparametric segmentation of curves into various representations. IEEE Transactions on Pattern Analysis and Machine Intelligence 17:1140–1153
32. Gander W, Golub GH, Strebel R (1994) Least square fitting of circles and ellipses. BIT 43:558–578
33. Fitzgibbon A, Pilu M, Fisher RB (1999) Direct least square fitting of ellipse. IEEE Transact Pattern Analysis and Machine Intelligence 21:476–480

Chapter 25
Mortality Modeling Perspectives

Hoang Pham

Department of Industrial and Systems Engineering,
Rutgers University,
96 Frelinghuysen Road,
Piscataway, NJ, 08854, USA

25.1 Introduction

As the human lifespan increases, more and more people are becoming interested in mortality rates at higher ages. Since 1909, the birth rate in the United States has been decreasing except for a major significant increase after World War II, between the years 1946 and 1964 [1], also known as the baby boom period. People born during the baby boom are now between the ages of 44 and 62. According to the National Center for Health Statistics, US Department Health and Human Services, in 1900 – 1902 [2, 3], one could expect to live for 49 years on average.

Today, an infant can expect to live about 77 years. As of recent years and in prediction, the life expectancy for an infant born may be even higher. With the human lifespan increasing and a large part of the United States population aging, many researchers in various fields have recently become interested in studying quantitative models of mortality rates [4]. Scientists in biological fields are not only interested in organisms and how they are made, they are also interested in what happens to organisms over time. A study of yeast, which would interest biologists, showed the effects of senescence as well as a model that accurately represents the experimental data. It has been shown that the addition of a *Sir2* gene can prolong life in yeast [5]. Once we can model human aging, we can look for ways to extend our lifespan and counteract the negative aspects of aging.

Researchers in the medical field are interested in the cost effectiveness of new medication with respect to old medication. The cost per life year saved can be evaluated for each of the two medications with respect to the mortality model. For this type of work, it is important to use a model that will fit the mortality data for the population in which the medication is to be distributed [6]. This implies that a mortality model be chosen that accurately describes the data in question and indeed very crucial, but that also depends on the modeling decision criteria [7].

In the mathematical and physical fields, researchers are interested in predicting the failure of machines, or other man-made objects [8]. Through a series of experiments and mathematical calculations, one can predict when an object will fail without a direct test on the object. This is useful because one can use available data to create a model that can predict the remaining lifetime of humans. A direct test would take the entire lifespan instead of the shorter time needed to create and use the model. Physicists and mathematicians are also interested in creating models that will depict how aging and environmental factors will effect the survival of the human population [9].

Social scientists in the United States are currently very interested in projections of life expectancy and aging due to the deteriorating system of social security. The current "Pay-As-You-Go" system, which provides benefits for the elderly, is breaking down as the life expectancy is increasing and a large part of the population is aging [10]. Many organizations, including insurance companies, today rely on mortality models and projections of human aging.

This chapter first introduces the motives as well as the reasons why studying mortality data is important. Some literature reviews on the mortality modeling and analysis are then discussed. Several distributions applicable and common used to human mortality studies are also mentioned.

25.2 Literature Discussions

Gavrilov and Gavrilova [8] discussed a brief overview of reliability theory, relating it to mechanical systems and biological systems. They stated that organisms tend to die according to the Gompertz law, while technical devices tend to die according to the Weibull law. They also brought up the idea that the individuality and uniqueness in humans is caused by different combinations of defects in the organism. Aging is a direct result of system redundancy in organisms, and initial flaws in organisms cause the tendency to follow the Gompertz law. Gavrilov and Gavrilova believed that when the Gompertz law and Weibull law both fail, then mortality is following a more general law in reliability theory. They believed that reliability theory provides a great predictive and explanatory power to theories of aging. Gavrilov and Gavrilova provided insight into several different distributions found in reliability theory and how they relate to organisms as well as machines, but did not compare the distributions against experimental data.

Thatcher [11] explored a model showing that the probability of dying increases with age, and believed that it is valid for both modern and historic data. First, he studied three existing models, the "fixed frailty" model, the stochastic process model, and a theory based on genetics, and tackled the question of why mortality seems to follow the logistic model by analyzing four models such as the logistic model, Gompertz model, Weibull model, and the law of mortality, for effectiveness of modeling mortality at high ages. Out of the four methods, Thatcher found that the logistic model out performed the other three models for data from 1980 to 1982, from England and Wales. He also found that the logistic model fit best for the historic data from these areas. In addition, he described several theories for predicting the probability distribution of the highest age. One view was that there is a fixed upper limit to the length of human life. The other is that there will continue to be a probability distribution for the highest age given the circumstances. Thatcher used the data from several different areas of the United Kingdom, such as England and Wales, but not from the United States. Since he was studying historic data, he chose an area where he would have access to mortality records from the past, such as the tenth and twelfth centuries. To analyze the models and fit them to the data, he used a modified version of the maximum likelihood estimation method to accommodate for using data from a life table.

Pletcher and Neuhauser [4] studied the creation of criteria for modeling aging, which they felt would create more consistency in the research of aging in different fields. They found five experimental results in the field of biodemography, which impacts the study of aging. They noted that in reliability theory, many models are useful in looking at biological aging, but felt that the models lack biological realism. They proceeded to examine models in the four research areas: molecular biology, physics, reliability engineering, and evolutionary biology/population genetics. They noted that the experimental results from one field of study may fail to model the same data in using another field's criterion. They also outlined a simple mechanistic model that built from all four fields and, through simulations, upheld the experimentally found results. They realized that with further investigation and simulations, their model would not hold up; however, their purpose was to open communication and research between the different fields. Pletcher and Neuhauser realized the need to find a way to join different scientific fields under common criteria.

Bongaarts [12, 13] used the data from the Human Mortality Database for females and males aged 25–109 in 14 different countries, to test the fit of logistic models for the force of mortality. He also proposed a new shifting logistic model that he hoped would better predict age-specific rates of adult mortality. He compared his model to the Lee–Carter method [14, 15] for modeling and forecasting mortality by age. Bongaarts found that his model addressed several weaknesses in the Lee–Carter model, and believed that it provided a basis for age-specific mortality projections. Higgins [16] examined several mathematical models including the Gompertz, Perks, Polynomial, and Wittstein models used to describe and explain human mortality. He discussed the importance of cohorts in models as well as requirements for the ideal model; however, it required a complicated function and is not yet defined.

Pletcher [17, 18] stated that "well-defined statistical techniques for quantifying patterns of morality within a cohort and identifying differences in age-specific mortality among cohorts are needed". He examined ways to find the parameters for each of the models using the maximum likelihood estimates. He also set a minimum for the number of individuals needed in a specific cohort to make the experiments effective, and noted that no fewer than 100–500 individuals are needed in each cohort, and that combining cohorts to achieve these numbers are much more helpful in accurately estimating mortalities for age groups, rather than keeping them separate. Pletcher explained how Gompertz parameters were traditionally estimated with linear regression, which has a much higher bias than the maximum likelihood method. He concluded that the extended use of the maximum likelihood methods provide for much better analysis of the mathematical models.

25.3 Mortality Modeling

This section briefly discusses six common used distributions in the area of aging and mortality modeling. Table 25.1 presents a list of common distributions such as Gompertz, Gompertz–Makeham, Logistic, log logistic, loglog and Weibull that commonly used on mortality modeling and analysis. The probability density function (pdf) and hazard rate of each distribution are listed in Table 25.1, column 2 and 3 respectively. The hazard rate function is also known as the failure rate in reliability engineering; the force of mortality in demography, the intensity function in stochastic processes, and the age-specific failure rate in epidemiology.

In the context of reliability modeling, the interrelationships between the pdf $f(t)$, failure rate function $h(t)$, cumulative failure rate function $H(t)$, and reliability function $R(t)$, for a continuous lifetime t can be summarized as

$$h(t) = \frac{f(t)}{R(t)}$$

$$H(t) = \int_0^t h(x)\,\mathrm{d}x$$

$$R(t) = \mathrm{e}^{-H(t)}$$

Note that the cumulative failure rate functions must be a non-decreasing function for all $t \geq 0$ and $\lim_{t \to \infty} H(t) = \infty$.

The characteristic of the hazard rate function for the occurrence of a particular event can be increased, decreased, constant, bathtub shaped [21], and Vtub shaped [20] that can help to describe the failure mechanism or symptoms. Distribution models with decreasing hazard rate functions are much less common but can find in the applications of learning behaviors. Models with increasing hazard rates often can be used in applications that reflect the aging or wear and tear. The bathtub-

Table 25.1 Probability density function and hazard rate of distributions

Distribution	pdf	Hazard rate
[19]	$f(t) = \beta\, \mathrm{e}^{\left[\theta t - \frac{\beta}{\theta}\left(\mathrm{e}^{\theta t}-1\right)\right]}$	$h(t) = \beta\, \mathrm{e}^{\theta t}$
Loglog [20]	$f(t) = \beta\, \ln(\theta)\, t^{\beta-1}\, \theta^{t^{\beta}}\, \mathrm{e}^{1-\theta^{t^{\beta}}}$	$h(t) = \beta\, \ln(\theta)\, t^{\beta-1}\, \theta^{t^{\beta}}$
[21]	$f(t) = \frac{\beta}{\theta}\left(\frac{t}{\theta}\right)^{\beta-1} \mathrm{e}^{-\left(\frac{t}{\theta}\right)^{\beta}}$	$h(t) = \frac{\beta}{\theta}\left(\frac{t}{\theta}\right)^{\beta-1}$
Gompertz–Makeham	$f(t) = \left(\gamma + \beta\,\mathrm{e}^{\theta t}\right) \mathrm{e}^{\left[-\frac{\beta}{\theta}\left(\mathrm{e}^{\theta t}-1\right)-\gamma t\right]}$	$h(t) = \gamma + \beta\,\mathrm{e}^{\theta t}$
Logistic	$f(t) = \beta\,\mathrm{e}^{\theta t}\left[1 + \frac{\gamma\beta}{\theta}\left(\mathrm{e}^{\theta t}-1\right)\right]^{-\frac{\gamma+1}{\gamma}}$	$h(t) = \dfrac{\beta\,\mathrm{e}^{\theta t}}{\left[1 + \frac{\gamma\beta}{\theta}\left(\mathrm{e}^{\theta t}-1\right)\right]}$
Log-logistic	$f(t) = \dfrac{\beta\,\theta\,t^{\beta-1}}{\left(1 + \theta\,t^{\beta}\right)^2}$	$h(t) = \dfrac{\beta\,\theta\,t^{\beta-1}}{\left(1 + \theta\,t^{\beta}\right)}$

shaped hazard rate function plays an important role in reliability applications such as human life and electronic devices [22].

In the Gompertz model, the two parameters β and θ are positive; β varies with the level of mortality and θ measures the rate of increase in mortality with age. The Gompertz model is used widely with biologists and demographers [18], and assumes that the mortality increases exponentially with age. This model was developed by Benjamin Gompertz and published in the *Philosophical Transactions of the Royal Society of London* in 1825 [19]. Although this distribution was developed almost 200 years ago, it is still commonly used in modeling biological mortality. In Messori's study [6] of cost-effectiveness analysis, he states that the Gompertz function best models the mortality data he used from England.

The Gompertz model has been extended with the addition of a constant γ to take into account the background mortality due to causes unrelated to age (see Table 25.1). This model is known as the Gompertz–Makeham model. The model likely represents an improvement over the Gompertz model at younger ages, but it still seems to over-estimate mortality at the oldest ages [11]. The Gompertz–Makeham mortality law is also a widely used distribution in gerontological investigation [23]. In Yu *et al.*'s conclusion, they do not recommend the model for human populations because the distribution needs to be corrected for sex differences.

The Weibull distribution listed in Table 25.1 is a two-parameter distribution; one parameter controls the scale and the other controls the shape of the distribution. Gavrilov and Gavrilova [8] indicate that a difference between the applications of the Weibull and Gompertz distributions is such that technical devices follow the Weibull distribution, while organisms follow the Gompertz. Even though Weibull may have been earmarked for technical devices, due to its flexibility depending on the parameters, it can also be applied to mortality and human population growth modeling. A recent study by Pham and Lai [22] on the generalization of the Weibull can also be applied in modeling the mortality and human population aspects. The Gompertz and Weibull distributions are both commonly used in reliability engineering.

Table 25.2 Mortality modeling and distribution functions and analysis literature

Group	References
General modeling and data analysis	[8, 9, 15, 24–37]
Age-specific mortality	[12, 13, 15, 16, 18, 38–41]
Gompertz and Weibull models	[6, 19, 21, 42–44]
Logistic, Loglog, and other common models	[7, 45]
Measurements, modeling fitting and criteria	[7, 18, 43, 44]
Aging	[4, 5, 8–10, 35, 42, 46–50]
Life expectancy	[13, 15, 29, 43, 44, 51]
Mortality modeling	[15–17, 19, 24, 34, 36–40, 43, 44, 47, 48, 52–59]
Population data	[6, 14, 38, 41, 60]
Forecasting	[24, 43, 60, 61]

The three-parameter logistic model is derived from the work of Pierre Verhulstin in the early 1800s and is commonly used in many disciplines including mortality human population, software reliability engineering, human factor analysis, cancer medicine applications and applied engineering statistics. In Thatcher's analysis of highest attained age [11], he noted how the logistic model better predicted higher ages for his data. In including this model, this study will be able to show whether or not this model can also accurately predict for the entire age range, not just the highest ages. A summary of references of research papers and books on mortality modeling and analysis is given in Table 25.2.

References

1. United States Center for Disease Control and Prevention. Live births, birth rates, and fertility rates, by race: United States. 1909–2000. 2000. http://www.cdc.gov/nchs/data/statab/t001x01.pdf
2. United States National Center for Health Statistics
3. US Census Bureau, Population Division. http://www.census.gov/population/cen2000/phc-t3/tab03.pdf
4. Pletcher SD, Neuhauser C (2000) Biological aging – criteria for modeling and a new mechanistic model. International Journal of Modern Physics C 11:525–546
5. Gillespie CS *et al.* (2004) A mathematical model of ageing in yeast. Journal of Theoretical Biology 229:189–196
6. Messori A (1997) Survival curve fitting using the Gompertz function: a methodology for conducting cost-effectiveness analysis on mortality data. Computer Methods and Programs in Biomedicine 53:175–164
7. Pham H, Deng C (2003) A predictive-ratio risk criterion for selecting software reliability models. The 9th ISSAT Proceedings on Reliability and Quality in Design, August
8. Gavrilov LA, Gavrilova NS (2001) The reliability theory of aging and longevity. Journal of Theoretical Biology 213:527–545
9. Raudys S (2002) An adaptation model for simulation of aging process. International Journal of Modern Physics C 13:1075–1086
10. Alders P, Broer P (2005) Ageing, fertility, and growth. Journal of Public Economics 89:1075–1095
11. Thatcher AR (1999) The long-term pattern of adult mortality and the highest attained age. Journal of the Royal Statistical Society Series A (Statistics in Society) 162:5–43
12. Bongaarts J, Feeney G (2002) How long do we live? Population and Development Review 28:13–29
13. Bongaarts J, Feeney G (2003) Estimating mean lifetime. Proc Natl Acad Sci 100:13127–13133
14. Lee ET (1992) Statistical Methods for Survival Data Analysis. Wiley, New York
15. Lee R, Carter LR (1992) Modeling and forecasting U.S. mortality. J Am Statist Assoc 87:659–671
16. Higgins T (2003) Mathematical models of mortality. Workshop on Mortality Modeling and Forecasting, Australian National University, 13–14 February (http://acpr.edu.au/Publications/Mortality%20talk%20-%20mathematical%20equations.pdf)
17. Pletcher S (1996) Age-specific mortality costs of exposure to inbred *Drosophila melanogaster* in relation to longevity selection. Experimental Gerontology 31:605–616
18. Pletcher SD (1999) Model fitting and hypothesis testing for age-specific mortality data. Journal of Evolutionary Biology 12:430–439

19. Gompertz B (1825) On the nature of the function expressive of the law of human mortality and on a new mode of determining the value of life contingencies. Philosophical Transactions of the Royal Society of London 115:513–585

20. Pham H (2002) A vtub-shaped hazard rate function with applications to system safety. International Journal of Reliability Applications 3:1–16

21. Weibull W (1951) A statistical distribution function of wide applicability. Journal of Applied Mechanics 18:293–297

22. Pham H, Lai CH (2007) On recent generalization of the Weibull distribution. IEEE Transactions on Reliability 56:454–458

23. Yu V, Pakin S, Hrisanov M (1984) Critical analysis of the applicability of the Gompertz–Makeham law in human populations. Gerontology 30:8–12

24. Bongaarts J (2005) Long-range trends in adult mortality: models and projection methods. Demography 42:23–49

25. Booth H, Maindonald J, Smith L (2002) Applying Lee–Carter under conditions of variable mortality decline. Population Studies 56:325–336

26. Carey JR (1993) Applied Demography for Biologists. Oxford University Press

27. Finch CE (1990) Longevity, Senescence and the Genome. University of Chicago Press

28. Fisher RA (1921) On the mathematical foundation of theoretical statistics. Philosophical Transactions of the Royal Society of London A 222:309–368

29. Gavrilov LA, Gavrilova NS (1991) The Biology of Life Span: a Quantitative Approach. Harwood Academic Publishers, Switzerland

30. Hughes KA, Charlesworth B (1994) A generic analysis of senescence in *Drosophila*. Nature 367:64–66

31. Kannisto V (1987) On the survival of centenarians and the span of life. Population Studies 42:389–406

32. Kannisto V, Lauritsen J, Thatcher AR, Vaupel JW (1994) Reductions in mortality at advanced ages: several decades of evidence from 27 countries. Population and Development Review 20:793–810

33. Keyfitz N (1977) Applied Mathematical Demography. Wiley, New York

34. Keyfitz N (1981) Choice of function for mortality analysis: effective forecasting depends on a minimum parameter representation. Theoretical Population Biology 21:329–352

35. Riggs JE (1990) Longitudinal Gompertzian analysis of adult mortality in the US, 1900–1986. Mechanisms of Ageing and Development 54:235–247

36. Riggs JE (2001) The dynamic of aging and mortality in the United States. Mechanisms of Ageing and Development 65:217–228

37. Thatcher AR, Kannisto V, Vaupel JW (1998) The Force of Mortality at Ages 80 to 120. Odense University Press, Odense

38. Brooks A, Lithgow GJ, Johnson TE (1994) Mortality rates in a genetically heterogeneous population of *Caenorhabditis elegans*. Science 263:668–671

39. McNown R, Rogers A (1989) Forecasting mortality: a parameterized time series approach. Demography 26:654–660

40. Olshansky SJ (1988) On forecasting mortality. The Milbank Quarterly 66:482–530

41. Pollard JH (1987) Projection of age-specific mortality rates. Population Bulletin of the United Nations, nos 21–22. United Nations, New York, pp. 55–69

42. Mueller LD, Nusbaum TJH, Rose MR (1995) The Gompertz equation as a predictive tool in demography. Experimental Gerontology 30:553–569

43. Wilson DL (1993) A comparison of methods for estimating mortality parameters from survival data. Mechanisms of Ageing and Development 66:269–281

44. Wilson DL (1994) The analysis of survival (mortality) data: fitting Gompertz, Weibull and logistic functions. Mechanisms of Ageing and Development 74:15–33

45. Pakin YV, Hrisanov SM (1984) Critical analysis of the applicability of the Gompertz-Makeham law in human populations. Gerontology 30:8–12

46. Lee R, Miller T (2001) Evaluating the performance of the Lee–Carter method for forecasting mortality. Demography 38:537–549

47. Manton KG, Stallard E, Vaupel JW (1981) Methods for comparing the mortality experience of heterogeneous populations. Demography 18:389–410
48. Manton KG, Stallard E, Vaupel JW (1986) Alternative models for the heterogeneity of mortality risks among the aged. Journal of the American Statistical Association 81:635–644
49. Nusbaum TJ, Mueller LD, Rose MR (1996) Evolutionary patterns among measures of aging. Experimental Gerontology 31:507–516
50. Taljapurkar S, Li N, Boe C (2000) A universal pattern of mortality decline in the G7 countries. Nature 405:789–792
51. Curtsinger JW, Fukui HH, Townsend DR, Vaupel JW (1992) Demography of genotypes: failure of the limited lifespan paradigm in *Drosophila melanogaster*. Science 258:461–463
52. Beard RE (1971) Some aspects of theories of mortality, cause of death analysis, forecasting and stochastic processes. In Brass W (ed) Biological Aspects of Demography. Barnes & Noble Inc., New York
53. Himes CL, Preston SH, Condran GA (1994) A relational model of mortality at older ages in low mortality countries. Population Studies 48:269–291
54. Horiuchi C, Coale AJ (1994) Age patterns of mortality for older women: an analysis using the age-specific rate of mortality change with age. Mathematical Population Studies 2:245–267
55. Pletcher S, Curtsinger JW (1998) Mortality plateaus and the evolution of senescence: Why are mortality rates so low? Evolution 52:454–464
56. Shyamalkumar ND (2002) Analysis of mortality data using smoothing spline Poisson regression. Department of Statistics, University of Iowa
57. Tabeau EA, Heathcote C (2001) Forecasting Mortality in Developed Countries. Kluwer Academic Publishers, Dordrecht
58. Vaupel JW (1986) How change in age-specific mortality affects life expectancy. Population Studies 40:147–157
59. Yashin AI, Vaupel JW, Iachine IA (1994) A duality in aging: the equivalence of mortality models based on radically different concepts. Mechanisms of Ageing and Development 74:1–14
60. Alho JM (1990) Stochastic methods in population forecasting. International Journal of Forecasting 6:521–530
61. Carey JR, Liedo P, Orozco D, Vaupel JW (1992) Slowing of mortality rates at older ages in large medfly cohorts. Science 258:457–461

Further Reading

Lee R (2000) The Lee–Carter method for forecasting mortality, with various extensions and applications. North American Actuarial Journal 4:80–93

About the Editor

Hoang Pham is currently Professor and chairman of the Department of Industrial and Systems Engineering at Rutgers University. Before joining Rutgers, he was a senior engineering specialist at the Boeing Company, Seattle, and the Idaho National Engineering Laboratory, Idaho Falls. He received his B.S. degree in mathematics, B.S. degree in computer science, both with high honors, from Northeastern Illinois University, Chicago, an M.S. degree in statistics from the University of Illinois, Urbana-Champaign, and M.S. and Ph.D. degrees in industrial engineering from the State University of New York at Buffalo. His research interests include software reliability, system reliability modeling, maintenance and biological risk assessment.

He is the author of four books, has published more than 90 journal articles, and edited more than 10 books including the *Handbook of Reliability Engineering* and *Springer Handbook of Engineering Statistics*. He is editor-in-chief of the *International Journal of Reliability, Quality and Safety Engineering*, associate editor of the *IEEE Transactions on Systems, Man and Cybernetics*, an editorial board member of a dozen journals. He is also the editor of Springer Series in Reliability Engineering and World Scientific Series of Industrial and Systems Engineering. He has been conference chair and program chair of over 30 international conferences and workshops. He is a fellow of IEEE.

Index

Printed in the United States
124891LV00001BA/112-135/P